T0230086

Lecture Notes in Artificial Intelligence   744

Subseries of Lecture Notes in Computer Science
Edited by J. Siekmann

Lecture Notes in Computer Science

Edited by G. Goos and J. Hartmanis

K.P. Jantke S. Kobayashi
E. Tomita T. Yokomori (Eds.)

Algorithmic
Learning Theory

4th International Workshop, ALT '93,
Tokyo, Japan, November 8-10, 1993
Proceedings

Springer-Verlag

Berlin Heidelberg New York
London Paris Tokyo
Hong Kong Barcelona
Budapest

K. P. Jantke   S. Kobayashi
E. Tomita   T. Yokomori (Eds.)

Algorithmic Learning Theory

4th International Workshop, ALT '93
Tokyo, Japan, November 8-10, 1993
Proceedings

Springer-Verlag

Berlin Heidelberg New York
London Paris Tokyo
Hong Kong Barcelona
Budapest

Series Editor

Jörg Siekmann
University of Saarland
German Research Center for Artificial Intelligence (DFKI)
Stuhlsatzenhausweg 3
D-66123 Saarbrücken, Germany

Volume Editors

Klaus P. Jantke
Fachbereich Informatik, Mathematik und Naturwissenschaften
Hochschule für Technik, Wirtschaft und Kultur Leipzig
Postfach 66, D-04251 Leipzig, Germany

Shigenobu Kobayashi
Interdisciplinary Graduate School of Science and Engineering
Tokyo Institute of Technology
4259 Nagatsuta, Midori-ku, Yokohama, 227 Japan

Etsuji Tomita
Department of Communications and System Engineering
University of Electro-Communications
1-5-1 Chofugaoka, Chofu, Tokyo, 182 Japan

Takashi Yokomori
Department of Computer Science and Inform. Math.
University of Electro-Communications
1-5-1 Chofugaoka, Chofu, Tokyo, 182 Japan

CR Subject Classification (1991): I.2.6, I.2.3, F.1.1

ISBN 3-540-57370-4 Springer-Verlag Berlin Heidelberg New York
ISBN 0-387-57370-4 Springer-Verlag New York Berlin Heidelberg

Typesetting: Camera ready by author
Printing and binding: Druckhaus Beltz, Hemsbach/Bergstr.
45/3140-543210 - Printed on acid-free paper

PREFACE

This volume contains all the papers that were presented at the Fourth Workshop on Algorithmic Learning Theory (ALT '93), which was held at University of Electro-Communications in Tokyo from November 8th to 10th, 1993. In addition to 3 invited papers, 29 papers were selected from among 47 submitted extended abstracts, which represent the highest number of papers submitted to an ALT workshop, exceeding 46 which were recorded in 1990.

This workshop was the fourth in a series of ALT workshops, whose focus is on theories of machine learning and the application of such theories to real world learning problems. The ALT workshops have been held annually since 1990, sponsored by the Japanese Society for Artificial Intelligence. In the past, ALT alternated between an English-language international conference and a Japanese-language domestic workshop. (ALT'90 was international whereas ALT'91 and ALT'92 were more or less domestic.) Starting with ALT'93, all the future ALT workshops will be English-language international conferences, and researchers from throughout the world will be invited to present papers and to attend the conference.

This year, we are fortunate to include three invited papers by distinguished researchers: "Identifying and Using Patterns in Sequential Data" by Dr. P. Laird, NASA Ames Research Center, "Learning Theory Toward Genome Informatics" by Prof. S. Miyano, Kyushu University, and "Optimal Layered Learning : A PAC Approach to Incremental Sampling" by Prof. S. Muggleton, Oxford University. It goes without saying that the three researchers are distinguished theoreticians, but at the same time the issues they addressed were highly practically relevant. It is our hope that the future ALT workshops will continue to bring together researchers from both theoretical and practical sides of machine learning to provide a forum for truly worthwhile research interactions.

We would like to extend our sincere gratitude to the many individuals who made this workshop possible. These include the invited speakers, all the presenters and participants at the workshop, the members of the steering committee, the members of the program committee, many referees who helped ensure the quality of the accepted papers, and many others.

Last but not least, we gratefully acknowledge the support of the Telecommunications Advancement Foundation.

Tokyo, November 1993

K.P. Jantke
S. Kobayashi
E. Tomita
T. Yokomori

Conference Chairman

Shigenobu Kobayashi

Program Committee

N. Abe	M. Harao	P. Laird	Masahiko Sato
K. Akama	Hideki Imai	M. Li	Masako Sato
D. Angluin	Hiroshi Imai	A. Maruoka	T. Sato
S. Arikawa	Y. Inagaki	D. Michie	T. Shinohara
J. Arima	B. Indurkhya	S. Miyano	Y. Shirai
H. Asoh	M. Ishikawa	K. Morik	C. Smith
A. Biermann	H. Ishizaka	H. Motoda	Y. Takada
J. Case	K. Jantke	S. Muggleton	H. Tsukimoto
R. Daley	E. Kinber	T. Nishida	E. Ukkonen
R. Freivalds	Y. Kodratoff	S. Nishio	O. Watanabe
K. Furukawa	A. Konagaya	M. Numao	R. Wiehagen
M. Hagiya	S. Kuhara	H. Ono	A. Yamamoto
M. Haraguchi	S. Kunifuji	L. Pitt	T. Yokomori (Chairman)

Local Arrangements Committee

T. Kasai	T. Nishizawa	H. Takahashi	M. Wakatsuki
Satoshi Kobayashi	A. Sakurai	E. Tomita (Chairman)	

Sponsored by Japanese Society for Artificial Intelligence (JSAI)

In Cooperation with

Information Processing Society of Japan (IPSJ)
Institute of Electronics, Information and Communication
    Engineers of Japan (IEICE)
Japanese Cognitive Science Society (JCSS)
Japan Society for Software Science and Technology(JSSST)
Japan Neural Network Society (JSNN)
Society of Instrument and Control Engineers of Japan (SICE)

List of Referees

N. Abe	Y. Kodratoff	L. Pitt
S. Akaho	A. Konagaya	Masahiko Sato
K. Akama	T. Koshiba	Masako Sato
T. Akutsu	S. Kuhara	T. Sato
D. Angluin	S. Kunifuji	A. Sakurai
S. Arikawa	T. Kurita	A. Sharma
J. Arima	P. Laird	S. Shimozono
H. Arimura	S. Lange	A. Shinohara
H. Asoh	M. Li	T. Shinohara
A. Biermann	W. Maass	Y. Shirai
J. Case	A. Maruoka	H. Simon
R. Daley	O. Maruyama	C. Smith
R. Freivalds	M. Matsuoka	M. Suraj
K. Furukawa	D. Michie	E. Suzuki
M. Hagiya	T. Miyahara	Y. Takada
M. Haraguchi	S. Miyano	H. Takahashi
M. Harao	K. Morik	J. Takeuchi
H. Iba	H. Motoda	K. Tanatsugu
Hideki Imai	T. Motoki	S. Tangkitvanich
Hiroshi Imai	S. Muggleton	N. Tanida
Y. Inagaki	Y. Mukouchi	M. Tatsuta
B. Indurkhya	A. Nakamura	A. Togashi
M. Ishikawa	T. Nishida	H. Tsukimoto
H. Ishizaka	T. Nishino	E. Ukkonen
K. Jantke	S. Nishio	O. Watanabe
Y. Katayama	T. Nishizawa	S. Weinstein
E. Kinber	M. Numao	R. Wiehagen
J. Kivinen	N. Ohtsu	S. Yamada
Satoshi Kobayashi	H. Onda	A. Yamamoto
Shigenobu Kobayashi	H. Ono	T. Yokomori

TABLE OF CONTENTS

[New Learning Paradigms]

Identifying and Using Patterns in Sequential Data

Philip Laird

NASA Ames Research Center, Moffett Field, CA 94035-1000, U.S.A.

Abstract. Whereas basic machine learning research has mostly viewed input data as an unordered random sample from a population, researchers have also studied learning from data whose input sequence follows a regular sequence. To do so requires that we regard the input data as a stream and identify regularities in the data values as they occur. In this brief survey I review three sequential-learning problems, examine some new, and not-so-new, algorithms for learning from sequences, and give applications for these methods. The three generic problems I discuss are:
- Predicting sequences of discrete symbols generated by stochastic processes.
- Learning streams by extrapolation from a general rule.
- Learning to predict time series.

1 Introduction

Algorithmic Learning Theory treats both the theory of learning and the design of practical algorithms. Over the years many useful and interesting algorithms have derived from the assumption of *data independence*, that is, that the observations be considered an unordered set of examples. Indeed, we consider an algorithm that learns successfully regardless of the presentation order of the examples to be more robust than one that depends on assumptions about the sequence of examples. For example, given an algorithm that infers the grammatical structure of a language from examples of sentences and non-sentences, we would prefer that the examples not have to be given in order of size or in some order depending on the structure of the grammar.

In practice, however, the requirement of order independence is a strong one: much information can be conveyed by carefully choosing the sequence of examples—*too much* information for some theoretical models, where the learning problem becomes trivial if the presenter is free to choose the presentation order. More significantly, many algorithms *depend* on the assumption that the examples come from random sampling of a fixed population; such algorithms usually turn in poor results when this assumption fails to hold.

In some applications the real problem is to learn the pattern responsible for generating the sequence, rather than to learn to attach a label to the individual examples. And in situations where examples arrive as a continual stream of symbols without any natural divisions or termination, we have little choice, at least initially, but to treat the incoming values as a sequence until enough is known to segment it into its parts. In the next section I relate some situations I have encountered where sequential effects cannot be overlooked in the learning problem.

This paper briefly surveys learning problems and algorithms for sequential data. I distinguish among three kinds of learning problems: stochastic sequence prediction, sequence extrapolation, and modeling time series. For each I review the characteristics of the problem, discuss some learning algorithms, and note several applications. In the conclusion I summarize some common features of these problems and the challenges for continuing research.

2 Some Sequential-Learning Applications

The descriptions below abstract some learning applications I have encountered where the sequential nature of the data is important to the problem.

Events in Telemetry Streams

Let $\{f(t), t \geq 1\}$ be a stream of numbers obtained by sampling a physical process at regular time intervals. This stream is our only way of observing the process since it is at a remote site. Most of the time this telemetry data conveys nothing of interest, but now and then an important event occurs. We recognize this fact by the pattern (or *signature* of successive values of $f(t)$ over some fixed time interval.

Humans learn quickly to recognize and identify these significant events visually by looking at a plot of $f(t)$. Suppose, however, we want to automate the process of signaling these interesting events. We obtain samples of $f(t)$ with labels by an expert of the interesting events. Thereupon the process of training a program to recognize and label interesting events is apparently straightforward: any number of static classification algorithms can be used—Bayesian discriminants, neural networks, decision trees, etc. After training, the system is turned on, and as expected, it performs quite well identifying these events.

As time continues, however, we discover that it reports successively fewer events of interest. To find out why, we examine the telemetry stream and discover a small but significant trend in the signatures, due perhaps to physical wear or some hysteresis effect. Unless this trend can be identified and corrected, no amount of random sampling will make our classifier work for more than a short time.

Engineering Monitoring and Maintenance

Replacing parts of a complex piece of machinery can be done by schedule or by need. Scheduled maintenance means that the part is replaced after a fixed time period, regardless of its actual condition. Other parts, especially expensive ones, are removed and examined frequently and replaced as soon as they show signs of deterioration. Sometimes for very critical parts, or for parts that are expensive to remove, sensors are attached to the part to indicate its condition without the need to remove it.

In practice, scheduled maintenance is expensive, and sensors are failure-prone. As a result the true condition of a critical part must often be inferred indirectly from a combination of sensors and observations. To automate the process of monitoring critical components, we need to be able to learn from sequences of multiple noisy observations $\{\mathbf{X}_t, t \geq 1\}$ when to replace the part.

Clustering

Clustering programs examine datasets and try to group the data into "meaningful" classes from which useful conclusions can be drawn. For example, in an attempt to discover some of the structure hidden in a huge database of spectral measurements from stars and galaxies, a clustering program groups the many thousands of observations into about fifty distinct classes. Astronomers then study these classes, looking for common physical properties of the objects in the classes.

The clustering program treats the observations as a set, and the time of the observation is not normally included as an attribute. But in actuality the observations occurred in time, and unexpected correlations between the measurement times of the objects in a class need to be explained. In the case of spectral imaging, temporal correlations could result from the fact that consecutive observations tend to be directed at objects physically proximate to each other and hence closely related. The correlations could, however, be a consequence of temporal effects in the use or adjustment of the apparatus rather than of any astrophysical phenomena.

Operator Training

Drivers, pilots, and other operators of machinery are often trained on simulators. By presenting a realistic set of scenarios on a model of the machine, the trainee can develops the motor and judgment skills to operate safely without any risk that a serious accident will result from his errors.

Often, however, the trainee also learns to predict the simulator. In one case a pilot learned that he could always avoid a collision in a simulated near-miss by reducing his altitude and banking to the right. This strategy might not be effective in a real encounter, so the designers of the simulator were worried that the pilots were learning the simulator rather than good piloting skills. They warned the trainee of the risk of relying on apparently consistent patterns in the simulator, and then reinforced this warning by "training" the pilot to expect a certain predictable response from the simulator—only to change this response suddenly after the pilot's expectations (and responses) had been conditioned. This tactic, however, does little to mitigate the human ability to learn sequential patterns, and this ability seriously limits the usefulness of simulators in training.

3 Models of Sequential Effects

The three models given below, while related, differ in the nature of the models and the learning objectives and hence call for different kinds of algorithms.

Stochastic Sequence Prediction

In the most basic form, the input sequence is an infinite stream of individual symbols with no meaning or internal structure. By assumption the symbols are produced by some unknown stochastic process that may or may not produce each symbol independently of its predecessors. The learning task is to construct a model of the process and to predict as accurately as possible each symbol in the stream.

For example, consider the stream fragment,

$$... \text{ T } + + + \text{ \$ \% T } + \text{ \$ \% T } + + + + \text{ \$ \% } ...$$

After reading this sequence from left to right, most people would agree that the most likely symbol to follow is T. The basis for this opinion is the heuristic that the future tends to follow the past, and in both previous occurrences the character % has been succeeded by T. The confidence in this prediction may be low since the evidence is meager, but in the absence of any knowledge about the process generating this stream, humans seem to default to this kind of statistical prediction.

Trying to predict beyond the next two or three symbols degenerates quickly into a simple frequency analysis—what one would predict with the assumption that all symbols were selected independently at random. Evidently the sequential information decays quite rapidly with time, something characteristic of state-based stochastic models. In general we can model the observed sequence as a deterministic or probabilistic function of some unobservable random state or "context". If the state were known, the prediction task would reduce to one where the symbols stream is a sequence of independent random variables. The task, then, is to model both the underlying state structure and the random process generating the symbols in each state.

Discrete-state Markov models are a popular way of modeling state dependencies. Hidden Markov models [31] consist of a discrete-state, discrete-time Markov process in which a multinomial process is attached to each state. Fairly efficient offline algorithms, known as *reestimation algorithms*, (e.g, [39]) are known for finding models from observations.

Markov trees are a different representation for Markov chains. Instead of representing the individual states of the process, the tree represents transitions from the steady state. (See Figure 1.) Each level of the tree represents a context: the root node ignores all previous input, nodes at the first level represent changes based on the single previous input symbol, and so forth. At any time a prediction can be based on any of these contexts, and different algorithms choose the appropriate context differently (e.g., [53]). Currently the best known

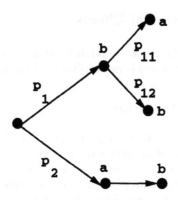

Fig. 1. A two-level Markov tree. The nodes are labeled by the input symbols, the arcs by probabilities. The root represents the steady-state or empty context.

general-purpose text compression algorithms are based on Markov trees ([5] surveys these algorithms), but exactly why they consistently outperform Markov models and other techniques is not clear.

Finding an optimal Markov process is in general an intractable problem, but when the underlying dynamics of the state-transition process are known, Kalman filters are an optimal procedure for estimating the state and making predictions ([9]). When the input observations are numeric, one can estimate the parameters of and quantify the effects of noise quite precisely.

We can point to a few of the many applications of stochastic sequence prediction. Some are based on the dual relationship between learning and compression. Learning algorithms become compression algorithms by forming a model of the sequence, predicting the next input, and encoding the difference between what is predicted and what actually occurs. Only these differences need be retained: there is no need to encode the model if the uncompression process is deterministic and undergoes exactly the same learning, forming the same models at the same point in the stream.

Conversely, a text-compression utility can be "taken from the box" and used as a learning algorithm Vitter and Krishnan [47, 48] used the Lempel-Ziv algorithm for data compression as the learning element in a database module that predicts record requests during queries. If they predict a request for a record that is not online, a prefetch is issued in anticipation thereof.

My colleague Ronald Saul and I [24] analyzed and generalized the Markov tree PPM algorithm for text compression [5]. Our variant, known as *TDAG*, is well suited to a variety of tasks, such as managing the cache memory of a mass storage system and dynamically optimizing Prolog programs [23]). Others have since applied TDAG to trend analysis and protein sequence prediction.

Hidden Markov models are widely used for speech recognition, and the carefully engineered system due to Kai-fu Lee and his colleagues [30] is one of the most successful examples. By drastically reducing the number of possibilities for the next input item, the learning element enabled them to advance substantially the state of the art in real-time speech recognition.

When the input stream is naturally segmented into "sentences", formal-language methods can also be used to predict shorter segments of the input stream. For example, Hermens and Schlimmer [19] used zero-reversible finite automata and a learning algorithm due to Angluin [2] to help predict the user's keyboard input in an electronic forms assistant. Stochastic context-free grammars are a more powerful model, and with new algorithms for inferring such grammars from the sequences they generate, they are finding applications in predicting the secondary structure of proteins [12].

Let us note some of the distinctive features of stochastic sequence prediction algorithms. First, we can usually characterize the number of degrees of freedom in the models by two numbers: the number of distinct symbols and the number of "states" (or the *order*) of the model. The models are usually not aware of "noise" (random disturbances of the input symbols) because the noise is incorporated into the model as modifications of the probabilities. On the other hand, the lack of a distinct noise model makes it hard to separate the signal from the noise, as sometimes can be done with other learning methods, e.g., the Kalman filter, where Gaussian white noise is assumed to affect both the dynamics and the observations. The ability of a model to compress the input stream serves as a useful comparative measure its predictive success. Finally, as for most forms of learning from examples, there is usually a tradeoff between the expressiveness of the model and the number of observations required to converge to a model.

Structured Sequence Extrapolation

When the symbols in an incoming stream are not atomic but instead have known relationships among them, we can make stronger models than statistical ones. For example, consider the following sequence of strings: $S = *, !**?, !!*!**??,$ $!!!*!!*!**???, \ldots$. Although we can ignore the string boundaries and predict symbol by symbol, we can also use what we know about the algebra of strings to look for regular substructure in the sequence. Observe, for example, that the simple sequence $R = *?^0, *?^1, *?^2, *?^3, \ldots$ forms a suffix of S. Hence we can write $S = L \cdot R$, and then focus on the subsequence L. After a few moments, most people find a reasonably simple rule for L and then are able to make a reasonable prediction for the next string in this sequence: $!!!!*!!!*!!*!**????$.

The knowledge that the elements of this sequence are strings, and that strings can be decomposed into substrings by the non-commutative concatenation operation, is the key to this approach. Other data types besides strings have similar properties: the sequence of integers 8, 9, 11, 15, 23, ... can be written as the sum of two simpler sequences 7, 7, ... and 1, 2, 4, 8, ...; and the sequence of binary lists

$$S = \langle a, [a, a], [[a, a], a], [[[a, a], a], a] \ldots \rangle,$$

(where [,] is the ordered-pair operation) can be represented by the recurrence expression $S_{n+1} = [S_n, a]$. The idea, then, is to extrapolate a sequence using prior knowledge about the data type of the elements in the sequence.

Humans exhibit remarkable skill at this task. "Tests of intelligence" often include integer extrapolation problems. Curiously, while there is usually no unique way to extrapolate a sequence (e.g., 1, 2, 3, 4, ... can continue with 5 or 97 or any other value depending on one's assumptions), there is typically a simplest or "most elegant" rule that most people will agree upon, or that, when shown to someone unable to find it, results in expressions of "Oh, of course!" In the case of integer sequences, the underlying assumptions are that each element S_{n+1} of the stream can be obtained from "a few" (one or two) of its predecessors (S_n, S_{n-1}) by means of only a "few" additions or multiplications, involving perhaps some small constants (0, ±1, ±2).

Moreover, when more than one rule of roughly equal complexity can account for the values seen so far, humans often hedge their prediction by offering more than one, along with a rough confidence estimate. As the number of possible explanations diminish, the confidence in the remaining rules increases correspondingly. Given that the input stream is deterministic, not stochastic, one can ask how these confidence estimates arise.

Most of all, that all this seems to be a common skill suggests this learning/generalization ability is somehow fundamental.

The number of principled algorithms for sequence extrapolation is small. Kotovsky and Simon [21] explored a model of human sequence extrapolation in the case of Thurstone letter sequences. Pivar and Finkelstein [37] implemented a Lisp program that extrapolates integer sequences based on the so-called *method of differences*, which reportedly goes back to Gauss. This idea is based on the fact that a polynomial sequence $\{f(n) \mid n = 1, 2, \ldots\}$ reduces to a constant sequence $\{c, c, c, \ldots\}$ by computing the successive k'th differences, defined recursively as

$$D_k f(n) = D_{k-1} f(n + 1) - D_{k-1} f(n)$$
$$D_0 f(n) = f(n),$$

where k is the degree of the polynomial f. Then the matrix of first, second, ..., k'th differences can be turned directly into a finite-difference representation or into a closed-form polynomial. Best of all, this requires only $k + 1$ consecutive examples (values) of the polynomial.

This simple method applies only for sequences that are polynomials, but some have tried to develop algorithms extending the method to incorporate factors, constants, and other options along with differences. But the need to search through a number of possible ways to decompose the numbers limits the approach. With some clever heuristics, Feenstra (cited in [10]) used an extended difference method to achieve an "IQ" of about 160 on a published intelligence test.

Genetic programming [22] has also been used to search for both polynomial forms and recurrence relations. While reasonably effective, this method has two drawbacks: it requires many more examples than necessary, and it comes up with rules that are more complex than necessary.

For other than numerical domains, few algorithms have been suggested than can be presented formally and analyzed. Many researchers, however, have been aware of the need to study sequence extrapolation in a way that is not restricted to just one data type (see, for example, [36, 18]). Intuition says that essentially the same algorithm should work for strings, integers, lists, queues, etc., with specific differences that depend on the properties of that type. Since general algorithms are more useful than specialized ones, we are encouraged to look for extrapolation algorithms not restricted to a single data type.

Recently my colleague Ronald Saul and I have developed such an algorithm [25]. The main features of this algorithm are the use of abstract data types, a recursive language for representing streams, and the ability to provide a confidence measure with each prediction. A data type is a set of expressions containing a finite set of generators (atoms) and closed under a finite set of operators. Congruences group the expressions into equivalence classes (e.g., $(2 * 3) = (3 + 3)$. The algorithm depends on an efficient algorithm for factoring elements into subelements—e.g., writing the string abc as $\lambda \cdot (abc)$, $(a) \cdot (bc)$, $(ab)(c)$, and $(abc) \cdot \lambda$. There are a few additional restrictions, but the model covers most of the commonly used data types, including multisorted ones like pairs of strings.

The representation language is called *elementary stream descriptions* and does not depend on the type. Streams are recursively defined in one of three forms:

- An *initial-value* form, giving the first element in the stream and an elementary description of its tail;
- A *functional* form, expressing the stream as a functional combination of two or more streams, which are in turn defined by elementary descriptions;
- A *recursive* form, stating that the stream is recursively equal to a stream within whose definition it occurs.

For example, the Fibonacci stream $F = \langle 1, 1, 2, 3, 5, \ldots \rangle$ is defined: $F = \langle 1 \mid F_2 \rangle$, $F_2 = \langle 1 \mid F_3 \rangle$, $F_3 = (F_4 + F_5)$, $F_4 = F$, $F_5 = F_2$.

The algorithm breaks each incoming example into its parts and uses these either to create new descriptions or to test existing ones. Descriptions that are inconsistent with the example are discarded. Hypotheses are stored in such as way that the simplest ones, especially the ones with the fewest initial values, can be found efficiently.

Another feature of the algorithm is that it provides confidence estimates with its prediction of the next value, estimates that agree qualitatively with our own. We make the simplifying assumption, that normally is not true, that any hypothesis H has a fixed probability p_H of incorrectly predicting a symbol in the next input example. p_H depends on H but not on the symbol or where it occurs in the input. The longer the string of correct predictions, the lower our estimate of p_H, which we can compute explicitly with Bayes' rule.

We have done some analysis of the algorithm, including a proof of correctness (identification in the limit). For the special case of freely generated types, the algorithm is efficient, running in time bounded by a polynomial in the sizes of the input values and revising its hypotheses in polynomial time. For more general types the algorithm is not efficient with respect to time or space. We have implemented the algorithm and are now experimenting with it.

Sample size analysis for elementary stream descriptions is difficult. Our only formal result so far [26] is to show that, for the family of streams whose type is dotted pairs (non-associative pair algebra) with at most one initial value, the number of input values required in the worst case is exactly $h + 3$, where h is the height of the first example. Thus if the first example is an atomic value, say a, then 4 input examples are necessary and sufficient to identify the stream uniquely. This tends to confirm our observation that sample sizes required for sequence extrapolation is extremely small compared to statistical models.

Because of the lack of good algorithms, we cannot point to many actual applications where sequence extrapolation has played an important role, but we can suggest applications where such an algorithm could be of value.

In the late 1960's and early 1970's there was a flurry of interest in automatic programming, including programming by examples [17, 42, 43]. Some methods looked for patterns in successive examples that could be turned into recurrence relations and thence into recursive programs. For example, given the following examples of the function f on lists:

$$f([a]) = a$$
$$f([a, b]) = b$$
$$f([a, b, c]) = c$$

Summers's program [43] finds two sequences: the sequence of input forms $\mathrm{cons}(a, \ \mathrm{nil}), \mathrm{cons}(a, \ \mathrm{cons}(b, \ \mathrm{nil})), \mathrm{cons}(a, \ \mathrm{cons}(b, \ \mathrm{cons}(c, \mathrm{nil})))$, and the sequence of value forms expressed in terms of the input x: $\mathrm{car}(x), \mathrm{car}(\mathrm{cdr}(x))$, $\mathrm{car}(\mathrm{cdr}(\mathrm{cdr}(x)))$. Then, using templates for certain kinds of recurrences, he derives the function (recall that nil is an atom in LISP):

$$f(\mathrm{cons}(X, Y)) = X \quad \text{if } Y \text{ is an atom}$$
$$= f(Y) \quad \text{otherwise}$$

The ability to generalize from $\mathrm{cons}(a, b)$ to $\mathrm{cons}(X, Y)$ (where X and Y are variables) derives from his (arbitrary) assumption that the functional form f depends only on the structure of the input list, not on the specific values.

Shapiro's Model Inference System shifted the approach to programming by examples away from finding recurrences. His algorithm is based on generalizing and specializing formulas by instantiating variables to terms and adding clauses to conjunctive-form logical sentences. Since then, most of the research—notably, the work on inductive logic programming [38, 32, 35]— has concentrated on inductive generalization—generalizing specific formulas until they are general enough to cover the examples—instead of looking for patterns and recurrences. While inductive generalization by itself is usually a sufficient technique, it does not take advantage of information about the target function or relation that is revealed by sequential patterns.

For example, consider the arithmetic function $F(X, Y)$ defined by the following examples:

X	0 1 0 2 1 0 3 2 1 0 4 3 ...
Y	0 0 1 0 1 0 0 1 2 3 0 1 ...
F	2 2 2 2 3 2 2 4 4 2 2 5 ...

The structure of this concept becomes much more discernible if arranged so that the sequential patterns emerge:

Y	0	1	2	3
X				
0	2	2	2	2
1	2	3	4	5
2	2	4	6	8
3	2	5	8	11

The patterns $F(X, 0) = 2$, $F(X, 1) = F(X-1, 1)+1$, $F(X, 2) = F(X-1, 2)+2$, ... are all easy extrapolations. Then by generalizing these patterns (instead of the original examples) we obtain

$$F(X, Y) = F(X - 1, Y) + Y,$$

with an initial value of $F(0, Y) = 2$. Inductive programming currently suffers from the need for a very large sample size; by contrast, extrapolation requires small sample sizes. In the preceding example, fewer than twenty values of F sufficed to infer a simple hypothesis for the concept. Whereas sequence extrapolation alone is not sufficient for programming by example, in combination with inductive generalization it may enable us to resurrect and incorporate the good ideas from some of the past research.

Numerical discovery is the term often used to describe the problem of finding a simple formula to account for numerical data. For example, given (noisy) measurements of pressure, volume, and temperature, one looks for a "simple" relation $f(P, V, T) = k$ that agrees "adequately" with the data. (Interpreting the quoted terms is the hardest part.) Current algorithms feature regression and knowledge-based discovery techniques, such as [27, 14].

At first glance sequence extrapolation seems not to apply, since the observations need not be in order and since they are noisy. But the formulas one seeks are generally rational functions with integer coefficients, and in many cases (e.g., economic measurements) one or more of the variables is evenly spaced (e.g., $t = 1, 2, \ldots$). Two kinds of noise affects sequence extrapolation: discrete noise, where an input symbol is occasionally changed to another, unrelated symbol, and continuous noise, where numerical input values differ from their true values by an amount whose probability decreases with its absolute value. The assumption of integral coefficients means, for example, that neither the formula $IR/E = 1$ nor $2IR/E = 1$ will fit the data exactly, but as more input values accrue, one's confidence in the one will outstrip that in the other. Moreover, one can quantify that confidence, as discussed above.

As a final potential application, let us mention the large class of formal program transformation techniques wherein a functional or logic program with undesirable features is transformed into one with desirable ones. For example, the unfold-fold technique suggested by Burstall and Darlington [7, 45] entails unfolding (i.e., partially evaluating) a recursive program some number of times, applying some equivalence transformations to the expanded form, and folding the program into a more efficient recursive form. Automating these transformations is difficult because a certain amount of insight into which transformations improve performance seems to be required. But many transformations, when applied to a sequence of unfoldings, lead to a sequence of forms that a sequence extrapolator can then fold back into a recursive form. Testing this resulting form for improved properties (speed, termination, correctness, etc.) is no small task, but the refolding process itself can be instantiated as sequence extrapolation.

Finally, we note some characteristics of the known sequence extrapolation algorithms. Like stochastic sequences, stream descriptions have two ways to characterize their degrees of freedom: the *order* (also called the delay, or latency), and what I shall call the breadth (not a standard term). The order measures how many preceding values determine the next value. The breadth depends on the representation and counts the number of substreams that are required to define the observed stream. (A close analogy would be the number of variables in a context-free grammar or the number of predicate symbols in a logic program.) As noted, there are two flavors of noise: metric and non-metric. Both are most damaging if they occur early in the stream, since then they are most likely to cause false hypotheses to be proposed and correct hypotheses to be rejected.

As with stochastic sequences the effectiveness of an extrapolation algorithm can again be measured by how well it compresses a data stream. Note that a perfect extrapolator can compress an infinite stream into a finite number of bits, but with noise or a weak representation language, the infinite stream can only be compressed into a smaller, but still infinite, stream.

Hypotheses are rules rather than descriptions of stochastic processes. Hence it is not clear how uniform convergence results can be applied to sequence extrapolation. But just as stochastic models typically exhibit uniform convergence of the likelihood of the possible hypotheses to their means, the confidence (Bayesian posterior) of our models likewise appears to converge uniformly, although we have not done any analysis on this.

Time Series

A time series is a vector function of time (or other continuous scalar variable), $X(t)$. Observations consist of samples of X at regular or irregular intervals, X_t, $X_{t+\tau_1}$, $X_{t+\tau_2}$, ... The learning task is to construct a model of $X(t)$ so that certain predictions can be made about the course of its future values.

The dream is that one can infer X_t and use it to predict stock prices, weather patterns, cardiovascular functions, and similarly important series. However, without strong assumptions about X this problem is hopelessly difficult.

Nevertheless a number of fairly general and useful algorithms are available for inferring the properties of time series.

Formal analysis of time series began about seventy-five years ago with efforts to separate the stationary (uncorrelated) component of the series from the trend (non-stationary, deterministic component). An elegant theory developed for a family of linear stationary models, known as ARMA (autoregressive moving-average) models, that fit the data to the form

$$\mathbf{X}_n = \sum_{i=1}^{m} a_i \mathbf{E}_{n-i} + \sum_{j=1}^{d} b_j \mathbf{X}_{n-j}.$$

Here $\mathbf{E}_n$ is a white-noise (uncorrelated) process that combines with a finite number of recurrent values of the series $\mathbf{X}$ to produce successive values. The "deconvolution" problem is to infer the coefficients of this process from the input examples. This turns out to be possible because from the spectrum of the series one can estimate the autocorrelation function, and from that one can solve for the coefficients. An iterative linear modeling procedure developed by Box and Jenkins [6] first applies differencing to subtract a polynomial non-stationary component of the series and then deconvolves, repeating the process until the best fit is obtained.

Despite its elegance, this "ARIMA" model often gives poor results with real data. Extensions to higher-order powers of $\mathbf{X}$ lead to integral equations that are difficult to solve. An analytical *theory of inverse problems* has arisen to study general mathematical issues of extracting a function f from discrete values of $\mathbf{B}f$, where $\mathbf{B}$ is an operator in a general family of operators [46]. Other nonlinear modeling methods such as regressive splines, Padé approximants, and maximum entropy are effective for specific problems; [13] contains good summaries of many of these.

Recently two research areas—chaos and neural networks—have contributed new ideas to the learning of time series. Statistical models decompose a stationary series into random and deterministic components:

$$X = A * R + D,$$

where R is a white-noise process and D is a deterministic process ($*$ is the convolution operator and A is a constant filter). A theorem due to Wold [54] states that, under very general conditions, any stationary sequence can be so decomposed. The novel insight is the "deterministic" and "predictable" are not the same: chaos theory has demonstrated that most nonlinear deterministic systems exhibit complex behavior that is difficult to predict, in the sense that system trajectories diverge exponentially in time no matter how close their initial position. Computing the behavior of such a system requires $\mathcal{O}(t)$ space and quickly exceeds the computational capacity of real machines. In the end such systems are as unpredictable as purely random ones.

Yet just as random processes have predictable properties (e.g., their moments), so do chaotic ones. For example, the well-known logistic recurrence,

$X_{n+1} = 4X_n(1 - X_n)$, yields a complex, uncorrelated time series. If $0 \le X_1 \le 1$, the values of X_n remain between 0 and 1. Given successive values of this sequence, how could we detect that it is generated by a simple deterministic process rather than a stochastic one? If we graph the *return map* X_{n+1} versus X_n, we obtain an extremely simple curve: a parabola. Since no random process would exhibit such regularity, we are sure that the process is deterministic. More generally, to distinguish a chaotic process from a random one, we can look for some deterministic function of the process. But how do we find such a property, and what can we do if the process is a mixture of random and chaotic processes?

As yet the answers to these questions are not fully known, but some intriguing hints have emerged. For sampled chaotic processes the *k-th order return map*, X_{n+1} as a function of $X_n, \ldots, X_{n-k+1}$ enjoys some remarkable topological properties. Roughly speaking, as k increases from 1, there exists a value k_0 such that for all $k \ge k_0$ the topological dimension of the surface of the return map embedded in the k-dimensional space R^k is much smaller than k. The conditions are (1) that the time interval, or *lag*, between the measurements be large enough that the chaos has a chance to eliminate most of the correlations between values, and (2) that the values $X(t)$ we observe be derived from those of the underlying process Y_t by a diffeomorphic coordinate transformation $X(Y(t))$. This so-called *embedding dimension* can be estimated numerically from the sample entropy of the sequence and serves as a measure of the inherent complexity ("degrees of freedom") of the underlying process. [44, 40]. Experimentally, graphical procedures are often effective in determining a bound on the embedding dimension for chaotic time series [33, 34]. One cannot help being impressed when an apparently random process—be it water dripping from a faucet or the planet Pluto moving in a gravitational potential—is coaxed into showing us its basic simplicity.

Scargle [41] has generalized Wold's theorem by further decomposing the deterministic component of a stationary time series:

$$X = A * R + B * Y + C,$$

where Y is an uncorrelated chaotic series and C is a non-chaotic deterministic series. He is developing deconvolution techniques whereby the components of X can be estimated from the observations.

Most algorithms for analyzing time series seek to predict the future course of the sequence, if not exactly, at least within specified ranges. Recently researchers have develop a number of new approaches. Their effectiveness evidently depends on the complexity of the underlying process and on whether the short-term or long-term behavior of the series is to be modeled. Weather-prediction methods, for example, are very different depending on whether one wants to know the weather a few hours or days hence or whether one is interested in three- to six month temperature trends. For short-term analysis, neural network methods have proved relatively successful, especially for high dimensional processes. The networks are trained to approximate the embedded surface $X_{t+1} = f(X_t, \ldots, X_{t-k+1})$ in "lag space" with a smooth, non-linear surface. Successful architectures have included sigmoid and radial-basis nets [28, 52] and *finite impulse response (FIR)* nets [49], in which simple connections are replaced by FIR filters and trained using a parallel backpropagation technique. As with other connectionist methods, the effectiveness of the algorithms is difficult to quantify, and the complexity increases rapidly with the size of the problem.

Recently a competition to predict the future of time series [51] received a number of connectionist entries, some quite successful and other not. Reportedly the successful ones required a lot of experimentation and analysis to determine the appropriate embedding delay, network structure, and training procedure for the task. In view of this per-problem empirical analysis, it is still a stretch to refer to these procedures as "algorithms."

Besides connectionist models, researchers have proposed algorithms that model small neighborhoods of the embedded surface as hyperplanes. These so-called *local-linear models* are based on the idea that the surface is approximately linear in a sufficiently small neighborhood. The hyperplane can be constructed using the nearest neighbors in R^k and used to project the vector $(X_n, \ldots, X_{n-k+1})$ to $(X_{n+1}, \ldots, X_{n-k+2})$ [15, 8]. (The ARMA model is a "global" linear model in that a single hyperplane represents the entire surface.) The larger the approximating surface, the greater the scale of the prediction: a local model based on a very small neighborhood is best for short-term predictions, whereas a longer time scale requires a coarser neighborhood model.

Let us review briefly some of the characteristics of time-series algorithms. The order (delay, etc.) of the series—i.e., the number of immediately preceding values upon which the next one depends—seems a fundamental value to establish. For a series obtained by sampling a continuous process, the *lag* or time between observations can also affect the results: if it is too short, one may overlook the effects of chaos that can eliminate any apparent predictability over short time intervals; if too long, one may not be able to extrapolate on a sufficiently short time scale. Related is the question of overfitting: if we fit the observations too closely, our model may incorporate noise and other spurious effects as non-random components of the series, leading to poor extrapolation results. One way that neural network models avoid overfitting is by halting training when cross-validation scores (obtained by testing the predictions against data withheld from training) stop improving and begin to worsen [50]. Another is to prune away elements that contribute little to reducing the error [29]. Still another is to build up the network incrementally, increasing the size only when doing so significantly improves the performance. The true issue here is how much generalization can justified by the data for a particular family of hypotheses; and while the fundamentals of this question are understood fairly well for concept-learning problems [3, 20], it is very much an open problem for time series.

Decomposition has been a recurring strategy for time-series analysis, whether separating the stationary from the non-stationary, the chaotic from the simple deterministic, or one local neighborhood from another. Finally, although we have not mentioned compression, the entropy of the underlying source process plays an important role in the ergodic theory of time series. Learning reduces the rate of increase in the information provided by the source process, and this in turn translates directly into greater compression of the sample.

4 Summary

Although the three types of sequential learning problems are very different one from the other, we should at least pick out some common aspects:

- They try in some fashion to reduce the sequence to simpler subsequences: by identifying common contexts or states, by expressing the sequence as a combination of other sequences, or by separating random, chaotic, and simple deterministic contributions.
- They predict or represent the next input value based on a finite number of its recent predecessors. Determining the number of those predecessors (the order of the sequence) seems to be a crucial part of any algorithm.
- The success of the learning algorithm is measured by prediction accuracy, or equivalently, by the ability to compress the information in the examples.
- The notion of convergence of random variables applies, but it has not been exploited as effectively as in concept learning. For stochastic sequences the likelihood of the hypothesis converges to its limit value, so that maximum likelihood strategies are effective [4, 1]. For sequence extrapolation—which is not a random process—our interpretation of the performance or likelihood of a hypothesis depends on our forming some probabilistic assumptions about the occurrences of prediction errors. To do so may seem rather arbitrary since these errors are deterministic rather than probabilistic, but evidently we humans do something like that in order to estimate the confidence in our hypotheses. For time series problems, Bayesian and maximum entropy techniques converge rapidly and as such discriminate with great sensitivity among different models when strong models about the underlying process are available. See [11], especially the article by Gull [16].

I conclude with these observations. Concept-learning and clustering research has had its greatest impact with relatively simple, general-purpose algorithms (decision trees, networks, hierarchical clustering, etc.) that apply broadly in the absence of strong models about the data. Similarly I expect that the most influential sequence learning algorithms to be simple even if naïve, effective if not rigorous, for divers types of data streams. For stochastic sequences we have such, but general-purpose algorithms are still lacking for the other two. Also, all the algorithms and procedures described here are first-order algorithms—in effect, search optimization algorithms. Aside from obtaining confidence estimates, little research is given to learning the properties of the solutions themselves.

Acknowledgment

Doug Fisher kindly reviewed a draft of this paper and returned helpful suggestions.

References

1. N. Abe and M. Warmuth. On the computational complexity of approximating distributions by probabilistic automata. *Machine Learning*, 9:205–260, 1992.

2. D. Angluin. Inference of reversible languages. *Journal of the Association for Computing Machinery*, 29:741–765, 1982.

3. E. Baum and D. Haussler. What size net gives valid generalization? *Neural Computation*, 1:151–160, 1989.

4. L. E. Baum and T. Petrie. Statistical inference for probabilistic functions of finite state markov chains. *Ann. Math. Stat.*, 37:1554–1563, 1966.

5. T. C. Bell, J. G. Cleary, and I. H. Witten. *Text Compression*. Prentice Hall, Englewood Cliffs, N.J., 1990.

6. G. Box and G. Jenkins. *Time Series Analysis, Forecasting, and Control*. Holden Day, 1976.

7. R. Burstall and J. Darlington. A transformation system for developing recursive programs. *Journal of the Association for Computing Machinery*, pages 44–67, 1977.

8. M. Casdagli. Nonlinear prediction of chaotic time series. *Physica D*, 35, 1989.

9. D. Catlin. *Estimation, Control, and the Discrete Kalman Filter*. Springer Verlag, New York, 1989.

10. A. K. Dewdney. Computer recreations. *Scientific American*, pages 14–21, 1986.

11. J. Skilling (ed.). *Maximum Entropy and Bayesian Methods*. Kluwer Academic, 1989.

12. Y. Sakakibara et al. Stochastic context-free grammars for modeling RNA. Technical Report UCSC-CRL-93-16, University of California, Santa Cruz, 1993. (submitted).

13. W. Press *et al. Numerical Recipes*. Cambridge University Press, 1992.

14. B. Falkenhainer and R. Michalski. Integrating qualitative and quantitative discovery: the ABACUS system. *Machine Learning*, 1:367–401, 1986.

15. J. D. Farmer and J. Sidorowich. Exploiting chaos to predict the future and reduce noise. In *Evolution, learning, and cognition*. World Scientific, 1989.

16. S. Gull. Developments in maximum entropy data analysis. In *Maximum Entropy and Bayesian Methods*, pages 53–71. Kluwer Academic, 1989.

17. S. Hardy. Synthesis of LISP programs from examples. In *Proceedings of 4th International Joint Conference on A.I.*, pages 268–273, 1975.

18. C. Hedrick. Learning production systems from examples. *Artificial Intelligence*, 7:21–49, 1976.

19. L. A. Hermens and J. C. Schlimmer. Applying machine learning to electronic form filling. In *Proc. SPIE Application of AI: Machine Vision and Robotics*, 1993.

20. G. Hinton and D. van Camp. Keeping neural networks simply by minimizing the description length of the weights. In *Proc. Sixth Annual ACM Conference on Computational Learning Theory*, 1993.

21. K. Kotovsky and H. Simon. Empirical tests of a theory of human acquision of concepts for sequential patterns. *Cognitive Psychology*, 4:399–424, 1973.

22. J. Koza. *Genetic Programming*. M.I.T. Press, 1992.

23. P. Laird. Dynamic optimization. In *Proc., 9th International Machine Learning Conference*. Morgan Kaufmann, 1992.

24. P. Laird and R. Saul. Discrete sequence prediction and its applications. *Machine Learning*, 1993. (To appear).

25. P. Laird and R. Saul. Sequence extrapolation. In *Proceedings, 13th International Joint Conference on Artificial Intelligence*, 1993.

26. P. Laird, R. Saul, and P. Dunning. A model of sequence extrapolation. In *Proceedings of the 6th Annual Conference on Computational Learning Theory*, 1993.

27. P. Langley. Rediscovering physics with BACON.3. In *Proc. IJCAI 6*, pages 505–507, 1977.

28. A. Lapedes and R. Farber. Non-linear signal processing using neural networks. Technical report, Los Alamos National Laborator, 1987.

29. Y. le Cun, J. Denker, and S. Solla. Optimal brain dammage. In *Advances in Neural Information Processing Systems 2*, pages 598–605. Morgan Kaufmann, 1990.

30. Kai-Fu Lee. *Large-vocabulary speaker-independent continuous speech recognition: the SPHINX System*. PhD thesis, Carnegie-Mellon University, Computer Science Department, 1988.

31. J. Levinson, L. Rabiner, and M. Sondhi. An introduction to the application of the theory of probabilistic functions of markov processes in automatic speech recognition. *Bell Sys. Tech. J.*, 62:1035–1074, 1983.

32. S. Muggleton, editor. *Inductive Logic Programming*. Academic Press, 1992.

33. N. Packard and J. Crutchfield *et al.* Geometry from a time series. *Physical Review Letters*, pages 712–716, 1980.

34. Jan Paredis. Learning the behavior of dynamical systems from examples. In *Proc. 6th International Workshop on Machine Learning*, pages 137 – 139. Morgan-Kaufmann, 1989.

35. M. Pazzani and D. Kibler. The utility of knowledge in inductive learning. *Machine Learning*, 9:57–98, 1992.

36. S. Persson. *Some Sequence Extrapolation Programs: A study of representation and modeling in inquiry system*. PhD thesis, University of California, Berkeley, 1966. Also printed as Stanford University Computer Science Department Technical Report # CS50, 1966.

37. M. Pivar and M. Finkelstein. Automation, using LISP, of induction on sequences. In E. Berkeley and D. Bobrow, editors, *The Programming Language LISP*. Information International, Inc., 1964.

38. R. Quinlan. Learning logical definitions from relations. *Machine Learning*, 5:239–266, 1990.

39. R. Redner and H. Walker. Mixture densities, maximum likelihood, and the EM algorithm. *SIAM Review*, 26:195–239, 1984.

40. T. Sauer, J. Yorke, and M. Casdagli. Embedology. *J. Statistical Physics*, pages 579–616, 1991.

41. J. Scargle. An introduction to chaotic and random time series analysis. *Int. J. of Imaging Systems and Technology*, pages 243–253, 1989.

42. D Shaw, W. Swartout, and C. Green. Inferring LISP programs from example problems. In *Proceedings of 4th International Joint Conference on A.I.*, pages 260–267, 1975.

43. P. Summers. A methodology for lisp program construction from examples. *J.ACM*, 24, 1977.

44. F. Takens. Detecting strange attractors in turbulence. In D. Rand and L.-S. Young, editors, *Dynamical Systems and Turbulence*. Springer Verlag, 1981.

45. H. Tamaki and T. Sato. Unfold/fold transformation of logic programs. In *2nd International Logic Programming Conf.*, 1984.

46. A. Tarantola. *Inverse Problem Theory*. Elsevier, 1987.

47. J. Vitter and P. Krishnan. Optimal prefetcching with data compression. Technical Report CS-91-46, Brown University Department of Computer Science, 1991.

48. J. Vitter and P. Krishnan. Optimal prefetching via data compression. In *Proceedings of the 32nd Annual IEEE Symposium on Foundations of Computer Science*, 1991.

49. E. Wan. Temporal backpropagation: An efficient algorithm for finite impulse response neural networks. In *Connectionist Methods: Proc. of the 1990 Summer School*, pages 131–140, 1990.

50. A. Weigend and B. Hubermann andD. Rumelhart. Predicting the future: a connectionist approach. *International Journal of Neural Systems*, 1:193–209, 1990.

51. A. Weigend and N. Gershenfeld. *Predicting the Future and Understanding the Past: A Comparison of Approaches*. Addison-Wesley, 1993.

52. A. Weigend, D. Rumelhart, and B. Hubermann. Backpropagation, weight elimination, and time series prediction. In *Connectionist Methods: Proc. of the 1990 Summer School*, pages 105–116, 1990.

53. Ross Williams. *Adaptive Data Compression*. Kluwer Academic Publishers, Boston, 1991.

54. H. Wold. *A study in the analysis of stationary time series*. Almqvst and Wiksell, Uppsala, 1938.

Learning Theory Toward Genome Informatics

Satoru Miyano

Research Institute of Fundamental Information Science
Kyushu University 33, Fukuoka 812, Japan
miyano@rifis.sci.kyushu-u.ac.jp

Abstract. This paper discusses some problems in Molecular Biology to
which learning paradigms may be applicable. As a case, we present our recent
study on knowledge discovery from amino acid sequences by PAC-learning
paradigm.

1 Introduction

Computer Science and Molecular Biology is now creating a rapidly evolving research
field called *Genome Informatics* or *Molecular Bioinformatics*. Rapid advances of
biotechonolgy in the last decade and the international Human Genome Project are
putting various databases of proteins and nucleic data into explosion. Everyday,
a considerable amount of genomic data from laboratories are compiled into the
databases via international networks.

One of the important issues in Genome Informatics is to establish technologies
for discovering knowledge from amino acid sequences and nucleic sequences that
may provide new directions of experimental investigations to biologists. Systematic
and efficient methods are strongly desired for discovering or acquiring important
biological knowledge from these genomic data.

The purpose of this paper is to show a way of getting into Genome Informatics from the side of learning theory in the above respect by (a) showing available
databases of genomic data, (b) defining problems on these data, (c) giving their
related works so far, and (d) presenting the trace of our approach [1, 31, 38, 39] to
some of the problems.

The research on gathering, analyzing, and managing amino acid sequences of
proteins and nucleic acid sequences has a considerably long tradition since the birth
of Molecular Biology. On the other hand, Computer Science seems to have made
relatively little effort on this topic while it has been contributing to various fields
with drastic changes. Genome Informatics involves nearly every aspect of Computer
Science. In particular, computational challenges of learning theory are really desired
for the next advance, and the potential payoff is very high both in Computer Science
and Molecular Biology. If algorithmic/computational learning theory could survive
through this big trend in Genome Informatics and develop soundly, it would establish
its firm identity in Science.

This paper is organized as follows: Section 2 surveys some databases which pro-
vides information about amino acid sequences of proteins and nucleic acid sequences.
In Section 3, we pick up some problems to which learning technologies are applicable.
Section 4 reviews the methods so far developed for these problems. In Section 5, we

present our research strategy using PAC-learning paradigm for the transmembrane domain identification problem and the signal peptide identification problem. We also sketch the machine discovery system BONSAI that is developed in this research and show its successful experimental results.

2 Databases for Genome Informatics

This section provides some useful databases which compile amino acid sequences and nucleic acid sequences with their associated information. All databases below are, for instance, accessible at Human Genome Center, Institute of Medical Science, The University of Tokyo, by ftp network service.

1. **GenBank**
   This database contains nucleic sequences from published articles and direct submissions from authors. Each record specifies the source of the data and related keywords. Translations of coding regions are also available in this database.

2. **PDB**
   Three-dimensional structures of proteins and nucleic acids are compiled in PDB. These are X-ray crystallographic and NMR data provided by direct submissions from their authors.

3. **PIR**
   PIR (Protein Identification Resources) contains amino acid sequences of proteins from published articles and their related information.

4. **EMBL**
   EMBL collects nucleic acid sequences from published articles and also by direct submission from authors.

5. **SWISS-PROT**
   Protein sequences and translations of coding regions from EMBL are compiled in this database.

6. **PROSITE**
   This database collects motifs for proteins. This is a kind of a dictionary of protein sites and patterns.

3 Problems

This section provides some problems in Genome Informatics to which learning technologies may be helpful. The final purpose is to establish techniques by employing learning paradigms for extracting biologically relevant information from amino acid sequences of proteins and nucleic acid sequences stored in the databases shown in Section 2. In practice, the problem is to determine regions or locations on a sequence which have specific functions or properties. Furthermore, it is also an important issue to discover biologically interesting knowledge from the databases by learning technologies.

In the following discussion, we may ignore biological correctness of explanations since there are lot of exceptions in biological data. Instead, we put priority on their understandability of the contents. The text books by Watson et al. [47] and Lewin [30] are very helpful to capture the general principles in Molecular Biology.

3.1 Protein Structure Prediction

The *primary structure* of a protein is described as a sequence of amino acid residues. It is generally accepted that the amino acid sequences determine their forms and functions.

The *secondary structure* of a protein describes the local structure of the protein. There are three kinds of major secondary structures; *α-helix*, *β-sheet*, and *turn*. The α-helix is a helical conformation of the polypeptide backbone that is stereochemically most satisfactory. Every turn of the helix consists of 3.6 amino acids. An α-helix comprises of 4 ~ 30 amino acids. The β-sheet is a regular lateral association of extended polypeptide chains into a sheet. The *turn* is a structure like a hairpin.

The *secondary structure prediction problem* is to predict the locations of α-helices, β-sheets and turns on the amino acid sequence of proteins.

The *tertiary structure* of a protein is the form in the three-dimensional space. Its structure is, in many cases, complex. It is also an important problem to predict possible tertiary structures of a protein form its primary structure. This problem is called the *tertiary structure prediction problem*.

PDB collects tertiary structures of proteins. Unfortunately, the number of proteins whose tertiary structures are known and open to public is small.

The following special issues have attracted many attentions.

Membrane Spanning Segments A membrane has a structure of lipid bilayer that surrounds the cytoplasm of a cell. A membrane of a cell usually contains various kinds of protein molecules directly inserted into the lipid bilayer. Some of these proteins have parts sticking out from both sides of the membrane and pass through this bilayer several times. Regions of a protein passing right through the membrane are called the *transmembrane domains*, each of which is a sequence of length about 20 constituting an α-helix structure. It has been accepted to assume that a protein containing these transmembrane domains is a membrane protein. The length of an α-helix in a membrane protein comes from the length to span the apolar center of a membrane [13]. In a typical membrane, the apolar center is approximately 30Å, and each amino acid residue has a thickness of 1.54Å. Hence 20 residues×1.54Å/residue = 31Å. If sequences corresponding to α-helices of transmembrane domains are found in an amino acid sequence of a protein, it is very likely that the protein is a membrane protein.

The *transmembrane domain identification problem* is, given an amino acid sequence of a protein, to decide if it contains transmembrane domains and to show the regions of transmembrane domains.

PIR provides amino acid sequences of proteins with FEATURE field where transmembrane domains are indicated as shown in Fig. 1.

Signal Peptides There are proteins that shall be conveyed across the membrane to the outside of the cell. Most of such proteins are synthesized in the form of longer precursors that consist of additional amino acid sequences of length 15 ~ 30 at their N-termini. A *signal peptide* is a sequence locating at this N-terminal region. This sequence usually begins with a Methionine (M).

```
ENTRY          MPHUPL     #Type Protein
TITLE          Myelin proteolipid protein - Human
ALTERNATE-NAME PLP lipophilin
DATE           30-Sep-1987 #Sequence 30-Sep-1987 #Text 30-Sep-1991
PLACEMENT      1081.0   1.0   1.0   1.0   1.0
SOURCE         Homo sapiens #Common-name man
ACCESSION      A24657
REFERENCE
    #Authors   Stoffel W., Giersiefen H., Hillen H., Schroeder W.,
               Tunggal B.
    #Journal   Biol. Chem. Hoppe-Seyler (1985) 366:627-635
    #Reference-number A24657
    #Accession   A24657
    #Molecule-type protein
    #Residues    1-276 [STO]
COMMENT        Five hydrophobic domains are present; three are
               transmembrane domains and two are intramembranous
               domains.
COMMENT        Most, and perhaps all, of the Cys residues are
               involved in disulfide bonds.
COMMENT        This is the major protein of myelin from the central
               nervous system.
SUPERFAMILY    #Name myelin proteolipid protein
KEYWORDS       membrane protein myelin oligodendrocyte
               structural protein
FEATURE
    9-34              #Domain transmembrane I [TM1]
    59-96             #Domain transmembrane II [TM2]
    151-190           #Domain hydrophobic intramembranous I
                      [HC1]
    205-216           #Domain hydrophobic intramembranous II
                      [HC2]
    232-267           #Domain transmembrane III [TM3]
    198               #Binding-site fatty acid
SUMMARY        #Molecular-weight 29890 #Length 276 #Checksum  559
SEQUENCE
             5      10     15     20     25     30
      1 G L L E C C A R C L V G A P F A S L V A T G L C F F G V A L
     31 F C G C E V E A L T G T E K L I E T Y F S K N Y Q D Y E Y L
     61 I N V I H A F Q Y V I Y G T A S F F F L Y G A L L L A X G F
     91 Y T T G A V R Q I F G D Y K T T I C G K G L S A T V T G G Q
    121 K G R G S R G Q H Q A H S L E R V C H C L G C W L G H P D K
    151 F V G I T Y A L T V V W L L V F A C S A V P V Y I Y F N T W
    181 T T C Q S I A A P C K T S A S I G T L C A D A R M Y G V L P
    211 W N A F P G K V C G S N L L S I C K T A E F Q M T F H L F I
    241 A A F V G A A A T L V S L L T F M I A A T Y N F A V L K L M
    271 G R G T K F
```

Fig. 1. An amino acid sequence in PIR containing four transmembrane domains.

GenBank provides nucleic sequences together with information about signal peptides. An example from GenBank is shown in Fig. 2 that is a DNA sequence containing a signal peptide.

The *signal peptide identification problem* is, given a DNA sequence, to decide if it contains a sequence representing a signal peptide at the N-terminal region and to identify the location of the signal peptide.

```
LOCUS       AFABCP       783 bp ds-DNA          BCT      15-MAR-1989
DEFINITION  A.faecalis blue copper protein gene, complete cds.
ACCESSION   M18267
KEYWORDS    anaerobic nitrate respiration system regulator;
            blue copper protein; electron carrier; pseudoazurin.
SOURCE      A.faecalis S-6 DNA, clone pAB101.
  ORGANISM  Alcaligenes faecalis
            Prokaryota; Bacteria; Gracilicutes; Scotobacteria;
            Aerobic rods and cocci; Alcaligenaceae.
REFERENCE   1  (bases 1 to 783)
  AUTHORS   Yamamoto,K., Uozumi,T. and Beppu,T.
  TITLE     The blue copper protein gene of Alcaligenes faecalis S-6 directs
            secretion of blue copper protein from Escherichia coli cells
  JOURNAL   J. Bacteriol. 169, 5648-5652 (1987)
  STANDARD  simple staff review
FEATURES            Location/Qualifiers
     CDS            249..689
                    /note="blue copper protein precursor"
                    /codon'start=1
                    /translation="MRNIAIKFAAAGILAMLAAPALAENIEVHMLNKGAEGAMVFEPA
                    YIKANPGDTVTFIPVDKGHNVESIKDMIPEGAEKFKSKINENYVLTVTQPGAYLVKCT
                    PHYAMGMIALIAVGDSPANLDQIVSAKKPKIVQERLEKVIASAK"
     sig'peptide    249..317
                    /note="blue copper protein signal peptide"
                    /codon'start=1
     mat'peptide    318..686
                    /note="blue copper protein"
                    /codon'start=1
BASE COUNT     207 a     204 c     216 g     156 t
ORIGIN      5 bp upstream of SphI site.
        1 gcatgcaggc ttgtcatgtc gcgcctaggc tggccggagg ctgcggcaaa agggctggcg
       61 ggcatatcga atggtgcgat ggtggaatgc agaaaatcaa acagttttat tctgagcgtc
      121 attatcatga ataacttcac ctgtttgatc cagatcaaag aggttggcag gcgacaggtc
      181 taaaccccgt tacgtggcgt gttgaggccg agacggcagg tgcgccaatc gttggagatc
      241 aggacaaaat gcgtaacatc gcgatcaaat ttgctgccgc aggcatcctc gccatgctgg
      301 ctgcccccgc tcttgccgaa aatatcgaag ttcatatgct caacaagggc gccgagggcg
      361 ccatggtttt cgagcctgcc tatatcaagg ccaatcccgg cgacacggtc acctttattc
      421 cggtggacaa aggacataat gtcgaatcca tcaaggacat gatccctgaa ggcgccgaaa
      481 agttcaaaag caagatcaac gagaactatg tgctgacggt tacccagccc ggcgcatatc
      541 tggtaaagtg cacaccgcat tatgccatgg gtatgatcgc gctcatcgct gtcggtgaca
      601 gcccggccaa tctcgaccag atcgtttcgg ccaagaagcc gaagattgtt caggagcggc
      661 tggaaaaggt catcgccagc gccaaataag agcgccaaat aagattgacc gaaaactctc
      721 gatgagccga acttgaaccg gcttcatgac gaggacatca tgaccagaca gccaggcctg
      781 cag
```

Fig. 2. A DNA sequence containing a signal peptide.

Epitopes This problem involves an aspect related to the tertiary structures of proteins. An *epitope* is also called an *antigenic determinant*. Each epitope consists of $5 \sim 8$ amino acids and locates on the protein surface. The expected number of antigenic determinants is at least $10,000$ and probably much more although the number of known epitopes is not large. PIR and GenBank provide information about eptiope sequences.

3.2 Coding Region Identification

For prokaryotic genes such as bacterial genes, there is the colinearity, i.e., a direct correspondence between the sequence of a gene, its mRNA and its protein product.

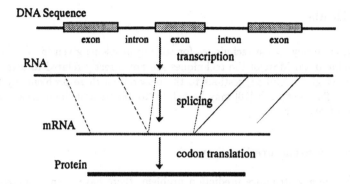

DNA Sequence

exon   intron   exon   intron   exon

RNA   transcription

splicing

mRNA

Protein   codon translation

Fig. 3. The gene is transcribed into a very large precursor RNA molecule, from which the introns are removed and the exons are concatenated to form a mRNA.

On the other hand, the nucleotides sequences in eukaryotes that specify amino acid sequences of proteins are usually interrupted by blocks of nucleotides that have no coding information. The segments that encode amino acids for proteins are called *exons* and the blocks that are not translated into amino acids and removed by the process called *RNA splicing* are called *introns* (Fig. 3). There is no specific range of the lengths of exons and introns. Moreover, the number of exons varies gene to gene. The *splicing site identification problem* is to find locations of exon/intron boundaries in a given DNA sequence. Information about these data are provided in GenBank.

The *coding region* is a region of a DNA sequence which is finally translated by ribosomes into a polypeptide. In prokaryotes, the coding region is a single contiguous sequence (*open reading frame* (ORF)), while the coding region of a eukaryotic gene comprises of several disjoint ORFs since the sequence is interrupted by introns. The problem of determining the coding region of a DNA sequence is one of the most challenging problems in Genome Informatics.

Translation of mRNA encoding a protein starts at some location and stops at some location. These locations are also encoded in a DNA sequence. Three codons (UAA, UAG, and UGA) are stop signals where translation stops. These codons do not correspond to any amino acids. The start signal is usually AUG that encodes Methionine (M). However, the converse is not true. Namely, the region starting with AUG is not necesarry the starting location of translation. The coding region is delimited by these start and stop signals.

In the upstream of the start signal, there is a region called a *promoter* that is recognized by RNA polymerase to locate the start site of RNA transcription on DNA template. The problem of searching for promoters in DNA sequences is still difficult. Promoter sites of E. coli RNA polymerase are compiled in [27]. Promoters in eukaryotes are collected in the Eukaryote Promoter Database (EPD) [8]. An important problem about promoters is to characterize promoter sequences and sites. GenBank is also useful to collect promoters.

4 Methods

This section surveys some techniques for solving problems given in Section 3 with bibliographic data. Most of the techniques are not directly related to learning theory, but may give useful suggestions to the research on these problems by learning paradigms. The text book [18] may be helpful to understand the overview and current state of research.

4.1 Protein Structure Prediction

For the secondary structure prediction problem, there have been proposed several methods to predict the locations of α-helices, β-sheets and turns from amino acid sequences of proteins. The prediction accuracy remains in 55–65%. The evaluation of accuracy strongly depends on the choice of data. Therefore, exact comparison is rather difficult by the data presented in published articles.

The most common strategy is the use of statistical information about amino acid residues appearing in α-helices, β-sheets and turns. Chou and Fasman [9] and Garnier et al. [15] employed this strategy. By analyzing the sequence patterns of the various types of secondary structure segments, Cohen et al. [10] proposed a turn prediction method. The Neural network approach is currently the most successful one. There are a bunch of neural network approaches and their experiments: Qian and Sejnowski [35], Holley and Karplus [21], Kneller at al. [26], etc. We do not enumerate those works here. The strategy of hidden Markov model also achieved good accuracy of prediction: [3, 19]. There are some interesting approaches to protein secondary structure prediction [1, 11, 39, 43]. Our approach by PAC-learning paradigm [1, 39] also attained the former accuracy but did not exceed drastically.

Transmembrane Domain Identification The most common method for prediction transmembrane domains is the hydropathy plot of the amino acid residues of the sequence. The hydropathy index due to Kyte and Doolittel [28] works quite well. Yanagihara [48] developed a very accurate method for this problem. Section 5 presents our approach by PAC-learning paradigm.

Signal Peptides Signal peptides have the hydrophobic nature [45, 46]. Ladunga et al. [29] used neural networks to predict signal peptide regions. There have been developed some software packages for this problem [14, 34]. Our approach by PAC-learning paradigm is shown in Section 5.

Epitopes Antigenic regions locate on the surfaces of proteins and are formed by two or more segments. Hopp and Woods [22] used the hydrophilicity for amino acids and developed a method to predict antigenic regions by plotting the hydrophilicity of the sequence. Jameson and Wolf [24] defined the antigenicity index by a linear combination of hydrophilicity, predicted side chain flexibility [25], surface probability [12], turn prediction by Chou-Fasman [9] and Garnier et al. [15].

4.2 Exon-Intron Boundary Prediction

In coding region identification problem, the most important problem is to predict the *splice junctions* which are boundaries of exons and introns. Recently, neural network approached have been developed [6, 7]. Brunak et al. [7] used neural networks to predict splice site location in human pre-mRNA. Most of the approaches are based on statistical analysis of patterns [16, 20, 23, 32, 37]. Senapathy et al. [37] have analyzed the base frequency distributions for the donor and acceptor sites for the the general categories in GenBank (for these categories, see Section 5.5.2). There are a lot of works on this problem including software packages [41, 42].

5 Machine Discovery by PAC-Learning Paradigm

In this section, we present our approach by PAC-learning paradigm and some algorithmic strategies [1, 39] to the signal peptide identification problem and the transmembrane domain identification problem.

5.1 Polynomial-Time PAC-Learnability

Let us begin with basic terminology about PAC-learning. We call a subset of Σ^* a *concept*. A *concept class* C is a nonempty collection of concepts. For a concept $c \in C$, a pair $\langle w, c(w) \rangle$ is called an example of c for $w \in \Sigma^*$, where $c(w) = 1$ $(c(w) = 0)$ if w is in c (is not in c). For an alphabet Σ and an integer $n \geq 0$, $\Sigma^{\leq n}$ denotes the set $\{w \in \Sigma^* \mid |w| \leq n\}$.

We assume a representation system R for concepts in C. We use a finite alphabet Λ for representing concepts. For a concept class C, a *representation* is a function $R : C \to 2^{\Lambda^*}$ such that $R(c)$ is a nonempty subset of Λ^* for c in C and $R(c_1) \cap R(c_2) = \emptyset$ for any distinct concepts c_1 and c_2 in C. For each $c \in C$, $R(c)$ is the set of *names* for c.

A concept class C is said to be *polynomial-time learnable* [5, 33, 44] if there is an algorithm $\mathcal{A}$ which satisfies (1) and (2).

(1) $\mathcal{A}$ takes a sequence of examples as an input and runs in polynomial-time with respect to the length of input.
(2) There exists a polynomial $p(\cdot, \cdot, \cdot)$ such that for any integer $n \geq 0$, any concept $c \in C$, any real numbers ε, δ $(0 < \varepsilon, \delta < 1)$, and any probability distribution P on $\Sigma^{\leq n}$, if $\mathcal{A}$ takes $p(n, \frac{1}{\varepsilon}, \frac{1}{\delta})$ examples which are generated randomly according to P, then $\mathcal{A}$ outputs, with probability at least $1 - \delta$, a name of a hypothesis h with $P(c \oplus h) < \varepsilon$.

5.2 Learning Decision Trees over Regular Patterns

Let us explain our approach [1, 39] to, for example, the signal peptide identification problem. Let $\mathcal{R} = \{A, ..., Z\}$ be the alphabet consisting of 20 symbols of amino acid residues and additional symbols for representing ambiguity. From GenBank, we have collected 1032 sequences of amino acids for primate signal peptides. Namely, these sequences are strings over $\mathcal{R}$. These data have been provided by experiments by

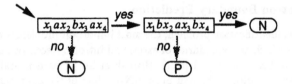

Fig. 4. Decision tree over regular patterns defining the language $\{a^m b^n a^l \mid m, n, l \geq 1\}$ over $\Sigma = \{a, b\}$.

biologists who have their own interests. Hence we may regard that the data are coming out by following some unknown probability distribution.

The objective is to find a "short explanation" of these sequences which may have a sense in Molecular Biology. Then by using the knowledge described in the "short explanation", we may predict signal peptides in the sequences. We also need to prepare negative examples for signal peptides. This is discussed in the sequel.

For applying PAC-learning paradigm to this problem, the first thing we have to do is to define a target concept class $\mathcal{C}$ in which there may be a concept representing the primate signal peptides. Discussions with molecular biologists would be helpful to define a target concept class. Then for this concept class $\mathcal{C}$, we have to find a learning algorithm running in feasible time. The following result is useful to show the polynomial-learnability of $\mathcal{C}$:

Theorem 1. [33] A concept class $\mathcal{C}$ is polynomial-time learnable if $\mathcal{C}$ is of polynomial dimension and there is a polynomial-time fitting for $\mathcal{C}$.

In analyzing of amino acid sequences and nucleic sequences, "motifs" are often searched that are common to the given sequences as complied in PROSITE [4]. For example, the expression C-x(2,4)-C-x(12)-H-x(3,5)-H is called the zinc finger motif. These "motifs" can be regarded as a kind of simple regular expressions. By this observation, we use *regular patterns* [40] as the view by which we analyze the sequences.

Formally, a regular pattern is an expression of the form $\pi = \alpha_0 x_1 \alpha_1 \cdots x_n \alpha_n$, where $\alpha_0, ..., \alpha_n$ are strings over an alphabet Σ and $x_1, ..., x_n$ are mutually distinct variables to which arbitrary strings in Σ^* are substituted. Thus it defines a regular language over Σ. We denote its defining language by $L(\pi)$ (we allow ε-substitutions). A regular pattern containing at most k variables is called a k-*variable regular pattern*.

Now we have established the basic view on the sequences. The next step is to combine these regular patterns in order to make a concept. The idea is to use a decision tree whose nodes have regular patterns as their labels for classification. We call such a tree a *decision tree over regular patterns*, which is a binary tree T such that each leaf is labeled with class name N (negative) or P (positive) and each internal node is labeled with a regular pattern (see Fig. 4). Let $L(T)$ be the set of strings that are classified as P. Obviously, $L(T)$ is also a regular language.

For integers $k, d \geq 0$, we denote by $DTRP(d, k)$ the class of languages defined by decision trees over k-variable regular patterns with depth at most d.

We have shown in [1] the following theorem:

Theorem 2. [1] $DTRP(d,k)$ is polynomial-time learnable for all $d, k \geq 0$.

We have discussed in [31] that the polynomial-time learnability requires the constant bound k on the number of variables.

5.3 Indexing of Amino Acid Residues

Kyte and Doolittle [28] have given the hydropathy index for amino acid residues in their study of membrane spanning segments. Each amino acid residue is assigned a real number between $-4.5 \sim 4.5$. According to the hydropathy index, we have classified 20 kinds of amino acid residues into three categories $\{*, +, -\}$ as shown in Table 1 and found that the transformation of amino acid residues by Table 1 makes learning drastically efficient in the experiments [2, 1]. Suprizingly, the transmembrane domain sequences transformed by Table 1 have only a few overlaps with nontransmembrane domain sequences. Hence, the information about positiveness and negativeness is not lost by this transformation.

Amino Acids	Hydropathy Index		New Symbol
A M C F L V I	$1.8 \sim 4.5$	$\rightarrow$	*
P Y W S T G	$-1.6 \sim -0.4$	$\rightarrow$	+
R K D E N Q H	$-4.5 \sim -3.2$	$\rightarrow$	-

Table 1. Categorization of amino acid residues by the hydropathy index

This observation led us to the following notion:

Definition 3. Let Σ be a finite alphabet and P and N be two disjoint subsets of Σ^*. Let Γ be a finite alphabet, called an *indexing alphabet*, satisfying $|\Sigma| > |\Gamma|$. An *indexing* ψ of Σ by Γ with respect to P and N is a mapping $\psi : \Sigma \rightarrow \Gamma$ such that $\tilde{\psi}(P) \cap \tilde{\psi}(N) = \emptyset$, where $\tilde{\psi} : \Sigma^* \rightarrow \Gamma^*$ is the homomorphism defined by $\tilde{\psi}(a_1 \cdots a_n) = \psi(a_1) \cdots \psi(a_n)$ for $a_1, \ldots, a_n \in \Sigma$.

In the above definition, the indexing ψ with respect to P and N must satisfy the condition $\tilde{\psi}(P) \cap \tilde{\psi}(N) = \emptyset$. However, this condition is too strong for practical use because we have the following theorem:

Theorem 4. [38] The problem of finding an indexing is NP-complete.

It should be also stated that the data from the PIR and GenBank databases contains some amount of errors. Therefore it is reasonable to relax the condition of indexing so that $\tilde{\psi}(P)$ and $\tilde{\psi}(N)$ have a few overlaps.

For the representation of a concept, we use a pair (T, ψ) of a mapping $\psi : \mathcal{R} \rightarrow \Gamma$ and a decision tree T over regular patterns, where constant strings appearing in the patterns are taken from Γ^*. Strings in $\mathcal{R}^*$ are first transformed by ψ to strings in Γ and then classified by T. The target concept class C consists of the concepts over $\mathcal{R}^*$ defined by such pairs (T, ψ).

5.4 Algorithms for Decision Trees and Indexing

Although the learning algorithm showing Theorem 2 runs in polynomial time, the running time is enormous and the algorithm does not have any sense in practical use. This is the reason why we developed in [1] a heuristic algorithm for learning a decision tree over regular patterns.

Based on these ideas, we have developed a machine learning system BONSAI that shall produce, from sets of positive and negative examples, a pair of an indexing and a decision tree over regular patterns.

This section gives algorithms for finding approximate indexings and constructing decision trees over regular patterns that are implemented in BONSAI.

Let POS and NEG be two disjoint subsets of $\mathcal{R}^*$, where POS is,in practice, the set of all collected positive examples and NOG is the set of suitably selected negative examples. From POS and NEG, two small subsets pos and neg are randomly chosen for training sets. Let Γ be a finite alphabet for indexing.

By following [1], we briefly review how to construct decision trees over regular patterns from pos and neg of positive and negative training examples. We employed the idea of ID3 algorithm [36] for constructing a decision tree. The ID3 algorithm assumes examples specified with explicit attributes in advance. On the other hand, our approach assumes a space of regular patterns which are simply generated by given pos and neg. The algorithm tries to find appropriate regular patterns from this space of attributes dynamically during the construction of a decision tree. Thus the view to the data determines the space of attributes. This is a point which is very suited for our empirical research.

Let P and N be the sets of strings obtained by transforming pos and neg of the training examples by some indexing ψ of $\mathcal{R}$ by Γ. Notice that P and N may have some intersection. Using P and N, we deal with regular patterns of the form $\alpha_0 x_1 \alpha_1 x_2 \cdots x_k \alpha_k$ such that $\alpha_0, ..., \alpha_k$ are substrings of some strings in $P \cup N$. Let $\Pi(P, N)$ be some family of such regular patterns made from P and N. The family $\Pi(P, N)$ is appropriately given and used as a space of attributes.

For a regular pattern $\pi \in \Pi(P, N)$, the cost $E(\pi, P, N)$ is the one defined in [36] by

$$E(\pi, P, N) = \frac{p_1 + n_1}{|P| + |N|} I(p_1, n_1) + \frac{p_0 + n_0}{|P| + |N|} I(p_0, n_0),$$

where p_1 (resp. n_1) is the number of positive examples in P (resp. negative examples in N) that match π, i.e., $p_1 = |P \cap L(\pi)|$, $n_1 = |N \cap L(\pi)|$, and p_0 (resp. n_0) is the number of positive examples in P (resp. negative examples in N) that do not match π, i.e., $p_0 = |P \cap \overline{L(\pi)}|$, $n_0 = |N \cap \overline{L(\pi)}|$, $\overline{L(\pi)} = \Sigma^* - L(\pi)$, and

$$I(x, y) = \begin{cases} 0 & \text{(if } x = 0 \text{ or } y = 0) \\ -\frac{x}{x+y} \log \frac{x}{x+y} - \frac{y}{x+y} \log \frac{y}{x+y} & \text{(otherwise)}. \end{cases}$$

Algorithm $MakeTree(P, N)$ sketches the decision tree algorithm for $\Pi(P, N)$, where $Create(\pi, T_0, T_1)$ returns a new tree with a root labeled with π whose left and right subtrees are T_0 and T_1, respectively.

The above algorithm is sufficiently fast and experiments show that small enough trees are very often produced. When we consider the class $DTRP(k, d)$, we have to specify k of variables in a pattern and d of the depth of a decision tree. However,

```
function MakeTree( P, N : sets of strings ): node;
  begin
    if N = ∅ then return( Create("P", null, null) )
    else if P = ∅ then return( Create("N", null, null) )
    else begin
      Find a shortest pattern π in Π(P, N) that minimizes E(π, P, N);
      P₁ ← P ∩ L(π);  P₀ ← P - P₁;
      N₁ ← N ∩ L(π);  N₀ ← N - N₁;
      if (P₀ = P and N₀ = N) or (P₁ = P and N₁ = N) then
                    /* No more division is possible */
            return( Create("P", null, null) )
      else return( Create(π, MakeTree(P₀, N₀), MakeTree(P₁, N₁) ) )
    end
  end
```

Algorithm 1: *MakeTree*

we can not know in advance which k and d are appropriate for the signal peptide problem. Therefore, as long as small hypotheses are produced, we may be satisfied with the results.

Now we describe the algorithm that finds indexings. It searches for a locally optimal indexing using a score function $Score(\psi)$ that calls *MakeTree* as a subroutine. Let Ψ be the set of all indexings of $\mathcal{R}$ by Γ. Let POS be the set of positive examples and NEG be the set of negative examples from $\mathcal{R}^*$. Let pos and neg be the sets of positive and negative training examples. To evaluate the "goodness" of an indexing $\psi \in \Psi$, it constructs a tree T by running the procedure $MakeTree(\tilde{\psi}(pos), \tilde{\psi}(neg))$ and then evaluates the success rate that T explains $\tilde{\psi}(POS)$ and $\tilde{\psi}(NEG)$ correctly. $Score(\psi)$ is determined by $Verify(MakeTree(\tilde{\psi}(pos), \tilde{\psi}(neg)), \tilde{\psi}(POS), \tilde{\psi}(NEG))$, where the function $Verify(T, Pos, Neg)$ returns the value $\sqrt{\frac{|L(T) \cap Pos|}{|Pos|} \cdot \frac{|L(T) \cap Neg|}{|Neg|}}$, which represents the geometric mean of the success rates of the decision tree T for positive examples Pos and negative examples Neg.

For indexings ψ and ϕ in Ψ, the distance between ψ and ϕ is defined by $d(\psi, \phi) = |\{\sigma \in \Sigma \mid \psi(\sigma) \neq \phi(\sigma)\}|$. The neighbors of ψ are the indexings whose distance from ψ is one. Algorithm *FindGoodIndex* begins with an indexing ψ which is selected randomly from Ψ. Then it searches an indexing ϕ from its neighbors such that its score is the best among the neighbors of ψ. This process continues until no better indexing is found from its neighbors. The strategy of the algorithm is sketched in Algorithm *FindGoodIndex*.

These two algorithms are combined and related as shown in Fig. 5. The following two subsections describe how to apply BONSAI to acquire knowledge about transmembrane domains and signal peptides.

5.5 Experiments

We have to specify two parameters in BONSAI. The first is the size of the small sets pos and neg which shall be chosen from POS and NEG at random. The second

```
function FindGoodIndex(POS,NEG: sets of strings) indexing;
    begin
        Select small subsets pos of POS and neg of NEG randomly;
        Select randomly ψ₀ from Ψ;
        repeat
            ψ ← φ;
            for each φ' ∈ Ψ with d(ψ, φ') = 1 do
                if Score(φ') > Score(φ) then φ ← φ';
        until ψ = φ;
        return ψ;
    end
```

Algorithm 2: *FindGoodIndex*

is the size of the indexing alphabet. With these parameters, the system with will produce a hypothesis (T, ψ) consisting of a decision tree T over regular patterns and an indexing ψ. The accuracy of (T, ψ) for POS and NEG is represented by a pair $(p\%, n\%)$, where $p\%$ (resp. $n\%$) of POS (resp. NEG) are recognized as positive (resp. negative).

In the following experiments, we set the size of pos and neg to be $|pos| = |neg| = 10$ and the indexing alphabet Γ has 2∼3 symbols.

Moreover, in order to avoid combinatorial explosions, we also assume that the regular patterns attached to the nodes of decision trees are of the form $x\alpha y$, where x and y are variables and α is a substring taken from pos and neg. There is no other special reason why we used only these regular patterns except that they are the simplest.

Transmembrane Domains The sequences corresponding to the transmembrane domains are positive examples. Negative examples are sequences of length around 30 taken from the parts other than transmembrane domains. For example, in the sequence shown in Fig. 1, there are three transmembrane domains as is indicated in its FEATURE field as below. These are taken as positive examples.

```
FEATURE
    9-34                #Domain transmembrane I
    59-96               #Domain transmembrane II
    232-267             #Domain transmembrane III
```

As negative examples, we use the amino acid sequences located in other parts than transmembrane domains. For instance, the sequence $w[100..130]$ is a possible negative example, while $w[46..75]$ is not, because the terminal segment of $w[46..75]$ is located in a transmembrane domain $w[59..96]$.

We collect all the positive examples from the PIR database. The number of positive examples is 689. We use 19256 negative examples randomly chosen. These sequences form the sets POS and NEG for the transmembrane domain identification problem.

32

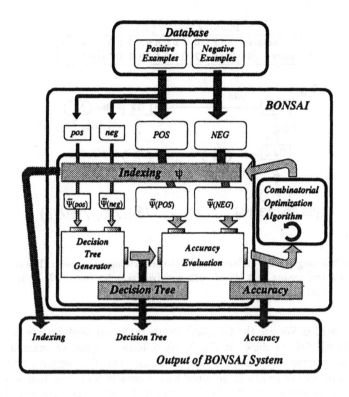

Fig. 5. BONSAI

We have reported in [39] that BONSAI has discovered a decision tree over regular pattern with just three internal nodes with accuracy more than 92%. More interestingly, the indexing associated with the decision tree exactly corresponds to the hydropathy index of amino acids due to Kyte and Doolittle [28] except Asparagine (N) as shown in Table 2. Biologists gave a comment that the exception of N is reasonable.

Table 2. Indexing discovered by BONSAI and its correspondence with the hydropahty index.

Amino Acid	A	C	D	E	F	G	H	I	K	L	M	N	P	Q	R	S	T	V	W	Y
New Symbol	0	0	1	1	0	0	1	0	1	0	0	0	0	1	1	1	0	0	0	0
Hydropathy	1.8	2.5	-3.5	-3.5	2.8	-0.4	-3.2	4.5	-3.9	3.8	1.9	-3.5	-1.6	-3.5	-4.5	-0.8	-0.7	4.2	-0.9	-1.3

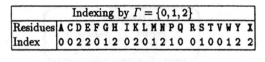

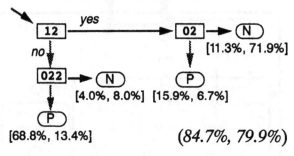

Fig. 6. The indexing and the decision tree for the primate signal peptides. The pair [p%, n%] attached to a leaf means that p% of 1032 positive examples (resp. n% of 3162 negative examples) reached the leaf.

Signal Peptides We use the GenBank database. Signal peptides are indicated in the FEATURE field as shown in Fig. 2. We collected as positive examples the signal peptide sequences beginning with a Methionine (M). The negative examples are N-terminal regions of length 30 obtained from complete sequences that have no signal peptide and begin with M. The following table shows the numbers of positive and negative examples taken from the files of viral, bacterial, invertebrate, primate, rodent, other mammalian, other vertebrate, and plant in GenBank.

Sequences	Positive	Negative
Viral	120	4882
Bacterial	495	7330
Invertebrate	263	1927
Primate	1032	3162
Rodent	1018	3158
Other Mammalian	235	588
Other Vertebrate	207	1056
Plant	370	3074

For each file, we made an experiment for finding a good decision tree and an indexing. We have reported the results in [39] that achieved good accuracy. The most unsuccessful is the case for the primate signal peptides shown in Fig. 6 (accuracy is (84.7%, 79.9%)), while the most successful is the case for the bacterial signal peptides whose accuracy is (86.3%, 90.6%) for positive and negative examples.

6 Conclusion

We have given an overview of problems for which learning paradigms are required. But the methods so far developed are just on the first stage of research. The next development is strongly expected for Genome Informatics. In the framework of PAC-learning, we sketched our research on knowledge acquisition from protein data. Based on the research, we have designed and implemented the system BONSAI, which produces a decision tree over regular patterns and an indexing of amino acid residues. We also reported some experiments on transmembrane domains and signal peptides by BONSAI that showed a large potential for discovering knowledge from biological data.

7 Acknowledgments

The author would like to thank S. Kuhara at Graduate School of Genetic Resources Technology, Kyushu University for guiding him into this new field. Special thanks are due to A. Shinohara and S. Shimozono. Without their contributions, the project could not achieve this high level of success. The author is also indebted to many people, especially, S. Arikawa, T. Shinohara and T. Uchida. This research is partly supported by Grant-in-Aid for Scientific Research on Priority Areas "Genome Informatics" from the Ministry of Education, Science and Culture, Japan.

References

1. Arikawa, S., Kuhara, S., Miyano, S., Mukouchi, Y., Shinohara, A., and Shinohara, T. [1993], Machine discovery of a negative motif from amino acid sequences by decision trees over regular patterns, *New Generation Computing* 11, 361–375.
2. Arikawa, S., Kuhara, S., Miyano, S., Shinohara, A., and Shinohara, T. [1992], A learning algorithm for elementary formal systems and its experiments on identification of transmembrane domains, *Proc. 25th Hawaii International Conference on System Sciences*, 675–684.
3. Asai, K., Hayamizu, S., and Onizuka, K. [1993], HMM with protein structure grammar, *Proc. 26th Hawaii International Conference on System Sciences*, 783–791.
4. Bairoch, A. [1991], PROSITE: a dictionary of sites and patterns in proteins, *Nucleic Acids Res.* 19, 2241–2245.
5. Blumer, A., Ehrenfeucht, A., Haussler, D., and Warmuth, M.K. [1989], Learnability and the Vapnik-Chervonenkis dimension, *JACM*, 36, 929–965.
6. Brunak, S., Engelbrecht, J., and Knudsen, S. [1990], Neural network detects erros in the assignment of mRNA splice sites, *Nucleic Acids Res.* 18, 4797–4801.
7. Brunak, S., Engelbrecht, J., and Knudsen, S. [1991], Prediction of human mRNA donor and acceptor sites from the DNA sequence, *J. Mol. Biol.* 220, 49–65.
8. Bucher, P. [1988], The eukaryote promoter database of the Weizmann Institute of Science, *EMBL Nucleiotite Sequence Data Library Release* 17, Heidelberg, Germany.
9. Chou, P.Y. and Fasman, G.D. [1978], Prediction of the secondary structure of proteins from their amino acid sequence, *Advances in Enzymology* 47, 45–147.
10. Cohen, R.E., Abarbanel, R.A., Kuntz, I.D., and Fletterick, R.J. [1986], Turn prediction in proteins using a pattern matching approach, *Biochemistry* 25, 266–275.
11. Dowe, D.L., Oliver, J., Dix, T.I., Allison, L., and Wallace, C.S. [1993], A decision graph explanation of protein secondary structure prediction, *Proc. 26th Hawaii International Conference on System Sciences*, 669–678.
12. Emini, E.A., Hughes, J.V., Perlow, D.S., and Boger, J. [1985], Induction of hepatitis A virus-neutralizing antibody by a virus-specific peptide, *J. Virol.* 55, 836–839.

13. Endgelman, D.M., Steiz, T.A., and Goldman, A. [1986], Identifying nonpolar transbilayer helices in amino acid sequences of membrane proteins, *Ann. Rev. Biophys. Chem.* **15**, 321–354.

14. Folz, R.J. and Gordon, J.I. [1987], Computer-assisted predictions of signal peptibase processing sites, *Biochem. Biophys. Res. Comm.* **146**, 870–877.

15. Garnier, J., Osguthorpe, D.J., and Robon, B. [1978], Analysis of the accuracy and implication of simple methods for predicting the secondary structure of globular proteins, *J. Mol. Biol.* **120**, 97–120.

16. Gelfand, M.S. [1989], Statistical analysis of mammalian pre-mRNA splicing sites, *Nucleic Acids Res.* **17**, 6369–6382.

17. GenBank, Genetic Sequence Data Bank, National Institute of General Medical Science, NIH by contract to Intelligenetics, Inc., and Los Alamos Laboratory.

18. Gribskov, M. and Devereux, J. [1991], *Sequence Analysis Primer*, UWBC Biotechnical Resource Series, Macmillan Publishers Inc.

19. Haussler, D., Krogh, A., Mian, I.S., and Sjölander, K. [1993], Protein modeling using hidden Markov models: analysis of globins, *Proc. 26th Hawaii International Conference on System Sciences*, 792–802.

20. Harris, N.L. and Senapathy, P. [1990], Distribution and consensus of branch point signals in eukaryotic genes: a computerized statistical analysis, *Nucleic Acids Res.* **18**, 3015–3019.

21. Holley, L.H. and Karplus, M. [1989], Protein secondary structure prediction with a neural network, *Proc. Nal. Acad. Sci. USA* **86**, 152–156.

22. Hopp, T.P. and Woods, K.R. [1981], Prediction of protein antigenic determinants from amino acid sequences, *Proc. Natl. Acad. Sci. USA* **78**, 3824–3828.

23. Iida, Y. and Sasaki, F. [1983], Recognition patterns for exon-intron junctions in higher organism as revealed by a computer search, *J. Biochem.* **94**, 1731–1738.

24. Jameson, B.A. and Wolf, H. [1988], The antigenic index: a novel algorithm for predicting antigenic determinants, *Comput. Appl. Biosci.* **4**, 181–186.

25. Karplus, P.A. and Schulz, G.E. [1985], Prediction of chain flexibility in proteins, *Naturwissenschaften* **72**, 212–213.

26. Kneller, D.G., Choen, F.E., and Langridge, R. [1990], Improvements in protein secondary structure prediction by an enhanced neural network, *J. Mol. Biol.* **214**, 171–182.

27. Kroeger, M., Wahl, R., and Rice, P. [1990], Compilation of DNA sequences of Escherichia coli (update 1990), *Nucleic Acids Res.* **18**, 2549–2587.

28. Kyte, J. and Doolittle, R.F. [1982], A simple method for displaying the hydropathic character of protein, *J. Mol. Biol.*, **157**, 105–132.

29. Ladunga, I., Czako, F., Csabai, I., and Geszti, T. [1991], Improving signal peptide prediction accuracy by simulated neural network, *Comput. Appl. Biosci.* **7**, 485–487.

30. Lewin, B. [1987], *Genes: Third Edition*, John Wiley & Sons, Inc.

31. Miyano, S., Shinohara, A., and Shinohara, T. [1993], Learning elementary formal systems and an application to discovering motifs in proteins, Technical Report RIFIS-TR-CS-37, Research Institute of Fundamental Information Science, Kyushu University, revised in April, 1993 (former version: Proc. 2nd Algorithmic Learning Theory, 139–150, 1991).

32. Nakata, K., Kanehisa, M., DeLisi, C. [1985], Prediction of splice junctions in mRNA sequences, *Nucleic Acids Res.* **13**, 5327–5340.

33. Natarajan, B.K. [1989], On learning sets and functions, *Machine Learning*, **4**, 67–97.

34. Pascarella, S. and Bossa, F. [1989], CLEAVAGE: a microcomputer program for predicting signal sequence cleavage sites, *Comput. Appl. Biosci.* **5**, 53–54.

35. , Qian, N. and Sejnowski, T.J. [1988], Predicting the secondary structure of globular proteins using neural network models, *J. Mol. Biol.* **202**, 865–884.
36. Quinlan, J.R. [1986], Induction of decision trees, *Machine Learning*, **1**, 81–106.
37. Senapathy, P., Shapiro, M.B., and Harris, N.L. [1990], Splice junctions, branch point sites, and exons: sequence statistics, identification, and applications to the genome project, *Meth. Enzym.* **183**, 252–278.
38. Shimozono, S. and Miyano, S. [1992], Complexity of finding alphabet indexing, Technical Report RIFIS-TR-CS-61, Research Institute of Fundamental Information Science, Kyushu University, August, 1992.
39. Shimozono, S., Shinohara, A., Shinohara, T., Miyano, S., Kuhara, S., and Arikawa, S. [1993], Finding alphabet indexing for decision trees over regular patterns: an approach to bioinformatical knowledge acquisition, *Proc. 26th Hawaii International Conference on System Sciences*, 763–772.
40. Shinohara, T. [1983], Polynomial time inference of extended regular pattern languages, *Proc. RIMS Symp. Software Science and Engineering* (Lecture Notes in Computer Science, **147**, 115–127.
41. Staden, R. [1990], An improved sequence handling package that runs on the Apple Macintosh, *Comput. Applic. Biosciences* **6**, 387–393.
42. Staden, R. [1990], Finding protein coding regions in genomic sequences, *Meth. Enzym.* **183**, 163–180.
43. Unger, R. and Moult, J. [1993], On the applicability of genetic algorithms to protein folding, *Proc. 26th Hawaii International Conference on System Sciences*, 715–725.
44. Valiant, L. [1984], A theory of the learnable, *Commun. ACM*, **27**, 1134–1142.
45. von Heijne, G. [1981], On the hydrophobic nature of signal sequences, *Eur. J. Biochem.* **116**, 419–422.
46. von Heijne, G. [1986], A new method for predicting signal sequences cleavage sites, *Nucleic Acids Res.* **14**, 4683–4690.
47. Watson,J.D., Hopkins, N.H., Robets, J.W., Steitz, J.A., and Weiner, A.M. [1987],*Molecular Biology of The Gene: Fourth Edition*, The Benjamin/Cummings Publishing Company, Inc.
48. Yanagihara, N., Suwa, M., and Mitaku, S. [1989], A theoretical method for distinguishing between soluble and membrane proteins, *Biophysical Chemistry*, **34**, No. 1, 69–77.

Optimal layered learning: a PAC approach to incremental sampling

Stephen Muggleton

Oxford University Computing Laboratory,
11 Keble Road, Oxford, OX1 3QD, United Kingdom. Email: steve@prg.oxford.ac.uk

Abstract. It is best to learn a large theory in small pieces. An approach called "layered learning" starts by learning an approximately correct theory. The errors of this approximation are then used to construct a second-order "correcting" theory, which will again be only approximately correct. The process is iterated until some desired level of overall theory accuracy is met. The main advantage of this approach is that the sizes of successive training sets (errors of the hypothesis from the last iteration) are kept low. General lower-bound PAC-learning results are used in this paper to show that optimal layered learning results in the total training set size (t) increasing linearly in the number of layers. Meanwhile the total training and test set size (m) increases exponentially and the error (ϵ) decreases exponentially. As a consequence, a model of layered learning which requires that t, rather than m, be a polynomial function of the logarithm of the concept space would make learnable many concept classes which are not learnable in Valiant's PAC model.

1 Introduction

Since the introduction [6] of the Probably-Approximately-Correct (PAC) model of learning many theoretical results indicate that it is not possible to learn a great deal at once. Indeed it is clear that weakly constrained languages either require too many examples to describe the target concept sufficiently accurately, or require an untractably large amount of time for concept formation and testing. In this paper an incremental approach called "layered learning" is investigated and analysed using general lower-bound results for PAC learning. Layered learning proceeds by first taking a small sample from a stream of data and using it to construct an approximately correct theory. A second approximately correct theory is then constructed based on the errors of the first theory in a new sample which is a supserset of the first. Further layers of correcting theories are then added using successively larger samples until a pre-specified level of overall theory accuracy is achieved. This approach is similar to what Quinlan [5] calls "windowing".

Clearly the minimal example requirements for layered learning will be limited by existing general lower-bound PAC results. However, it is also clear that most of these examples will simply be used for testing the present stage of the theory. Only a small number of examples, related to the present error-rate, will be used for any layer of the training set. In this paper it is shown using lower-bound PAC learning results that an optimal use of examples leads to the following.

- the total training set increases linearly with the number of layers,
- the total test set increases exponentially with the number of layers and
- the error decreases exponentially with the number of layers.

2   Lower bound PAC result

Valiant [6] introduced what has now become a widely studied stochastic model of machine learning. In this model positive and negative examples of some unknown concept, chosen from a concept class C, are presented to a learning algorithm. These examples are drawn according to a fixed but arbitrary probability distribution. From the examples drawn, the learning algorithm must, with high probability, produce a hypothesised concept that is a good approximation to the target.

Suppose the concept class C is of size 2^n. Then if a uniform prior distribution over the concept class is assumed each concept $c \in C$ can be expressed in n bits. According to [3] the following is an expression of the minimum number of examples, m, required to allow construction of an hypothesis H for which the probability of $error(H) \leq \epsilon$ is at least $1 - \delta$.

$$m = \frac{n + ln(\frac{1}{\delta})}{-ln(1 - \epsilon)}$$

Existing algorithms for learning monomials, kDNF formulae, kCNF formulae and symmetric functions all use the optimal number of examples (within a constant factor).

The result can be re-expressed as follows to give the accuracy $(1 - \epsilon)$ expected in terms of m, n and δ.

$$(1 - \epsilon) = e^{-\frac{1}{x}}$$

where

$$x = \frac{m}{n + ln(\frac{1}{\delta})}$$

Since n is measured in bits, for a fixed value of δ, x is proportional to the number of examples required per bit of the concept learned. For fixed values of δ and n, x is simply proportional to m. Figure 1 shows the increase of accuracy with increasing x. Note the following properties of this curve.

1. $d(1 - \epsilon)/dx \to 0$ as $x \to 0$.
2. $d(1 - \epsilon)/dx \to 0$ as $x \to \infty$.
3. $(1 - \epsilon)$ increases monotonically with x.

These observations correspond to a law of diminishing returns in machine learning. When only a small number of examples have been observed accuracy increases slowly with each example. The same occurs when a large number of

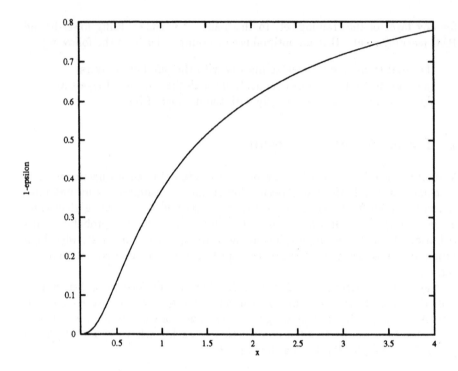

Fig. 1. Increase of $(1 - \epsilon)$ with x

examples have been observed. The maximum rate of accuracy increase occurs somewhere between these two extremes. By setting the double differential to zero we find that the maximum value of $e^{-\frac{1}{x}}$ occurs when $x = \frac{1}{2}$, i.e.

$$(1 - \epsilon) = e^{-2} = 0.135$$

Note that this maximum rate of accuracy increase is independent of m, n and δ. Since the lower-bound PAC results on which this are based are also independent of both the example distribution and the concept language it can be considered that an accuracy of $(1 - \epsilon) = e^{-2}$ has a fundamental significance throughout inductive learning.

3   Maximising performance increase per example

In the last section it was demonstrated that when using lower-bound PAC learning results maximum performance increase occurs when $(1 - \epsilon) = e^{-2}$. However the point of maximum increase in accuracy at

$$m = \frac{n + ln(\frac{1}{\delta})}{2}$$

does not provide the optimal number of examples required for the first stage of layered learning. In order to make best use of the training set it is necessary to maximise the increase in accuracy per training example. This can be achieved by solving

$$\frac{d}{dm}\frac{(1-\epsilon)}{m} = 0$$

This gives

$$m = n + ln(\frac{1}{\delta})$$

for which

$$(1-\epsilon) = e^{-1} = 0.368$$

The increase of $(1-\epsilon)/x$ with x is shown in Figure 2.

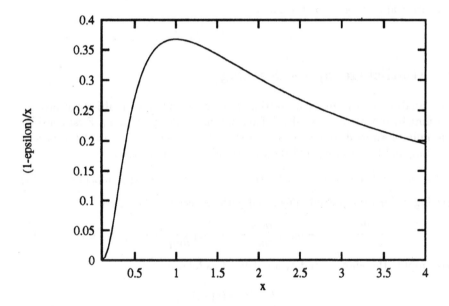

Fig. 2. Increase of $(1-\epsilon)/x$ with x

4 Two stage layered learning

In the last section we found that the optimal number of training examples for the first stage of layered learning is that which produces an accuracy of $(1-\epsilon_1) = e^{-1}$. Suppose that an accuracy of e^{-1} is produced from a training set of size m_1 in at least $(1 - \delta)$ retrials. It is now possible to predict that the size, in bits, of the target concept to be learned is

$$n = m_1 - ln(\frac{1}{\delta})$$

Now suppose that the total sample for training and testing in two staged layered learning is m_2. Since only the errors in the second sample are used for training, the total training set t_2 (stage 1 + stage 2) is

$$t_2 = m_1 + \epsilon_1(m_2 - m_1)$$

Solving

$$\frac{dt_2}{dm_1} = 0$$

gives

$$m_2 = 2m_1 = 2(n + ln(\frac{1}{\delta}))$$

The error after stage two will thus be

$$(1 - \epsilon_2) = e^{-\frac{1}{2}} = 0.607$$

5 Multi-stage layered learning

The analysis of the previous section can be extended to multi-stage layered learning by repeatedly partially differentiating with respect to m_i. First consider the infinite series defining the size of the training set t in terms of the partial training and test sets m_i and the corresponding error-rates ϵ_i.

$$t = m_1 + \epsilon_1(m_2 - m_1) + \ldots + \epsilon_i(m_{i+1} - m_i) + \epsilon_{i+1}(m_{i+2} - m_{i+1}) + \ldots \quad (1)$$

Setting to zero the partial differential with respect to m_{i+1} gives

$$\frac{\partial t}{\partial m_{i+1}} = \epsilon_i + m_{i+2}\frac{\partial \epsilon_{i+1}}{\partial m_{i+1}} - (m_{i+1}\frac{\partial \epsilon_{i+1}}{\partial m_{i+1}} + \epsilon_{i+1}) = 0 \quad (2)$$

To simplify the above b and x_i are defined as follows

$$b = (n + ln(\frac{1}{\delta}))$$

$$x_i = \frac{m_i}{b}$$

ϵ_i can now be approximated using series expansion as follows.

$$\epsilon_i = 1 - e^{-\frac{b}{m_i}}$$

$$= 1 - e^{-\frac{1}{x_i}}$$

$$= 1 - (1 - \frac{1}{x_i} + \frac{1}{x_i^2} + \ldots)$$

$$\simeq \frac{1}{x_i} \qquad \text{for large } x_i$$

Simplifying equation (2) and rearranging gives

$$m_{i+2} = \frac{m_{i+1}^2}{b(1-\epsilon_{i+1})}(\epsilon_i - \epsilon_{i+1} + \frac{b}{m_{i+1}}(1 - \epsilon_{i+1}))$$

$$\simeq \frac{bx_{i+1}}{x_{i+1}-1}(\frac{x_{i+1}^2}{x_i} - 1)$$

$$bx_{i+2} \simeq \frac{bx_{i+1}^2}{x_i} \qquad \text{for large } x_{i+1}$$

$$x_{i+2} \simeq \frac{x_{i+1}^2}{x_i}$$

It can be shown that x_i is an exponential series as follows. Let $x_1 = a$ and $x_2 = ad$. Then it follows from the above that

$$x_3 \simeq ad^2$$
$$x_4 \simeq ad^3$$
$$x_5 \simeq ad^4$$
$$\dots$$

and in general

$$\frac{x_{i+1}}{x_i} \simeq d$$

The general term in equation (1) can now be expressed as

$$t_{i+1} = \epsilon_i(m_{i+1} - m_i)$$
$$= \epsilon_i b(x_{i+1} - x_i)$$
$$\simeq \frac{b}{x_i}(x_{i+1} - x_i)$$
$$\simeq b(d - 1)$$

Thus for fixed n and δ, and large x_i the size of each successive training set remains constant at $b(d-1)$ for each value of i. The total training and test set increases exponentially in i since $x_i \simeq ad^{i-1}$. Similarly the error decreases exponentially since $\epsilon_i \simeq \frac{1}{x_i}$. Without approximations, equation (2) can be rearranged to show that

$$x_{i+2} = (1 + \frac{1}{x_{i+1}} - e^{\frac{1}{x_{i+1}} - \frac{1}{x_i}})x_{i+1}^2$$

This recurrence formula is used in Figure 3 to give a tabulation of t_i/b (relative training set size), x_i (relative test set size) and ϵ_i (error) for i ranging between 1 and 10. Note that although t_i/b is not a constant (it asymptotes to one), this tabulation shows that the three general trends arrived at in the analysis above hold for large x_i.

i	t_i/b	x_i	ϵ_i
1	1	1.00	.63
2	.63	2.00	.39
3	.62	3.57	.24
4	.62	6.10	.15
5	.62	10.17	.09
6	.61	16.73	.06
7	.61	27.32	.03
8	.61	44.42	.02
9	.61	72.03	.01
10	.61	116.62	.009

Fig. 3. Relative size of training set, test set and error for various values of i

6 Discussion

Layered learning offers a general approach to incremental machine induction in which successive layers of constructed knowledge decrease the error exponentially. Since successive training sets remain constant in size a learner can reach arbitrarily low values of ϵ without increasing memory requirements.

For instance, layered learning of a 10,000 bit theory requires around 6,100 training examples in each layer to produce arbitrarily low error (see Figure 3). However, the cumulative training and test set size at layer 10 would be 1,166,200 examples.

Suppose that the memory limit of the inductive learner is $l \leq 6,100$. The same general effect can be achieved as that in Figure 3 by letting $l = b(d-1)$, i.e. $d = \frac{l}{b} + 1$. In this case errors would still reduce in $O(d^{-i})$.

Layered learning provides a basis in computational complexity for what has been termed "predicate invention" within the field of Inductive Logic Programming [4]. Predicate invention involves the decomposition of predicates being learned into useful sub-concepts. Such sub-concepts can be viewed as modifiers to predicate definitions which would otherwise be both incomplete and incorrect. This approach is similar to that taken in [2, 1] and [7].

A model of layered learning which requires t, rather than m, to be a polynomial function of the number of bits in the target concept would make learnable many concept classes which are not learnable in Valiant's PAC model. The computational complexity of learning such concept classes is not explored in detail in this paper. However, it is believed that this will be a fruitful topic for future research.

Acknowledgements
The author would like to thank Ashwin Srinivasan, Donald Michie, Tony Hoare and Brian Ripley for helpful discussions on the topics in this paper. This work was supported by the Esprit Basic Research Action ILP, project 6020.

References

1. M. Bain. Experiments in non-monotonic first-order induction. In *Proceedings of the Eighth International Machine Learning Workshop*, San Mateo, CA, 1991. Morgan-Kaufmann.
2. M. Bain and S. Muggleton. Non-monotonic learning. In D. Michie, editor, *Machine Intelligence 12*. Oxford University Press, 1991.
3. A. Ehrenfeucht, D. Haussler, M. Kearns, and L. Valiant. A general lower bound on the number of examples needed for learning. In *COLT 88: Proceedings of the Conference on Learning Theory*, pages 110–120, San Mateo, CA, 1988. Morgan-Kaufmann.
4. S. Muggleton. Inductive Logic Programming. *New Generation Computing*, 8(4):295–318, 1991.
5. J.R. Quinlan. Discovering rules from large collections of examples: a case study. In D. Michie, editor, *Expert Systems in the Micro-electronic Age*, pages 168–201. Edinburgh University Press, Edinburgh, 1979.
6. L. Valiant. A theory of the learnable. *Communications of the ACM*, 27(11):1134–1142, 1984.
7. S. Wrobel. On the proper definition of minimality in specialization and theory revision. In P.Brazdil, editor, *EWSL-93*, pages 65–82, Berlin, 1993. Springer-Verlag.

Reformulation of Explanation by Linear Logic — Toward Logic for Explanation –

Jun ARIMA and Hajime SAWAMURA

FUJITSU Labs. LTD.
140 Miyamoto, Numazu-shi, Shizuoka, 410-03, Japan
TEL.: + 81 - 559 - 24 - 7210, FAX.: + 81 - 559 - 24 - 6180
arima@iias.flab.fujitsu.co.jp, hajime@iias.flab.fujitsu.co.jp

Abstract. The use of the concept of "explanation" spreads extensively over fields of *Artificial Intelligence*: EBG, analogy, abduction, natural language understanding, diagnosis, etc. Their formalisms, however, suffer inconveniences from the nature of the logic underlying them – *classical logic*. This paper explores one of the crucial inconveniences stemming from classical logic and attempts newly to construct an adequate logic for "explanation" based on *linear logic*.

1 Introduction

The use of the concept of "explanation" spreads extensively over fields of *Artificial Intelligence* (AI): EBG, analogy, abduction, natural language understanding, diagnosis, etc. Their purposes of the use of "explanation" are fundamentally common in that it is used to extract some useful information by *explaining* a certain phenomenon, though used, in EBG and analogy, to extract a relevant *fact* with the explained phenomenon, while used, in abduction and natural language understanding, to propose a *hypothesis* by which the phenomenon is made explicable.

Their formalisms, however, suffer unavoidable dubiousness and/or crucial fetters from the nature of the logic underlying them – *classical logic*. After a brief survey on "explanation" in AI, this paper makes the following contributions:

i) to point out and explore a logically serious problem arising from the choice of classical logic as an underlying logic for "explanation". A fallacy of classical logic, called the *positive paradox*, is its direct cause.

ii) to construct a preliminary logic, EL^2, [1] for "explanation" based on *linear logic*. EL^2 has the following features: a) being free from the positive paradox, b) partly embedding *Occam's razor* and *coherence* in its inference rules which are general criteria for the choice of "good" explanation, and c) treating a trinomial relation, < a set of used logical formulas in explaining, a phenomenon to be explained, its explanation >, which leads to extraction of a good "explanation" from the proof and verification for such a "natural" explanation, especially here, as same sub-explanations are omitted from.

---

[1] EL^2 is an acronym of *ELementary Explanatory Logic*.

General definition of "explanation"

Though the concept of "explanation" is extensively used over fields of AI, terminologies on it are different in every field. To avoid confusion, we will coherently use the following terminologies in this paper.

Definition 1. For an adequate logical system $\vdash$ and $\models$, let *knowledge* $\mathcal{K}$ be a set of its formulas and a *phenomenon* to be explained G be its formula. A set of formulas $\mathcal{E}$ will be called the *(explanatory) factors* of a phenomenon G under knowledge $\mathcal{K}$ if, for some set of formulas $\mathcal{H}$, $\mathcal{E} = k \cup \mathcal{H}$ and $\mathcal{E} \models G$, where $\mathcal{K} \models k$. $\mathcal{H}$ is called a set of *hypotheses*, while a formula k s.t. $\mathcal{K} \models k$ called a *fact*.

Then, an *explanation* of G under $\mathcal{K}$ is an adequate representation of a proof of G from $\mathcal{E}$ ($\mathcal{E} \vdash G$). Moreover, an explanation is said to be *good* if it satisfies some conditions, called *(explanatory) criteria*.

Here, given *knowledge* and a *phenomenon*, we intend to choose a *good explanation* (or the *explanatory factors* for it) by introducing *hypotheses* adequately.

"Explanation" which has been treated so far in AI is limited in the case of either ($\mathcal{K} \models G$ and $\mathcal{K} \models \mathcal{E}$) or ($\mathcal{K} \not\models G$ (therefore, $\mathcal{K} \not\models \mathcal{E}$)). The former case is mainly treated in studies of EBG and analogy, which will be introduced in "Empirical Approach" in the next section, and the latter case is in studies of abduction and natural language understanding, which will be introduced in "Logical Approach".

This paper describes the first report on an attempt to construct a theory for "explanation" which treats uniformly the above two cases and, additionally, the remaining last case — $\mathcal{K} \models G$ and $\mathcal{K} \not\models \mathcal{E}$.

2 "Explanation" and Inadequacy of Classical Logic

2.1 Definitions on "Explanation" and Criteria for "Good" One

We survey definitions on "explanation" proposed so far in two approaches to AI, taking formalization by classical logic apart from criteria for a good explanation. Then, we describe logical problems arising from the choice of classical logic as a underlying logic in each approach.

"Explanation" in Empirical Approach – This approach is characterized by the case of $\mathcal{K} \models G$ and $\mathcal{K} \models \mathcal{E}$.

Many works have been done vigorously on EBG and analogy [5, 1]. The original work of EBG [6] can be representative of this approach w.r.t. the concept of "explanation".

i) Definition

"An *explanation* of how an instance is an example of a concept is a *proof* that the example satisfies the concept definition." by T.Mitchell et. al.[6].

Though the purpose of EBG is to extract a sufficient condition for the concept by generalizing an explanation which satisfies a criterion, called *operational criterion*, let us focus our attention on the explanation instantiated by a training example which is expected to be generalized.

In this approach, *proof* is implicitly assumed to be a proof of classical logic. Provided that we chose classical logic as a logic underlying this approach, we specialize our general terminologies to EBG's as follows.

$\vdash$ and $\models$ express *provability* and *logical consequence* relation of *classical logic* respectively (we write $\vdash_C$ and $\models_C$). Knowledge $\mathcal{K}$ expresses a set of *concept definition, training example* and *domain theory* in EBG. A phenomenon G corresponds to the *goal concept*.

Let a conjunction of atoms C be the *sufficient condition* to be extracted in EBG. Then, the following relations hold in this approach.

a) A training example satisfies the sufficient conditions (by using *training example* and *domain theory*): $\mathcal{K} \models_C C$.

b) A fact that the training example satisfies the goal concept is derivable from the fact that it satisfies the sufficient conditions (by using *concept definition* and *domain theory*): $\mathcal{K} \models_C C \supset G$.

Thus, we can correspond an "explanation" in EBG with a proof of $\mathcal{E} \vdash_C G$ (the explanatory factors $\mathcal{E} = C, (C \supset G)$).

ii) Criterion

A syntactical restriction is imposed on C, which is called *operational(ity) criterion*. It is usually that C (more strictly, $C \supset G$) must be expressed in terms of the predicates used to describe example or other selected, easily evaluated predicates from the *domain theory*.

Logical Problem – A paradox of explanation by arbitrary facts:

Example 1. $\mathcal{K} = \mathcal{D} \cup Te$, $G = Cup(A)$,
  concept definition + domain theory: $\mathcal{D} = \{\forall x.(Concave(x) \supset Cup(x))\}$
  training examples.: $Te = \{Concave(A), Flowery(A), \cdots\}$

This example is particularly simple in that a concept definition is embedded into $\mathcal{D}$ and domain theory has a sufficient concept definition itself which satisfies the operational criterion. It is, however, enough to show a key point of a logical problem.

The intended explanation in EBG is a series of proof process, $Concave(A)$, $(Concave(A) \supset Cup(A))$, $Cup(A)$, that is, the intended sufficient condition C is $Concave(A)$. However, classical logic makes the following explanation that satisfies the operational criterion: $Flowery(A)$, $(Flowery(A) \supset Cup(A))$, $Cup(A)$, which implies that "A training example A is a cup because it is flowery". This strange explanation is possible in classical logic because *any implicational sentence holds if its consequence holds* ($\mathcal{K} \models_C anyP \supset Cup(A)$, where $anyP$ is an arbitrary formula) and, even if its precondition is restricted by the operational criterion, any *facts* in training example are possible to be relativized with the goal concept $(\mathcal{K} \models_c Flowery(A) \supset Cup(A))$ [2]. The key of the paradox originates from the nature of classical logic, and has direct relationship with a famous fallacy of the logic, called the *positive paradox*[3].

"Explanation" in Logical Approach – It is characterized by the case of $\mathcal{K} \not\models G$.

We survey formalization of "explanation" and its criteria in studies of abduction, natural language understanding, diagnosis, etc.

i) Definition

This approach deals with a set of formulas $\mathcal{H}$, which satisfies the following [4]:
$$\mathcal{K} \models_c \mathcal{H} \supset G \quad (\mathcal{K} \cup \mathcal{H} \models_c G), \quad \text{where} \quad \mathcal{K} \cup \mathcal{H} \text{ is consistent.}$$
Thus, $\mathcal{H}$ is a set of hypotheses and $\{\mathcal{H}, (\mathcal{H} \supset G)\}$ is the explanatory factors, $\mathcal{E}$.

ii) Criteria

Criteria will be introduced by using the following example.

Example 2   A modified Pearl's example [8].

$\mathcal{B} = \{Rain \supset WG, \; Sprinkler \supset WG,$
$\qquad Rain \supset SA, \; WornTire \supset LowFriction, \; LowFriction \supset SA\}$
$G = WG \wedge SA,$

where they are read as "Rain causes wet grass (WG).", "Sprinkler causes wet grass.", "Rain causes slip accidents (SA).", "A worn tire causes low friction." and "Low friction causes slip accidents.".

a) **Occam's Razor:** One of the most often-used criterion.

A hypothesis should not be introduced more than it is needed. For example, as a set of hypotheses for the explanation of G, $\{Rain\}$ is better than $\{Rain, Sprinkler\}$ and $\{Rain, anyP\}$, where $anyP$ is an arbitrary formula.

b) **Level of specificity:** Many works have been done on the specificity level of an explanation. The adequate level, in general, depends on the domain where an explanation is made: the most specific level is used in the diagnosis work [2], the least specific for the natural language interpretation [12], the selected level based on the weighted rules is proposed in [4] and syntactical restriction is used in the diagnosis work, planing, common sense reasoning [9, 10] and, additionally, EBG.

---

[2] In this example, though the explanation does not use the domain theory, it is not the cause of this strange explanation. The following another strange explanation can be made:
$Concave(A), Flowery(A), (Flowery(A) \supset (Concave(A) \supset Cup(A))), Cup(A)$
(Note that $\mathcal{K} \models_c \forall x, y.(Flowery(y) \supset (Concave(x) \supset Cup(x)))$
Also here, an irrelevant fact is relativized.

[3] $\models_c p \supset (q \supset p)$, that is, "If p is true, then any q implies p." is a theorem of classical logic, which is something unreasonable.

[4] $\mathcal{H}$ is usually called "explanation" in this approach.

c) **Coherence:** An explanation which "ties" the various phenomena is better [7]. *Occam's razor isn't sharp enough* because, for example, Occam's razor does not distinguish between minimal sets of hypotheses, $\{Rain\}$ and $\{Sprinkler, WornTire\}$. However, it is generally accepted that $\{Rain\}$ is more preferable for the explanation of WG and SA. The viewpoint of coherence is that it is because the explanation from $\{Sprinkler, WornTire\}$ consists of two independent sub-explanations for WG and SA, while the explanation from $\{Rain\}$ ties the both phenomena. Coherence is general because it often meets the concept of Occam's razor.

Logical problems

Obviously from the definition, this approach treats nothing about an explanation to be made for a known phenomenon ($\mathcal{K} \models G$). This situation, if we look this approach as a general frame work for "explanation" or "inference for cause", is something strange. Let us see the previous Pearl's example with adding new modifications. In this case, let the system know the sprinkler worked and their tires were worn indeed: $\mathcal{B}' = \mathcal{B} \cup \{Sprinkler, WornTire\}$, and be ordered to explain a phenomenon, "Grass is wet": $G' = WG$. Then, though the system knows the direct cause of G' which happened indeed, the system can explain nothing because it does need no hypothesis to explain G'. This situation seems more regrettable, considering the generality of criteria proposed in this approach. Then, is it easy to give a formal definition for explanation when a hypothesis is unnecessary? The fact is not so easy.

If we try to extend the usual definition to cover the explanation for known phenomenon ($\mathcal{B}' \models G'$), we must soon face the same problem, the *positive paradox*. From $\mathcal{B}' \vdash_C WG$, $\mathcal{B}' \vdash_C (WornTire \supset WG)$ holds, which makes an explanation, "Wet grass was caused by a worntire". In this approach, additionally, a strange explanation with a reversal order in causality can be made. For example, from $\mathcal{B}' \vdash_C WornTire$, $\mathcal{B}' \vdash_C (LowFriction \supset WornTire)$ also holds, which, by additionally using the above explanation, makes an explanation, "Wet grass was caused by a worn tire and a worn tire was caused by low friction". This difficulty which occurs in extending the usual definition is probably the most influential cause why the works in this approach have been locked up in the case where hypotheses are necessary.

As mentioned above, classical logic has a crucial fallacy from the view point of "a logic for explanation". The next section gives a closer look at the very core of the fallacy and proposes a logic which aims at removing the core and being a logic for explanation.

3 Reformulation of Explanation by Linear Logic

3.1 Adequacy of Linear Logic

i) The Positive Paradox

Let us see why the paradox happens. For this purpose, let us trace a Gentzen-style proof of $\mathcal{K} \vdash_C (Flowery(A) \supset Cup(A))$, which relativizes a phenomenon to be explained, $Cup(A)$, with an arbitrary fact, $Flowery(A)$.

$$\cfrac{\cfrac{\mathcal{K} \vdash Cup(A)}{\mathcal{K}, Flowery(A) \vdash Cup(A)} \; W}{\mathcal{K} \vdash Flowery(A) \supset Cup(A)} \; \supset Right$$

This proof shows that an inference rule called *weakening* (the above rule labeled 'W') introduce the irrelevant fact (indeed, it is not necessary to be a "fact"). Therefore, a reasonable way to avoid the paradox is to reject weakening.

Linear logic, proposed by Girard[3], is its instance. Another fundamental character of linear logic is to reject an inference rule called *contraction*. With forward reading these two inference rules, it could be said that weakening introduces an arbitrary formula and contraction crushes two identical formulas into one. A fact that linear logic rejects the both implies that it is denied to increase or decrease a formula arbitrarily in the logic. By this fact, it can be said that linear logic provides a new view, "formulas as resources"[5].

ii) Occam's Razor and Coherence

This resource-sensitiveness yields another important advantage in dealing with explanation. Occam's razor, which is the commonest criterion for a good explanation, is one that rejects more hypotheses than necessary. Unnecessary hypotheses can be divided into three types. The first type of hypothesis is quite irrelevant to an explanation, for example, $anyP$ of hypotheses $\{Rain, anyP\}$ in the Pearl's example. The second is introduced by explaining the same set of phenomena more than *twice*, for example, *Rain* and *Sprinkler* in the case of $G = WG$. The third is one which explains, not the same set of, but each of phenomena explained by other hypotheses; *Sprinkler* of $\{Rain, Sprinkler\}$ in the case of $G = WG \wedge SA$. The first type of hypotheses can be rejected by removing the nature of weakening, which allows the introduction of irrelevant and/or redundant formula in an explanation, as we have seen. The nature of weakening lies, not only in weakening, but among classical logical connectives. This also allows the second hypotheses to slip into an explanation. With classical logic, an explanation of $G = WG$ could be made by introducing hypotheses *Rain* and *Sprinkler* as follows:

$$\cfrac{\cfrac{\cfrac{\cfrac{Rain \vdash Rain \quad WG \vdash WG}{(Rain \supset WG), Rain \vdash WG} \supset Left}{B, Rain, Sprinkler \vdash WG} W \quad \cfrac{\cfrac{Sprinkler \vdash Sprinkler \quad WG \vdash WG}{(Sprinkler \supset WG), Sprinkler \vdash WG} \supset Left}{B, Rain, Sprinkler \vdash WG} W}{B, Rain, Sprinkler \vdash WG \wedge WG} \wedge Right \quad \cfrac{WG \vdash WG}{WG \wedge WG \vdash WG} \wedge Left}{B, Rain, Sprinkler \vdash WG} Cut$$

---

[5] To express arbitrariness in the number of formulas, two modalities (!, ?) are introduced in linear logic.

Reading the above proof *backward*, we easily see that ∧Left makes it possible to duplicate the phenomenon to be explained and, by explaining them, an unnecessary hypothesis is introduced.

Rejection of both weakening and contraction, which characterizes *linear logic*, leads to a change in the very nature of the logical connectives. For a remarkable instance, classical 'and ($\wedge$)' is divided into the two linear 'and'; the *multiplicative* one ($\otimes$) and the *additive* one ($\&$). The latter is closer to classical 'and' and the former is more resource-sensitive than the latter in that $A \otimes A \nvdash A$ while $A \& A \vdash A$. Therefore, we will also block the second type of hypotheses by dealing with phenomena in a strict resource-sensitive manner; for this example, using the multiplicative 'and' for the conjunction of phenomena and by rejecting weakening.

For the third, we would neither propose a complete way nor, indeed, intend to do so, but we could show a partial solution for it by using linear logic in a subtle manner. The key is to adopt the coherence criterion. Let us show it by the Pearl's example ($G = WG \wedge SA$). In the example, even if the duplication of phenomenon is inhibited, it is still possible to introduce two hypotheses; *Sprinkler* for an explanation of WG and *Rain* for an explanation of SA [6]. Without introduction of *Sprinkler*, G is explicable by the remaining hypothesis *Rain*, which is against Occam's razor. This problem seems to be unavoidable just by the resource-sensitive manner. However, it can be avoided if we take another principle, to use common hypotheses – or, more generally, *formulas* – for explanations of different phenomena when it is possible. This idea could be identified with the coherent criterion. 'A common hypothesis' implies that it is *used more than once* in an explanation, therefore we should allow contraction w.r.t. hypotheses, which is possible by using linear modalities.

To adopt the idea of coherence is necessary in a more active sense. "Explanation" which we intend to deal with involves the case when a phenomenon can be explained without hypothesis. In this case, Occam's razor, as it is a criteria for a choice of a set of *hypotheses*, is of no use, therefore we must adopt another general criterion, coherence.

As we will see soon, we can construct a logic which embeds the idea of coherence in itself by controlling use of linear logic with modalities. The logic will be called EL^2. Proving by EL^2 with a certain deterministic strategy makes

---

[6] Strictly, speaking, it is *not* possible if the logic drops weakening throughout. However, it is necessary to adopt weakening (and contraction) w.r.t. knowledge as a set of proper axioms so that $\mathcal{K}$ might involve unnecessary formulas for an explanation (for example, formulas of $\mathcal{B}$ except ($Sprinkler \supset WG$) in the left sub-proof of WG in the following proof). If weakening is allowed in order to delete such irrelevant facts (not hypotheses), this case could happen.

$$
\cfrac{
  \cfrac{
    \cfrac{
      Sprinkler \vdash Sprinkler \quad WG \vdash WG
    }{(Sprinkler \supset WG), Sprinkler \vdash WG} \supset \text{Left}
  }{\mathcal{B}, Sprinkler \vdash WG} W\text{Left}
  \quad
  \cfrac{
    \cfrac{
      Rain \vdash Rain \quad SA \vdash SA
    }{(Rain \supset SA), Rain \vdash SA} \supset \text{Left}
  }{\mathcal{B}, Rain \vdash SA} W\text{Left}
}{
  \cfrac{
    \mathcal{B}, \mathcal{B}, Rain, Sprinkler \vdash WG \wedge SA
  }{\mathcal{B}, Rain, Sprinkler \vdash WG \wedge SA} C\text{Left}
} \wedge^* \text{Right},
$$

where $\wedge^*$ is multiplicative 'and'.

an preferred explanation which meets common criteria such as Occam's razor and coherence.

Before going ahead, it would be worth summarizing and specifying a little more how to deal with knowledge, hypotheses and phenomena in the EL^2. *A hypothesis must be used 'more than once', while use of an axiom of knowledge should be arbitrary.* Therefore, weakening (Left) is allowed only when the extinguished formula with backward reading is a member of knowledge (*proper axioms*). Contraction (Left) is, on the other hand, allowed freely in the left-hand side of $\vdash$, namely unlimited use of knowledge and hypotheses are allowed (with backward reading). *To a formula representing phenomena, strict resource-sensitiveness sticks.* W.r.t. phenomena, weakening is not allowed for the purpose of blocking a positive paradox, and contraction is not allowed, either, because it makes a constructive explanation possible.

In addition to these two features – being free from the positive paradox and partly embedding Occam's razor and coherence, – EL^2 has another feature which directly "extracts explanation" by proof.

iii) Extraction of Explanation

EL^2 deals with the following form of expression:
$$\Sigma \ \vdash \ \alpha \ ; \ \mathcal{F},$$
where α expresses a (multi-)set of phenomena, Σ expresses a used set of facts and hypotheses in explaining phenomena, that is, the explanatory factors, and $\mathcal{F}$ expresses its explanation.

A proof of EL^2 is an attempt to extract a more "natural" explanation from a formal proof, and, from another point of view, to verify such an explanation. The next example intuitively and roughly illustrates what explanation is considered "natural" in EL^2.

Example 3. Let us consider an explanation of a phenomenon, $Unsafe$, under knowledge $\mathcal{B}$, with using hypotheses, $\{YA, Rich\}$.

$$\mathcal{B} = \{(Car \wedge Drink \supset Unsafe), \ (YA \wedge GF \supset Drink),$$
$$(YA \wedge Car \supset GF), \ (YA \wedge Rich \supset Car)\}$$

Then, the following two explanation are identified in EL^2.

Explanation 1 : A rich ($Rich$) and young adult (YA) is likely to own his car (Car). A young adult who owns a car is likely to have a girl friend (GF). A young adult who has a girl friend is likely to drink alcohol ($Drink$). (Again,) a rich and young adult is likely to own his car. He who owns a car and drinks alcohol is likely to be unsafe ($Unsafe$).

Explanation 2 : A rich and young adult is likely to own his car. Therefore, he is likely to have a girl friend. And therefore, he is likely to drink alcohol. Consequently, he is likely to be unsafe.

What difference between these explanations we mean is that the Explanation 1 retains an original proof structure while the Explanation 2 loose sub-explanations which have once been made. EL^2 is constructed so that both explanations could have an identical expression, that is, both explanations are equally verified[7]. Such a shortening, though it might not occur in every explanations, is very common in a natural conversation, and therefore we would take a standpoint that such a shorten explanation is equally valid.

3.2 A logic for Explanation: EL^2

Greek capital letters: a set of Horn clause $(A \Leftarrow A_1 \wedge \cdots \wedge A_n \ (n \geq 0)$, where $A, A_1 \cdots A_n$ are atoms and we will write simply A when $n = 0$), Greek small letters: a conjunction of atoms, V: a set of atoms, $[A]$: a *hypothetical* atom, where A is an atom, $\mathcal{K}$: a set of clauses, $\mathcal{F}$: (Explanatory) forest, $\mathcal{T}$: (Explanatory) tree and $\bar{A}$: (Explanatory) atom s.t.

$$\mathcal{F} = \mathcal{T} \mid \mathcal{T} + \mathcal{F} \qquad \mathcal{T} = \bar{A} \mid \bar{A} \circ \mathcal{F} \qquad \bar{A} = \hat{A} \mid \tilde{A} \mid \bot,$$

where $\hat{A}$, $\tilde{A}$ and $\bot$ are atoms.

The intended meaning of an atom with brackets '[ ]' is that the atom is a hypothesis. For an explanatory forest, $\hat{A}$ expresses "a fact A", $\tilde{A}$ "a hypothesis A" and $\bot$ is an empty or 'silence'. $A \circ \mathcal{F}$ expresses "A because $\mathcal{F}$" and $\mathcal{F}_1 + \mathcal{F}_2$ is "$\mathcal{F}_1$ and $\mathcal{F}_2$". V is a vocabulary for explanatory forest.

EL^2 under $\mathcal{K}$ is expressed by the following axioms and rules (Its provability relation is expressed by $\vdash_{EL}^{\mathcal{K}}$, or simply $\vdash$ if it is obvious from the context).

i) Axioms
  a) for $A \in \mathcal{K}$, $A \vdash A$; $\hat{A}$ $(A \in V)$, and
  b) for $A \notin \mathcal{K}$, $[A] \vdash A$; $\tilde{A}$ $(A \in V)$,
      where, when $A \notin V$, $\hat{A}$ and $\tilde{A}$ are replaced with $\bot$ in a) and b)).

ii) Branching
$$\frac{\Gamma, \Sigma_1 \vdash A; \ \mathcal{T} \quad \Gamma, \Sigma_2 \vdash \alpha; \ \mathcal{F}}{\Gamma, \Sigma_1, \Sigma_2 \vdash A \wedge \alpha; \ \mathcal{T} + \mathcal{F}} \ \text{B}$$
Especially, if $\Gamma \neq \phi$, this inference rule is called *Coherent Branching* and expressed by CB.

iii) Expanding
For $(A \Leftarrow \alpha) \in \mathcal{K}$, if $A[t/x] \in V$,
$$\frac{\Gamma \vdash \alpha[t/x]; \ \mathcal{F}}{\Gamma, (A \Leftarrow \alpha) \vdash A[t/x]; \ \widehat{A[t/x]} \circ \mathcal{F}} \ \text{E} \ ,$$
otherwise $\widehat{A[t/x]}$ is replaced with $\bot$ in the above rule.

Reduction Rules to the Shortened Explanatory Forest:
Shortening is done by the following reductions.
  a) $\bot \circ \mathcal{F} \longrightarrow \mathcal{F}$, $\mathcal{F} \circ \bot \longrightarrow \mathcal{F}$ **b)** $\mathcal{F} + \bot \longrightarrow \mathcal{F}$, $\bot + \mathcal{F} \longrightarrow \mathcal{F}$
  c) $\mathcal{T} + \mathcal{F}[\mathcal{T}] \longrightarrow \mathcal{F}[\mathcal{T}]$, where $\mathcal{F}[\mathcal{T}]$ is a forest where a tree $\mathcal{T}$ appears.

Moreover, let $+$ be *commutative* but $\circ$ be not. (These meanings are given in linear logic.)

---

[7] Though this paper does not mention, it is easy to modify EL^2 so that it might distinguish them.

Example 2 (continued). Let $LowFriction \notin V$ and $\mathcal{K} = \mathcal{B}^E \cup G$, where $\mathcal{B}^E$ is obtained by the replacement of $\subset$ with $\Leftarrow$ in $\mathcal{B}$. Then, the following are instances of proofs of EL^2.

Proof 1:

$$\frac{WG \vdash WG; \ \widehat{WG} \quad SA \vdash SA; \ \widehat{SA}}{WG, SA \vdash WG \wedge SA; \ \widehat{WG} + \widehat{SA}} \ \text{B}$$

Proof 2:

$$\frac{WG \vdash WG; \ \widehat{WG} \quad \dfrac{\dfrac{[WornTire] \vdash WornTire; \ \widetilde{WornTire}}{[WornTire], (LowFriction \Leftarrow WornTire), \vdash \ LowFriction; \widetilde{WornTire}} \ \text{E}}{[WornTire], (LowFriction \Leftarrow WornTire), (SA \Leftarrow LowFriction) \vdash SA; \ \widehat{SA} \circ \widetilde{WornTire}} \ \text{E}}{WG, [WornTire], (LowFriction \Leftarrow WornTire), (SA \Leftarrow LowFriction) \vdash WG \wedge SA; \ \widehat{WG} + \widehat{SA} \circ \widetilde{WornTire}} \ \text{B}$$

Proof 3:

$$\frac{\dfrac{[Rain] \vdash Rain; \ \widetilde{Rain}}{[Rain], (WG \Leftarrow Rain) \vdash WG; \ \widetilde{WG} \circ \widetilde{Rain}} \ \text{E} \quad \dfrac{[Rain]' \vdash Rain; \ \widetilde{Rain}}{[Rain], (SA \Leftarrow Rain) \vdash SA; \ \widetilde{SA} \circ \widetilde{Rain}} \ \text{E}}{[Rain], (WG \Leftarrow Rain), (SA \Leftarrow Rain) \vdash WG \wedge SA; \ \widehat{WG \circ Rain} + \widehat{SA \circ Rain}} \ \text{CB}$$

Thus, an explanatory forest is a kind of representation of a proof structure and extracted by a proof of EL^2. Generally, EL^2 proof allows introduction of hypotheses, which is expressed by $\tilde{A}$ in its explanatory forests ($WornTire$ in Proof 2 and $Rain$ in Proof 3 are instances). In Proof 2, for $LowFriction$ is not in a vocabulary V, an explanation without $LowFriction$ is extracted consequently. Now, we are ready to describe shortening in explanation a little more precisely. In EL^2, an explanatory tree T can be deleted only if it appears in the forest which belongs to the same forest that T belongs to. It corresponds to the reduction rule c). In Proof 3, each occurrences of tree $\widetilde{Rain}$ is not allowed to be deleted because two occurrences belong to different forests from each other, that is, $\widetilde{WG} \circ \widetilde{Rain}$ and $\widetilde{SA} \circ \widetilde{Rain}$[8].

Of course, all of these proofs does not produce good explanations. The next section describes strategies to extract a good explanations.

3.3 Adequacy of EL^2 – Embedding Criteria into Strategies

a) **The Positive Paradox:** An implicational formula can be used only if it is in knowledge (See the Expanding rule). It implies that EL^2 is free from the positive paradox in itself.

b) **Occam's Razor and Coherence:** The concept of "formulas as resources" of linear logic is applied only to a (multi-)set of phenomena to be explained. Therefore, similarly to arguments on linear logic in the previous section, it suppresses introduction of unnecessary hypotheses.

---

[8] Try to explain $G'' = WG \wedge Rain$ (or, of course, *Example 3*), and readers will find the case of shortening and obtain an explanation, "Grass is Wet because it *Rained*.", which would be interesting if it is taken as information obtained from utterance, "Grass is Wet. It *Rained*. (G'')"

Introduction of unnecessary hypotheses might make a coherent explanation, thus we have two main strategies. If coherence is given a higher priority than Occam's razor, application of coherent branching (CB) should be tried as many as possible. Otherwise, use of axiom b) should be avoided as much as possible. The former strategy yields Proof 3 and the latter Proof 1 (Note that $\mathcal{K}$ involves WG and SA themselves in this example. Therefore, it shows that explanations are made even if hypotheses are unnecessary).

c) **Level of specificity:** Syntactical restriction on explanation can be done by selecting vocabulary V, which would be useful in improving "naturality" of an explanation.

3.4 EL^2 and Linear Logic

This section assumes knowledge on linear logic [3]. An interpretation of EL^2 can be given by linear logic. In this section, we introduce two new notation, $\iota S = S \otimes !S$ and $\L S = S \,\mathscr{V}\, ?S$ as abbreviations, where S is a formula.

For (multi-)set of formulas in the left side of $\vdash$, we interpret as follows. Every clause in knowledge $\mathcal{K}$ is prefixed by ! followed by a universal quantifier $\forall$ for each of its free variables, which will represented by $!\mathcal{K}$, because a formula in $\mathcal{K}$ is allowed to be *used arbitrarily many times* ($n \geq 0$) and, therefore, weakening and contraction are both allowed to it. An explanatory factor of Γ including used facts and hypotheses is, on the other hand, expected to be *used more than once* ($n \geq 1$), that is, weakening is inhibited while contraction is allowed, which can be expressed by prefixing ι for every formula in Γ, written by $\iota\Gamma$.

For a formula, we introduce an interpretation·in two steps, which might make it easy to see.

1. Translation of EL^2 without Explanatory Forest
   Every formula is interpreted in a strictly resource-sensitive manner, therefore, multiplicative operators are used. A translation e is given by the following:
   $$(A)^e = A, \quad (A \wedge \alpha)^e = (A)^e \otimes (\alpha)^e,$$
   $$([A])^e = A, \quad (A \Leftarrow \alpha)^e = (\alpha)^e \multimap (A)^e.$$

2. Translation of EL^2 with Explanatory Forest
   We can get a translation for EL^2 with explanatory forest by some generalization of the above translation e. The above e is a special case when every atom A is not in the vocabulary V.
   It begins with modification of e. We replace the above last line of e with:
   $$(A)^{ehf} = A \,\mathscr{V}\, \hat{A}, \quad ([A])^{ehf} = A \,\mathscr{V}\, \tilde{A}, \quad (A \Leftarrow \alpha)^e = (\alpha)^e \multimap (A)^{ehf},$$
   so that a head of clause could absorb its explanatory atom, $\bar{A} = \hat{A} \mid \tilde{A} \mid \perp$. Then, we introduce a translation ef for $\mathcal{F}$, which is given by the following:
   $$(\bar{A})^{ef} = \L \bar{A}, \qquad\qquad\qquad (\bar{A})^{ef'} = ?\bar{A},$$
   $$(\mathcal{T} + \mathcal{F})^{ef} = (\mathcal{T})^{ef} \,\mathscr{V}\, (\mathcal{F})^{ef}, \qquad (\mathcal{T} + \mathcal{F})^{ef'} = (\mathcal{T})^{ef'} \,\mathscr{V}\, (\mathcal{F})^{ef'},$$
   $$(\bar{A} \circ \mathcal{F})^{ef} = \L \bar{A} \,\mathscr{V}\, (\mathcal{F})^{ef} \,\mathscr{V}\, ?(\L \bar{A} \,\mathscr{V}\, (\mathcal{F})^{ef'}), \quad (\bar{A} \circ \mathcal{F})^{ef'} = (\mathcal{F})^{ef'} \,\mathscr{V}\, ?\bar{A}.$$

In sequent calculus of linear logic, a formula $A \invamp B$ in the right side of $\vdash$ is equivalent to a sequent A, B in the right. Thus, in the following, we take the result of ef-translation as a multi-set of formulas.

Example 4 Translation of a forest.

$$(A \circ B \circ (C + D \circ E))^{ef}$$
$$= \imath A, \; ?(\imath A, (B \circ (C + D \circ E))^{ef'}), \; (B \circ (C + D \circ E))^{ef}$$
$$= \imath A, \; ?(\imath A, ?B, ?C, ?D, ?E), \; \imath B, \; ?(\imath B, ?C, ?D, ?E),$$
$$\imath C, \; \imath D, \; ?(\imath D, ?E), \; \imath E.$$

The result of ef-translation of $\bar{A} \circ \mathcal{F}$ can be seen to have two components: $\imath \bar{A} \invamp ?(\imath \bar{A} \invamp (\mathcal{F})^{ef'})$ and $(\mathcal{F})^{ef}$. Roughly speaking, the former expresses that A is surely explained *at least once*, $(\imath \bar{A})$, and obtained through explanations already done for $(\mathcal{F})^{ef'}$, which can express a whole process of the explanation together with the corresponding parts in the latter component. The latter implies that $(T)^{ef} \in (\mathcal{F}[T])^{ef}$ holds, as the above example shows. That is, the whole, $\mathcal{F}[T]$, involves its part, T as its element. Every $(T)^{ef}$ is allowed to be duplicated by contraction from the nature, therefore, it makes *shortening* of EL^2 possible (the following c)).

Proposition 2 Guarantee of shortening. *Under translation ef to linear logic, the following properties hold w.r.t. a forest.*
a) $(\bot \circ \mathcal{F})^{ef} = (\mathcal{F} \circ \bot)^{ef} = (\mathcal{F})^{ef}$ **b)** $(\mathcal{F} + \bot)^{ef} = (\bot + \mathcal{F})^{ef} = (\mathcal{F})^{ef}$
c) Let $\mathcal{F}[T]$ be a forest where a tree T appears.
$$\frac{\Gamma \vdash_{\mathcal{L}} \Sigma, (T + \mathcal{F}[T])^{ef}}{\Gamma \vdash_{\mathcal{L}} \Sigma, (\mathcal{F}[T])^{ef}}$$

Then, the following relation with linear logic provability $\vdash_{\mathcal{L}}$ holds.

Theorem 3. *Let $\mathcal{K}$ and Γ be a set of Horn clauses, $\mathcal{H}$ be a set of all the hypothetical atoms in Γ, α be a conjunction of atoms, and $\mathcal{F}$ be an explanatory forest. Then,*

$$i(\Gamma)^e \vdash_{\mathcal{L}} (\alpha)^e, (\mathcal{F})^{ef} \quad and \quad !(\mathcal{K})^e, i(\mathcal{H})^e \vdash_{\mathcal{L}} (\alpha)^e, (\mathcal{F})^{ef}, \quad if \quad \Gamma \vdash_{EL}^{\mathcal{K}} \alpha; \mathcal{F}.^9$$

4   Concluding Remarks and Future Works

This is the first report on an attempt to construct a logic for explanation which is seen in many AI-fields. Classical logic is not enough adequate for it. The first version which rejects inadequacies stemming from the choice of classical logic is proposed. Additionally to explanation dealt with in AI, it underlies 'good' explanation introducing hypotheses even when the phenomenon is provable just from knowledge (proper axioms). It is intended to be a foothold to an intelligent system which gives an adequate explanation for *every item* that it knows.

Many works are left in future.

---

^{9} This theorem can easily proved from the fact that every axioms and rules of EL^2 translated by e and ef are derived rules of linear logic.

1) Various proofs: especially, completeness w.r.t. linear logic.
2) Relaxation in restriction of Horn clause: we will deal with one with negation as failure and multiple-heads (disjunctive-heads).
3) Improvements of EL^2, where the concept of Coherence will be embedded more precisely.
4) Considerations on how the concept of inconsistency should be dealt with in "explanation".
5) Construction of model theory.

Acknowledgment
We are grateful to Takeshi Ohtani, our colleague, and anonymous referees of ALT93 for helpful comments.

References

1. Arima, J.: A logical analysis of relevance in analogy, in *Proc. of Workshop on Algorithmic Learning Theory (ALT'91)*, Japanese Society for Artificial Intelligence, 255-265 (1991).
2. Cox P.T. and Pietrzykowski T.: Causes for events: their computation and applications, in: *Proc. of Eighth International Conference on Automated Deduction*, Lecture Notes in Computer Science **230** (Springer-Verlag, Berlin, 1986) pp. 608-621.
3. Girard,J-Y: Linear Logic, *Theoretical Computer Science 50*, North-Holland, pp. 1-102, (1987).
4. Hobbs,J.R., Stickel,M.E., Martin,P., & Edwards,D.: Interpretation as abduction, In *Proc. of the 26th Annual Meeting of the Association for Computational Linguistics*, Philadelphia, PA, pp. 32-37, (1986).
5. Kedar-Cabelli,S.: Purpose-directed analogy, in *the 7th Annual Conference of the Cognitive Science Society*, Hillsdale, NJ: Lawrence Erlbaum Associates, pp.150-159 (1985).
6. Mitchell, T., Keller, R. & Kedar-Cabelli,S.: Explanation-Based Generalization: A Unifying View, in *Machine Learning 1*, Kluwer Academic Publishers, Boston, pp.47-80 (1986).
7. Ng,H.T. & Mooney,R.J.: On the Role of Coherence in Abductive Explanation, in *Proc. of the 8th National Conference on AI (AAAI-90)*, pp. 337-342 (1990).
8. Pearl,J.: Embracing causality in default reasoning, *Artificial Intelligence* **35**, pp.259-271 (1988).
9. Poole D., Goebel R. and Aleliunas R.: Theorist: a logical reasoning system for defaults and diagnosis, in: N. Cercone and G. McCalla (eds.), *The Knowledge Frontier: Essays in the Representation of Knowledge* (Springer-Verlag, New York, 1987) 331-352.
10. Poole,D.: A logical framework for default reasoning, *Artificial Intelligence* **36**, pp.27-47 (1988).
11. Stickel M.E.: Rationale and methods for abductive reasoning in natural-language interpretation, in: R. Studer (ed.), *Natural Language and Logic, Proceedings of the International Scientific Symposium*, Hamburg, Germany, Lecture Notes in Artificial Intelligence **459** (Springer-Verlag, Berlin, 1990) 233-252.

Towards Efficient Inductive Synthesis of Expressions from Input/Output Examples

Jānis Bārzdiņš

Guntis Bārzdiņš

Kalvis Apsītis

Uģis Sarkans

Institute of Mathematics and Computer Science
University of Latvia
Rainis blvd. 29, Riga, LV-1459, Latvia
E-mail: jbarzdin@mii.lu.lv, guntis@mii.lu.lv

Abstract. Our goal through several years has been the development of efficient search algorithm for inductive inference of expressions using only input/output examples. The idea is to avoid exhaustive search by means of taking full advantage of semantic equality of many considered expressions. This might be the way that people avoid too big search when finding proof strategies for theorems, etc. As a formal model for the development of the method we use arithmetic expressions over the domain of natural numbers. A new approach for using weights associated with the functional symbols for restricting search space is considered. This allows adding constraints like the frequency of particular symbols in the expression. Additionally the current state of the art of computer experiments using this methodology is described. An example that is considered is the inductive inference of the formula for solving quadratic equations, the finding of which by pure exhaustive search would be unrealistic.

1 Introduction

Inductive synthesis of recursive functions from input/output examples is a very well studied problem in the recursive-theoretic framework [6] [7]. At the same time few works have been devoted to the problem in the practical perspective because it was considered impossible to synthesize non-trivial functions from input/output examples in the reasonable time (without exhaustive search).

One of the methods used to synthesize functions from input/output examples is the Occam razor principle stating that we have to search for the simplest hypothesis which complies with all available examples. This method could produce very reliable results, but it is difficult to be implemented without exhaustive search.

Nevertheless people are able to guess quite complicated functions from several input/output examples. Many such functions in the form of number sequences are collected in the Angluin's paper "Easily inferred sequences" [1]. The question is: what allows us to generalize such sequences so easily?

In [4] an idea to use algebraic axioms to synthesize functions from input/output examples was suggested. The main advantage of axioms is that they can be synthesized independently of each other and that the complexity of a separate axiom is much smaller than that of the whole program computing the function. If sufficiently many axioms are found, they can describe the function completely. The techniques known in the theory of term rewriting systems can be used to construct an executable program. The computer experiments have shown that algorithms for adding and multiplying binary numbers can be synthesized in this way. The most time-consuming part in such synthesis happens to be the synthesis of expressions in fixed signature which satisfy several input/output examples.

The problem of efficient synthesis of expressions from input/output examples presents interest also by itself. We might wish to be able to induce some formula, like the one for the volume of the frustum of a square pyramid:

$$V(h, a, b) = h \frac{a^2 + ab + b^2}{3}$$

using as input only the results of several measurements, e.g.:

$$V(6, 4, 2) = 56$$
$$V(3, 4, 3) = 37$$
$$V(9, 2, 1) = 21$$
$$V(6, 1, 3) = 26$$

(This formula is particularly interesting because it was known in the ancient Egypt a long time before Euclid's deductive method in geometry was introduced [8]).

The studies on the problem of efficient inductive synthesis of expressions were initiated in [3] where a reasonably efficient method for such synthesis was proposed. The results of computer experiments illustrating this approach were described in [5]. These experiments have shown that the formulas like the one of the volume of the frustum of the square pyramid can be synthesized on the 33MHz Sparc workstation in about 10 minutes. The ultimate goal was to improve the method by an order of magnitude so that the formulas like the one for solving quadratic equations could be synthesized from input/output examples. In this case the method might become practically interesting.

In this paper a new, improved algorithm for inductive synthesis of expressions from input/output examples is presented. In many cases it might be more efficient than the one described in [3]. The first experimental results with this algorithm are given at the end of the paper.

2 Definitions and notations

Let the signature Σ be a finite set of functional symbols $\{f_1, \ldots, f_m\}$ where any symbol f_i has a fixed arity. Let $\mathcal{D}$ be a finite domain set. For the sake of simplicity we shall assume that $\mathcal{D}$ is a subset of natural numbers. We will say that Σ *is interpreted on* $\mathcal{D}$, if with all functional symbols $f \in \Sigma$ there is associated a partially defined function with domain and range in $\mathcal{D}$. By $K_{\Sigma,\mathcal{D}}$ we will denote a particular interpretation of Σ on $\mathcal{D}$. Since we are considering a finite domain $\mathcal{D}$, an interpretation $K_{\Sigma,\mathcal{D}}$ can be specified completely by a finite number of *equalities*. By equality we mean an expression

$$f(a_1, \ldots, a_n) = a_0$$

where $f \in \Sigma$ is a functional symbol of arity n and $a_0, a_1, \ldots, a_n \in \mathcal{D}$.

Example. Let $\Sigma_0 = \{z, s, +\}$ and $\mathcal{D}_0 = \{0, 1, 2, 3\}$. Then a particular interpretation $K_{\Sigma_0, \mathcal{D}_0}$ can be described by the following equalities:

$$\{z = 0, s(0) = 1, s(1) = 2, s(2) = 3,$$
$$+(0,0) = 0, +(0,1) = 1, +(1,0) = 1, +(1,1) = 2, +(2,0) = 2,$$
$$+(0,2) = 2, +(1,2) = 3, +(2,1) = 3, +(3,0) = 3, +(0,3) = 3\}.$$

For other parameter values functions are undefined in this interpretation.

Further by Σ-*algebra* $\mathcal{K}$ we shall understand the set of equalities as in the Example, which describes an interpretation $K_{\Sigma,\mathcal{D}}$ over fixed domain $\mathcal{D}$. We assume that $\mathcal{D}$ contains only those domain elements which appear on left or right side of some equality from $\mathcal{K}$. The *volume* of Σ-algebra $\mathcal{K}$ is defined to be the number of equalities in $\mathcal{K}$ and will be denoted $|\mathcal{K}|$. Σ_0-algebra $\mathcal{K}_0$ in the Example has volume $|\mathcal{K}_0| = 14$.

Let there be a fixed alphabet $\{x_1, x_2, \ldots\}$ of *term variables*. *Open terms* are expressions made of term variables and functional symbols from the signature Σ.

By *weight function* we shall understand a mapping from the set of all open terms to the set of all nonnegative integers: $w : T(\Sigma) \to N$ with certain properties. Namely, weight function w is defined by means of *auxiliary weight functions* $\{\tilde{f}_1, \ldots, \tilde{f}_m\}$ where $\tilde{f}_j : N^n \to N$ is of the same arity as f_j:

- Any variable x_i is an open term of weight $w(x_i) = 0$.
- Any 0-arity functional symbol is an open term of constant weight

$$w(f()) = \tilde{f}$$

- If f is a functional symbol of arity n ($n > 0$) and $t_1, \ldots, t_n$ are open terms, then the expression $f(t_1, \ldots, t_n)$ is an open term of weight

$$w(f(t_1, \ldots, t_n)) = \tilde{f}(w(t_1), \ldots, w(t_n))$$

Auxiliary weight functions $\tilde{f}_j$, $j \in \{1, \ldots, m\}$, must satisfy monotonicity axioms:

A1. $\tilde{f}_j(p_1, \ldots, p_n) > \max(p_1, \ldots, p_n)$.
A2. If $p'_i \geq p_i$ for each $i \in \{1, \ldots, n\}$ then $\tilde{f}_j(p'_1, \ldots, p'_n) \geq \tilde{f}_j(p_1, \ldots, p_n)$.

Besides that functions $\tilde{f}_j$, $j \in \{1, \ldots, m\}$, must be computable from its arguments in constant time.

Some examples of weight function w (over signature $\Sigma_0 = \{z, s, +\}$):

- $\tilde{z} = 1, \tilde{s}(p) = p + 1, \tilde{+}(p_1, p_2) = \max(p_1, p_2) + 1$ (in this case $w(t)$ describes the number of levels in the term),
- $\tilde{z} = 1, \tilde{s}(p) = p + 1, \tilde{+}(p_1, p_2) = p_1 + p_2 + 1$ (in this case $w(t)$ is number of functional symbols in the term).

Let weight function w is arbitrary fixed. Further by *weight* of open term t we will understand $w(t)$.

The *size* of the open term t is defined to be the number of instances of the functional and variable symbols in t and will be denoted $|t|$.

The term obtained from the open term t by replacing its variables by elements of domain set $\mathcal{D}$ will be called a *closed term*, and its weight and size are defined to be the same as the weight and size of the open term t.

We will say that a closed term t can be computed in Σ-algebra $\mathcal{K}$ if its value can be derived by means of elementary equations of $\mathcal{K}$.

Let there be given a tuple $\langle a_1, \ldots, a_k \rangle \in \mathcal{D}^k$, $a_i \neq a_j$ if $i \neq j$, and $b \in \mathcal{D}$.[1] Such pair

$$(\langle a_1, \ldots, a_k \rangle, b)$$

we will call *input/output example*. We say that an open term t *satisfies* the I/O-example $(\langle a_1, \ldots, a_k \rangle, b)$ in the Σ-algebra $\mathcal{K}$ if:

- t contains no other variables than $x_1, \ldots, x_k$,
- the value of a closed term t' obtained from t by replacing variables $x_1, \ldots, x_k$ by $a_1, \ldots, a_k$ respectively, can be computed in $\mathcal{K}$,
- the value of t' is equal to b.

Let l be a natural number $(l > 0)$. We will denote by $A_{\mathcal{K},l}^{(\langle a_1, \ldots, a_k \rangle, b)}$ a set of all open terms such that:

- they satisfy the input/output example $(\langle a_1, \ldots, a_k \rangle, b)$ in $\mathcal{K}$,
- they have weight no more than l.

If we consider the Σ_0-algebra $\mathcal{K}_0$ and the weight function $w(t)$ equal to the number of functional symbols in t, then the set $A_{\mathcal{K}_0,2}^{(\langle 1,2 \rangle, 3)}$ is

$$\{+(x_1, x_2), \ +(x_2, x_1), \ +(x_1, s(x_1)), \ +(s(x_1), x_1), \ +(x_1, +(x_1, x_1)),$$
$$+(+(x_1, x_1), x_1), \ s(x_2), \ s(s(x_1)), \ s(+(x_1, x_1))\}.$$

3 Synthesis from one example

We will say that an algorithm having received the input U enumerates the set of objects $\{w_1, w_2, \ldots, w_s\}$ in *setup time* T and ith *step time* T_i, if this algorithm outputs ("prints" on the output tape) the first object w_1 in $T + T_1$ time, and the ith object $w_i (i = 2, 3, \ldots)$ in T_i time from the moment when the previous object w_{i-1} was output. In this paper by *algorithm* we mean the RAM-machine.

---

[1] The restriction that $a_i \neq a_j$ if $i \neq j$ is not essential — it is added only to simplify the algorithm.

Theorem 1. *Let signature Σ and weight function w be fixed. There exists an algorithm which, given any Σ-algebra $\mathcal{K}$, any input/output example $(\langle a_1, \ldots a_k \rangle, b)$ and any natural number $l > 0$, enumerates without repeating the set of terms $A_{\mathcal{K},l}^{(\langle a_1, \ldots a_k \rangle, b)}$ in setup time $O(|\mathcal{K}| \log l)$ and ith-step time $O(|t_i| \log l)$, where t_i is the output term in the ith step $(i = 1, 2, \ldots, |A_{\mathcal{K},l}^{(\langle a_1, \ldots a_k \rangle, b)}|)$.*

Proof. For Σ-algebra $\mathcal{K}$ we define a corresponding Σ-algebra graph $G_\mathcal{K}$. Σ-algebra graph $G_\mathcal{K}$ contains nodes of two types: domain nodes denoted by $D_\mathcal{K}$ and functional nodes denoted by $F_\mathcal{K}$. To distinguish nodes of these two types, in the figures we will show domain as dots and functional nodes as small circles. Domain nodes will correspond to the elements of domain set $\mathcal{D}$ of Σ-algebra $\mathcal{K}$. Functional nodes will correspond to elementary equations of Σ-algebra $\mathcal{K}$ and will be marked by the functional symbol on the left-hand side primitive term in this equation. For any equation

$$f(a_1, \ldots, a_n) = b$$

of Σ-algebra $\mathcal{K}$ the following arcs are added in graph $G_\mathcal{K}$. From a functional node v corresponding to this equation an arc is drawn to the domain node b; the node b is called the *upper node* for the functional node v. From domain nodes $a_1, \ldots, a_n$ arcs (marked by numbers $1, 2, \ldots, n$ respectively) are drawn to the functional node v; domain nodes $a_1, \ldots, a_n$ are called *lower nodes* for the functional node v.

The Σ_0-algebra graph $G_{\mathcal{K}_0}$ which corresponds to the Σ_0-algebra $\mathcal{K}_0$ given above is shown in Fig.1.

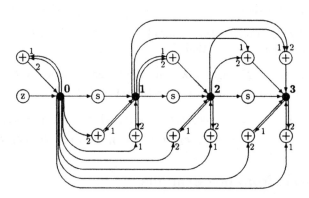

Fig. 1. Σ_0-algebra graph $G_{\mathcal{K}_0}$.

Let there be some tuple $\langle a_1, \ldots, a_k \rangle \in \mathcal{D}^k$ such that $a_i \neq a_j$ if $i \neq j$. Let there be some natural number l. In this case we define weights for nodes of the graph $G_\mathcal{K}$ according to the following conditions (weight will not be defined for all nodes):

- Domain nodes $a_1, \ldots, a_k$ have weight 0.
- Any functional node which corresponds to 0-arity functional symbol f has weight $\tilde{f}$.
- If a functional node v corresponds to some n-arity functional symbol f and $d_1, \ldots, d_n$ are lower nodes of the node v and for all of them weights $h_1, \ldots, h_n$ are defined, then the weight of the node v is $\tilde{f}(h_1, \ldots, h_n)$; otherwise the weight for the node v is not defined.
- Let $v_1, \ldots, v_n$ be functional nodes for which the upper domain node is d and weights $h_1, \ldots, h_n$ are defined. In this case the weight of the domain node d is $\min(h_1, \ldots, h_n)$; otherwise the weight of the domain node d is not defined (i.e. if no such functional nodes v_i do exist).
- All weights do not exceed l.

It is easy to see that a weight related this way to some domain node d is exactly the smallest weight among weights of closed terms which in Σ-algebra K satisfy the pair $(\langle a_1, \ldots, a_k \rangle, d)$, if it does not exceed l. (According to axiom A2, if $t = f(t_1, \ldots, t_k)$ and minimal weights of terms $t_1, \ldots, t_k$ are, respectively, $\tilde{t}_1, \ldots, \tilde{t}_k$, minimal weight of t is

$$\tilde{t} = \tilde{f}(\tilde{t}_1, \ldots, \tilde{t}_k).)$$

Additional arcs, called *dotted arcs* (they are denoted by dotted lines), are added to the graph G_K in the following way. Let d be some domain node for which weight is defined. Let $v_1, \ldots, v_n$ be functional nodes whose upper node is d and for which weights are defined and let these nodes $v_1, \ldots, v_n$ be already ordered according to their weights (weight(v_i)$\leq$weight(v_{i+1})). The dotted arc is drawn to connect the node d to v_1, the node v_1 to v_2, ..., the node v_{n-1} to v_n.

The graph G_K which is completed by weights and dotted arcs according to the fixed tuple $\langle a_1, \ldots, a_k \rangle$ and fixed l will be called *an annotated Σ-algebra graph* and will be denoted $G_{K,l}^{(a_1, \ldots, a_k)}$.

The graph $G_{K_0,3}^{(1,2)}$ which corresponds to the Σ_0-algebra graph G_{K_0} is shown in Fig.2.

3.1 Construction of annotated sample graph

Lemma 2. *The annotated Σ-algebra graph $G_{K,l}^{(a_1, \ldots, a_k)}$ can be built from Σ-algebra K in time $O(|K| \log l)$.*

Proof. The algorithm consists of initial step, iterative step and final step.

Initial step. For every domain node $a_i, i = 1, \ldots, k$, we introduce an additional 0-arity functional node with the upper node a_i and set its weight equal to zero. For every "proper" 0-arity functional node f_i we set the weight to $\tilde{f}_i$, if and only if $\tilde{f}_i \leq l$.

Iterative step. First let us examine the graph after the ith iterative step. (We will refer to the initial step as the 0-th step.) There are four kinds of functional nodes.

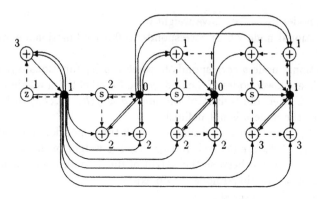

Fig. 2. Annotated Σ_0-algebra graph $G_{\mathcal{K}_0,3}^{\langle 1,2 \rangle}$.

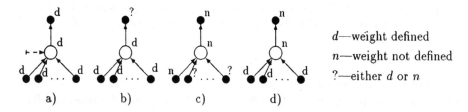

d—weight defined
n—weight not defined
?—either d or n

Fig. 3. Types of functional nodes after iterative step.

- Node with a dotted arc entering it. Weights of its lower nodes, upper node and itself are defined (Fig. 3a). We denote the set of all such nodes by A_i.
- Node without a dotted arc entering it, but with defined weight. Weights of the lower nodes are defined (Fig. 3b). The set of all such nodes is denoted by B_i.
- Node with undefined weight. The weight of the upper node is also undefined. At least one lower node has undefined weight (Fig. 3c). The set of all such nodes will be called C_i.
- Node with undefined weight, but all lower nodes has defined weight (Fig. 3d). The corresponding set is denoted by D_i.

After the initial step only B_0 and C_0 are not empty.

And now the iterative step itself.

While B_i is not empty we take from it the functional node with minimal weight (any, if there are several ones) f_i. Two cases should be distinguished.

1. The weight of f_i upper node d_i is defined. Then we add f_i to the end of the dotted path associated with d_i.

$$B_{i+1} = B_i \setminus \{f_i\}, A_{i+1} = A_i \cup \{f_i\}, C_{i+1} = C_i, D_{i+1} = D_i$$

2. The weight of f_i upper node d_i is not defined. We set it equal to the weight of f_i, draw a dotted arc from d_i to f_i.

$$A_{i+1} = A_i \cup \{f_i\}$$

Afterwards we examine all functional nodes with d_i as a lower node (they belong to C_i). Suppose there are k_i functional nodes $g_{i,0}, \ldots, g_{i,k_i}$ that have only lower nodes with defined weights. (For any node $g_{i,j}, j = 0, \ldots, k_i$, d_i was the last node with undefined weight.) Now we can compute $\tilde{g}_{i,0}, \ldots, \tilde{g}_{i,k_i}$. Let us suppose that there is l_i such that for any $j = 0, \ldots, l_i$, $\tilde{g}_{i,j} \leq l$, and for any $j = l_{i+1}, \ldots, k_i$, $\tilde{g}_{i,j} > l$. We set weights for nodes $g_{i,0}, \ldots, g_{i,l_i}$.

$$B_{i+1} = (B_i \setminus f_i) \cup \{g_{i,j} | j = 0, \ldots, l_i\},$$

$$C_{i+1} = C_i \setminus \{g_{i,j} | j = 0, \ldots, k_i\},$$

$$D_{i+1} = D_i \cup \{g_{i,j} | j = l_i + 1, \ldots, k_i\}.$$

It is important to notice that, if every node in A_i has weight not greater than any node in B_i has, the same property holds also for A_{i+1} and B_{i+1} (because f_i is the node with the minimal weight and for every $j = 0, \ldots, k_i$, $\tilde{g}_{i,j} > \tilde{f}_i$ — axiom A1). Therefore every dotted path is correct, i. e. nodes along a path are ordered according to their weights.

Final step. We remove additional 0-arity functional nodes that were added during initial step. (It is necessary to delete them correctly from the dotted paths.)

We must be able to find a functional node in B_i with minimal weight. We will organize set B_i so that it would be possible to insert, delete and find a functional node with minimal weight in time $O(\log l)$. For this purpose a priority queue [2] will be used.

Let us analyze time complexity of our algorithm.

Initial step. The number of domain nodes and 0-arity functional nodes is $O(|\mathcal{K}|)$, adding to B_0 takes time $O(\log l)$ for every node. Therefore initial step takes time $O(|\mathcal{K}| \log l)$.

Iterative step. The number of iterative steps is $O(|\mathcal{K}|)$. Although an individual step may require time greater than $O(\log l)$ (if there are many functional nodes with d_i as a lower node), total time required by iterative steps is $(O(|\mathcal{K}| \log l)$. (Every node can be moved from C_i to B_{i+1} only once and from B_i to A_{i+1} also only once.)

Final step. It takes time $O(|\mathcal{K}|)$. $\Box$

3.2 Term output

Let there be a graph $G_{\mathcal{K},l}^{(a_1, \ldots, a_k)}$ placed in memory so that its nodes can be accessed by their addresses.

A closed term will be called an α-*term* if:

– it contains no other domain symbols than $a_1, \ldots, a_k$,
– it belongs to the Σ-algebra $\mathcal{K}$.

An α-term will be called *annotated* if, for each of its symbols, the address of some node in the graph $G_{\mathcal{K},l}^{(a_1,\ldots,a_k)}$ is attached according to the following conditions:

– to domain symbols there are attached the addresses of the corresponding domain nodes,
– if $f(g_1, \ldots, g_n)$ is a subterm of a term t and $v_1, \ldots, v_n$ are domain nodes which represent values of closed terms $g_1, \ldots, g_n$, then to the symbol f there is attached the address of the functional node marked by f and having lower nodes $v_1, \ldots, v_n$.

It is easy to see that any α-term can be annotated.

We shall use two representations of α-terms— ordered tree form and linear form (in prefix notation). Subterms of the same level which are included in brackets of the same functional symbol (their parent in the tree) will be called siblings. Siblings are ordered according to the number of argument they represent. Therefore we can speak of right siblings for given subterm. (Such trees and linear layouts for current α-terms will be stored apart from the Σ-algebra graph $G_{\mathcal{K},l}^{(a_1,\ldots,a_k)}$.)

We define the *minimal α-term* of a domain node d in the following way. If $d \in \{a_1, \ldots, a_k\}$, then the symbol d itself is the minimal α-term of the node d. Otherwise we consider the functional node v which is connected to the node d by a dotted arc (if there is no dotted arc from the node d, then the minimal term is not defined). If the node v corresponds to 0-arity functional symbol f, then this symbol itself is the minimal term. If the node v corresponds to n-arity ($n > 0$) functional symbol f, then the minimal term is the term $f(t_1, \ldots, t_n)$ where $t_1, \ldots, t_n$ are minimal α-terms of the lower nodes for the node v.

It is easy to see that the so defined minimal α-term of the node d is a closed term which satisfies the pair $(\langle a_1, \ldots, a_k \rangle, d)$ in Σ-algebra $\mathcal{K}$ and has the minimal weight which is the weight of the node d in the Σ-algebra graph.

Now we also define the minimal α-term for functional nodes. Let v be some functional node for which weight is defined (otherwise the minimal term for v is not defined). Let functional node v correspond to the functional symbol f of arity n. If $n = 0$, then the symbol f itself is the minimal term of node v. Otherwise there are lower nodes $d_1, \ldots, d_n$ for the node v and minimal terms $t_1, \ldots, t_n$ for those domain nodes. Then the term $f(t_1, \ldots, t_n)$ is the minimal term of the functional node v. It is easy to see that the weight of the minimal term of the functional node v is equal to the weight of the node v in the Σ-algebra graph.

Lemma 3. *Inequality*

$$\tilde{f}(p_1, \ldots, p_{i-1}, x, p_{i+1}, \ldots, p_n) \leq l$$

can be solved in time $O(\log l)$ and solution, if it exists, is in form

$$x \leq l'$$

$(p_1, \ldots, p_n, l, l', x$ are natural numbers, $l' < l)$.

Indeed, A2 implies that together with l' all $x < l'$ suit inequality as well. A1 implies that l' (if it exists at all) should be smaller than l. The exact value of l' can be found in $O(\log l)$ steps of dichotomy.[2]

For the representation of a given α-term we attach *actual* and *maximal weights* to every ordered tree node (or, equivalently, to every instance of functional or variable symbol) in the following way:

- for the root symbol maximal weight l is given,
- if a parent node f has its maximal weight l', then its i-th child node has maximal weight l'', where $x \leq l''$ is the solution of inequality

$$\tilde{f}(p_1, \ldots, p_{i-1}, x, p_{i+1}, \ldots, p_n) \leq l'$$

$(p_1, \ldots, p_n$ are actual weights of sibling nodes). In the case solution does not exist we assume $l'' = -1$.

Maximal weight shows how far we can raise the weight of a given subterm without making other changes.

Actual weights are computed upwards in $O(|t|)$ time, and after that — maximal weights are computed downwards in time $O(|t| \log l)$.

We will say that there exists an l'-*alternative* for annotated α-term t in the graph $G_{K,l}^{(a_1, \ldots, a_k)}$ if from the node q_1 there is a dotted arc to the node q_2 of weight l'. In this case the minimal α-term of the node q_2 is called the l'-*alternative* of term t.

The following *ordering of subterms* for a term t is defined: we say that a subterm t_1 is *left* from the subterm t_2 if the root symbol of t_1 is left from the root symbol of t_2 in the usual linear layout of the term t.

We mark the current α-term in the following way. A special symbol, say "$*$", is associated with the rightmost subterm t_1 of the term t (according to the ordering of subterms defined above) for which the l'-alternative exists, where l' is not greater than the maximal weight assigned to this subterm in the tree representation of t.

It is easy to see that for marking of the annotated α-term t, or detecting it as unmarkable, $O(|t|)$ steps are necessary.

For marked α-term t we define *succeeding α-term t'* in the following way. Let t_1 be the subterm marked by "$*$" and t_2 be the alternative of the term t_1. Succeeding term is obtained from the term t by replacing its subterm t_1 by the subterm t_2 and all its right siblings by the minimal terms of the same value.

The next lemma follows easy from the facts already proved.

Lemma 4. *The number of the steps necessary to construct, from a marked α-term t, its succeeding α-term t' and to mark it is $O(|t'| \log l)$.*

---

[2] Some important weight functions allow solving the inequality in constant time

The algorithm mentioned in Theorem 1 works as follows. Having received as input the Σ-algebra $\mathcal{K}$, the pair $(\langle a_1, \ldots, a_k \rangle, b)$ and $l > 0$, it constructs first the graph $G_{\mathcal{K},l}^{\langle a_1, \ldots, a_k \rangle}$. It takes $O(|\mathcal{K}| \log l)$ steps. After that the algorithm outputs the minimal term t_1 of the node b (in which the different domain values $a_1, \ldots, a_k$ are replaced by variables $x_1, \ldots, x_k$ respectively); this takes $O(|t_1|)$ steps. Then for the term t_2, succeeding t_1, the corresponding open term is output (according to the last lemma it takes $O(|t_2|)$ steps), etc. We proceed while marking is possible. It is easy to see that in this way the algorithm enumerates the set $A_{\mathcal{K},l}^{(\langle a_1, \ldots, a_k \rangle, b)}$ without repeating in the time mentioned in Theorem 1.$\square$

4 Synthesis from several examples

In the previous section the case where an open term satisfies one input/output example

$$(\langle a_1, \ldots, a_k \rangle, b)$$

was considered. Now we will consider the case where an open term has to satisfy several input/output examples simultaneously. More precisely, let there be a set of input/output examples belonging to Σ-algebra $\mathcal{K}$:

$$Q = \{(\langle a_1^1, \ldots, a_k^1 \rangle, b^1), (\langle a_1^2, \ldots, a_k^2 \rangle, b^2), \ldots, (\langle a_1^s, \ldots, a_k^s \rangle, b^s)\}.$$

We will say that an open term t *satisfies* a set Q in Σ-algebra $\mathcal{K}$, if t satisfies all elements of Q in $\mathcal{K}$.

Let us denote by $A_{\mathcal{K},l}^{Q}$ the set of all open terms which in Σ-algebra $\mathcal{K}$ satisfy a set Q and have weight no more than l ($l > 0$). In other words,

$$A_{\mathcal{K},l}^{Q} = A_{\mathcal{K},l}^{(\langle a_1^1, \ldots, a_k^1 \rangle, b^1)} \cap \ldots \cap A_{\mathcal{K},l}^{(\langle a_1^s, \ldots, a_k^s \rangle, b^s)}.$$

In real situations it is typical that separate sets

$$A_{\mathcal{K},l}^{(\langle a_1^1, \ldots, a_k^1 \rangle, b^1)}, A_{\mathcal{K},l}^{(\langle a_1^2, \ldots, a_k^2 \rangle, b^2)}, \ldots$$

are of relatively large size but their intersection is of relatively small size. Therefore an important question arises: can elements of $A_{\mathcal{K},l}^{Q}$ be found directly (i.e. enumerated sufficiently fast) without constructing all s sets?

The next theorem gives a positive answer to this question in some sense.

Theorem 5. *Let signature Σ and weight function w be fixed. There exists an algorithm which, given any Σ-algebra $\mathcal{K}$, any set Q of input/output examples and any natural number $l > 0$, enumerates without repeating the set of terms $A_{\mathcal{K},l}^{Q}$ in setup time $O(|\mathcal{K}|^{|Q|})$ and ith-step time $O(|t_i|)$ where t_i is the output term in the step i $(i = 1, 2, \ldots, |A_{\mathcal{K},l}^{Q}|)$.*

Proof. For the sake of simplicity we will consider the case when the set Q contains only two examples: $((a_1^1, \ldots, a_k^1), b^1)$ and $((a_1^2, \ldots, a_k^2), b^2)$.

For Σ-algebra $\mathcal{K}$ we define a *square Σ-algebra* $\mathcal{K}^2$ which describes an interpretation $K_{\Sigma, \mathcal{D}^2}$:

$$(f_i((a_1, a_1'), \ldots, (a_n, a_n')) = (b, b')) \in \mathcal{K}^2$$

if and only if

$$(f_i(a_1, \ldots, a_n) = b) \in \mathcal{K} \text{ and } (f_i(a_1', \ldots, a_n') = b') \in \mathcal{K}.$$

It is easy to deduce the next important equality:

$$A_{\mathcal{K},l}^{\{((a_1^1, \ldots, a_k^1), b^1), ((a_1^2, \ldots, a_k^2), b^2)\}} = A_{\mathcal{K},l}^{((a_1^1, \ldots, a_k^1), b^1)} \cap A_{\mathcal{K},l}^{((a_1^2, \ldots, a_k^2), b^2)} =$$

$$= A_{\mathcal{K}^2,l}^{(((a_1^1, a_1^2), \ldots, (a_k^1, a_k^2)), (b^1, b^2))}$$

To enumerate the set $A_{\mathcal{K},l}^{\{((a_1^1, \ldots, a_k^1), b^1), ((a_1^2, \ldots, a_k^2), b^2)\}}$ the algorithm described in Theorem 1 can be applied with respect to Σ-algebra $\mathcal{K}^2$. Because $|\mathcal{K}^2| \le |\mathcal{K}|^2$ the correctness of Theorem 5, when $|Q| = 2$, follows immediately. $\square$

5   Computer experiments

The advantage of the given algorithm (if compared with the algorithm described in [3]) is that it allows to enumerate effectively not only terms having limited depth, but it also permits to enumerate terms according to more sophisticated criteria, like the number of functional symbols etc. In the case some function (like square root) is not likely to appear in the expression more than once, then the new method allows to restrict that by assigning higher weight to the particular function.

First some implementation details. Setup time $O(|\mathcal{K}|^{|Q|})$ and corresponding volumes of graphs still are too large for practical implementations. Another approach was used — not only the weight of terms was restricted with some w, but also the term level was said not to exceed some l (it is easy to see that the method described above works also for such a pair of restrictions). Besides, the annotated sample graph was constructed dynamically, it contained only reachable domain nodes. Reachability of some node here means that

- there is a term of depth not greater than l such that, substituting its variables with values from input example, the value of the term corresponds to the node;
- the weight of this term does not exceed w;
- there is a term of depth not greater than l such that, substituting some of its variables with the value corresponding to the node and other variables with values corresponding to reachable nodes, the value of the term equals to the value of output example;

– the weight of this term also does not exceed w. (Note that there is only one input example consisting of several tuples and only one output example (tuple) — see Section 4).

Using described approach a formula for the volume of the frustum of the square pyramid was synthesized. The experiment showed that the new algorithm finds the formula several times faster than the algorithm described in [5].

The next formula we tried to synthesize was the formula for finding the greatest root of quadratic equation of the form $x^2 + bx + c = 0$. Unfortunately we were not able to synthesize the formula from input/output examples using 33MHz Sparc workstation in the reasonable time, assuming $\Sigma = \{+, -, *, /, \sqrt{}, 1, 2, 3, 4\}$ and using weight function which counts instances of functional symbols . Therefore the algorithm of Theorem 2 was further modified taking into account the following observation: the outermost function of the formula is one of the 9 functions appearing in Σ and having fixed one of them we can reduce the depth of unknown term by 1 level (we have to consider all 9 possible cases of the outermost function). In fact it is reasonable to fix functions appearing in the two outermost levels what yields us 377 cases, but reduces the depth of the unknown term by 2 levels. In this way the formula for greatest root of the quadratic equation

$$x = \frac{\sqrt{b^2 - 4c} - b}{2}$$

was synthesized on the 33MHz Sparc workstation in 20 minutes from the following examples:

$$F(3, -4) = 1$$
$$F(-2, -3) = 3$$
$$F(5, 4) = -1$$
$$F(-4, 3) = 3$$
$$F(-3, -4) = 4$$
$$F(-3, 2) = 2$$
$$F(3, 2) = -1$$

Limit on the number of functional symbols — 8, limit on term depth — 6, domain — $[-16,25]$. Analogical attemt without weight restriction failed.

Then we changed domain for $[-24,36]$, and examples were as follows:

$$F(-1, -6) = 3$$
$$F(4, -5) = 1$$
$$F(5, 6) = -2$$
$$F(-4, -5) = 5$$
$$F(6, 5) = -1$$
$$F(-5, 4) = 4$$

We expected to obtain greater annotated sample graphs, as the domain was greater, but surprisingly the correct formula was synthesised in 10 minutes with much smaller graph volumes. Appearently these examples were "better" if compared to previous ones. Therefore it seems to be hard to make any theoretical estimations of the behavior of annotated sample graphs on various examples.

Further experiments are intended in order to more precisely understand the power of the described method.

References

1. D.Angluin: Easily inferred sequences. Memorandum No. ERL-M499, University of California, 1974
2. P.van Emde Boas, R.Kaas and E.Zijlstra: Design and Implementation of an Efficient Priority Queue. Mathematical Systems Theory 10 (1977), pp. 99—127
3. J.M.Barzdins and G.J.Barzdins: Rapid construction of algebraic axioms from samples. Theoretical Computer Science 90 (1991), pp. 199—208
4. G.Barzdins: Inductive synthesis of term rewriting systems. Lecture Notes in Computer Science, vol.502 (1991), pp. 253—285
5. J.Barzdins and G.Barzdins: Towards efficient inductive synthesis: rapid construction of local regularities. Lecture Notes in Computer Science, vol.659 (1993), pp. 132 — 140
6. R.Freivalds: Inductive inference of recursive functions. Qualitative theory. Lecture Notes in Computer Science, vol.502 (1991), pp. 77—110
7. R.Freivalds, J.Barzdins, K.Podnieks: Inductive inference of recursive functions: Complexity bounds. Lecture Notes in Computer Science, vol.502 (1991), pp. 565—613
8. E.Howard: An Introduction to the History of Mathematics. New York: Holt, Reinhard and Winston 1961

A Typed λ-Calculus for Proving-by-Example and Bottom-Up Generalization Procedure

Masami Hagiya

Department of Information Science, University of Tokyo
Hongo 7-3-1, Bunkyo-ku, Tokyo 113, JAPAN
hagiya@is.s.u-tokyo.ac.jp

Abstract. A generalization procedure which reconstructs mathematical inductions that are expanded in a proof of an example is formulated under a typed λ-calculus that is designed for increasing the applicability of the procedure. The λ-calculus, an extension of Logical Framework, allows recursions and inductions on natural numbers, and inferences on linear arithmetical terms are built into its type system. The generalization procedure is iterated in a bottom-up fashion to construct nested inductions. Consequently, it can also find inductions whose induction formula is a limited form of bounded quantification.

1 Introduction

Arguments using concrete examples often result in friendly communication between humans who exchange knowledge with each other. This must also be true for human-computer interaction. In this paper, we discuss the use of examples in the field of proof-checking or theorem-proving, in which a human and a machine communicate mathematical proofs with each other. We investigate principles for constructing a proof-checker that allows arguments beginning with "For example..." A user of such a proof-checker writes a proof of a theorem using a concrete example, and the proof-checker automatically generalizes the proof of the example to that of the general case.

In this paper, we focus on the problem of reconstructing mathematical inductions that are expanded in a proof of an example. A mathematical induction of form

$$
\cfrac{A[0] \quad \cfrac{[A[i]] \\ \vdots \\ A[i+1]}{}}{A[n]}
\quad \text{is expanded to} \quad
\begin{matrix} A[0] \\ \vdots \\ A[1]\,, \\ \vdots \\ A[2] \end{matrix}
$$

if n is equal to 2 in a concrete proof. Reconstructing a mathematical induction means recovering the former inductive proof given the latter one. In the machine learning community, this problem is called *generalization-to-N* or *generalizing number*, and has been studied by several researchers in the field of explanation-based learning [7, 12, 13, 3].

Since proof generalization as discussed above is a kind of proof transformation, we need a framework in which proofs can be treated as first-class objects. The framework based on typed λ-calculi [1, 8] is probably the best for the purpose. The aim of this paper is to propose a typed λ-calculus suitable for proof generalization or *proving-by-example*, and rigorously formulate the generalization procedure under the calculus. The most important benefit of using typed λ-calculi is that various syntactical constructs can be treated in a uniform fashion. For our purpose, a single generalization procedure suffices to generalize programs, propositions, types and proofs.

In this paper, we extend Logical Framework [8], one of the typed λ-calculi of Lambda Cube [1]. The extended calculus allows recursions and inductions on natural numbers, and inferences on linear arithmetical terms are built into its type system. The extension corresponds to that of logic programming by constraints [4]. The constraint solver checks constraints between arithmetical terms while cooperating with the ordinary type-checking algorithm.

We then formulate a bottom-up procedure for generalizing a concrete proof by recovering inductions that are expanded in the concrete proof. The procedure can reconstruct inductions whose induction formula $A[i]$ is of form $\exists x B[i, x]$, where $B[i, x]$ is atomic. The generalization procedure is iterated in a bottom-up fashion to construct nested inductions. Consequently, it can also find inductions whose induction formula $A[i]$ is of form $\forall j (j \leq i \supset B[i, j])$, and whose induction step consists of proofs of two propositions $\forall i B[i, i]$ and $\forall i \forall j (j \leq i \wedge B[i, j] \supset B[i+1, j])$. This kind of induction often appears in a proof about an array, in which output of a function is constructed as elements of an input array are sequentially scanned. The correctness proof of a sorting algorithm is a typical example. We call such an induction *a triangular induction* here, because it takes a triangular shape if it is expanded in a concrete proof.

In the next section, we examine the correctness proof of the selection sort algorithm. The proof is an instance of a triangular induction, and is a typical example that motivates the study of this paper. In the following section, we present the typed λ-calculus, giving its inference rules and some of its properties. We then formulate the generalization procedure. We finally examine how a triangular induction can be reconstructed from a concrete proof with the generalization procedure.

2 Selection Sort

Let **swap** $a \, i \, j$ denote the result of swapping the i-th and j-th elements of array a, and **minindex** $a \, i$ denote the index of the least element of $a[i], a[i+1], a[i+2], \cdots$.[1]

---

[1] In this formalization, arrays do not have an upper bound.

Let a_0 be an arbitrary array. We successively make the following definitions.

$$k_0 = \text{minindex } a_0\; 0, \quad a_1 = \text{swap } a_0\; 0\; k_0,$$
$$k_1 = \text{minindex } a_1\; 1, \quad a_2 = \text{swap } a_1\; 1\; k_1,$$
$$k_2 = \text{minindex } a_2\; 2, \quad a_3 = \text{swap } a_2\; 2\; k_2,$$
$$k_3 = \text{minindex } a_3\; 3, \quad a_4 = \text{swap } a_3\; 3\; k_3, \;\cdots$$

We can then prove

$$a_4[0]\leq a_4[1] \wedge a_4[1]\leq a_4[2] \wedge a_4[2]\leq a_4[3]$$

as follows. Using the properties of swap and minindex, we can first prove $a_2[0] \leq a_2[1]$. From this, $a_3[0] \leq a_3[1]$ is derived because $a_3[0] = a_2[0]$ and $a_3[1] = a_2[1]$ hold. From this, $a_4[0] \leq a_4[1]$ is further derived. As we proved $a_2[0] \leq a_2[1]$, we can prove $a_3[1] \leq a_3[2]$. From this, $a_4[1] \leq a_4[2]$ is derived. We can finally prove $a_4[2] \leq a_4[3]$. Taking the conjunction of the facts proved so far, we obtain a proof that looks like Figure 1.

$$\vdots$$
$$a_2[0] \leq a_2[1]$$

$$a_3[0] \leq a_3[1] \qquad a_3[1] \leq a_3[2]$$

$$\underline{a_4[0] \leq a_4[1] \qquad a_4[1] \leq a_4[2] \qquad a_4[2] \leq a_4[3]}$$
$$a_4[0]\leq a_4[1] \wedge a_4[1]\leq a_4[2] \wedge a_4[2]\leq a_4[3]$$

Fig. 1. Selection Sort

Since the proof has a simple iterative structure, it is expected that one can construct an inductive proof by *folding* an iteration of similar proofs into an induction, and obtain a proof of $\forall i \forall j (j\leq i \supset a_{i+2}[j]\leq a_{i+2}[j + 1])$. Under the framework of ordinary first-order arithmetic, however, the inductive proof cannot be obtained so easily. First of all, the above proposition must be translated to the bounded quantification $\forall j (j\leq 2 \supset a_4[j]\leq a_4[j + 1])$. Since proofs of *bounds* such as $j \leq 2$ do not appear in the concrete proof, they must be appropriately generated and inserted. Moreover, as the index j in $a_4[j] \leq a_4[j + 1]$ increases, the number of iteration decreases. This requires the ability to do arithmetical reasoning on bounds. These are the motivations for proposing a new typed λ-calculus in this paper. The calculus allows implicit reasoning on bounds, because arithmetical inferences of certain forms are built into its type system.

A more fundamental consequence of using a typed λ-calculus is that recursions and inductions are treated uniformly because proofs and programs are both represented by λ-terms. In particular, proofs and programs are uniformly generalized by our generalization procedure.

3 An Extension of Logical Framework

For the purpose of the paper, we extend Logical Framework [8] as follows. We first introduce the notion of arithmetical variables and terms.

- In addition to ordinary variables, we introduce *arithmetical variables* that range over **nonnegative** integers. Arithmetical variables are denoted by Greek letters such as χ, η, etc.
- Expressions of the following form are allowed as *arithmetical terms*.

$$k_0 + k_1\chi_1 + \cdots + k_n\chi_n$$

Notice that as arithmetical terms, we only allow linear expressions of arithmetical variables with integer coefficients. In the above arithmetical term, k_i denotes an integer constant and χ_i denotes an arithmetical variable. Arithmetical terms are denoted by Greek letters such as μ, ν, etc.

Terms and types of Logical Framework are then extended. As is explained again, M and F denote terms, and A denotes types.

- We allow λ-abstractions by an arithmetical variable of form $\lambda\chi<\mu.M$, and Π-abstractions by an arithmetical variable of form $\Pi\chi<\mu.A$, where χ is an arithmetical variable and μ is an arithmetical term. The abstraction $\lambda\chi<\mu.M$ denotes a function that returns M for χ such that $0 \leq \chi < \mu$. Abstractions of form $\lambda\chi<\infty.M$ or of form $\Pi\chi<\infty.M$ are also allowed.
- Expressions of forms $\mu<\nu$, $\mu=\nu$, $\mu\leq\nu$, etc., are allowed as types (i.e., propositions), where μ and ν are arithmetical terms.
- An expression of form $?_{\mu<\nu}$ is allowed as a proof of proposition $\mu<\nu$. Expressions of forms $?_{\mu=\nu}$, $?_{\mu\leq\nu}$, etc., are also allowed. They are called *anonymous proofs*.
- We introduce *the recursion operator* **R**. If μ is an arithmetical term, and F and M are terms, then $\mathbf{R}\mu F M$ is a term. Terms of form $\mathbf{R}\mu F M$ are called *recursion terms*. The term M in $\mathbf{R}\mu F M$ is called *the recursion base* of $\mathbf{R}\mu F M$ and F *the recursion step*. $\mathbf{R}\mu F M$ denotes the result of applying the recursion step F to recursion base M for μ times. $\mathbf{R}\mu F M$ is reduced to M if $\mu = 0$, and to $F(\mu-1)(\mathbf{R}(\mu-1)FM)$ if $\mu > 0$.

According to the above extension, the type system of Logical Framework is extended as follows. Remember that in the following explanations, μ, ν, etc., may also denote the symbol ∞.

- As an element of a context, we allow an expression of form $\chi<\mu$, where χ is an arithmetical variable and μ is an arithmetical term or ∞.

- We introduce judgements of form $\Gamma \vdash \mu < \nu$, which means that $\mu < \nu$ always holds under the range restrictions for the arithmetical variables in context Γ. Judgements of the following forms are also introduced: $\Gamma \vdash \mu = \nu$, $\Gamma \vdash \mu \leq \nu$, $\Gamma \vdash \mu \neq \nu$, and $\Gamma \vdash 0 \leq \mu < \nu$. Judgements of these forms are called *arithmetical judgements*.
- The equality between ordinary terms is also explicitly derived as a judgement [9, 6]. We introduce judgements of form $\Gamma \vdash M = N$, where M and N are terms that are not arithmetical.

In Figures 4–8 in the appendix, the extended type system is summarized by a set of inference rules that derive judgements. In the inference rules as well as in the following explanations, M, N, K, L, A, B, F, G, etc., denote ordinary terms that are not arithmetical. In particular, A, B, etc., denote types, and F, G, etc., denote functions. Ordinary variables that are not arithmetical are denoted by x, y, etc. Contexts are denoted by Γ. The so called *sorts*, i.e., **Type** (*) and **Kind** ($\square$), are denoted by s. We use i, j, k, l, m, n, etc., for integer constants. A term with a designated variable is denoted by an expression such as $M[x]$, where x is the designated variable. The expression $M[N]$ denotes the result of replacing x with N in $M[x]$.

Arithmetical judgements are defined as in Figure 7. Judgements of forms $\Gamma \vdash \mu = \nu$, $\Gamma \vdash \mu \leq \nu$, etc., are defined similarly. If $\chi_1 < \mu_1, \cdots, \chi_n < \mu_n$ are all the elements in context Γ of form $\chi_i < \mu_i$, the validity of the arithmetical judgement $\Gamma \vdash \mu < \nu$ is equivalent to that of the arithmetical formula $\forall \chi_1 \cdots \forall \chi_n (0 \leq \chi_1 < \mu_1 \wedge \cdots \wedge 0 \leq \chi_n < \mu_n \supset \mu < \nu)$. Since it is a Presburger formula [5], its validity is decidable. Moreover, because all the quantifiers are universal and placed at the outermost position of the formula, the SUP-INF method [2, 14] can be used to test its validity.

By the inference rules, the formula $0 \leq \chi_1 < \mu_1 \wedge \cdots \wedge 0 \leq \chi_n < \mu_n$ is guaranteed to be consistent, i.e., there exist some $\chi_1, \cdots, \chi_n$ that satisfy the formula. This implies that two judgements $\Gamma \vdash 0 < \mu$ and $\Gamma \vdash \mu = 0$ cannot be simultaneously valid. Therefore, a recursion term $\mathbf{R}\mu FM$ is not simultaneously reducible to $F(\mu-1)(\mathbf{R}(\mu-1)FM)$ and to M. This is the key point when the calculus is shown to be Church-Rosser.

The calculus has the following basic properties for typed λ-calculi: Church-Rosser, type unicity[2], subject reduction, strong normalization, and decidability of typechecking. The Church-Rosser property is proved by the standard parallel reduction method [15] with a slight extension that the parallel reduction relation is now indexed by a context. The strong normalization property is also proved by the standard computability method with a slight extension that the reducibility predicate is indexed by a context.[3]

The extension we have made to typed λ-calculi is similar to that of logic programming by constraints. Arithmetical judgements correspond to *constraints* in

---

[2] We do not have coercion in this version of the calculus.

[3] Since this version of the calculus does not have η-conversion, the Church-Rosser property can be proved independently from others.

a goal, while judgements of other kinds are considered as ordinary atoms. The initial goal consists of the judgement to be derived, which is of form $\Gamma \vdash M \in A$. Ordinary typechecking algorithm cooperates with the constraint solver in the sense that the constraint solver solves arithmetical judgements while other judgements are resolved with inference rules. Notice that reduction is also controlled by the constraint solver.

4 Generalization Procedure

The generalization procedure is roughly of the following form.

> **while** a recursion step is found from a repetition of similar terms **do**
> > **while** a subterm that is an iteration of the recursion step is found **do**
> > > replace the iteration with a recursion term
> > **end while**
> **end while**

In the rest of this section, we explain how to find a recursion step and how to replace a subterm with a recursion term.

4.1 Join Operation

As is explained in the next subsection, we use the *join operation*, denoted by $Join(\{M_1, \cdots, M_n\})$, for constructing a recursion step in the generalization procedure. It compares terms $M_1, \cdots, M_n$, ignoring the difference of constant parts of arithmetical terms. Moreover, it is defined such that $Join(\{\mathbf{R}\mu FM, M\}) \equiv M$.[4] The formal definition of $Join(\{M_1, \cdots, M_n\})$ follows those of two relations $\sim$ and $\leadsto$ between terms.

If two terms differ only in the constant part of arithmetical terms, they are said to be *approximately equal*. For example, $4 - 2\chi + \eta$ is approximately equal to $3 - 2\chi + \eta$. We write $M \sim N$ when M and N are approximately equal. The reflexive and transitive relation $M \leadsto N$, which is compatible with the structure of terms, is then defined as follows.

- If $M \sim N$, then $M \leadsto N$. If $\mu \sim \nu$, then $\mu \leadsto \nu$.
- $\mathbf{R}\mu FM \leadsto M$.

Notice that $\leadsto$ contains the approximate equality $\sim$.

Let $\{M_1, \cdots, M_n\}$ be a nonempty finite set of terms. By the following rules, we compute a term denoted by $Join(\{M_1, \cdots, M_n\})$, which satisfies

$$Join(\{M_1, \cdots, M_n\}) \leadsto M_i.$$

Note that the term $Join(\{M_1, \cdots, M_n\})$ may not be well-typed, even if $M_1, \cdots, M_n$ are.

---

[4] As is usual, $\equiv$ denotes syntactical identity.

- If $M_1 \sim M_2 \sim \cdots \sim M_n$, then $Join(\{M_1, \cdots, M_n\}) \equiv M_1$.[5] If $\mu_1 \sim \mu_2 \sim \cdots \sim \mu_n$, then $Join(\{\mu_1, \cdots, \mu_n\}) \equiv \mu_1$.
- If $M_i \equiv R\mu_i F_i N_i$ for $i = 1, \cdots, m$ $(1 \leq m \leq n)$ and M_i does not begin with R for $i = m + 1, \cdots, n$, then

$$Join(\{M_1, \cdots, M_n\})$$
$$\equiv R\,Join(\{\mu_1, \cdots, \mu_m\})\,Join(\{F_1, \cdots, F_m\})\,Join(\{N_1 \cdots N_m, M_{m+1}, \cdots M_n\}).$$

- If $M_i \equiv \lambda\chi < \mu_i . N_i$, then

$$Join(\{M_1, \cdots, M_n\}) \equiv \lambda\chi < Join(\{\mu_1, \cdots, \mu_n\}).\,Join(\{N_1, \cdots, N_n\}).$$

- Similarly for ordinary λ-abstractions and Π-abstractions.
- If $M_i \equiv F_i \mu_i$, then

$$Join(\{M_1, \cdots, M_n\}) \equiv Join(\{F_1, \cdots, F_n\})\,Join(\{\mu_1, \cdots, \mu_n\}).$$

- Similarly for ordinary applications.
- If $M_i \equiv \mu_i < \nu_i$, then

$$Join(\{M_1, \cdots, M_n\}) \equiv Join(\{\mu_1, \cdots, \mu_n\}) < Join(\{\nu_1, \cdots, \nu_n\}).$$

- Similarly for $=$, $\leq$, etc.
- In other cases, $Join(\{M_1, \cdots, M_n\})$ is not defined.

4.2 Finding a Recursion Step

In order to reconstruct an induction from the given concrete proof, we first search the proof for two iterations of some recursion step, i.e., we search for a term $\bar{M}$ such that $\bar{M} = M[1, M[0, M_0]]$ for some terms $M[\chi, x]$ and M_0. This in general requires us to solve the equation $f1(f0z) = \bar{M}$, where f and z are unknowns [10, 11]. In order to solve the equation, we need to do higher-order unification in the extended calculus, whose complete procedure is difficult to implement because the calculus has recursion terms.

Instead, we take the following practical approach here. We search for a term $\bar{M}$ of form $M_2[M_1[M_0]]$, where x occurs exactly once in $M_1[x]$ and $M_2[x]$, and $Join(\{M_1[x], M_2[x]\})$ is defined. We then put $M'[x] \equiv Join(\{M_1[x], M_2[x]\})$.

At this moment, we only have a naïve procedure for searching for $\bar{M}$. For each pair of nodes in the tree representation of term M, where one node is an ancestor of the other, the procedure looks down the latter node to check if the same path from the former to the latter is repeated (Figure 2). If it finds a repetition, it calls the join operation.

After $M'[x]$ is obtained as above, we generalize $M'[x]$ by introducing a new arithmetical variable χ, and replacing each occurrence of an arithmetical term in $M'[x]$ of form $k_0 + k_1\chi_1 + \cdots + k_n\chi_n$ with an arithmetical term of form $k + l\chi + k_1\chi_1 + \cdots + k_n\chi_n$, where k and l are unknown integers. These unknown

---

[5] M_1 is chosen arbitrarily from the set $\{M_1, \cdots, M_n\}$.

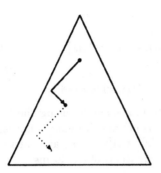

Fig. 2. Searching for a Repetition

integers are distinct for each occurrence of an arithmetical term in $M'[x]$. The result of generalization is denoted by $M[\chi, x]$. Note that $M[\chi, x] \rightsquigarrow M_i[x]$ holds for $i = 1, 2$.

Let Γ be the context for the term $M_2[M_1[M_0]]$ and assume $\Gamma \vdash M_0 \in A_0$, $\Gamma \vdash M_1[M_0] \in A_1$ and $\Gamma \vdash M_2[M_1[M_0]] \in A_2$. As we obtained $M[\chi, x]$ from $Join(\{M_2[x], M_1[x]\})$, we also generalize $Join(\{A_0, A_1, A_2\})$ to $A[\chi]$ by introducing unknown integers. Note that $A[\chi] \rightsquigarrow A_i$ holds for $i = 0, 1, 2$.

It remains to instantiate the unknown integers in $M[\chi, x]$ and $A[\chi]$. We do it in the next subsection.

4.3 Introducing a Recursion Term

Assume that we have obtained $M[\chi, x]$ and $A[\chi]$ as in the previous section. We then search the given concrete proof for a term $\bar{N}$ of form

$$N_n[N_{n-1}[\cdots [N_2[N_1[N_0]]] \cdots]]$$

such that

- $n \geq 1$,
- $M[\chi, x] \rightsquigarrow N_i[x]$ for $i = 1, \cdots, n$,
- $A[\chi] \rightsquigarrow B_i$ for $i = 0, 1, \cdots, n$, where $\Gamma \vdash N_i[N_{i-1}[\cdots [N_2[N_1[N_0]]] \cdots]] \in B_i$.

The term $M_2[M_1[M_0]]$, which were used to compute $M[\chi, x]$ and $A[\chi]$, may be taken as $\bar{N}$, but we look for a maximal term satisfying the above conditions, and make n as large as possible. We also apply $M[\chi, x]$ to terms in the proof that do not contain $M_2[M_1[M_0]]$. This means that we *propagate* the recursion step throughout the entire proof.

After fixing the term $\bar{N}$, we introduce another unknown integer m, and instantiate m and the unknown integers in $M[\chi, x]$ and $A[\chi]$ so that the following judgements become valid.

$$\Gamma, x{:}A[i{-}1] \vdash M[i{-}1, x] = N_i[x] \qquad \text{(for } i = 1, \cdots, n) \qquad (1)$$
$$\Gamma \vdash A[i] = B_i \qquad \text{(for } i = 0, 1, \cdots, n) \qquad (2)$$
$$\Gamma, \chi{<}m, x{:}A[\chi] \vdash M[\chi, x] \in A[\chi{+}1] \qquad (3)$$

We can usually instantiate unknowns by using the facts $M[\chi, x] \rightsquigarrow N_i[x]$ and $A[\chi] \rightsquigarrow B_i$ so that equations (1) and (2) hold. That is, if $n \geq 2$, we can usually obtain from $M[\chi, x] \rightsquigarrow N_i[x]$ or $A[\chi] \rightsquigarrow B_i$ at least two equations for each pair of unknowns k and l, i.e., equations of the forms

$$0 \times k + l = \text{some constant}$$
$$1 \times k + l = \text{some constant},$$

from which k and l are immediately instantiated. In some cases, e.g., when $n = 1$, we must also use the judgement (3).[6]

We finally replace the term $\bar{N}$ in the proof with the recursion term

$$\mathbf{R}n(\lambda\chi{<}m.\lambda x{:}A[\chi].M[\chi, x])N_0,$$

if m has been instantiated, or with the recursion

$$\mathbf{R}n(\lambda\chi{<}\infty.\lambda x{:}A[\chi].M[\chi, x])N_0,$$

if m need not be instantiated.

Notice that we find a recursion step from **two** iterations of the step, while we introduce a recursion term even for **one** iteration. Moreover, the join operation allows **zero** iterations (i.e., recursion base) to be matched with a recursion term.

4.4 Bottom-Up Procedure

The generalization procedure indefinitely iterates the two operations explained so far, i.e., finding a recursion step and introducing a recursion term. It is of the following form.

> **while** term $M[\chi, x]$ is found **do**
> **while** term $\bar{N}$ is found **do**
> replace $\bar{N}$ with a recursion term
> **end while**
> **end while**

In order to effectively iterate the operations, the procedure must be applied in a bottom-up fashion from inner subterms to outer ones in the given proof. In particular, we apply it to terms representing programs before terms representing proofs.

---

[6] One of the problems for typechecking $M[\chi, x]$ is that we cannot decide how to reduce a recursion term $\mathbf{R}\mu GK$ if μ contains unknown integers. We must try to reduce $\mathbf{R}\mu GK$ under the hypothesis $\mu > 0$ or $\mu = 0$.

5 Reconstructing Triangular Inductions

Assume that the given concrete proof contains proofs of the three propositions $B[2,0]$, $B[2,1]$ and $B[2,2]$, and the three proofs take the forms in Figure 3. $B[1,0]$ is a proof from $B[0,0]$ and $0 < 1$, $B[2,0]$ is a proof from $B[1,0]$ and $0 < 2$, etc. This is the result of expanding an induction which we call *triangular induction*.

$$
\begin{array}{ccc}
\vdots & & \\
B[0,0] \quad 0 < 1 & & \\
\vdots & \vdots & \\
B[1,0] \quad 0 < 2 & B[1,1] \quad 1 < 2 & \\
\vdots & \vdots & \vdots \\
B[2,0] & B[2,1] & B[2,2]
\end{array}
$$

Fig. 3. Triangular Induction

The proof $\vdots$ in Figure 3 is assumed to be uniform in the sense that it can be represented by the term $M[\chi, \eta, x, y]$ such that

$$\Gamma, \chi < \infty, \eta < \infty, x{:}B[\chi, \eta], y{:}\eta < \chi + 1 \vdash M[\chi, \eta, x, y] \in B[\chi+1, \eta].$$

The proof $\vdots$ is also uniform and represented by the term $N[\chi]$ such that

$$\Gamma, \chi < \infty \vdash N[\chi] \in B[\chi, \chi].$$

Therefore, the three proofs are represented by the following three terms M_0, M_1 and M_2.

$$
\begin{aligned}
M_0 &\equiv M[1, 0, M[0, 0, N[0], ?_{0<1}], ?_{0<2}] \\
M_1 &\equiv M[1, 1, N[1], ?_{1<2}] \\
M_2 &\equiv N[2]
\end{aligned}
$$

If we apply the generalization procedure, we can find the recursion step in M_0. Let $F_{i,j}$ denote the term $\lambda x{:}B[i,j].M[i,j,x,?_{j<i+1}]$. By the generalization procedure, the three terms are rewritten as follows.

$$
\begin{aligned}
M_0 &\equiv \mathbf{R}2(\lambda \chi < \infty.F_{\chi,0})N[0] \\
M_1 &\equiv \mathbf{R}1(\lambda \chi < \infty.F_{\chi+1,1})N[1] \\
M_2 &\equiv N[2]
\end{aligned}
$$

Now, assume that $B[i,j]$ is of form $\mathbf{T}(b[i,j])$, where $b[i,j]$ is a term of type **bool** under the following context.

  bool : **Type**

   t : **bool**

    T : **bool** $\rightarrow$ **Type**

   tt : **T** t

   and : **bool** $\rightarrow$ **bool** $\rightarrow$ **bool**

   andI : $\mathit{\Pi}w_1$:**bool**.$\mathit{\Pi}w_2$:**bool**.$(\mathbf{T}w_1) \rightarrow (\mathbf{T}w_2) \rightarrow (\mathbf{T}(\text{and}w_1w_2))$

By **andI** and the above three terms M_0, M_1 and M_2, we can construct a proof of proposition

$$\mathbf{T}(\text{and}(b[2,2])(\text{and}(b[2,1])(\text{and}(b[2,0])\mathbf{t})))$$

as follows.

    $\text{andI}(b[2,0])(\text{and}(b[2,1])(\text{and}(b[2,2])\mathbf{t})))$

    M_0

    $(\text{andI}(b[2,1])(\text{and}(b[2,2])\mathbf{t})$

     M_1

     $(\text{andI}(b[2,2])\mathbf{t}$

      $M_2\,\mathbf{tt}))$

This proof is generalized as follows.

From the term $\text{and}(b[2,1])(\text{and}(b[2,2])\mathbf{t})$, we first find the recursion step $\lambda\eta{<}3.\lambda z$:**bool**.$\text{and}(b[2,2{-}\eta])z$. Let $r_{i,j}$ denote the recursion term

$$\mathbf{R}i(\lambda\eta{<}j.\lambda w\text{:bool}.\text{and}(b[2,2{-}\eta])w)\mathbf{t}.$$

Using $r_{i,j}$, the above proof is rewritten as follows.

    $\text{andI}(b[2,0])r_{2,3}$

    M_0

    $(\text{andI}(b[2,1])r_{1,3}$

     M_1

     $(\text{andI}(b[2,2])\mathbf{t}$

      $M_2\,\mathbf{tt}))$

We finally find the following recursion step.

$$\lambda\eta{<}3.\lambda w\text{:}\mathbf{Tr}_{\eta,3}.\text{andI}(b[2,2{-}\eta])r_{\eta,3}(\mathbf{R}\eta(\lambda\chi{<}\infty.F_{\chi+2{-}\eta,2{-}\eta})N[2{-}\eta])w$$

The entire proof is generalized to

  $\mathbf{R}\,3\ (\lambda\eta{<}3.\lambda w\text{:}\mathbf{Tr}_{\eta,3}.$

     $\text{andI}(b[2,2{-}\eta])r_{\eta}(\mathbf{R}\eta(\lambda\chi{<}\infty.F_{\chi+2{-}\eta,2{-}\eta})N[2{-}\eta])w)$

  $\mathbf{tt}.$

Since the result is of type $Tr_{3,3}$, by simply replacing number 2 by a variable ζ and number 3 by $\zeta+1$, we obtain the proof

$\mathbf{R}\ (\zeta+1)\ (\lambda\eta{<}\zeta{+}1.\lambda w{:}Tr_{\eta,\zeta+1}.$

$\qquad\qquad \mathtt{andI}(b[\zeta,\zeta-\eta])r_\eta(\mathbf{R}\eta(\lambda\chi{<}\infty.F_{\chi+\zeta-\eta,\zeta-\eta})N[\zeta-\eta])w)$

$\quad$ tt,

whose type is $Tr_{\zeta+1,\zeta+1}$,

6   Concluding Remarks

Representing proofs as λ-terms is not only theoretically important but is also practically useful, when one wants to manipulate formal proofs. In this paper, reconstructing inductions and reconstructing recursions can be realized by a single procedure because proofs and programs are both represented as terms.

However, as logic programming has been extended with domain-specific knowledge formulated as in the form of *constraints*, if type theory is to find more applications, it is also necessary to extend it with application-specific knowledge as we did in this paper. We consider that it is important to build a general theory for incorporating constraints into type theory. One of the research issues is to investigate whether basic properties such as Church-Rosser or strong normalization hold in a calculus with constraints.

Bruynooghe, De Raedt and De Schreye in [3] use a concrete proof to guide unfold/fold program transformation, but a concrete proof is not directly generalized. Since the program obtained by their program transformation corresponds to an inductive proof whose induction formula is a conjunction, we must extend our approach to cope with conjunctive induction formulas in order to directly generalize a concrete proof to obtain a transformed program.

Whether a proof is written by hand or generated by a theorem prover, automatically reconstructing inductions is an important and interesting research issue. However, it would be nice if an environment for writing formal proofs allows one to write a proof about a concrete example first and then manually generalize the concrete proof by explicitly specifying appropriate induction steps in it. Although automatic reconstruction of inductions requires more researches, implementing such an environment is not a very difficult task and will immediately improve the user-interface for writing formal proofs.

Acknowledgements

The author would like to thank Masahiko Sato, Susumu Hayashi, Hiroshi Nakano, and other many people for fruitful discussions. He also thanks the referees for valuable comments to improve the paper.

References

1. Barendregt,H.: Introduction to generalized type systems, *Journal of Functional Programming*, Vol.1, No.2 (1991), pp.125–154.

2. Bledsoe,W.W.: A new method for proving certain Presburger formulas, *Proceedings of IJCAI*, 1975, pp.15–21.

3. Bruynooghe,M., De Raedt,L., De Schreye,D.: Explanation based program transformation, *Proceedings of IJCAI*, 1989, pp.407–412.

4. Cohen, J.: Constraint logic programming languages, *Communications of the ACM*, Vol.33, No.7 (1990), pp.52–68.

5. Cooper,D.C.: Theorem proving in arithmetic without multiplication, *Machine Intelligence*, No.7 (1972), pp.91–99.

6. Coquand,T.: An algorithm for testing conversion in type theory, *Logical Framework* (Huet,G., Plotkin,G., eds.), 1991, pp.255–279.

7. Ellman,T.: Explanation-based learning: A survey of programs and perspectives, *ACM Computing Surveys*, Vol.21, No.2 (1989), pp.163–221.

8. Harper,R., Honsell,F., Plotkin,G.: A framework for defining logics, *Symposium on Logic in Computer Science*, 1987, pp.194–204.

9. Harper,R.: An equational formulation of LF, ECS-LFCS-88-67, Edinburgh, 1988.

10. Hagiya,M.: Programming by example and proving by example using higher-order unification, *10th International Conference on Automated Deduction* (Stickel,M., ed.), Lecture Notes in Artificial Intelligence, Vol.449 (1990), pp.588–602.

11. Hagiya,M.: From programming-by-example to proving-by-example, *Theoretical Aspects of Computer Software* (Ito, T., Meyer,A.R., eds.), Lecture Notes in Computer Science, Vol.526 (1991), pp.387–419.

12. Shavlik,J.W., DeJong,G.F.: An explanation-based approach to generalizing number, *Proceedings of IJCAI*, 1987, pp.236–238.

13. Shavlik,J.W., DeJong,G.F.: Acquiring general iterative concepts by reformulating explanations of observed examples, *Machine Learning Volume III* (Kodratoff,Y., Michalski,R., eds.), 1990, pp.302–350.

14. Shostak,R.E.: On the SUP-INF method for proving Presburger formulas, *Journal of the ACM*, Vol.24, No.4 (1977), pp.529–543.

15. Takahashi,M.H.: Parallel reductions in λ-calculus, *Journal of Symbolic Computation*, Vol.7 (1989), pp.113–123.

Appendix — Inference Rules

$$\frac{\Gamma \text{ is an empty context}}{\Gamma \text{ context}}$$

$$\frac{\Gamma \vdash A \in s}{\Gamma, x{:}A \text{ context}}$$

$$\frac{\Gamma \vdash 0 < \mu}{\Gamma, \chi < \mu \text{ context}}$$

Fig. 4. Rules for Contexts

$$\frac{\Gamma \text{ context}}{\Gamma \vdash \textbf{Type} \in \textbf{Kind}}$$

$$\frac{\Gamma \text{ context} \quad x{:}A \text{ belongs to } \Gamma}{\Gamma \vdash x \in A}$$

$$\frac{\Gamma \vdash A \in \textbf{Type} \quad \Gamma, x{:}A \vdash B \in s}{\Gamma \vdash \Pi x{:}A.B \in s}$$

$$\frac{\Gamma \vdash A \in \textbf{Type} \quad \Gamma, x{:}A \vdash M \in B \in s}{\Gamma \vdash \lambda x{:}A.M \in \Pi x{:}A.B}$$

$$\frac{\Gamma \vdash M \in \Pi x{:}A.B[x] \quad \Gamma \vdash N \in A}{\Gamma \vdash MN \in B[N]}$$

$$\frac{\Gamma \vdash A = B \quad \Gamma \vdash M \in B}{\Gamma \vdash M \in A}$$

Fig. 5. Rules for Ordinary Terms

Γ context.

$\chi_1 < \mu_1, \cdots, \chi_n < \mu_n$ are all the elements in Γ of form $\chi_i < \mu_i$.

μ is an arithmetical term whose variables are among χ_i.

ν is such an arithmetical term or ∞.

$\mu < \nu$ holds for all the integers $\chi_1, \cdots, \chi_n$ such that $0 \le \chi_1 < \mu_1, \cdots, 0 \le \chi_n < \mu_n$.

$$\frac{}{\Gamma \vdash \mu < \nu}$$

Fig. 7. Rules for Arithmetical Judgements

$$\frac{\Gamma, \chi < \mu \vdash B \in s}{\Gamma \vdash \Pi \chi < \mu.B \in s}$$

$$\frac{\Gamma, \chi < \mu \vdash M \in B \in s}{\Gamma \vdash \lambda \chi < \mu.M \in \Pi \chi < \mu.B}$$

$$\frac{\Gamma \vdash M \in \Pi \chi < \mu.B[\chi] \quad \Gamma \vdash 0 \le \nu < \mu}{\Gamma \vdash M\nu \in B[\nu]}$$

$$\frac{\Gamma \vdash \mu < \infty \quad \Gamma \vdash \nu < \infty}{\Gamma \vdash \mu < \nu \in \textbf{Type}}$$

$$\frac{\Gamma \vdash \mu < \nu}{\Gamma \vdash ?_{\mu < \nu} \in \mu < \nu}$$

$$\frac{\Gamma \vdash \nu < \infty \quad \Gamma \vdash \nu \le \mu \quad \Gamma \vdash F \in \Pi\chi < \mu.A[\chi] \to A[\chi+1] \quad \Gamma \vdash M \in A[0]}{\Gamma \vdash \textbf{R}\nu FM \in A[\nu]}$$

Fig. 6. Rules for Arithmetical Terms

$$\frac{\Gamma \vdash M \in A \quad \Gamma \vdash N \in A \quad M \to_\beta N}{\Gamma \vdash M = N}$$

$$\frac{\Gamma \vdash 0 < \mu}{\Gamma \vdash \textbf{R}\mu FM = F(\mu-1)(\textbf{R}(\mu-1)FM)}$$

$$\frac{\Gamma \vdash \mu = 0}{\Gamma \vdash \textbf{R}\mu FM = M}$$

Fig. 8. Rules for Equality

Case-Based Representation and Learning
of
Pattern Languages

Klaus P. Jantke and Steffen Lange *

Hochschule für Technik, Wirtschaft und Kultur Leipzig (FH)

Fachbereich Informatik, Mathematik & Naturwissenschaften

P.O.Box 66

04251 Leipzig

Germany

jantke@informatik.th-leipzig.de

steffen@informatik.th-leipzig.de

Abstract

Pattern languages seem to suit case-based reasoning particularly well. Therefore, the problem of inductively learning pattern languages is paraphrased in a case-based manner. A careful investigation requires a formal semantics for case bases together with similarity measures in terms of formal languages. Two basic semantics are introduced and investigated. It turns out that representability problems are major obstacles for case-based learnability. Restricting the attention to so-called proper patterns avoids these representability problems. A couple of learnability results for proper pattern languages are derived both for case-based learning from only positive data and for case-based learning from positive and negative data. Under the so-called competing semantics, we show that the learnability result for positive and negative data can be lifted to the general case of arbitrary patterns. Learning under the standard semantics from positive data is closely related to monotonic language learning.

1 Introduction

This paper is dealt with problems of case-based learning in a particular area where we can exploit a remarkable amount of inductive learning results. This is the area of pattern languages as introduced in [Ang80]. This area has attracted enormous attention in learning theory (cf. [Ang80], [Shi82], [Nix83], [Jan84], [LW91], [Jan91b], [Lan91], and others). A key reason for the intensive research work dedicated to the

---

*The work has been partially supported by the German Federal Ministry for Research and Technology (BMFT) within the Joint Project (BMFT-Verbundprojekt) **GOSLER** on **Algorithmic Learning for Knowledge-Based Systems** under contract no. 413-4001-01 IW 101 A.

learning of pattern languages is the naturalness of the general learning problem as well as the closeness of individual texts to the general underlying pattern structures. From this insight, there is outgrowing a particular motivation of the investigations presented here.

Here, we are briefly illustrating what will be considered in more detail below. Given any text structure like

$$x_{author}, x_{title}, x_{journal}\ x_{volume}\ (x_{year}), x_{pages}$$

one may easily imagine a number of typical instances. Vice versa, from some typical cases like

> Dana Angluin and Carl H. Smith, A Survey of Inductive Inference: Theory and Methods, Computing Surveys 15 (1983), 237-269

> Reinhard Klette and Rolf Wiehagen, Research in the Theory of Inductive Inference by GDR Mathematicians - A Survey, Information Sciences 22 (1980), 149-169

most people will infer underlying patterns like the one above. In this particular domain, there is an easy concept of cases, and humans are usually able to learn from a small number of those cases (cf. [Nix83] for experiments and measurements).

This consideration motivated the following intention. First, if pattern inference is an area where we have a natural and easy to understand concept of cases, we should be able to develop and illustrate basic ideas of case-based learning. Second, if there are general difficulties of case-based learning in such a nice area, this could be understood as testbed for problems we are faced to in a large number of areas where formal considerations may be of a considerably greater complexity. In a sense, the results about case-based learning of pattern languages developed in the sequel may be interpreted as *lower bounds* for the difficulties of case-based learning in a huge variety of further areas.

1.1 Case-Based Learning

Case-based reasoning is a recently booming subarea of artificial intelligence. One important reason is that human experts tend to use knowledge in the form of particular cases or episodes rather frequently than generalized knowledge as described by rules, e.g. Therefore, there is some hope that case-based reasoning may help to widen the bottleneck of knowledge acquisition. The reader is directed to [Kol92] for a recent introduction in and survey of case-based reasoning. Within case-based reasoning, case-based learning as investigated in [Aha91] and [AKA91], for instance, is a rather natural way of designing learning procedures. Recent formalizations (cf. [Jan92]) have exhibited the remarkable power of case-based learning algorithms. In the particular setting of learning total recursive functions, which covers the problem of learning effective classifiers in formalized areas, everything learnable inductively turns out to be learnable in a case-based manner. This may be understood as a normal form result for inductive inference algorithms.

This general result raised the question how to be interpreted in particular settings where there may be or not a natural concept of cases. In some areas, cases seem to be conceptually far from the target objects to be learnt. For example, in the area of learning number-theoretic functions from input/output examples, those input/output examples may be considered as cases specifying the intended target behaviour. In despite of any particular choice of an underlying programming language, there is usually a considerable syntactical difference between programs and examples of their corresponding behaviours. Usually, there is no position in a program where some input/output examples occur syntactically. There is a minor class of exceptions including textual dialogue components.

In contrast, domains like containment decision lists (cf. [SS92], e.g.) or text patterns look quite promising. First, let us briefly consider containment decision lists. If those lists are understood to accept formal languages, and if cases are formalized as labelled words, those decision lists are obviously constructed directly from the best cases describing the language accepted. This yields an immediate syntactic correspondence between the information to be processed and the hypotheses to be generated within an inductive learning process. Because of these extraordinary formal assumption, one might expect further insights into the nature of case-based learning when investigating both the power and the limitations of case-based learning applied to containment decision lists. Motivated by problems of incremental learning, [Jan93] contains a first investigation of containment decision lists, in this regard. Second, pattern languages seem to be similarly tailored to case-based learning. This will be investigated in more detail troughout this paper.

1.2 Text Patterns

In this section, we formally define the concept of a text pattern as introduced in [Ang80].

Assume any finite alphabet A with at least two different symbols. By A^+ we denote the set of all finite non-empty strings of symbols from A. $X = \{x_1, x_2, \ldots\}$ is a countable set of symbols disjoint from A. Elements of X are called variables. A pattern p is any string from $(A \cup X)^+$. By $\mathcal{P}$ we denote the set of all patterns, i.e. $\mathcal{P} = (A \cup X)^+$.

For a pattern p, we denote by $\mathcal{L}(p)$ the corresponding pattern language defined by p. $\mathcal{L}(p)$ contains all strings which can be obtained by substituting non-empty strings for the variables of p, where the same variables have to be substituted by the same strings. For example, the pattern language $\mathcal{L}(ax_1bx_2x_1)$ contains among others the strings $aaabbaaa$ and $ababbba$, whereas $abbba$ as well as $bbaaaa$ do not belong to this pattern language.

This quite simple concept reflects very well the intuitive notion of text patterns as explained in the introduction.

During the last decade, learnability of pattern languages has been intensively investigated within different learning models (cf. [Ang80], [Shi82], [Mar88], [LW91], and others). Besides that, pattern languages form the basis of a couple of applications in different fields, e.g. in the intelligent text processing system EBE (cf. [Nix83]) or in a classification system for transmembrane proteins (cf. [AKM+92]).

1.3   Inductive Pattern Inference

Inductive inference is the process of hypothesizing a general rule from eventually incomplete data. It has its origins in philosophy of sciences. During the last three decades, it received much attention in computer science (cf. [AS83]).

The general situation investigated in language learning can be described as follows: There is some target language to be learnt (identified, ...) inductively. Given more and more possibly incomplete information concerning the language to be learnt, an inference device has to produce in every step a hypothesis about the phenomenon to be inferred. The set of all admissible hypotheses is called space of hypotheses. The given information may contain only *positive examples*, i.e. exactly all the strings contained in the language to be recognized, or both *positive and negative examples*, i.e. the learner is fed with arbitrary strings over the underlying alphabet which are classified with respect to their containment to the unknown language. The sequence of hypotheses has to converge to a hypothesis correctly describing the object to be learnt. To sum up, the inference process as a whole is a limiting one.

By $N = \{0, 1, 2, 3, ...\}$ we denote the set of all natural numbers. By A we denote any fixed finite alphabet of symbols. Any subset $L \subseteq A^*$ is called a language. By $\overline{L}$ we denote the complement of L, i.e. $\overline{L} = A^* \setminus L$. Let L be a language and $t = (s_0, 1), (s_1, 1), (s_2, 1), ...$ an infinite sequence of elements from $A^* \times \{1\}$ such that $range(t) = \{s_k \mid k \in N\} = L$. Then t is said to be a *text* for L or, synonymously, a *positive presentation*. By $Text(L)$ we denote the set of all positive presentations of L. Furthermore, let $i = (s_1, d_1), (s_2, d_2), ...$ be a sequence of elements of $A^* \times \{1, 0\}$ such that $range(i) = \{s_k \mid k \in N\} = A^*$, $i^+ = \{s_k \mid (s_k, d_k) = (s_k, 1), k \in N\} = L$ and $i^- = \{s_k \mid (s_k, b_k) = (s_k, 0), k \in N\} = \overline{L}$. (Usually, the empty word ε is excluded from consideration.) Then we refer to i as an *informant*. If L is classified via an informant, then we also say that L is presented by *positive and negative data*. By $Inf(L)$ we denote the set of all positive and negative presentations of L. Moreover, let t and i be a text and an informant, respectively, and let x be any natural number. Then t_x and i_x denote the initial segment of t and i of length $x + 1$, respectively. The notation $A \subset_{fin} B$ indicates that A is a finite subset of B.

We define an *inductive inference machine* (abbr. IIM) to be an algorithmic device which works as follows in order to learn a language L from a class of languages $\mathcal{L}$: The IIM takes as its input larger and larger initial segments of any text t (any informant i) and outputs hypotheses, accordingly. We write $M(t_x) = j_x$ ($M(i_x) = j_x$, respectively) to indicate that the IIM M has produced the hypothesis j_x when fed with t_x (or i_x). Furthermore, a hypothesis j_x will be interpreted as a grammar G_{j_x} in some underlying space of hypotheses $\mathcal{G} = (G_j)_{j \in N}$ satisfying $\mathcal{L} \subseteq \{L(G_j) \mid j \in N\}$. A sequence $(j_x)_{x \in N}$ of numbers is said to be convergent in the limit if and only if there is a number j such that $j_x = j$ for all numbers x past some point. This is abbreviated by $lim_{x \to \infty} j_x = j$.

The learnability concept introduced by the following definition is an immediate adaptation of the classical identification types in recursion-theoretic inductive inference (cf. [AS83], [KW80]). It is reflecting the approaches underlying [Ang80], [Shi82], [Mar88], [LW91], e.g.

Definition 1

Let Q be a class of pattern languages, i.e. $Q \subseteq P$, and let $\mathcal{G} = (G_j)_{j\in N}$ be a space of hypotheses.
Q is identifiable in the limit from text (resp. from informant)
iff
there is an IIM M such that for all $p \in Q$ and for all $t \in Text(L(p))$ (resp. for all $i \in Inf(L(p))$):
(1) $\forall x \in N : M(t_x)$ (resp. $M(i_x)$) is defined.
(2) $lim_{x\to\infty}M(t_x) = j$ (resp. $lim_{x\to\infty}M(i_x) = j$)
(3) $L(p) = L(G_j)$

According to identification types in inductive inference (cf. [AS83], [KW80]), learnability as above is denoted by $Q \in LIM.TXT$ and $Q \in LIM.INF$, respectively. If an IIM M identifies a class Q from text or from informant, this is abbreviated by $Q \subseteq LIM.TXT(M)$ and $Q \subseteq LIM.INF(M)$, respectively.

It is well known that the class of all pattern languages P is identifiable in the limit from positive examples (cf. [Ang80]). This directly implies that P can be identified in the limit from positive and negative examples, too.

Theorem 1

(1) $P \in LIM.TXT$
(2) $P \in LIM.INF$

Both results above may help to relate our approach to scientific work already done in formal language learning.

In [LZ93], the notion of monotonic language learning has been introduced. An inductive inference machine M works monotonically on a text t, if $L(G_{j_x}) \subseteq L(G_{j_{x+1}})$, for any two consecutive hypotheses j_x and j_{x+1} output by M. The corresponding approach to learning pattern languages has been introduced in [Jan91b].

Definition 2

Let Q be a class of pattern languages, i.e. $Q \subseteq P$, and let $\mathcal{G} = (G_j)_{j\in N}$ be a space of hypotheses.
$Q \in MON.TXT$
iff
there is an IIM M such that for all $p \in Q$ and for all $t \in Text(L(p))$:
(1) $\forall x \in N : M(t_x)$ is defined.
(2) $\forall x \in N : L(G_{M(t_x)}) \subseteq L(G_{M(t_{x+1})})$
(3) $lim_{x\to\infty}M(t_x) = j$ (resp. $lim_{x\to\infty}M(i_x) = j$)
(4) $L(p) = L(G_j)$

In [LZ93], the notion of monotonic language learning has been introduced. An inductive inference machine M works monotonically on a text t, if $L(G_{j_x}) \subseteq L(G_{j_{x+1}})$, for any two consecutive hypotheses j_x and j_{x+1} output by M. The corresponding approach to learning pattern languages has been introduced in [Jan91b].

In [JSSY93] it is shown that the inclusion problem for pattern languages is undecidable, in general. Exploiting the characterization theorem for monotonic language learning (cf. [LZ92]) one obtains the following result.

Theorem 2

Assume any space of hypotheses $\mathcal{G}$ with $L(\mathcal{P}) = \{L(G_j) \mid j \in N\}$. Then there does not exists any IIM M such that $\mathcal{P} \subseteq MON.TXT(M)$.

On the other hand, in order to design pattern inference algorithms, an appropriate choice of the space of hypotheses is necessary. The following result is due to [LZ93].

Theorem 3

There is some space of hypotheses $\mathcal{G}$ with $L(\mathcal{P}) \subset \{L(G_j) \mid j \in N\}$ such that there exists an IIM M satisfying $\mathcal{P} \subseteq MON.TXT(M)$.

2 Case-Based Representation of Pattern Languages

If some algorithm is expected to learn any member of some class of objects in a case-based manner by processing information about particular target objects to come up with some finite case-base and some similarity measure describing the particular target object, this obviously assumes some interpretation of pairs built from case-bases and similarity functions in terms of the objects under consideration. Formally spoken, one needs some well-defined semantics. In general, there is no standard semantics. [Jan92] is introducing three slightly different semantics, in a particular setting. Similarly, the reader will find below two slightly different approaches used in the paper on hand. It is especially surprising that a remarkable number of papers does not make the chosen semantics explicit. But for a formally correct treatment, the choice of some precise semantics is inevitable.

2.1 Semantics

There is assumed some finite, non-empty alphabet A. Cases about some formal language are labelled words indicating whether or not some word provided belongs to the language to be represented or even to be learnt. For labelling words, we choose 0 and 1 meaning *no* and *yes*, respectively. Certain papers in the area of case-based reasoning provide some rough concept of semantics as follows (cf. [Aha91], [AKA91], for example). If there is some finite case-base CB and some given similarity measure σ, this classifies words w according to the following procedure: Search CB for some labelled word (v, d) where $\sigma(v, w)$ is maximal. Return d to classify w. There may obviously arise some ambiguity, if there are conflicting classifications by cases $(v_1, 0)$ and $(v_2, 1)$ where both v_1 and v_2 are of maximal similarity to w. There are several ways to resolve those conflicts. Two of them are chosen for the formal semantics introduced in the sequel. The first approach is called *standard* and assumes implicitly that there is no proper conflict. If there is any similar candidate in the case-base, this is considered as sufficient information for including some word into the language specified. The second approach considers all other examples as *competing* for classifying words. The existence of some most similar case is explicitly required. Formal semantic concepts in the area of case-based pattern representation and inference should specify how a case-base CB together with a similarity measure σ defines a languages denoted by $\mathcal{L}(CB, \sigma)$. The *standard* approach and the *competing* approach will be denoted by $\mathcal{L}_{st}(CB, \sigma)$ and $\mathcal{L}_c(CB, \sigma)$, respectively. There may be further approaches, but the two considered seem to be basic.

Definition 3

$$\mathcal{L}_{st}(CB, \sigma) = \{w/\exists(u, 1) \in CB\,[\,\sigma(u, w) > 0 \land \forall(v, 0) \in CB\,[\sigma(u, w) > \sigma(v, w)]\,]\}$$

$$\mathcal{L}_c(CB, \sigma) = \{w/\exists(u, 1) \in CB[\,\sigma(u, w) > 0 \land \\ \forall(v, d) \in CB[u \neq v \Rightarrow \sigma(u, w) > \sigma(v, w)]\,]\}$$

The reader may easily recognize that the standard semantics is satisfying some monotonicity property. If any similarity measure σ is fixed, $\mathcal{L}_{st}(CB, \sigma)$ turns out to be monotonous in CB for sets of positive cases. Obviously, this does not hold for the competing semantics.

Under some chosen semantics, one can investigate the problem of learning target pattern languages by learning case bases and similarity measures describing them. The problem turns out to be posed improperly, if already the problem of representing certain languages under the assumed semantics is unsolvable. Therefore, we are considering representability first.

2.2 Representability Results

In the results listed below, the notation $\mathcal{L}_*$ refers to both the standard semantics and the competing semantics as introduced above. For the readers convenience, every theorem will be paraphrased (in italics), first.

Under both semantics, there is no universal similarity measure σ which allows to represent every pattern language by a finite number of its elements considered as positive cases.

Theorem 4

$$\neg\exists\sigma\forall p \in \mathcal{P}\exists CB \subset_{fin} \mathcal{L}(p) \times \{1\}\,[\mathcal{L}(p) = \mathcal{L}_*(CB, \sigma)]$$

In contrast, if one does no longer require a case base to contain only words of the language to be described, every languages is representable. Naturally, this generalization is senseless from the viewpoint of case-based reasoning. It is mentioned here for two reasons. First, it illustrates the borderlines of representability problems. Second, its proof illuminates certain encoding techniques. From the viewpoint of case-based reasoning, those techniques seem to be unfair tricks which should be prohibited. As usual, it is quite difficult to guarantee that some formal framework does not allow undesirable recursive tricks and effects.

Under both semantics, there is a universal similarity measure σ which allows to represent every pattern language by a finite number of cases, if the words of these cases are not restricted to represent the target language correctly.

Theorem 5

$$\exists \sigma \forall p \in \mathcal{P} \exists CB \subseteq_{fin} A^+ \times \{1\} \, [\mathcal{L}(p) = \mathcal{L}_*(CB, \sigma)]$$

The proof of **Theorem 4** exhibited the variable-free patterns to be the main obstacles for representability. As these objects are quite useless for generating pattern languages, it seems reasonable to restrict the investigations to patterns containing at least one variable. This is done by introducing the class $\mathcal{PP}$ of so-called *proper patterns*. $\mathcal{PP} =_{def} \mathcal{P} \backslash A^*$. For proper patterns, there is a universal methodology for representing arbitrary pattern languages.

Under both semantics, there is a universal similarity measure σ which allows to represent every proper pattern language by a finite number of its words considered as positive cases.

Theorem 6

$$\exists \sigma \forall p \in \mathcal{PP} \exists CB \subseteq_{fin} \mathcal{L}(p) \times \{1\} \, [\mathcal{L}(p) = \mathcal{L}_*(CB, \sigma)]$$

Again, the proof is based on certain suspicious encoding tricks. Therefore, one may try to admit only symmetric similarity measures which would not allow encodings as invoked in the proof above. But it turns out that symmetry does not help at all.

Under both semantics, there is no universal similarity measure σ being symmetric which allows to represent every proper pattern language by a finite number of its words considered as positive cases.

Theorem 7

$$\neg \exists \sigma \, [\sigma \; symmetric \wedge \forall p \in \mathcal{PP} \exists CB \subseteq_{fin} \mathcal{L}(p) \times \{1\} \, [\mathcal{L}(p) = \mathcal{L}_*(CB, \sigma)]]$$

To sum up the investigations into representability of pattern languages in a case-based manner using positive cases, only, one should restrict the considerations to only proper patterns, and one should give up the hope for symmetric similarity measures. Under these assumptions, the results about representability are setting the stage for learnability investigations.

The situation is slightly different, if we allow to use positive and negative cases within the case-base.

Under standard semantics as well as under competing semantics, there is a universal similarity measure σ allowing to represent every pattern language by a finite case-base CB of both examples and counter-examples considered as positive and negative cases,

Theorem 8

$$\exists \sigma \forall p \in \mathcal{P} \exists CB \subset_{fin} \mathcal{L}(p) \times \{1\} \cup \overline{\mathcal{L}(p)} \times \{0\} \left[\mathcal{L}(p) = \mathcal{L}_*(CB, \sigma) \right]$$

Again, the proof is based on some encoding tricks.

At the very moment, it is still open whether or not **Theorem 8** is valid, if it is required that the corresponding similarity measure σ is symmetric. We conjecture that there does not exist any symmetric similarity measure σ which allows to represent the class of all pattern languages using positive and negative cases under any of the two semantics investigated.

3 Case-Based Learning of Pattern Languages

There is the basic decision whether or not to consider counter-examples in the course of learning pattern languages. This is known to be the distinction between text and informant according to [Gol67]. The first definition of the following chapter is providing a direct formalization.

3.1 Learning Scenario

The following definitions are intended to characterize inductive inference of pattern languages as motivated and introduced above.

As our first results above illustrate the applicability of certain undesirable encoding tricks, and as particular results (cf. [Jan91a]) exhibit the possibilities of encoding knowledge into similarity measures, we are initially interested in the problem of learning on an a priori fixed similarity measures. There is not any restriction of the type of similarity concepts assumed.

First, one needs to specify admissible specifications of pattern languages. Every specification to be processed has actually to be finite. As finite samples may be insufficiently incomplete, one needs to consider the behaviour of learning devices on growing information sequences. This motivates the following limiting concepts.

Learnability is based on the technical concepts defined before. The reader may consult similar approaches in several related publications (cf. [AS83] and [KW80] for an overview, [Jan89] for an easy introduction, and [Jan92] for case-based approaches). The following two concepts are both distinguished by the type of admissible information according to **Definition 1** and the underlying semantics. There is formalized the idea of collecting cases in a computable manner.

Definition 4

Let $\mathcal{Q}$ be a class of pattern languages, i.e. $\mathcal{Q} \subseteq \mathcal{P}$.
$\mathcal{Q} \in S - CBL.TXT$ (resp. $\mathcal{Q} \in C - CBL.TXT$)
iff
there is an IIM M and a similarity measure σ such that for all $p \in \mathcal{Q}$ and for all
respectively.

$t \in Text(L(p))$ there exists some CB:
(1) $\forall x \in N : M(t_x) = CB_x$ is defined.
(2) $\forall x \in N : \emptyset \subseteq CB_0 \subseteq \{(s_0, 1)\} \wedge CB_x \subseteq CB_{x+1} \subseteq CB_x \cup \{(s_{x+1}, 1)\}$
(3) $\lim_{x \to \infty} M(t_x) = CB$
(4) $\mathcal{L}(p) = \mathcal{L}_{st}(CB, \sigma)$ (resp. $\mathcal{L}(p) = \mathcal{L}_c(CB, \sigma)$)

The prefixes "$S-$" and "$C-$" are used to indicated the underlying semantics, respectively. The notations are similar to those in a large variety of related publications. [JB81] is an early paper introducing and using notations of identification types similarly. The reference to the type of admissible information is expressed by the extension and should be self-explaining.

If we consider case-based learning from informant, a learning algorithm may collect positive as well as negative cases within a case base.

Definition 5

Let $\mathcal{Q}$ be a class of pattern languages, i.e. $\mathcal{Q} \subseteq \mathcal{P}$.
$\mathcal{Q} \in S - CBL.INF$ (resp. $\mathcal{Q} \in C - CBL.INF$)
iff
there is an IIM M and a similarity measure σ such that for all $p \in \mathcal{Q}$ and for all $i \in Inf(L(p))$ there exists some CB:
(1) $\forall x \in N : M(i_x) = CB_x$ is defined.
(2) $\forall x \in N : \emptyset \subseteq CB_0 \subseteq \{(s_0, d_0)\} \wedge CB_x \subseteq CB_{x+1} \subseteq CB_x \cup \{(s_{x+1}, d_{x+1})\}$
(3) $\lim_{x \to \infty} M(i_x) = CB$
(4) $\mathcal{L}(p) = \mathcal{L}_{st}(CB, \sigma)$ (resp. $\mathcal{L}(p) = \mathcal{L}_c(CB, \sigma)$)

Under these formalizations, we are going to investigate case-based learnability of pattern languages.

3.2 Learnability Results

Theorem 6 and **Theorem 8** circumscribe the possibilities of case-based learning of pattern languages. Again, every theorem will be paraphrased for the readers convenience. The following result comprises the general positive answer concerning the learnability of pattern languages from text.

For the class of proper pattern languages, there are universal case-based learning algorithms based on text for both semantics considered.

Theorem 9

(1) $\mathcal{PP} \in S - CBL.TXT$
(2) $\mathcal{PP} \in C - CBL.TXT$

The first result above has some special features distinguishing it from the other results within this section. It is easy to verify that every learning algorithm M inferring pattern languages from text in a cased-based manner under standard semantics is working monotonically, i.e. $\mathcal{L}_{st}(CB_x, \sigma) \subseteq \mathcal{L}_{st}(CB_{x+1}, \sigma)$, for any two case bases produced subsequently by M. (Recall the corresponding remark above.) Because of

Theorem 2, a learning algorithm witnessing $\mathcal{PP} \in S - CBL.TXT$ has to be based on a similarity measure σ being able to represent languages not contained in $\mathcal{PP}$. The proof is exhibiting a class of sufficiently expressive similarity measures. For the design of a case-based learning algorithm as desired, it is sufficient to use a similarity measure σ which allows to represent all finite non-empty intersections of pattern languages.

Next to, we consider case-based learning of pattern languages from informant, the situation changes drastically. First, there is an immediate corollary of the theorem above.

Corollary

(1) $\mathcal{PP} \in S - CBL.INF$
(2) $\mathcal{PP} \in C - CBL.INF$

The additional strength of counter-examples suggests to attempt an extension of the learnability result to all pattern languages.

Under competing semantics, the whole class of pattern languages $\mathcal{P}$ is case-based learnable from positive and negative examples.

Theorem 10

$\mathcal{P} \in C - CBL.INF$

Until now, it is still open whether a similar result can be achieved under standard semantics, too. This completes the list of problems and solutions attacked in the present paper.

4   Conclusions

There is a considerable number of related problems and further questions. Here, we are going to mention only some of them.

- What about further semantics of interest? The two semantics introduced are different, in some sense, but they are of the same expressiveness. Does there exist a lattice of some reasonable semantics naturally ordered?

- Some proofs of our results above invoke tricky constructions of similarity measures which contradict human intuition, i.e. they do not express any syntactical relation humans would call similarity. Does there exist abstract properties characterizing reasonable classes of similarity measures?

- Case-based learnability is basically characterized by collecting suitable cases. In more advanced approaches, learning of similarity measures is considered, too. How does the power of basic resp. advanced approaches relate to the underlying semantics?

- Learnability of similarity concepts is known to be crucial. How to extend the results above by learning similarity measures without allowing undesired encoding tricks?

- How does certain semantics interact with learning similarity concepts? Are there certain semantics particularly useful when learning similarity?

- Case-based learning from positive data under the standard semantics is monotonic. Are there further considerably different semantics implying certain monotonicity effects in case-based learning? How does this relate to the rich area of results in monotonic and non-monotonic language learning?

- What about the impact of further natural properties in inductive inference like consistency, e.g.? There is a huge amount of knowledge about the relation of several those properties. How do they depend on the chosen semantics?

5 Appendix

Two of the referees suggested to explain in some more detail at least one of the constructions of a tricky similarity measure invoked. For the readers convenience, we decided to choose a representation theorem, only. We want to avoid any confusion of representability and learnability issues. Our **Theorem 8** seems appropriate. We have to construct a similarity measure σ satisfying

$$\forall p \in \mathcal{P} \; \exists CB \subseteq_{fin} \mathcal{L}(p) \times \{1\} \cup \overline{\mathcal{L}(p)} \times \{0\} \; [\; \mathcal{L}(p) = \mathcal{L}_*(CB, \sigma) \;]$$

For our construction, we need any enumeration of all non-empty words over the alphabet A, say $w_1, w_2, w_3, \cdots$. Furthermore, assume $p_1, p_2, p_3, \cdots$ to be any enumeration of the class $\mathcal{PP}$ of all proper patterns. We may assume that both enumerations are repetition-free. Now, we are going to define an infinite sequence of words $u_{i,j}$ exhausting A^+. The definition proceeds in steps as follows, where U is used to collect all words $u_{i,j}$ specified so far:

$u_{1,0} := w_1 \; ; \; U := \{u_{1,0}\}$

In the ith step ($i > 1$), there are defined $u_{i,0}$ and $u_{1,i-1}, \cdots, u_{i-1,1}$ as follows:

$u_{i,0} := w_{\mu k [w_k \notin U \wedge w_k \in \mathcal{L}(p_i)]} \; ; \; U := U \cup \{u_{i,0}\}$

For $k = 1$ until $i - 1$ do

$u_{k,i-k} := w_{\mu l [w_l \notin U]} \; ; \; U := U \cup \{u_{k,i-k}\}$

The similarity measure σ is defined by the standard definition

$$\forall w \in A^+ : \sigma(w, w) = 1$$

and the following additional requirements, for arguments of σ different from each other.

$$\sigma(u_{i,0}, w) = \begin{cases} 1 & : \quad w \in \mathcal{L}(p_i) \\ 0 & : \quad otherwise \end{cases}$$

$$\sigma(u_{i,j+1}, w) = \begin{cases} 1 & : \quad w \in \mathcal{L}(p_i) \ \wedge \ w \neq u_{i,j} \\ 0 & : \quad otherwise \end{cases}$$

Obviously, σ is effectively computable, if the underlying enumerations are so. This uses the decidability of membership for pattern languages (cf. [Ang80]). For every $p_i \in \mathcal{PP}$, the appropriate case base is $\{ (u_{i,0}, 1) \}$. Every variable-free pattern w occurs as some $u_{i,j}$. Its case base w.r.t. σ is $\{ (u_{i,j}, 1), (u_{i,j+1}, 0) \}$. Note that the similarity measure constructed is $\{0, 1\}$-valued, only. Although this deemed to be not very expressive, also the necessary case bases are extremely simple. Every case base contains either exactly one positive or exactly one positive and one negative case.

References

[Aha91]   David W. Aha. Case-based learning algorithms. In *DARPA Workshop on Case Based Reasoning*, pages 147–157. Morgan Kaufmann, 1991.

[AKA91]   David W. Aha, Dennis Kibler, and Marc K. Albert. Instance-based learning algorithms. *Machine Learning*, 6:37–66, 1991.

[AKM+92]   Setsuo Arikawa, Satoru Kuhara, Satoru Miyano, Yasuhito Mukouchi, Ayumi Shinohara, and Takeshi Shinohara. A machine discovery from amino acid sequences by decision trees over regular patterns. In *Intern. Conference on Fifth Generation Computer Systems, June 1-5, 1992*, volume 2, pages 618–625. Institute for New Generation Computer Technology (ICOT), Tokyo, Japan, 1992.

[Ang80]   Dana Angluin. Finding patterns common to a set of strings. *Journal of Computer and System Sciences*, 21:46–62, 1980.

[AS83]   Dana Angluin and Carl H. Smith. A survey of inductive inference: Theory and methods. *Computing Surveys*, 15:237–269, 1983.

[Gol67]   E Mark Gold. Language identification in the limit. *Information and Control*, 14:447–474, 1967.

[Jan84]   Klaus P. Jantke. Polynomial time inference of general pattern languages. In M. Fontet/ K. Mehlhorn, editor, *STACS'84*, Lecture Notes in Computer Science 166, pages 314–325. Springer-Verlag, 1984.

[Jan89]   Klaus P. Jantke. Algorithmic learning from incomplete information: Principles and problems. In J. Dassow and J. Kelemen, editors, *Machines, Languages, and Complexity*, Lecture Notes in Computer Science, pages 188–207. Springer-Verlag, 1989.

[Jan91a]   Klaus P. Jantke. Monotonic and non-monotonic inductive inference. *New Generation Computing*, 8(4):349–360, 1991.

[Jan91b]   Klaus P. Jantke. Monotonic and nonmonotonic inductive inference of functions and patterns. In K.P. Jantke J. Dix and P.H. Schmitt, editors, *Nonmonotonic and Inductive Logic, 1st Int. Workshop*, Lecture Notes in Artificial Intelligence 543, pages 161–177. Springer-Verlag, 1991.

[Jan92]   Klaus P. Jantke. Case based learning in inductive inference. In *Proc. of the 5th ACM Workshop on Computational Learning Theory, (COLT'92), July 27-29, 1992, Pittsburgh, PA, USA*, pages 218–223. ACM Press, 1992.

[Jan93]   Klaus P. Jantke. Nonstandard concepts of similarity in case-based reasoning. In Bock, Lenski, and Richter, editors, *17. Jahrestagung der deutschen Gesellschaft für Klassifikation e.V.* Kaiserslautern, (Proc. in prep.), 1993.

[JB81]   Klaus P. Jantke and Hans-Rainer Beick. Combining postulates of naturalness in inductive inference. *EIK*, 17(8/9):465–484, 1981.

[JSSY93] Tao Jiang, Arto Salomaa, Kai Salomaa, and Sheng Yu. Inclusion is undecidable for pattern languages. In *Intern. Conference Automata, Languages and Programming, ICALP'93*, volume to appear of *Lecture Notes in Computer Science*, page ? Springer-Verlag, 1993.

[Kol92]   Janet L. Kolodner. An introduction to case-based reasoning. *Artificial Intelligence Review*, 6:3–34, 1992.

[KW80]   Reinhard Klette and Rolf Wiehagen. Research in the theory of inductive inference by GDR mathematicians - a survey. *Information Sciences*, 22:149–169, 1980.

[Lan91]   Steffen Lange. A note on polynomial-time inference of k-variable pattern languages. In K.P. Jantke J. Dix and P.H. Schmitt, editors, *NIL'91, Nonmonotonic and Inductive Logic, 1st Int. Workshop*, volume 543 of *Lecture Notes in Artificial Intelligence*, pages 178–183. Springer-Verlag, 1991.

[LW91]   Steffen Lange and Rolf Wiehagen. Polynomial-time inference of arbitrary pattern languages. *New Generation Computing*, 8(4):361–370, 1991.

[LZ92]   Steffen Lange and Thomas Zeugmann. Types of monotonic language learning and their characterization. In *Proc. of the 5th ACM Workshop on Computational Learning Theory, (COLT'92), July 27-29, 1992, Pittsburgh, PA, USA*, pages 377–390. ACM Press, 1992.

[LZ93]   Steffen Lange and Thomas Zeugmann. Monotonic versus non-monotonic language learning. In K.P. Jantke G. Brewka and P.H. Schmitt, editors, *NIL'91, Nonmonotonic and Inductive Logic, 2nd Int. Workshop*, volume 659 of *Lecture Notes in Artificial Intelligence*, pages 254–269. Springer-Verlag, 1993.

[Mar88]   Assaf Marron. Learning pattern languages from a single initial example and from queries. In *Proc. of the 1988 Workshop on Computational Learning Theory*, pages 345–358, 1988.

[Nix83]   Robert P. Nix. Editing by examples. Technical Report 280, Yale University, Dept. Comp. Sci., 1983.

[Shi82]   Takeshi Shinohara. Polynomial time inference of pattern languages and its applications. In *7th IBM Symposium on Mathematical Foundations of Computer Science*, pages 191–209, 1982.

[SS92]   Yasubumi Sakakibara and Rani Siromoney. A noise model on learning sets of strings. In *Proc. of the 5th ACM Workshop on Computational Learning Theory, (COLT'92), July 27-29, 1992, Pittsburgh, PA, USA*, pages 295–302. ACM Press, 1992.

Inductive Resolution

Taisuke SATO[†] and Sumitaka AKIBA[‡]

Electrotechnical Laboratory 1-1-4 Umezono Tsukuba
Ibaraki Japan 305
{Email: sato@etl.go.jp†, akiba@etl.go.jp‡}

ABSTRACT:
A clausal system **IDR** for inductive inference is proposed. Given background knowledge B and an observed fact O, each represented in terms of clauses, it infers a set H of clauses such that $B \cup H \vdash O$. It is *inductively complete*, i.e., can find every possible H regardless of whether $O = \square$ or not. When hypothesis generation is suppressed, it simulates OL-deduction [1]. **IDR** thus integrates resolution and induction in a single framework.

1 Introduction

A clausal system **IDR** for inductive inference is proposed. Given background knowledge B and an observed fact O, each represented in terms of clauses, it infers a set H of clauses such that $B \cup H \vdash O$. It is *inductively complete*, i.e., can find every possible H regardless of whether $O = \square$ or not. When hypothesis generation is suppressed, it simulates OL-deduction [1]. **IDR** thus integrates resolution and induction in a single framework.

It has four inference rules: the assumption rule, the expansion rule, the truncation rule and the factoring rule. The latter three are incidentally almost isomorphic, if not identical, to OL-deduction. We emphasize, however, that **IDR** is not a refutational system and logical operations working behind greatly differs from that of OL-deduction. To prove a clause α from a set Σ of clauses, we do not negate α and derive $\square$ from $\Sigma \cup \{\neg\alpha\}$. Instead **IDR** directly infers the logical relationship $\Sigma \vdash \alpha$.

We briefly state related work. Technically, **IDR** is a macro system developed on top of a basic inductive inference system R[16]. Although R can infer any Σ such that $\Sigma \vdash \alpha$ from a clause α, the inference made by R does not take Σ into account until it happens to produce a clause belonging to Σ. **IDR** came out of an effort to remedy this apparent lack of use of Σ during the inference process by organizing R's rules into a macro rule so that it always produces clauses one of which belongs to Σ.

There have already been a number of inductive inference systems in various fields such as ML (Machine Learning) [6, 10, 13], ILP (Inductive Logic Programming) [8, 9] and abduction [2, 3, 11, 12, 14, 15]. Among them however, SOL-deduction developed by Inoue [3] seems the closest to our **IDR**. It is OL-deduction augmented with Skip operation for selecting hypotheses. When SOL-deduction is applied to $\Sigma \cup \{\neg\alpha\}$ to find a set of hypothesis $\mathcal{H}$ such that $\Sigma \cup \mathcal{H} \vdash \alpha$, it nondeterministically transfers an "abducible" literal L from a center clause during the refutation and adds to $\mathcal{H}$ as $\neg L$. While generated hypotheses are syntactically limited to literals, more complex ones, say $\phi(x_1, \ldots, x_n)$, can be generated by adding $p(x_1, \ldots, x_n) \to \phi$ to Σ. Here $p(x_1, \ldots, x_n)$ is a new atom to name ϕ [11]. Hence when we expect complex hypotheses but do not know well about their precise form, we have no way but to add all possible candidates (in the form of atoms as above) to Σ beforehand because there is no inventing new predicates or new literals during the refutation. If Σ is empty for example, SOL-deduction will end up with A as the only hypothesis for A.

IDR's assumption rule covers Skip operation. It also allows free invention of literals and clauses during inference, thereby increasing flexibility of hypothesis selection. Also **IDR** allows "inconsistent hypotheses" in pursuit of logical uniformity and generality. Thus $\{\neg B, B\}$ is not precluded as a hypothesis explaining A where A, B are atoms. By contrast, search for consistent hypotheses is norm in almost all abductive systems [2, 3, 11, 12, 14, 15] and it remains unclear how to to derive $\{\neg B, B\}$ from A. Another notable difference between **IDR** and SOL-deduction (OL-deduction) is that, given an input clause αL and a center clause $\neg L \beta^1$, **IDR**'s expansion rule infers $\alpha L \beta$ whereas resolution, used by SOL-deduction, infers $\alpha\beta$ in which L simply disappears (though, actually, both keep L as a framed literal).

Demolonbe and Cerro proposed an inference rule for hypothesis generation called L-deduction based on Hyper resolution [2] instead of OL-resolution. However it neither allows predicate invention nor inconsistent hypotheses.

SOL-deduction and L-deduction are not intended to deal with specific models, or specific semantics. Neither is Poole's approach to default reasoning [11, 12]. Satoh and Iwayama on the other hand took a semantic approach to abductive logic programming [15]. They investigated a problem of finding a set of hypothesis that has a generalized stable model for a normal logic program and proposed a correct query evaluation procedure.

The contents of this paper are organized as follows. In Section 2, we introduce basic terminology. After explaining how the expansion rule, the truncation rule and the factoring rule work at ground level in Section 3, we present the whole inference rules of **IDR** in Section 4 and prove the completeness and soundness

---

[1] We use resolution terminology here.

of **IDR** (Theorem 4.1). Section 5 includes conclusion. The reader is assumed to be familiar with basic notions of logic and resolution [1, 17].

2 Preliminaries

We use first-order language $\mathcal{L}$ equipped with denumerably many function symbols and predicate symbols. Those symbols are denoted by strings beginning with a lower case letter. Variables on the other hand are those beginning with an upper case letter like $X, Y, \ldots$ Basic terminology such as literal, clause, interpretation, model, substitution, unification, etc. related to logic and resolution is defined as usual [1, 17].

L, M are used as meta-variables for literals. For a literal L, $\neg L$ stands for A if $L = \neg A$, or $\neg A$ otherwise where A is an atom. α, β, γ are meta-variables for clauses. They may be empty. The empty clause is specifically represented by $\square$ and considered false. We regard clauses as ordered, i.e. "ordered as presented" and treat them just as sequences of literals when the context is clear. For a clause α, $|\alpha|$ denotes the length of α and if θ is a substitution, $\alpha\theta$ denotes the result of applying θ to α. $L \in \alpha$ means that a literal L occurs in α. Let α and β be clauses. $\alpha \setminus \beta$, the *subtraction of β from α*, is defined as an ordered clause obtaining from α by dropping any literal $L \in \alpha$ such that $L \in \beta$, with the order of remaining literals in α preserved. $\alpha \subseteq \beta$ means every literal in α occurs in β also. *Complementary literals* are those two literals such as A and $\neg A$ that have a common atom with opposite sign. A clause is said to be *tautology* if it contains complementary literals.

Although our final goal is to derive $\mathcal{H}$ from $\mathcal{O}$ using background knowledge $\mathcal{B}$ such that $\mathcal{B} \cup \mathcal{H} \vdash \mathcal{O}$, we first concentrate on a special case where $\mathcal{B}$ and $\mathcal{H}$ are ground set of clauses, $\mathcal{O}$ is a single ground clause and no distinction is made between $\mathcal{B}$ and $\mathcal{H}$.

3 IDR derivation at ground level

Our setting here is that we have a set Σ of ground clauses and a ground clause α such that $\Sigma \vdash \alpha$. Σ may be inconsistent and α can be $\square$. In this section, we introduce three inference rules and prove that they are enough at ground level to infer $\Sigma \vdash \alpha$ from Σ and α. Before proceeding, we need to define some terminology and notation.

A *framed literal* is a literal surrounded by a frame like $\boxed{L}$ and $\boxed{\neg L}$. A *framed clause* is a clause that may contain framed literals whereas a *normal clause* is a clause containing no framed literals. A clause hereafter refers to both a framed one and a normal one. If a clause is presented as $\alpha L \beta$, it is to be understood that it is the concatenation of clause α, non-framed literal L and clause β. Similarly

$\alpha\boxed{L}\beta$ means that L is a framed literal. An *i-clause* (inductive clause) is a clause of the form $\beta|\alpha$ where α is a normal clause and β is a framed clause. Since the sole purpose of introducing i-clauses is to record the history of inference performed on clauses, they should be treated just as normal clauses from a logical point of view. Logically speaking, for example, an i-clause $\beta|\alpha$ is equivalent to a clause $\beta\alpha$ (without frames).

We now introduce three inference rules: the *expansion rule*, the *truncation rule* and the *factoring rule*. They all work on i-clauses.

- The *expansion rule* is applied to an i-clause $\delta_1 = L\alpha_1|\alpha_0$ whose leftmost literal L is non-framed and a normal clause of the form $\gamma\neg L\beta \in \Sigma$ (called an *input clause*). It produces a new i-clause $\delta_2 = \gamma\beta\boxed{\neg L}\alpha_1|\alpha_0$. We say that δ_2 *is inferred from* δ_1 by the expansion rule using $\gamma\neg L\beta$. As a special case, when $|\gamma\beta| = 0$, δ_2 takes the form $\boxed{\neg L}\alpha_1|\alpha_0$. Note a big difference between expansion and resolution. In expansion, every literal in the input clause is preserved in the inferred clause δ_2, while in resolution, $\neg L$ resolved upon just disappears.

- The *truncation rule* truncates an i-clause $\boxed{L}\alpha_1|\alpha_0$ whose leftmost literal is framed to $\alpha_1|\alpha_0$. We say that $\alpha_1|\alpha_0$ *is inferred from* $\boxed{L}\alpha_1|\alpha_0$ by the truncation rule. If α_1 is empty, the result is $|\alpha_0$.

- The *factoring rule* factors away a non-framed literal of an-iclause $L\alpha_1|\alpha_0$ if $\boxed{L}$ occurs in α_1 or if L occurs in α_0, and gives $\alpha_1|\alpha_0$. We say that $\alpha_1|\alpha_0$ *is inferred from* $L\alpha_1|\alpha_0$ by the factoring rule.

An **IDR** *derivation of* $\alpha_1|\alpha_0$ *from* Σ *with a top-clause* β is a sequence $\delta_1,\ldots,\delta_n$ $(n \geq 1)$ of i-clauses satisfying the following conditions:

1. $\delta_1 = \beta|\alpha_0$ and $\beta \in \Sigma$.

2. δ_i $(2 \leq i \leq n)$ is inferred from δ_{i-1} by the expansion rule using a clause in Σ as an input clause, by the truncation rule or by the factoring rule.

3. $\delta_n = \alpha_1|\alpha_0$

When $|\alpha_1| = 0$, the sequence is called an **IDR** *derivation of* α_0 *from* Σ.

Lemma 3.1 *Let* $\delta_1,\ldots,\delta_n$ *be an* **IDR** *derivation of* α. *If* δ_i $(1 \leq i \leq n)$ *takes the form* $\beta\boxed{\neg L}\gamma|\alpha$, *there is* $j(< i)$ *such that* δ_j *is* $L\gamma|\alpha$.

Proof By induction on i. Details omitted.

Proposition 3.1 *Suppose there is an* **IDR** *derivation* $\delta_1,\ldots,\delta_n$ $(n \geq 1)$ *of* α *from* Σ. *Then* $\Sigma \vdash \alpha$.

Proof We prove $\Sigma \vdash \delta_i$ $(1 \leq i \leq n)$ by induction on i. For $i = 1$, this is obvious because $\delta_1 = \beta | \alpha_0$ and β is a clause in Σ. Suppose the proposition holds for less than i. We prove it also holds for i. If δ_i is inferred by the expansion rule, δ_i contains an input clause from Σ again. So $\Sigma \vdash \delta_i$. Suppose δ_i is inferred from $\delta_{i-1} = \boxed{L} \delta_i$ by the truncation rule. By Lemma 3.1, there is an i-clause $\neg L \, \delta_i$ in the derivation prior to δ_{i-1}. So we have by the induction hypothesis $\Sigma \vdash \boxed{L} \delta_i$ and $\Sigma \vdash \neg L \, \delta_i$, respectively. Hence $\Sigma \vdash \delta_i$ (by resolution). The remaining case where δ_i is inferred by the factoring rule is obvious. Q.E.D.

For a set Σ of clauses such that $\Sigma \vdash \alpha$, we say Σ is *minimal* w.r.t. α if no proper subset of Σ logically entails α.

Lemma 3.2 *Let Σ be* minimal *w.r.t. a clause α. If $L \in \alpha$, $\neg L$ does not appear in Σ.*

Proof Suppose $L \in \alpha$. Suppose also that $\neg L$ appears in Σ. There are clauses containing $\neg L$ in Σ. Delete those clauses and let Σ' be the remaining set. Since Σ' is a proper subset of Σ, we have $\Sigma' \not\vdash \alpha$. So $\Sigma' \cup \{\neg \alpha\}$ has a model, in which $\neg L$ is true. Hence we have a model of $\Sigma \cup \{\neg \alpha\}$, but this contradicts $\Sigma \vdash \alpha$. So $\neg L$ does not appear in Σ. Q.E.D.

Lemma 3.3 *Suppose $\Sigma = \{L\beta\} \cup \Sigma' \vdash \alpha$ is minimal w.r.t. α. Then $\Sigma' \not\vdash \beta\alpha$.*

Proof Suppose $\Sigma' \vdash \beta\alpha$. Since $\Sigma' \not\vdash \alpha$, there is a model of $\Sigma' \cup \{\neg \alpha\}$. It follows from $\Sigma' \vdash \beta\alpha$ that the model satisfies β, and hence $L\beta$ as well. Accordingly it is a model of $\Sigma \cup \{\neg \alpha\}$, which contradicts $\Sigma \vdash \alpha$ though. Q.E.D.

Lemma 3.4 *Let Σ be a set of unit clauses which is minimal w.r.t. a clause α. If α is not tautology, there exists an **IDR** derivation of α from Σ with an arbitrary clause in Σ as a top-clause.*

Proof First note that Σ is non-empty as α is not tautology. Suppose Σ is inconsistent. Since Σ is minimal, it must be of the form $\Sigma = \{A, \neg A\}$ for some atom A. Let L be A or $\neg A$ chosen as a top-clause. If $L \in \alpha$, $L|\alpha, |\alpha$, is a required derivation (L is factored away). If not, take $L|\alpha, \boxed{\neg L}\alpha, |\alpha$. Suppose Σ is consistent. α is non-empty again. If no literal in Σ occurs in α, they have no atom in common by Lemma 3.2, and hence there is a model of $\Sigma \cup \{\neg \alpha\}$ (recall that α is not tautology). But this contradicts $\Sigma \vdash \alpha$. So some literal L occurs both in Σ and in α. The minimality of Σ on the other hand means $\Sigma = \{L\}$. Hence we have an **IDR** derivation $L|\alpha, |\alpha$ (recall that $L \in \alpha$). Q.E.D.

Proposition 3.2 *Let Σ be a set of clauses minimal w.r.t. a clause α. If α is not tautology, there exists an **IDR** derivation of α from Σ with an arbitrary clause in Σ as a top-clause.*

Proof We prove by induction on the number N of occurrences of $\vee$, the disjunction symbol, in Σ. The base case where $N = 0$ is already proved by Lemma 3.4.

Assume the proposition holds for less than N and let $L\beta$ be a clause in Σ chosen as a top-clause. Write $\Sigma = \{L\beta\} \cup \Sigma' \vdash \alpha$. We first deal with the case where β is non-empty.

We show that there is an **IDR** derivation I of $\beta|\alpha$ from $\{L\beta\} \cup \Sigma'$ with a top-clause $L\beta$. The case for $L \in \beta$ or $L \in \alpha$ is easy. Suppose not. First, $\{L\} \cup \Sigma' \vdash \beta\alpha$ is straightforward. Second, we have $\Sigma' \nvdash \beta\alpha$ from Lemma 3.3. So there is a subset $\Sigma'' \subseteq \Sigma'$ such that $\{L\} \cup \Sigma'' \vdash \beta\alpha$ and $\{L\} \cup \Sigma''$ is minimal w.r.t. $\beta\alpha$. By applying the induction hypothesis to $\{L\} \cup \Sigma''$, we have an **IDR** derivation $I' = L|\beta\alpha, \gamma\boxed{\neg L}|\beta\alpha, \ldots, |\beta\alpha$ of $\beta\alpha$ from $\{L\} \cup \Sigma''$. We may assume L is not used by the expansion rule in I'[2]. Therefore $I = L\beta|\alpha, \gamma\boxed{\neg L}\beta|\alpha, \ldots, \beta|\alpha$ becomes an **IDR** derivation from $\{L\beta\} \cup \Sigma'$.

On the other hand, from $\{\beta\} \cup \Sigma' \vdash \alpha$ and $\Sigma' \nvdash \alpha$ it follows that there is an **IDR** derivation $J = \beta|\alpha, \ldots, |\alpha$ of α from $\{\beta\} \cup \Sigma'$ with a top-clause β. Suppose in J, there is a sequence $M\delta|\alpha, \beta'\boxed{\neg M}\delta|\alpha$ in which β is used as input clause by the expansion rule. We replace this inference with $M\delta|\alpha, L\beta'\boxed{\neg M}\delta|\alpha, \ldots, \beta'\boxed{\neg M}\delta|\alpha$ [3]. With such modification to J, we get an **IDR** derivation J' of α from Σ with a top-clause β. The concatenation of I to J' gives an **IDR** derivation of α from Σ with a top-clause $L\beta$.

We finally consider the case where β appearing in $\{L\beta\} \cup \Sigma'$ is empty. But this case is rather straightforward, and omitted. Q.E.D.

Theorem 3.1 *Let Σ be a set of ground clauses and α a ground clause. Suppose α is not tautology. Then there exists an **IDR** derivation of α from Σ with a top-clause clause $\beta \in \Sigma$ iff $\Sigma \vdash \alpha$.*

Proof If-part by Proposition 3.2 and Lemma 3.2. Only-if-part by Proposition 3.1. Q.E.D.

We remark that if Σ is consistent and α is neither tautology nor empty, the top-clause β can be chosen so that α and β share at least one literal (proof omitted).

4 Rules for IDR derivation at general level

In this section, we move up to general level where clauses may contain variables, and discuss how to derive a set of hypotheses $\mathcal{H}$ from an observed fact $\mathcal{O}$ such

---

[2] Owing to $\boxed{\neg L}$ in $\gamma\boxed{\neg L}\beta\alpha$, expansion using L can be replaced by factoring followed by truncation.

[3] These first two i-clauses represent an expansion of M using $L\beta$ as input clause. The rest is obtained by modifying $I = L\beta|\alpha, \ldots, \beta|\alpha$ to $L\beta'\boxed{\neg M}\delta|\alpha, \ldots, \beta'\boxed{\neg M}\delta|\alpha$. This is legal because $\beta = \beta'\neg M$ as a whole and adding δ to the right of β still keeps I as an **IDR** derivation.

that $\mathcal{B} \cup \mathcal{H} \vdash \mathcal{O}$, using background knowledge $\mathcal{B}$. Unless otherwise stated, $\mathcal{O}$ is assumed to be a single ground clause. First we (re-)state various definitions at general level.

- The *expansion rule* infers from an i-clause $L\alpha_1|\alpha_0$ and an input (normal) clause $\gamma\neg M\beta$ (renaming assumed) such that L and M have mgu θ an i-clause $(\gamma\beta\boxed{\neg M}\alpha_1|\alpha_0)\theta$.

- The *truncation rule* infers from an i-clause $\boxed{L}\alpha_1|\alpha_0$ an i-clause $\alpha_1|\alpha_0$.

- The *factoring rule* infers from an i-clause $L\alpha_1|\alpha_0$ an i-clause $(\alpha_1|\alpha_0)\theta$ if there is a literal M such that L and M has mgu θ and $\boxed{M}$ occurs in α_1 or M occurs in α_0.

- There are two assumption rules:

  1. The *t-assumption rule* creates an i-clause $\beta|\alpha_0$ assuming a normal clause β as a top-clause. β is called the *generated assumption* by the t-assumption rule. When $\alpha_0 \subseteq \beta$, we say the assumption β is *conservative*.

  2. The *i-assumption rule*, infers from an i-clause $L\alpha_1|\alpha_0$ an i-clause $(\beta\boxed{\neg M}\alpha_1|\alpha_0)\theta$, assuming a normal clause $\beta\neg M$ such that L and M has a unifier θ. $\beta\neg M$ is called the *generated assumption* by the i-assumption rule. When $\alpha_0 \subseteq \beta\neg M$, we say the assumption $\beta\neg M$ is *conservative*. If the i-assumption rule only assumes $\neg L$ such that L is the leftmost literal an i-clause $L\alpha_1|\alpha_0$, it coincides with Skip operation in SOL-deduction [3].

Let α be a non-tautological normal ground clause. $\langle\delta_1, H_1\rangle,\ldots,\langle\delta_n, H_n\rangle$ $(n \geq 1)$, an **IDR** *derivation of α from background knowledge $\mathcal{B}$ generating a set of assumptions $\mathcal{H}$*, is a sequence constructed by the **IDR** derivation procedure as follows:

IDR Derivation Procedure

[**Initialization**]: Set $\delta_1 = \beta|\alpha$ where β is an assumption generated by the t-assumption rule and $H_1 = \{\beta\}$, or $\beta \in \mathcal{B}$ and $H_1 = \{\}$.

[**Rule application** $(i > 1)$]: Infer δ_i $(2 \leq i \leq n)$ from δ_{i-1} nondeterministically by the expansion rule using an input clause in $\mathcal{B} \cup H_{i-1}$, by the truncation rule, by the factoring rule or by the i-assumption rule. Set $H_i = H_{i-1} \cup \{\beta\}$ if the i-assumption rule is used and an assumption β is generated. Else $H_i = H_{i-1}$.

[**Termination**]: Stop if we get $\delta_n = |\alpha$. Set $\mathcal{H} = H_n$.

Theorem 4.1 (Completeness of IDR) *Let B be a set of clauses representing background knowledge and α be a ground clause. Suppose $B \not\vdash \alpha$. If there is a set of clauses $\mathcal{H}$ such that $B \cup \mathcal{H} \vdash \alpha$, there is an* **IDR** *derivation of α from B generating a set of assumptions $\mathcal{H}'$ such that $\mathcal{H}' \subseteq \mathcal{H}$ and $B \cup \mathcal{H}' \vdash \alpha$. Besides, if $B \not\vdash \neg\alpha$ and all assumptions are conservative, $B \cup \mathcal{H}'$ is consistent.*

Proof From $B \not\vdash \alpha$ and $B \cup \mathcal{H} \vdash \alpha$, α is not tautology and there is a set Σ of ground instances from $B \cup \mathcal{H}$ such that $\Sigma \vdash \alpha$. Without loss of generality, we may assume Σ is minimal w.r.t. α. Then by Proposition 3.2, there is an **IDR** derivation $\beta' | \alpha = \delta'_1, \ldots, \delta'_n$ $(n \geq 1)$ of α from Σ with a clause $\beta' \in \Sigma$. Then it is not very difficult to construct, with the help of the assumption rule, an **IDR** derivation $\langle \delta_1, H_1 \rangle, \ldots, \langle \delta_n, H_n \rangle$ from background knowledge B such that $\delta_1 = \beta | \alpha$ where $\beta \in B \cup \mathcal{H}$, δ'_i is an instance of δ_i $(1 \leq i \leq n)$, Σ is an instance of $H_n \subseteq \mathcal{H}$, thereby $B \cup H_n \vdash \alpha$. Suppose all assumptions in H_n are conservative. Then their conjunction is written as $F \vee \alpha$ for some formula F. Consequently, if $B \cup H_n$ is inconsistent, we have $B \vdash \neg\alpha$. By contraposition, if $B \not\vdash \neg\alpha$, $B \cup H_n$ is consistent. Q.E.D.

A couple of comments are in order.

- **Universally quantified α:** We have only been considering ground α as a fact to be explained. In the case that α contains variables $x_1, \ldots, x_m$ (by convention they are all universally quantified), and we are searching for a hypothesis $\mathcal{H}$ such that $B \cup \mathcal{H} \vdash \forall x_1, \ldots, x_m \alpha$, all we need to do is to introduce new constants $c_1, \ldots, c_m$ and look for $\mathcal{H}$ which does not contain any of those constants but $B \cup \mathcal{H} \vdash \alpha[c_1/x_1, \ldots, c_n/x_n]$. Here $\alpha[c_1/x_1, \ldots, c_n/x_n]$ is α in which x_i is replaced by c_i $(1 \leq i \leq m)$.

- **Existentially quantified α:** Suppose a fact to be explained is existentially quantified. Then we have only to treat as if it were an infinite disjunction made up of all ground instance of α. For example, we can factor $p(f(a), f(b))q(a) | \exists x, y \, p(f(x), y)$ to $q(a) | \exists x, y \, p(f(x), y)$ (detail omitted).

- **OL-refutation:** It is very apparent that if $\alpha = \square$ and if we never apply the assumption rule, an **IDR** derivation becomes "isomorphic" to an OL refutation except the sign of a framed literal and minor differences concerning reduction and factoring. We however would like to emphasize again that **IDR** is essentially a forward reasoning system. What expansion does is, for example, to expand an input clause using an inference from A to $A \vee B$. Also we should notice that in an **IDR** derivation, truncation is a real logical operation, i.e. ancestor resolution (this is obvious from the proof of Propositon 3.1) while truncation in an OL-deduction has no logical meaning.

5 Conclusion

We have presented **IDR**, a clausal inference system for inductive reasoning to find a set of clauses $\mathcal{H}$, given background knowledge $\mathcal{B}$ and a fact $\mathcal{O}$ to be explained, such that $\mathcal{B} \cup \mathcal{H} \vdash \mathcal{O}$. It has four inference rules: the assumption rule, the expansion rule, the truncation rule and the factoring rule. It is shown that they constitutes a complete and sound[4] inference system (Theorem 4.1). From an operational point of view, **IDR** looks like OL-deduction with the assumption rule added that "can create" assumptions $\mathcal{H}$ needed to explain an observed fact $\mathcal{O}$. In spite of operational similarities however, logical mechanism working behind should not be confused with that of OL-deduction.

    IDR presented in this paper is of theoretical nature and there remains much to be done to make **IDR** useful. In particular, establishing criteria for selecting (creating) assumptions is an urgent need. We are currently looking into this problem.

References

[1] Chang,C.C. and Lee,R.C.T., Symbolic Logic and Mechanical Theorem Proving, Academic Press Limited, (1973).

[2] Demolombe,R. and Cerro,L.F., "An Inference Rule for Hypothesis Generation," Proc. of IJCAI'91, (1991) pp.152-157.

[3] Inoue,K., "Consequence-Finding Based on Ordered Linear Resolution," Proc. of IJCAI'91, (1991) pp.158-164.

[4] Ishizaka,H. and Arimura,H., "Efficient Inductive Inference of Primitive Prologs from Positive Data," Proc. of ALT'92 (Workshop on Algorithmic Learning Theory) Tokyo (1992) pp.135-146.

[5] Kijsirikul,B. et al., "Discrimination-Based Constructive Induction of Logic Programs," Proc. of AAAI'92, San Jose, (1992) pp.44-49.

[6] Kodratoff,Y. and Michalski (ed.), Machine Learning III, Morgan Kaufmann Publishers, Inc., (1990).

[7] Lapointe,S. and Matwin,S., "Sub-unification: A Tool for Efficient Induction of Recursive Programs," Proc. of ML'92, (1992) pp.273-281.

[8] Muggleton,S. and Feng,C., "Efficient Induction of Logic Programs," Proc. of ALT'90, Tokyo (1990) pp.368-381.

[9] Muggleton,S.(ed.), Inductive Logic Programming, Academic Press Limited, (1992).

---

[4]The soundness is obvious.

[10] Natarajan,B., Machine Learning, Morgan Kaufmann Publishers, Inc., (1991).

[11] Poole,D., "A Logical Framework for Default Reasoning," Artificial Intelligence 36 (1988) pp.27-47.

[12] Poole,D., "Compiling a Default Reasoning System into Prolog," New Generation Computing 9 (1991) pp.3-38.

[13] Quinlan,J.R., "Induction of Decision Trees," Machine Learning 1 (1986) pp.81-106.

[14] Roulveirol,C., "Completeness for inductive procedures," Proc. of ML'91, (1991) pp.452-456.

[15] Satoh,K. and Iwayama,N., "A Query Evaluation Method for Abductive Logic Programming," JICSLP'92 (Joint Int'l Conf. and Symposium on Logic Programming), Washington, (1992) pp.671-685.

[16] Sato,T., "A Complte Set of Rules for Inductive Inference," Electrotechncial Laboratory report TR-92-44 (1992).

[17] Shoenfield,J.R., Mathematical Logic (2nd ed.) Addison-Wesley, (1973).

[18] Valiant,L.G., "Learning Disjunctions of Conjunctions," Proc. of IJCAI'85, (1985) pp.560-566.

Generalized Unification as Background Knowledge in Learning Logic Programs

Akihiro YAMAMOTO

Department of Electrical Engineering
Hokkaido University
N 13 W 8, Kita-ku, Sapporo 060, JAPAN
Phone : Int.+81-11-716-2111 ext.6473
Fax : Int.+81-11-707-9750
Email : yamamoto@huee.hokudai.ac.jp

Abstract. In this paper we investigate the roles of generalized unification as background knowledge in learning logic programs. Our framework of learning is PAC-learning. We treat logic programs in which function symbols and recursions appear. We generalize the hereditary programs, which Miyano et. al have defined to investigate the learnability of elementary formal systems, by introducing generalized unification as the background knowledge of the learning algorithm. As a consequence, we succeed to revise Miyano's algorithm so that it treats another class of logic programs. Our algorithm is superior to the algorithm given by Džeroski et. al in the point that it uses no queries on target predicates. We also define the size of a sample S not as the number of atoms in S, but as the number of symbols in S. This becomes possible because the evaluation of destructors in generalized unification corresponds to the use of background predicates in Džeroski's algorithm.

1 Introduction

Learning systems that use logic programming are attracting much attentions. The research area concerning such systems is called Inductive Logic Programming. In the area *background knowledge* acts important roles [3, 4, 12, 14]. *Generalized unification*, which is sometimes called *semantic unification*, has been one of the most attractive subjects in researches on logic programming languages and theories. Since the earlier work by Plotkin[15], many contributions were made(see, for example, [2, 5, 6, 18]). In this paper we investigate the roles of generalized unification as background knowledge in learning logic programs. Our framework of learning is PAC-learning[17]. We treat logic programs in which function symbols and recursions appear.

Džeroski et al.[4] showed polynomial-time learnable classes of logic programs with no function symbols or no recursions, assuming some background knowledge. However, if any function symbol or any recursion is allowed to appear in programs, their learning algorithm needs membership queries and equivalence queries about the *target* predicate, the predicate which the algorithm is learning. Another research on learning logic programs was made by Miyano et al.[10, 11].

They used Elementary Formal Systems (EFSs, for short), which are programs of a logic programming language useful to treat formal languages on alphabets [1, 19]. Their learning algorithm uses no queries, but the EFSs must be hereditary. A hereditary EFS is a set of hereditary clauses. A clause is hereditary if, all terms appearing in its body are subterms in its head. Hereditary programs can be defined for ordinary logic programs in the same manner. For example, a famous program

$$P_0 = \{append([\,], X, X) \leftarrow , \ append([A|X], Y, [A|Z]) \leftarrow append(X, Y, Z)\}$$

is hereditary. It is easy to revise the learning algorithm in [10, 11] for ordinary logic programs, but hereditary programs are narrow from the viewpoint of programming. For example, we cannot easily imagine how to write a hereditary program which reverses the lists taken as inputs.

In this paper we generalize hereditary programs by introducing generalized unification as the background knowledge of the learning algorithm. As a consequence, we succeed to revise Miyano's algorithm so that it treats another class of logic programs. Our algorithm is superior to the algorithm in [4] in the point that it uses no queries on target predicates. We also define the size of a sample S not as the number of atoms in S, but as the number of symbols in S. This becomes possible because the evaluation of destructors in generalized unification corresponds to the use of background predicates in Džeroski's algorithm.

It is not easy to introduce generalized unification into logic programming, because there may be infinitely many maximal unifiers for two terms [15]. We overcome the problem as the following manner. At first we partition the set of function symbols into two disjoint sets, the set of *constructors* and the set of *destructors*. The constructors are used to represent datum in the form of ground terms, while each destructor is used as a function in generalized unification. This assumption is natural because, in many situations, we regard already developed logic programs as functions which take tuples of ground terms as inputs and output tuples of ground terms. In the generalized unification, two ground terms are unifiable if they are identical after all destructors in them are evaluated. For example, if a destructor *app* is defined as a function which concatenates two lists, then two terms $app([[\,]], [[[\,]]])$ and $[[\,], app([[[\,]]], [\,])]$ are unifiable. We represent the generalized unifcation as an equality theory, and give some conditions to it. The equality theory satisfying the conditions are called regular background knowledge. It has the property required in [4]. At last we introduce innermost simple programs to define SLD-resolution with generalized unification in the same way as the ordinary logic programming. Off course, recursions and function symbols may appear in any innermost simple programs.

We show a class SIM-IST(k, l, m, n) of concepts is polynomial-time learnable for $k, l, m, n \geq 0$. It is the class of concepts defined by innermost simple programs for each P of which P consists at most k clauses, the maximum length of clauses in P is less than or equal to l, and at most m variables and n destructors appear in P.

The paper is organized as follows: in Section 2, we give some preliminaries on logic programming and PAC-learning. In Section 3, we explain generalized unification and equality theories. Then we discuss the conditions which the equality theory should satisfy when we use generalized unification as background knowledge in learning logic programs. We also generalize the hereditary programs. In Section 4 we define simple programs and show SLD-resolution is definable for simple programs. In Section 5, we give the polynomial-time learnable classes of programs. In Section 6, we discuss some relations between our result and other works.

2 Preliminaries

We use fundamental terminologies and concepts on logic programming by referring Lloyd[9].

For a set S, $|S|$ denotes the number of all elements in S. Let Π be a finite set of predicate symbols, Σ a finite set of function symbols, and V a set of variables. We assume these sets are mutually disjoint. Each element in Σ and Π is denoted by a small letter, and each element in V is denoted by a capital letter. They are sometimes subscripted. For a function symbol or a predicate symbol f, $arity(f)$ denotes the arity of f. We prepare a constant symbol $[]$ and a binary function symbol $[\_|\_]$ to represent an empty list and a list constructor, respectively. A list $[t_1|[t_2|[\cdots[t_n|[]]\cdots]]]$ is also written as $[t_1, t_2, \cdots, t_n]$.

Let Σ' be a subset of $\Sigma \cup V$, and Π' be a subset of Π. Then $T(\Sigma')$ denotes the set of all first order *terms* constructed from the elements in Σ', and $A(\Pi', \Sigma')$ denotes the set of all *atoms* constructed from the elements in Π' and Σ'. Each element of $T(\Sigma)$ is called a *ground term*, and each element of $A(\Pi, \Sigma)$ is called a *ground atom*. For a term t in $T(\Sigma \cup V)$, $|t|$ denotes the *size* of t, which is the number of all symbols appearing in t, and $var(t)$ denotes the set of variables occurring in t. For an atom $p(t_1, \ldots, t_n)$, we let

$$|p(t_1, \ldots, t_n)| = |t_1| + \cdots + |t_n|,$$
$$var(p(t_1, \ldots, t_n)) = var(t_1) \cup \cdots \cup var(t_n).$$

For a set S of atoms and an integer n, we define

$$S_n = \{A \in S \; ; \; |A| \le n\}.$$

Example 1. Let $\Sigma = \{[\_|\_], [], a, b\}$ and $p \in \Pi$. Then $|[a, X, b]| = 7$, $|p([a, b], [])| = 6$, and $v(p([X, b|X], Y)) = \{X, Y\}$.

Let $A, B_1, \ldots, B_n$ be atoms. A *definite clause* is a clause of the form $A \leftarrow B_1, \ldots, B_n$, and a *goal clause* is of the form $\leftarrow B_1, \ldots, B_n$. A *program* P is a finite set of definite clauses. For a definite clause $A \leftarrow B_1, \ldots, B_n$, we define

$$loa(A \leftarrow B_1, \ldots, B_n) = arity(p) + arity(q_1) + \cdots + arity(q_n)$$

where $p, q_1, \ldots,$ and q_n be the predicate symbols of $A, B_1, \ldots,$ and B_n, respectively. For a program P, we put $loa(P) = max_{C \in P}(loa(C))$.

Now we partition Σ into two disjoint sets Γ and Δ. We call each element of Γ a *constructor*, and each element of Δ a *destructor*. A *constructor term* is a term in $T(\Gamma \cup V)$ and a *ground constructor term* is a term in $T(\Gamma)$. We sometimes write $A(\{p\}, \Gamma)$ as $B(p)$, and $A(\Pi, \Gamma)$ as $B(\Pi)$.

We assume a special predicate symbol c in Π to use it as the name of the target predicate for our learning algorithm.

Definition. A target predicate is a subset I of $B(c)$. It is also called a *concept*. A *concept class* $\mathcal{C}$ is a subset of $2^{B(c)}$. For a concept class $\mathcal{C}$, we define $\mathcal{C}_n = \{I \cap B(c)_n ; I \in \mathcal{C}\}$ and $dim\ \mathcal{C}_n = log_2 |\mathcal{C}_n|$.

Definition. A concept class $\mathcal{C}$ is *polynomial dimension* if there is a polynomial $d(n)$ such that $dim\ \mathcal{C}_n \leq d(n)$ for all $n \geq 0$.

Definition. An *example* is a tuple $\langle A, b \rangle$ where $A \in B(c)$ and $b = 1$ or 0. It is *positive* if $b = 1$ and *negative* if $b = 0$. A concept I is *consistent* with a given sequence of examples $\langle A_1, b_1 \rangle$, $\langle A_2, b_2 \rangle$, ..., $\langle A_n, b_n \rangle$ if $A_i \in I$ is equivalent to $b_i = 1$.

Definition. A *polynomial-time fitting* for a concept class $\mathcal{C}$ is a polynomial-time algorithm that takes a sequence of examples and returns a concept in $\mathcal{C}$ consistent with the sequence.

We do not use the original definition of PAC-learnability because it is proved (see, for example, [13]) that a concept class $\mathcal{C}$ is polynomial-time learnable if $\mathcal{C}$ is polynomial dimension and there is a polynomial-time fitting for $\mathcal{C}$.

3 Generalized Unification as Background Knowledge

3.1 Generalized Unification

Unification is often represented as an equality theory. For example, Clark's equality theory (see [9]) represents the ordinary unification. We use an equality theory $\mathcal{B}$ to represent generalized unification. At first we give some conditions to $\mathcal{B}$ in order that each destructor f is defined as a function $f_\mathcal{B} : T(\Gamma)^n \longrightarrow T(\Gamma)$.

Definition ([7]). The following pair of conditions is called the *definition principle* for an equality theory $\mathcal{B}$:

1. For every pair of terms t and s in $T(\Gamma)$,

$$\mathcal{B} \models t = s \iff t \text{ and } s \text{ are syntactically identical.}$$

2. For every term t in $T(\Sigma)$, there is a unique constructor term s in $T(\Gamma)$ such that $\mathcal{B} \models t = s$.

Under an equality theory $\mathcal{B}$ satisfying the definition principle, we can define a function $f_\mathcal{B}$ for each destructor f as follows:

$$\mathcal{B} \models f(s_1, \ldots, s_n) = t \iff f_\mathcal{B}(s_1, \ldots, s_n) = t,$$

where $s_1, \ldots, s_n, t$ are ground constructor terms in $T(\Gamma)$.

Remark. A destructor f in Δ and the function f_B are different. The symbol f represents the function f_B. Thus $f(s_1, \ldots, s_n)$ and t are not always the same even if $B \models f(s_1, \ldots, s_n) = t$. These two become equal when we use the generalized unification w.r.t B defined below.

Example 2. Let $\Gamma = \{[\_|\_], []\}$, $\Delta = \{app\}$. If app_B is a function which concatenates two lists, then it holds that $B \models app([[], []], [[]]) = [[], [], []]$. However, $app([[], []], [[]])$ and $[[], [], []]$ are different terms, so $|app([[], []], [[]])| = 9$ and $|[[], [], []]| = 7$.

Remark. We intend that B is implemented by some developed logic program P_B. But B need not coincide with P_B. We require that B is a specification of the least Herbrand model of P_B.

Example 3. The function app_B in Example 2 might be implemented by the program P_0 shown in Section 1 with some equality axioms. We could define the function app_B so that $app_B(s_1, s_2) = t$ is equivalent to $P_0 \models append(s_1, s_2, t)$. Another implementation of B might be an equality theory including two axioms

$$\forall X (app([], X) = X),$$
$$\forall AXY (app([A|X], Y) = [A|app(X, Y)]).$$

Now we describe the generalized unification.

Definition. Let t and s be two terms. A *unifier* of t and s w.r.t. B is a substitution θ satisfying

$$B \models t\theta = s\theta.$$

Two terms are *unifiable* if their unifier exists. A substitution θ is a unifier of two atoms A and B if $A = p(t_1, \ldots, t_n)$ and $B = p(s_1, \ldots, s_n)$ and

$$B \models t_1\theta = s_1\theta \wedge \cdots \wedge t_n\theta = s_n\theta.$$

Two atoms are *unifiable* if their unifier exists.

We prepare a useful class of unifiers.

Definition. Let t and s be two terms. A *solution* of an equation $t = s$ w.r.t. B is a unifier $\theta = \{X_1 := s_1, \ldots, X_n := s_n\}$ of t and s w.r.t. B such that $\{X_1, \ldots, X_n\} = var(t) \cup var(s)$ and $s_1, \ldots, s_n$ are all ground constructor terms.

Example 4. Let Γ, Δ, and B be given as in Example 2. Let $t = app(X, Y)$ and $s = [[[]], []]$. Then $\theta = \{X := [], Y := app([[[]]], [[]])\}$ is a unifier of t and s. The solutions for the equation $t = s$ are the following three:

$$\theta_1 = \{X := [], \qquad Y := [[[]], []]\},$$
$$\theta_2 = \{X := [[[]]], \qquad Y := [[]] \quad\},$$
$$\theta_3 = \{X := [[[]], []], Y := [] \quad\}.$$

Next we define the least model of a program P when B is given. Note the following lemma.

Lemma 1. *If B satisfies the definition principle, the congruence over $T(\Sigma)$ defined as*

$$B \models t = s \iff t \sim s$$

is the finest among the congruences which are models of B. Moreover, the set $T(\Sigma)/\sim$ of congruence class is isomorphic to $T(\Gamma)$.

Proof. Let a congruence R over $T(\Sigma)$ satisfy B. The congruence R is not finer than $\sim$, because of the second condition of definition principle. So we consider the case that there is a term $t \in T(\Sigma)$ and $s \in T(\Gamma)$ such that $t \, R \, s$ but $t \not\sim s$. Let $u \in T(\Gamma)$ be the term satisfying $u \sim t$. Then $t \, R \, u$ and $u \, R \, s$ because R satisfies B. Thus from the first condition of the definition principle, $\sim$ is finer than R. The second statement of the theorem is clear from the definition of the definition principle.

Jaffer et. al [8] showed that if B generates a finest congruence $\sim$ over $T(\Sigma)$, there is a model $M(P)$ of P w.r.t. B such that

$$P \cup B \models A \iff M(P) \models A,$$

for every ground atom A. The model $M(P)$ is the least among the models which are subsets of a set

$$\{p(\mathbf{t}_1, \ldots, \mathbf{t}_n) \, ; \, p \in \Pi \text{ and } \mathbf{t}_1, \ldots, \mathbf{t}_n \in T(\Sigma)/\sim\}.$$

Because $T(\Sigma)/\sim$ is isomorphic to $T(\Gamma)$ from Lemma 1, we give the following definitions.

Definition. Let B satisfy the definition principle. An *Herbrand interpretation*, is a subset of $B(\Pi)$. An *Herbrand model* of P w.r.t. B is an Herbrand interpretation I w.r.t. B such that $I \models P$. The Herbrand model $M(P)$ w.r.t. B is called the *least Herbrand model* of P w.r.t. B.

Example 5. In the same setting as Example 2, we define the following definite clause:

$$P_1 = \left\{ \begin{array}{l} reverse([\,],[\,]) \leftarrow \\ reverse([A|X], app(Y,[A])) \leftarrow reverse(X,Y) \end{array} \right\}.$$

Then two atoms $reverse([\,],[\,])$ and $reverse([[[\,]],[\,]],[[\,],[[\,]]])$ are in $M(P_1)$, but $reverse([[[\,]],[\,]], app([[\,]],[[[\,]]]))$ is not because the second argument is not a constructor term.

Definition. A *background knowledge* is an equality theory B satisfying the following conditions:

1. The theory B satisfies the definition principle.
2. For every destructor $f \in \Delta$ there is an algorithm that takes $s_1, \ldots, s_n \in T(\Gamma)$ as inputs and returns $f_B(s_1, \ldots, s_n)$ in polynomial time in $|s_1| + \cdots + |s_n|$.

3.2 Regular Background Knowledge

We give more restrictions for the background knowledge $\mathcal{B}$ to use it in our learning algorithm. The first one is determinacy in the same sense as [4, 12]. The reader may think that a kind of determinacy has already introduced, because a destructor assumed to be defined as a function. Unfortunately, such determinacy is not sufficient. Consider the case that a ground constructor goal $\leftarrow reverse([[[\,]], [\,]], [[\,], [[\,]]]])$ is given to the program P_1 in Example 5. Then the restriction we want is that the equation $app(X, [A]) = [[\,], [[\,]]]$ has a unique solution. This requires that all destructors are interpreted as injections. The second restriction is that $\mathcal{B}$ should return the solution of $f(X_1, \ldots, X_n) = t$ in polynomial-time. We also require that destructors are defined as increasing functions in order to define the size of a sample for our learning algorithm as the number of symbols in it. Thus we give the following definitions.

Definition. Let $g : T(\Gamma)^n \longrightarrow T(\Gamma)$ be a function. Then the function g is *solvable* if there is an algorithm which takes as input a term $t \in T(\Gamma)$ and outputs all tuples $\langle s_1, \ldots, s_n \rangle$ satisfying $g(s_1, \ldots, s_n) = t$ if exists, or reports *fail* otherwise. The function g is *polynomial-time solvable* if the algorithm runs in polynomial-time of $|t|$.

Definition. A function $g : T(\Gamma)^n \longrightarrow T(\Gamma)$ is *increasing* if $|s_1| + \cdots + |s_n| < |g(s_1, \ldots, s_n)|$ for $s_1, \ldots, s_n$ in $T(\Gamma)$.

Example 6. Let $\Gamma = \{[\_|\_], [\,]\}$, $\Delta = \{alt\}$, and $\mathcal{B}$ defines the destructor alt as the following function:

$$\begin{aligned} \mathcal{B} &\models \forall X(alt(X, [\,]) = [X]), \\ \mathcal{B} &\models \forall AXY(alt(X, [A|Y]) = [A|alt(X, Y)]). \end{aligned} \tag{1}$$

Then $alt_{\mathcal{B}}$ is injective, polynomial-solvable, and increasing.

Example 7. The function $app_{\mathcal{B}}$ in Example 2 is neither injective nor increasing.

Now we generalize hereditariness. For every function symbol f in Σ, an integer i $(1 \leq i \leq arity(f) = m)$, and two terms t, s in $T(\Gamma)$, an expression $p_f^i(t, s)$ denotes the fact that there are $s_1, \ldots, s_{i-1}, s_{i+1}, \ldots, s_m$ in $T(\Gamma)$ which satisfy

$$\mathcal{B} \models t = f(s_1, \ldots, s_{i-1}, s, s_{i+1}, \ldots, s_n).$$

Definition. We define the *dependent set* $D(t)$ of a ground constructor term t as

$$D(t) = \left\{ s \in T(\Gamma) \;\middle|\; \begin{array}{l} \text{There are } f_1, \ldots, f_m \in \Sigma, \, s_1, \ldots, s_{m-1} \in T(\Gamma), \\ \text{and } i_1, \ldots, i_m \text{ such that } p_{f_1}^{i_1}(t, s_1), \, p_{f_2}^{i_2}(s_1, s_2), \ldots, \\ \text{and } p_{f_m}^{i_m}(s_{m-1}, s) \text{ hold.} \end{array} \right\}.$$

The dependency defined above is a generalization of the subterm relation.

Example 8. Consider the case $\Sigma = \Gamma = \{[\_|\_], [\,]\}$. For the constructor $[\_|\_]$, $p^1_{[\_|\_]}$ and $p^2_{[\_|\_]}$ are respectively called *car* and *cdr* (or sometimes called *head* and *tail*). Thus the dependency set $D(t)$ of $t \in T(\Gamma)$ is the set of all subterms of t.

Definition. A background knowledge $\mathcal{B}$ has *polynomial dependency property* if $|D(t)|$ is bounded by a polynomial in $|t|$.

Definition. A background knowledge $\mathcal{B}$ is *regular* if it has polynomial dependency property and, for any destructor f in Δ, $f_\mathcal{B}$ is injective, polynomial-time solvable, and increasing.

The following theorem shows the existence of interesting regular background knowledge.

Theorem 1. *Let* $\Gamma = \{[\_|\_], [\,]\}$, $\Delta = \{alt\}$. *Then the background knowledge* $\mathcal{B}$ *satisfying two conditions of (1) has polynomial dependency property.*

Proof. For every $s \in D(t)$ there are ground constructor terms $s_1, s_2, \ldots s_m$ such that s_1 is a subterm of t, $s_m = s$, $p^2_{alt}(s_1, s_2)$, $p^2_{alt}(s_2, s_3)$, $\cdots$, and $p^2_{alt}(s_{m-1}, s_m)$ hold. Note that $m \leq |t|$ because $alt_\mathcal{B}$ is increasing. Since the total amount of subterm of t is $|t|$, $|D(t)|$ is bounded by $|t|^2$.

4 Simple Programs

In generalized unification, there may be infinitely maximal unifiers of two terms, and thus the set of all the resolvents of a goal may be infinite. This makes it difficult to define SLD-resolutions in logic programming with generalized unification[18]. If destructors are solvable, we can overcome the problem by introducing a class of programs.

Definition. A definite clause $A \leftarrow B_1, \ldots, B_n$ is *simple* if all arguments of $B_1, \ldots, B_n$ are variables in A. The clause is *innermost simple* if it is simple and all arguments of destructors in A are variables. A program P is *simple* (*innermost simple*) if each clause in P is simple (innermost simple, respectively).

Example 9. The program P_1 in Example 5 is simple. But it is not innermost simple because the destructor *app* has $[A]$ as its arguments.

Example 10. Let Γ, Δ, $\mathcal{B}$ be defined as Example 6. The following program P_2 is innermost simple:

$$P_2 = \left\{ \begin{array}{l} reverse([\,], [\,]) \leftarrow \\ reverse([A|X], alt(A, Y)) \leftarrow reverse(X, Y) \end{array} \right\}.$$

Definition. A *ground constructor goal* is a goal clause in which no terms other than constructor ground terms appear.

Lemma 2. *Let G be a ground constructor goal. If $\mathcal{B}$ is solvable and P is simple, then every resolvent of G is a ground constructor goal, and the set of all the resolvents of G is finite and computable.*

Proof. Let A be an atom in G and $B \leftarrow B_1, \ldots, B_n$ be a clause in P. It is suffcient to consider the case that A and B are unifiable. We put $A = p(t_1, \ldots, t_n)$ and $B = p(s_1, \ldots, s_n)$. Then, from the solvability of $\mathcal{B}$, we can decide the equations $t_1 = s_1$, ..., $t_n = s_n$ have a common solution or not. If the solution exists, it substitutes each variable in $B_1, \ldots, B_n$ by a ground constructor term. Thus the resolvent of G is a ground constructor goal.

From the lemma above we can formalize SLD-derivations and SLD-refutations from ground constructor goals just as in the ordinary logic programming[9]. The SLD-derivations have an interesting property.

Lemma 3. *Let $\leftarrow p(t_1, \ldots, t_n)$ be a ground constructor goal. If $\mathcal{B}$ is solvable and P is simple, then all terms appears in an SLD-derivation from $\leftarrow p(t_1, \ldots, t_n)$ are in $D(t_1) \cup \cdots \cup D(t_n)$.*

Proof. Let t be a ground constructor term and f be a constructor or a destructor. If $f(X_1, \ldots, X_m) = t$ has a solution $\{X_1 := s_1, \ldots, X_m := s_m\}$, $p_f^i(t, s_i)$ holds for $i = 1, \ldots, m$. By applying this proposition recursively (this is possible from Lemma 2), the conclusion of this lemma is proved.

5 Polynomial-time Learnable Classes

We give classes of programs and prove that they are of polynomial-dimension, in very similar way to that of Proposition 1 of [10] and Lemma 4 of [11].

Definition. For $k, l \geq 0$, SIM(k, l) is the class of concepts defined by simple programs each P of which satisfies $|P| \leq k$ and $loa(P) \leq l$.

Lemma 4. *If $f_{\mathcal{B}}$ is increasing for any destructor f in Δ, then SIM(k, l) is of polynomial-dimension.*

Proof. We evaluate $|\{I \cap B(c)_n ; I \in \text{SIM}(k, l)\}|$ for $n \geq 0$. Let P be a simple program with $|P| \leq k$ and $loa(P) \leq l$, and $C = q(t_1, \ldots, t_m) \leftarrow B_1, \ldots, B_h$ is a clause in P. If $t_i \geq n$ for some i, $M(P) \cap B(c)_n = M(P - \{C\}) \cap B(c)_n$ because $\mathcal{B}$ is increasing and P is simple. If there is a predicate symbol p which occurs in C but does not occur in the head of any clause in P, $M(P) = M(P - \{C\})$. So we can assume that $\Pi = \{p_1, \ldots, p_k\}$, $c = p_1$, and the arity of p_i is less than or equal to l for $i = 1, \ldots, k$. Since the number of variables in the head of C are at most $n \cdot l$, we may let the set of variables in P be $\{X_1, \ldots, X_{n \cdot l}\}$. Thus it is sufficient to consider only the programs consisting of at most k simple clauses whose heads are in $A(\Pi, \Sigma \cup \{X_1, \ldots, X_{n \cdot l}\})_{n \cdot l}$, and whose bodies are sequences of at most l atoms in $A(\Pi, \{X_1, \ldots, X_{n \cdot l}\})$. The number of such programs is roughly bounded by

$$(k \cdot (|\Sigma| + n \cdot l)^n \cdot (k \cdot (n \cdot l)^l + 1)^l)^k.$$

Hence $SIM(k, l)$ is of polynomial-dimension.

Now we give the main result.

Definition. For $k, l, m, n \geq 0$, $SIM\text{-}IST(k, l, m, n)$ is the class of concepts defined by innermost simple programs each P of which satisfies $|P| \leq k$ and $loa(P) \leq l$ with at most m variables and n destructors.

Theorem 2. *If $\mathcal{B}$ is regular, then there is a polynomial-time fitting of SIM-IST(k, l, m, n) for any $k, l, m, n \geq 0$.*

Proof. Let P be a program in $SIM\text{-}IST(k, l, m, n)$. Without loss of generality, we can assume $\Pi = \{p_1, \ldots, p_{k \cdot l}\}$, $arity(p_m) = (m \bmod l) + 1$, and $c = p_j$ for some j.

The learning algorithm is the same one as that of [10, 11]. The algorithm is founded on the so-called *generate and test* method. It works as follows: it takes a sequence of examples S. Let S^+ be the set of positive examples in S. If $S^+ = \phi$ it returns $P = \phi$ as a fitting. Otherwise, it generates a set

$$\mathcal{S} = \{t \in T(\Gamma) ; t = t_i \text{ for some } c(t_1, \ldots, t_k) \in S\}.$$

Then for each $p_i \in \Pi$, the algorithm generates a set

$$H_i = \{p_i(t_1, \ldots, t_h) ; h = (i \bmod l) + 1 \text{ and } t_1, \ldots, t_h \in \bigcup_{t \in \mathcal{S}} D(t)\}.$$

Note that $\mathcal{B}$ has the polynomial dependency property, $|H_i|$ is polynomial in $\sum_{s \in S} |s|$ for every $p_i \in \Pi$.

Next the learning algorithm generates a set $\mathcal{P}$ of all programs P satisfying:

1. P is an innermost simple program, $|P| \leq k$, and $loa(P) \leq l$,
2. at most m variables and n destructors appear in each clause C in P, and
3. if the clause C is $A \leftarrow B_1, \ldots, B_h$, and the predicate symbol of A is p_i, there is B in H_i which is unifiable with A.

The set $\mathcal{P}$ is sufficient as the sets candidates of fitting, because we can claim from Lemma 3 that if a program P which satisfies $S^+ \subset M(P) \cap B(c)$, for every clause $A \leftarrow B_1, \ldots, B_n$, there is B in H_i which is unifiable with A. $|\mathcal{P}|$ is polynomial because k, l, m, n is fixed, Δ is finite, and $\mathcal{B}$ is increasing.

Then the algorithm checks for each $P \in \mathcal{P}$ whether $s \in M(P)$ or not for every $s \in S$, by trying to construct a bottom-up proof of s. Since all atoms appearing in the proof are in the set $\bigcup_{1 \leq i \leq k \cdot l} H_i$, this procedure works in polynomial of $\sum_{s \in S} |s|$.

Example 11. The program P_2 in Example 10 is in $SIM\text{-}IST(2, 4, 3, 1)$. Since $\mathcal{B}$ is regular from Theorem 1, the class is polynomial-time learnable.

6 Concluding Remarks

We start this research in order to find the reason why the algorithm in [10, 11] need not any queries or any background knowledge. By regarding the subterm relation as the background knowledge, we reached the generalization of hereditariness.

To delete destructors from programs, we can use the flattening [2]. For example, program P_2 is transformed into an ordinal logic program

$$\left\{ \begin{array}{l} reverse([],[]) \leftarrow \\ reverse([A|X], W) \leftarrow addlast(A, Y, W), reverse(X, Y) \end{array} \right\}.$$

Note that internal variables appears in the program as 'pipes' between background knowledge and predicates defined by the program. This role of internal variables is just as that in[4]. From the observation above, our work is intermediate between Miyano's work and Džeroski's work.

Moreover, we could solve another problem with flattening. The problem is that Γ should be exactly equal to the set $\{[\_|\_],[]\}$ to use the destructor alt. This comes from the condition that $\mathcal{B}$ should satisfy the definition principle, that is, $alt_\mathcal{B}$ should be total functions. In usual cases, Γ might be a proper superset of the set $\{[\_|\_],[]\}$, and $alt_\mathcal{B}$ might not be a total function. However, after flattening program P_2, the problem would disappear because $addlast$ is a predicate, not a function. Sato and Tamaki[16] called such a predicate a *functional part*. It would not be so difficult to revise our theory of this paper by using functional parts instead of functions.

References

1. Arikawa, S. , Shinohara, T. , and Yamamoto, A. , Learning Elementary Formal Systems, *Theoretical Computer Science* 95(1):97–113 (1992).
2. Bosco, P. G. , Giovannetti, E. , and Moiso, C. , Narrowing vs. SLD-Resolution, *Theoretical Computer Science* 59(1,2):3–23 (1988).
3. Buntine, W. , Generalized Subsumption and Its Applications to Induction and Redundancy, *Artificial Intelligence* 36:149–176 (1988).
4. Džeroski, S. , Mugglton, S. , and Russell, S. , PAC-Learnability of Determinate Logic Programs, in *Proceedings of the 5th Annual Workshop on Computational Learning Theory*, 128–135, Academic Press, 1992.
5. Fribourg, L. , Prolog with Simplification, in *Proceedings of France-Japan Artificial Intelligence and Computer Science Symposium*, 239–266, ICOT, 1986.
6. Hölldobler, S. , From Paramodulation to Narrowing, in Kowalski, R. A. and Bowen, K. A. (eds.), *Proceedings of the 5th International Conference and Symposium on Logic Programming*, 327–342, The MIT Press, 1988.
7. Huet, G. and Hullot, J.-M. , Proofs by Induction in Equational Theories, *Journal of Computer and System Science* 23(1):239–226 (1982).
8. Jaffar, J. , Lassez, J.-L. , and Maher, M. J. , Logic Programming Scheme, in DeGroot, D. and Lindstrom, G. (eds.), *Logic Programming: Functions, Relations, and Equations*, 441–467, Prentice-Hall, 1986.

9. Lloyd, J. W. , *Foundations of Logic Programming : Second, Extended Edition*, Springer - Verlag, 1987.

10. Miyano, S. , Shinohara, A. , and Shinohara, T. , Which Classes of Elementary Formal Systems are Polynomial-time Learnable?, in S. Arikawa and A. Maruoka and T. Sato (ed.), *Proceedings of the Second International Workshop on Algorithmic Learning Theory*, 139–150, JSAI, 1991.

11. Miyano, S. , Shinohara, A. , and Shinohara, T. , Learning Elementary Formal Sytems and an Application to Discovering Motifs in Proteins, Technical Report RIFIS-TR-CS-37, Research Institute of Fundamental Information Science, Kyushu University, 1991, revised 1993.

12. Mugglton, S. and Frish, A. M. , Efficient Induction of Logic Programs, in S. Arikawa and S. Goto and S. Ohsuga and T. Yokomori (ed.), *Proceedings of the First International Workshop on Algorithmic Learning Theory*, 368–381, JSAI, 1990.

13. Natarajan, B. K. , *Machine Learning : A Theoretical Approach*, Morgan-Kaufmann, 1991.

14. Page Jr., C. D. and Frish, A. M. , Generalization and Learnability : A study of Constrained Atoms, in S. Mugglton (ed.), *Inductive Logic Programming*, 29–61, Academic Press, 1992.

15. Plotkin, G. D. , Building in Equational Theories, in *Machine Intelligence 7*, 132–147, Edinburgh University Press, 1972.

16. Sato, T. and Tamaki, H. , Transformational Logic Program Synthesis, in *Proceedings of International Conference on FGCS 1984*, 195 – 201, ICOT, 1984.

17. Valiant, L. G. , A Theory of Learnable, *Communications of the ACM* 27(11):1134–1142 (1984).

18. Yamamoto, A. , A Theoretical Combination of SLD-Resolution and Narrowing, in Lassez, J.-L. (ed.), *Proceedings of the 4th International Conference on Logic Programming*, 470–487, The MIT Press, 1987.

19. Yamamoto, A. , Procedural Semantics and Negative Information of Elementary Formal System, *The Journal of Logic Programming* 13(1):89–97 (1992).

Inductive Inference Machines That Can Refute Hypothesis Spaces*

Yasuhito Mukouchi and Setsuo Arikawa

Research Institute of Fundamental Information Science
Kyushu University 33, Fukuoka 812, Japan
e-mail: {mukouchi, arikawa}@rifis.sci.kyushu-u.ac.jp

Abstract. This paper intends to give a theoretical foundation of machine discovery from examples. We point out that the essence of a logic of machine discovery is the refutability of the entire spaces of hypotheses. We discuss this issue in the framework of inductive inference of length-bounded elementary formal systems (EFS's, for short), which are a kind of logic programs over strings of characters and correspond to context-sensitive grammars in Chomsky hierarchy.

We first present some characterization theorems on inductive inference machines that can refute hypothesis spaces. Then we show differences between our inductive inference and other related inferences such as in the criteria of reliable identification, finite identification and identification in the limit. Finally we show that for any n, the class, i.e. hypothesis space, of length-bounded EFS's with at most n axioms is inferable in our sense, that is, the class is refutable by a consistently working inductive inference machine. This means that sufficiently large hypothesis spaces are identifiable and refutable.

> A scientist, whether theorist or experimenter, puts forward statements, or systems of statements, and test them step by step. In the field of the empirical sciences, more particularly, he constructs hypotheses, or systems of theories, and test them against experience by observation and experiment.
>
> *Karl R. Popper: The Logic of Scientific Discovery*

1 Introduction

In the middle of this century the logic of scientific discovery was deeply discussed by philosophers[20, 21]. Recently in Artificial Intelligence, especially in Cognitive Science, researchers are extensively discussing frameworks for scientific discovery from various viewpoints[27].

Before going into the detailed discussions we need to set up a computational logic of scientific discovery in a mathematical way so that we can precisely discuss what kinds of machine discovery can work. One of the best ways to this should be to reexamine the philosophical results from computational viewpoints. In the present paper we start with making the Popperian logic of scientific discovery[20, 21] computational.

---

* This work is partly supported by Grant-in-Aid for Scientific Research on Priority Areas, "Knowledge Science" and "Genome Informatics" from the Ministry of Education, Science and Culture, Japan. This paper is an extended abstract version of [19].

The Popperian logic of scientific discovery concentrated on the testability, falsifiability or refutability of hypotheses or scientific theories. Popper also asserted that scientific theory should have been refuted by observed facts and any such theory could by no means be verified[20]. Thus we tentatively believe the current theory until we face with an observation which is inconsistent with the theory.

Hence the consistent and conservative inductive inference can be viewed as a computational realization of the Popperian notion of refutability. In the inductive inference, the inference machine requests data or facts from time to time and produces hypotheses from time to time. The hypotheses produced by the machine are to be consistent with the facts read so far, and each of them is to be refuted when the machine faces with inconsistent data or facts.

Thus the Popperian logic of scientific discovery can be viewed as a basis of the modern inductive inference studies. The inductive inference is thus a mathematical basis of machine learning. *Then what should be a logic of machine discovery or a computational logic of scientific discovery?*

The machine discovery we are concerned with in this paper is to make computers discover some scientific theories from given data or facts. Hence machine learning should be a key technology for machine discovery. In machine learning first we must select a hypothesis space from which the learning machine proposes theories or hypotheses. The space is naturally required to be large, but to make the learning efficient it is required to be small. As far as data or facts are presented according to a hypothesis that is unknown but guaranteed to be in the space as in the ordinal inductive inference, the machine will eventually identify the hypothesis, and hence no problem may arise. In machine discovery, however, we can not assume this.

If the hypothesis is not in the space, most learning machines will continue for ever to search the space for a new hypothesis. Usually we can not know the time when to stop such an ineffective searching. This is the most crucial problem we must solve in realizing machine discovery systems. In machine discovery the sequences of data or facts are given at first independently of the space. We can not give in advance the space that includes the desired theory. If the learning machine can explicitly tell us that there are no theories in the space which explain the given sequence, the machine will work for machine discovery.

Hence the essence of a computational logic of scientific discovery should be that the entire hypothesis space is refutable by a sequence of observed data or facts. If there exist rich hypothesis spaces that can be refuted, we can give a space and a sequence to the machine, and then we can just wait for an output from it. The machine will discover a hypothesis which is producing the sequence if it is in the space, otherwise it will refute the whole of the space and stop. When the space is refuted, we may give another space to the machine and try to make such a discovery in the new space.

In this paper we choose the inductive inference as the framework for machine learning. Then the machine discovery system is an inductive inference machine that can refute hypothesis spaces. In order to make our discussion clearer we also choose, as the hypothesis spaces, classes of elementary formal systems (EFS's, for short) which is a kind of logic programs over strings of characters. Moreover we assume as in the ordinal inductive inference of languages that every class in question be an indexed family of recursive languages. This assumption is quite natural to make a grammar, i.e. a hypothesis or theory, refutable by an observation and also to generate grammars as hypotheses automatically and successively.

If the class is a finite set of recursive languages, then it is trivially refutable from complete data. Also if the class contains all finite languages, then it is easily shown not to be refutable. *Then are there any meaningful classes, i.e., hypothesis spaces, that are identifiable and refutable?* We give a positive answer to this question. We will say such classes to be refutably inferable.

We first show some characterizations of such inductive inference. Then we show that some sufficiently large classes of formal languages, i.e. the classes definable by length-bounded EFS's with at most n axioms, are refutably inferable from complete data.

2 Preliminaries

We start with basic definitions and notions on inductive inference of indexed families of recursive concepts.

Let U be a countable set to which we refer as a *universal set*. Then we call $L \subseteq U$ a *concept* or a *language*.

Definition 1. Let $N = \{1, 2, \cdots\}$ be the set of all natural numbers. A class $C = \{L_i\}_{i \in N}$ of concepts is said to be an *indexed family of recursive concepts*, if there is a recursive function $f : N \times U \to \{0, 1\}$ such that

$$f(i, w) = \begin{cases} 1, & \text{if } w \in L_i, \\ 0, & \text{otherwise.} \end{cases}$$

In what follows, we assume that a class of concepts is an indexed family of recursive concepts without any notice, and identify a class with a hypothesis space.

Definition 2. A *positive presentation*, or a *text*, of a *nonempty* concept L is an infinite sequence $w_1, w_2, \cdots$ of elements in U such that $\{w_1, w_2, \cdots\} = L$. A *complete presentation*, or an *informant*, of a concept L is an infinite sequence $(w_1, t_1), (w_2, t_2), \cdots$ of elements in $U \times \{+, -\}$ such that $\{w_i \mid t_i = +, i \geq 1\} = L$ and $\{w_i \mid t_i = -, i \geq 1\} = L^c (= U \setminus L)$. By σ and δ, we denote positive or complete presentations, and by $\sigma[n]$, we denote the σ's initial segment of length $n \geq 0$. For a presentation σ, each element in σ is called a *fact*.

A set T is said to be *consistent* with a concept L, if $T \subseteq L$. A pair $\langle T, F \rangle$ of sets is said to be *consistent* with a concept L, if $T \subseteq L$ and $F \subseteq L^c$. For a positive presentation σ and $n \geq 0$, the finite sequence $\sigma[n]$ is said to be *consistent* with a concept L, if all facts in $\sigma[n]$ are members of L. For a complete presentation σ and $n \geq 0$, the finite sequence $\sigma[n]$ is said to be *consistent* with a concept L, if $\{w_i \mid (w_i, +) \in \sigma[n]\} \subseteq L$ and $\{w_i \mid (w_i, -) \in \sigma[n]\} \subseteq L^c$.

An *inductive inference machine* (*IIM*, for short) is an effective procedure that requests inputs from time to time and produces positive integers from time to time. An *inductive inference machine that can refute hypothesis spaces* (*RIIM*, for short) is an effective procedure that requests inputs from time to time and either (i) produces positive integers from time to time or (ii) refutes the class and stops after producing some positive integers. The outputs produced by the machine are called *guesses*.

For an IIM M and a finite sequence $\sigma[n] = w_1, w_2, \ldots, w_n$, by $M(\sigma[n])$, we denote the last guess of M which is successively presented $w_1, w_2, \ldots, w_n$ on its input requests. For an RIIM M and a finite sequence $\sigma[n] = w_1, w_2, \ldots, w_n$, by $M(\sigma[n])$, we denote the last guess or the 'refutation' sign produced by M which is successively presented $w_1, w_2, \ldots, w_n$ on its input requests.

An IIM M or an RIIM M is said to *converge* to a positive integer i for a presentation σ, if there is a positive integer m such that for any $n \geq m$, $M(\sigma[n])$ is defined and equal to i. An RIIM M is said to *refute* a class C from a presentation σ, if there is a positive integer n such that $M(\sigma[n])$ is the 'refutation' sign. In this case we also say that M refutes the class C from $\sigma[n]$.

Definition 3. Let $C = \{L_i\}_{i \in N}$ be a class. For a concept $L_i \in C$ and a presentation σ of L_i, an IIM M or an RIIM M is said to *infer the concept L_i w.r.t. C from σ*, if M converges to a positive integer j with $L_j = L_i$ for σ.

(a) An IIM M is said to *infer a class $C = \{L_i\}_{i \in N}$ in the limit from positive data* (resp., *complete data*), if for any concept $L_i \in C$, the M infers L_i w.r.t. C from any positive presentation σ (resp., complete presentation σ) of L_i. A class C is said to be *inferable in the limit from positive data* (resp., *complete data*), if there is an IIM M which infers the class C from positive data (resp., complete data).

(b) An IIM M is said to *reliably infer a class C from positive data* (resp., *complete data*), if for any nonempty concept L (resp., any concept L) and any positive presentation σ (resp., complete presentation σ) of L, (i) if $L \in C$, then M infers L w.r.t. C from σ, (ii) otherwise M does not converge to any positive integer for σ. A class C is said to be *reliably inferable from positive data* (resp., *complete data*), if there is an IIM M which reliably infers the class C from positive data (resp., complete data).

(c) An RIIM M is said to *refutably infer a class C from positive data* (resp., *complete data*), if for any nonempty concept L (resp., any concept L) and any positive presentation σ (resp., complete presentation σ) of L, (i) if $L \in C$, then M infers L w.r.t. C from σ, (ii) otherwise M refutes the class C from σ. A class C is said to be *refutably inferable from positive data* (resp., *complete data*), if there is an RIIM M which refutably infers the class C from positive data (resp., complete data).

In what follows, we simply say 'inferable' to mean 'inferable in the limit', when it is clear from the context. Note that when we consider inductive inference from positive data, we restrict a class to an indexed families of *nonempty* recursive concepts, because we can not make any positive presentation of the empty concept.

The notion of reliable inference was introduced by Minicozzi[15] and Blum&Blum[8] for function learning, and it was adapted to language learning by Sakurai[22]. By definition, it is clear that if a class C is *refutably* inferable from positive data (resp., complete data), then C is *reliably* inferable from positive data (resp., complete data). However the converse does not hold as shown in Section 4.

For reliable inference, the following Theorem 4 holds.

Theorem 4. *(a) A class C is reliably inferable from positive data, if and only if C contains no infinite concept (Sakurai[22]).*

(b) Every class is reliably inferable from complete data.

In this paper, a finite set $F(x)$ with a parameter x is said to be *recursively generable*, if there is an effective procedure which produces all elements in $F(x)$ and then halts.

3   Characterizations

In order to characterize the refutable inferability, we need the following Lemma 5.

Lemma 5. *Let M be an RIIM which refutably infers a class C from positive data (resp., complete data). For a nonempty concept L (resp., a concept L) and for a positive presentation σ (resp., a complete presentation σ) of L, if M refutes the class C from $\sigma[n]$, then $\sigma[n]$ is inconsistent with any concept $L_i \in C$.*

In characterizing the refutable inferability, the notion of consistency plays an important role.

Definition 6. An RIIM M which refutably infers a class C from positive data (resp., complete data) is said to be *consistently working*, if M satisfies the following condition: For any nonempty concept L (resp., any concept L), for any positive presentation σ (resp., any complete presentation σ) of L and for any $n \geq 1$, (i) if there is a concept $L_i \in C$ such that the initial segment $\sigma[n]$ is consistent with the concept L_i, then $\sigma[n]$ is consistent with the concept L_j with $j = M(\sigma[n])$, (ii) otherwise M refutes the class C from $\sigma[n]$. A class C is said to be *refutably and consistently inferable* from positive data (resp., complete data), if there is a consistently working RIIM which refutably infers the class C from positive data (resp., complete data).

As shown later, if a class C is refutably inferable from *positive data*, then it can be achieved by a consistently working RIIM. However, even though a class C is refutably inferable from *complete data*, it is not known at present whether it can be achieved by a consistently working RIIM or not.

Definition 7. Let $T, F \subseteq U$ be finite sets. We define two 01-valued functions $econs_p(T)$ and $econs_c(T, F)$ as follows: $econs_p(T) = 1$ if and only if there exists a concept $L_i \in C$ such that T is consistent with L_i, and $econs_c(T, F) = 1$ if and only if there exists a concept $L_i \in C$ such that $\langle T, F \rangle$ is consistent with L_i.

For any concept $L_i \in C$, whether $T \subseteq L_i$ and $F \subseteq L_i^c$ or not is recursively decidable, because L_i is recursive, and T and F are explicitly given finite sets. Therefore in general, the above functions are recursively enumerable functions.

For the refutable inferability from positive data, we have the following characterizations.

Theorem 8. *For a class C, the following four statements are equivalent:*
- *C is refutably inferable from positive data.*
- *C is refutably and consistently inferable from positive data.*
- *C satisfies the following three conditions:*

   (3.1) C is inferable from positive data.

   (3.2) For any nonempty concept $L \notin C$, there is a finite set $T \subseteq L$ such that T is inconsistent with any concept $L_i \in C$.

   (3.3) The function $econs_p$ for C is recursive.
- *C satisfies the following three conditions:*

   (3.4) For any concept $L_i \in C$, all nonempty subsets of L_i are also members of C.

   (3.5) There is no infinite sequence of concepts $L_{i_1}, L_{i_2}, \cdots \in C$ such that $L_{i_1} \subsetneq L_{i_2} \subsetneq \cdots$.

   (3.3) The function $econs_p$ for C is recursive.

We can easily show that if a class C satisfies the conditions (3.4) and (3.5), then C contains no infinite concept.

Theorem 9. *(a) A class C is refutably and consistently inferable from complete data, if and only if C satisfies the following two conditions:*

(3.6) For any concept $L \notin C$, there are finite sets $T \subseteq L$ and $F \subseteq L^c$ such that $\langle T, F \rangle$ is inconsistent with any concept $L_i \in C$.

(3.7) The function $econs_c$ for C is recursive.

(b) A class C is refutably inferable from complete data, then C satisfies the above condition (3.6).

Corollary 10. *If a class C contains all nonempty finite concepts, then C is not refutably inferable from positive data or complete data.*

4 Comparisons with Other Identifications

In this section, by some distinctive examples of classes, we compare the criterion of refutable inference with some other criteria. This is motivated by the following question: What should we do if we face with facts that are inconsistent with a finitely inferred hypothesis?

In what follows, for the criterion of finite inference for a class of recursive concepts, please refer to Mukouchi[18], Lange&Zeugmann[12] and Kapur[10]. For the purpose of comparing the inferability from positive data with the inferability from complete data, we assume that all concepts are *nonempty* throughout this section.

Proposition 11. *If a class C is inferable in the limit from positive data, then C is inferable in the limit from complete data.*

Furthermore, the above assertion is still valid, if we replace the phrase 'inferable in the limit' with the phrase 'finitely inferable', 'reliably inferable' or 'refutably inferable'.

Theorem 12 (Lange&Zeugmann[11]). *If a class C is finitely inferable from complete data, then C is inferable in the limit from positive data.*

We can summarize the obtained comparisons as in Figure 1. In the illustration, the prefix 'LIM', 'FIN', 'REL' or 'REF' means the collection of classes that are 'inferable in the limit', 'finitely inferable', 'reliably inferable' or 'refutably inferable', respectively, and the postfix 'TXT' or 'INF' means from positive data or from complete data, respectively. For example, REF-TXT is the collection of all classes that are refutably inferable from positive data.

The classes $\mathcal{FC}_1$, $\mathcal{FC}_n$, $\mathcal{FC}_{\leq n}$ and $\mathcal{FC}_*$ consist of all finite concepts each of which cardinality is just 1, $n \geq 2$, at most n and unrestricted finite, respectively. The class $\mathcal{SFC}$ is the so-called superfinite class[9], that is, a class contains all finite concepts and at least one infinite concept.

The class $\mathcal{PAT}$ is the class of pattern languages (cf. Angluin[1, 2], Mukouchi[17]).

The classes $\mathcal{PR}_1$, $\mathcal{PR}_n$ and $\mathcal{PR}_{\leq n}$ consist of all multiples of a prime number, $n \geq 2$ distinct prime numbers and at most n prime numbers, respectively. For example, the concept $\{2, 4, 6, \ldots, 7, 14, 21, \ldots\}$ is a member of $\mathcal{PR}_{\leq i}$ with $i \geq 2$ and also a member of $\mathcal{PR}_2$, but is not a member of $\mathcal{PR}_i$ with $i = 1$ or $i \geq 3$. Note that $\mathcal{PR}_1 = \mathcal{PR}_{\leq 1}$ holds.

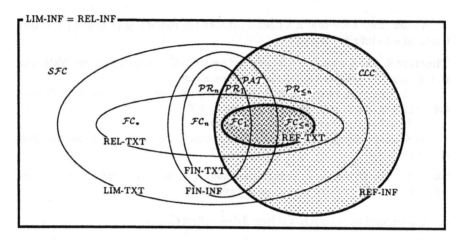

Fig. 1. Comparisons with other identifications

The class $CLC = \{L_i\}_{i\in N}$ is organized as follows: Put $\Sigma = \{a\}$, $L_1 = \{a^j \mid j \geq 1\}$ and $L_i = \{a^j \mid 1 \leq j \leq i-1\}$ for any $i \geq 2$. Note that in Lange&Zeugmann[13], this class was shown to be inferable from complete data with one mind change but not inferable from positive data.

By the above illustration, we see that a subclass of a refutably inferable class is not always refutably inferable.

5   Union of Some Classes

In this section, we consider two types of union classes. First we take a class as a collection of concepts from n classes.

Definition 13. Let $C_1,\ldots,C_n$ be classes of concepts ($n \geq 1$). For integers i with $1 \leq i \leq n$ and $j \geq 1$, by $L_{(i,j)}$, we denote the j-th concept L_j of the class C_i. Then the union class of $C_1,\ldots,C_n$ is represented:

$$\bigcup_{i=1}^{n} C_i = \{L_{(i,j)}\}_{1\leq i\leq n,\ j\in N}.$$

By assuming a bijective coding from $\{1,\ldots,n\} \times N$ to N, the above new class becomes an indexed family of recursive concepts.

Theorem 14. *Let $C_1,\ldots,C_n$ be classes each of which is refutably inferable from positive data (resp., complete data) ($n \geq 1$). Then the class $\bigcup_{i=1}^{n} C_i$ is refutably inferable from positive data (resp., complete data).*

Now we consider a class of concepts each of which is a union of at most n concepts from n classes (cf. Wright[29]).

Definition 15. Let $C_1,\ldots,C_n$ be classes of concepts ($n \geq 1$). For integers i with $1 \leq i \leq n$ and $j \geq 0$, by $L_{(i,j)}$, we denote the empty concept if $j = 0$, otherwise the j-th concept L_j of the class C_i. Then we define a class generated by $C_1,\ldots,C_n$ as follows:

$$\bigsqcup_{i=1}^{n} C_i \overset{\text{def}}{=} \left\{ \bigcup_{i=1}^{n} L_{(i,j_i)} \ \middle| \ (j_1,\ldots,j_n) \in N^n \right\},$$

where $\mathcal{N}^n$ is the set of all n-tuples of nonnegative integers, that is, $\mathcal{N}^n \stackrel{\text{def}}{=} \{0, 1, 2, \cdots\}^n$.

By assuming a bijective coding from $\mathcal{N}^n$ to N, the above new class becomes an indexed family of recursive concepts.

Theorem 16. *Let $C_1, \ldots, C_n$ be classes each of which satisfies the following condition (5.1) ($n \geq 1$):*

(5.1) For any $w \in U$, the set of indices $\{j \mid w \in L_j \in C_i\}$ is a recursively generable finite set.
Then the class $\bigsqcup_{i=1}^{n} C_i$ is refutably and consistently inferable from complete data.

Example 1. We consider the class of pattern languages. We can easily show that this class satisfies the condition (5.1). Therefore by Theorem 16, for any $n \geq 1$, the class of unions of at most n pattern languages is refutably and consistently inferable from complete data.

By Corollary 10, we see that if the number of patterns is not bounded by a finite number, then the class is not refutably inferable from complete data, because it contains all nonempty finite languages.

Note that the class of unions of at most n pattern languages is shown to be inferable from positive data (cf. Wright[29], Shinohara[25, 26]).

6 EFS Definable Classes

In this section, we consider the so-called model inference (cf. Shapiro[23]) and language learning using *elementary formal systems* (*EFS*'s, for short).

The EFS's are originally introduced by Smullyan[28] to develop his recursion theory. In a word, EFS's are a kind of logic programming language which uses strings instead of terms in first order logic[30], and is shown to be a natural device to define languages[3].

In this paper, we briefly recall EFS's. For detailed definitions and properties of EFS's, please refer to Smullyan[28], Arikawa[3], Arikawa et al.[4] and Yamamoto[30].

Let Σ, X and Π be mutually disjoint nonempty sets. We assume that Σ is finite, and fix it throughout this section. Elements in Σ, X and Π are called constant symbols, variables and predicate symbols, respectively. By $p, q, p_1, p_2, \cdots$, we denote predicate symbols. Each predicate symbol is associated with a positive integer which we call an *arity*. In general, for a set S, by S^+, we denote the set of all nonempty finite strings over S, and by $|S|$, we denote the cardinality of S.

Definition 17. A *term*, or a *pattern*, is an element in $(\Sigma \cup X)^+$, that is, it is a nonnull string over $(\Sigma \cup X)$. By $\pi, \pi_1, \pi_2, \cdots$, we denote terms. A term π is said to be *ground*, if $\pi \in \Sigma^+$. By $w, w_1, w_2, \cdots$, we denote ground terms.

An *atomic formula* (*atom*, for short) is an expression of the form $p(\pi_1, \ldots, \pi_n)$, where p is a predicate symbol with arity n, and $\pi_1, \ldots, \pi_n$ are terms. By $A, B, A_1, A_2, \cdots$, we denote atoms. An atom $p(\pi_1, \ldots, \pi_n)$ is said to be *ground*, if $\pi_1, \ldots, \pi_n$ are ground terms.

We define well-formed formulas and clauses in the ordinal ways[14].

Definition 18. A *definite clause* is a clause of the form

$$A \leftarrow B_1, \ldots, B_n,$$

where $n \geq 0$, and $A, B_1, \ldots, B_n$ are atoms.

Then an *EFS* is a finite set of definite clauses, each of which is called an *axiom*.

A *substitution* is a homomorphism from terms to terms which maps each symbol $a \in \Sigma$ to itself.

In the world of EFS's, the *Herbrand base* (*HB*, for short) is the set of all ground atoms, and we also define *Herbrand interpretation*, *Herbrand model*, and *the least Herbrand model* in the ordinal ways[14].

For an EFS Γ, by $\mathcal{M}(\Gamma)$, we denote the least Herbrand model, and for an EFS Γ and a predicate symbol p with arity n, we define the set of n-tuples of ground terms as follows:

$$L(\Gamma, p) \stackrel{\text{def}}{=} \{(w_1, \ldots, w_n) \in (\Sigma^+)^n \mid p(w_1, \ldots, w_n) \in \mathcal{M}(\Gamma)\}.$$

In case the arity of p is 1, we regard $L(\Gamma, p)$ as a language over Σ.

Now we put a syntactical restriction on EFS's, because the least Herbrand model $\mathcal{M}(\Gamma)$ for an unrestricted EFS Γ may not be recursive, that is, for a ground atom A, we can not recursively decide whether $A \in \mathcal{M}(\Gamma)$ or not.

For a term π, by $\|\pi\|$, we denote the length of π, and by $o(x, \pi)$, the number of all occurrences of a variable x in π. For an atom $p(\pi_1, \ldots, \pi_n)$, we define the length of the atom and the number of variable's occurrences in the atom as follows:

$$\|p(\pi_1, \ldots, \pi_n)\| \stackrel{\text{def}}{=} \|\pi_1\| + \cdots + \|\pi_n\|,$$
$$o(x, p(\pi_1, \ldots, \pi_n)) \stackrel{\text{def}}{=} o(x, \pi_1) + \cdots + o(x, \pi_n).$$

Definition 19. A clause $A \leftarrow B_1, \ldots, B_n$ is said to be *length-bounded*, if

$$\|A\theta\| \geq \|B_1\theta\| + \cdots + \|B_n\theta\|$$

for any substitution θ.

An EFS Γ is said to be length-bounded, if all axioms of Γ are length-bounded.

The notion of length-bounded clauses is characterized by the following Lemma 20.

Lemma 20 (Arikawa et al.[4]). *A clause $A \leftarrow B_1, \ldots, B_n$ is length-bounded, if and only if $\|A\| \geq \|B_1\| + \cdots + \|B_n\|$ and $o(x, A) \geq o(x, B_1) + \cdots + o(x, B_n)$ hold for any variable x.*

From now on, we only consider length-bounded EFS's. For length-bounded EFS's, the following Theorem 21 holds.

Theorem 21 (Arikawa et al.[4], Yamamoto[30]). *For a length-bounded EFS Γ, the least Herbrand model $\mathcal{M}(\Gamma)$ is recursive.*

Furthermore, the following Theorem 22 shows the power of length-bounded EFS's.

Theorem 22 (Arikawa et al.[4]). *A language $L \subseteq \Sigma^+$ is context-sensitive, if and only if the language L is definable by a length-bounded EFS.*

We devote the rest of this section to investigating the refutable inferability of length-bounded EFS definable classes.

Definition 23. For an atom A, by $pred(A)$, we denote the predicate symbol of A.

For a set $\Pi_0 \subseteq \Pi$, by $HB[\Pi_0]$, we denote the set of all ground atoms whose predicate symbols are members of Π_0, that is,

$$HB[\Pi_0] \stackrel{\text{def}}{=} \left\{ p(w_1, \ldots, w_n) \;\middle|\; \begin{array}{l} p \text{ is a predicate symbol in } \Pi_0 \text{ with arity } n, \text{ and} \\ w_1, \ldots, w_n \in \Sigma^+ \end{array} \right\}.$$

For a set $\Pi_0 \subseteq \Pi$ and a set S of atoms, by $S|_{\Pi_0}$, we denote the set of atoms restricted to Π_0, that is,

$$S|_{\Pi_0} \stackrel{\text{def}}{=} \{A \in S \mid pred(A) \in \Pi_0\}.$$

Then we adapt the definitions of inferability to the case of EFS's as follows, but as easily seen, the essential part is kept unchanged from Definition 3. In what follows, we assume that outputs from an RIIM are EFS's, and that an RIIM works on complete data, if not specified.

Definition 24. A *predicate-restricted presentation* of a set $I \subseteq HB[\Pi]$ w.r.t. $\Pi_0 \subseteq \Pi$ is an infinite sequence $(A_1, t_1), (A_2, t_2), \cdots$ of elements in $HB[\Pi_0] \times \{+, -\}$ such that $\{A_i \mid t_i = +, i \geq 1\} = I|_{\Pi_0}$ and $\{A_i \mid t_i = -, i \geq 1\} = HB[\Pi_0] \setminus I|_{\Pi_0}$.

An RIIM M is said to *converge to an EFS Γ* for a presentation σ, if there is a positive integer n such that for any $m \geq n$, $M(\sigma[m])$ is defined and equal to Γ.

Let $\mathcal{EC}$ be a class of EFS's. For an EFS $\Gamma \in \mathcal{EC}$ and a predicate-restricted presentation σ of $\mathcal{M}(\Gamma)$ w.r.t. $\Pi_0 \subseteq \Pi$, an RIIM M is said to *infer the EFS Γ w.r.t. $\mathcal{EC}$ from σ*, if M converges to an EFS $\Gamma' \in \mathcal{EC}$ with $\mathcal{M}(\Gamma')|_{\Pi_0} = \mathcal{M}(\Gamma)|_{\Pi_0}$ for σ.

A class $\mathcal{EC}$ is said to be *theoretical-term-freely and refutably inferable*, if for any nonempty finite subset Π_0 of Π, there is an RIIM M which satisfies the following condition: For any set $I \subseteq HB[\Pi]$ and any predicate-restricted presentation σ of I w.r.t. Π_0, (i) if there is an EFS $\Gamma \in \mathcal{EC}$ such that $\mathcal{M}(\Gamma)|_{\Pi_0} = I|_{\Pi_0}$, then M infers Γ w.r.t. $\mathcal{EC}$ from σ, (ii) otherwise M refutes the class $\mathcal{EC}$ from σ.

Theoretical terms are supplementary predicates that are necessary for defining some goal predicates. In the above definition, the phrase 'theoretical-term-freely inferable' means that *from only the facts on the goal predicates*, an RIIM generates some suitable predicates and infers an EFS which explains the goal predicates.

Definition 25. For EFS's Γ and Γ' and for a set $\Pi_0 \subseteq \Pi$, we write $\Gamma \equiv_{\Pi_0} \Gamma'$, if we can make Γ' identical to Γ by renaming predicate symbols other than those in Π_0 and by renaming variables in each axiom. We assume some canonical form of an EFS w.r.t. Π_0, and by $canon(\Gamma, \Pi_0)$, we denote the representative EFS for the set of EFS's $\{\Gamma' \mid \Gamma \equiv_{\Pi_0} \Gamma'\}$.

For an EFS Γ, by $HPRED(\Gamma)$ (resp., $BPRED(\Gamma)$), we denote the set of all predicate symbols appearing in the heads (resp., the bodies) of the axioms of Γ. We also define various sets as follows:

$$LB^{[n]} \stackrel{\text{def}}{=} \{\Gamma \mid \Gamma \text{ is a length-bounded EFS with } n \text{ axioms}\},$$
$$LB^{[n]}[\![\Pi_0]\!] \stackrel{\text{def}}{=} \{canon(\Gamma, \Pi_0) \mid \Gamma \in LB^{[n]}\},$$

$$MLB^{[n]}[\![\Pi_0]\!] \stackrel{\text{def}}{=} \{\Gamma \in LB^{[n]}[\![\Pi_0]\!] \mid BPRED(\Gamma) \subseteq HPRED(\Gamma)\},$$

$$MLB^{[n]}[\![\Pi_0]\!](l) \stackrel{\text{def}}{=} \left\{\Gamma \in MLB^{[n]}[\![\Pi_0]\!] \;\middle|\; \begin{array}{l}\text{the head's length of each axiom}\\ \text{of } \Gamma \text{ is not greater than } l.\end{array}\right\},$$

$$LB^{[\leq n]} \stackrel{\text{def}}{=} \bigcup_{i=0}^{n} LB^{[i]}, \qquad MLB^{[\leq n]}[\![\Pi_0]\!] \stackrel{\text{def}}{=} \bigcup_{i=0}^{n} MLB^{[i]}[\![\Pi_0]\!],$$

$$MLB^{[\leq n]}[\![\Pi_0]\!](l) \stackrel{\text{def}}{=} \bigcup_{i=0}^{n} MLB^{[i]}[\![\Pi_0]\!](l),$$

where $l, n \geq 0$, and $\Pi_0 \subseteq \Pi$.

Note that for any EFS's Γ and Γ' and for any set $\Pi_0 \subseteq \Pi$, whether $\Gamma \equiv_{\Pi_0} \Gamma'$ or not is effectively decidable, and that we can effectively obtain the EFS $canon(\Gamma, \Pi_0)$.

We prepare some basic lemmas.

Lemma 26 (Shinohara[26]). *Let Γ be a length-bounded EFS and A be a member of $\mathcal{M}(\Gamma)$. Then if Γ have an axiom D whose head is longer than A, then A is also a member of $\mathcal{M}(\Gamma \setminus \{D\})$.*

Lemma 27. *For any EFS $\Gamma \in LB^{[\leq n]}$ and any set $\Pi_0 \subseteq \Pi$, there is an EFS $\Gamma' \in MLB^{[\leq n]}[\![\Pi_0]\!]$ such that $\mathcal{M}(\Gamma')|_{\Pi_0} = \mathcal{M}(\Gamma)|_{\Pi_0}$.*

Lemma 28. *Let $n \geq 1$ and $\Pi_0 \subseteq \Pi$, and let $T \subseteq HB[\Pi_0]$ be a nonempty finite set and $F \subseteq HB[\Pi_0]$ be a finite set. Assume that $\langle T, F \rangle$ is inconsistent with $\mathcal{M}(\Gamma)|_{\Pi_0}$ for any EFS $\Gamma \in MLB^{[n-1]}[\![\Pi_0]\!]$.*

Then for any EFS $\Gamma \in MLB^{[n]}[\![\Pi_0]\!] \setminus MLB^{[n]}[\![\Pi_0]\!](l)$, $\langle T, F \rangle$ is inconsistent with $\mathcal{M}(\Gamma)|_{\Pi_0}$, where $l = \max\{\|A\| \mid A \in T\}$.

Lemma 29. *For any $l, n \geq 0$ and any finite subset Π_0 of Π, the set $MLB^{[n]}[\![\Pi_0]\!](l)$ is a recursively generable finite set.*

Lemma 30. *Let $L, L_1, \ldots, L_n$ $(n \geq 1)$ be concepts with $L \neq L_i$ $(1 \leq i \leq n)$.*

Then for any complete presentation σ of L, there is a positive integer m such that $\sigma[m]$ is inconsistent with any concept L_i $(1 \leq i \leq n)$.

Theorem 31. *For any $n \geq 0$, the class $LB^{[\leq n]}$ is theoretical-term-freely, refutably and consistently inferable.*

Proof. Let $\Pi_0 \subseteq \Pi$ be a nonempty finite set. We consider the procedure RIIM(n, Π_0) as shown in Figure 2.

Claim: In the procedure RIIM, for any m with $0 \leq m \leq n$, if T_m and F_m are defined, then $\langle T_m, F_m \rangle$ is inconsistent with $\mathcal{M}(\Gamma)|_{\Pi_0}$ for any EFS $\Gamma \in MLB^{[m]}[\![\Pi_0]\!]$.

The proof of this claim is given by a mathematical induction on m.

(I) In case $m = 0$. It is clear because T_0 is nonempty and the least Herbrand model of the empty EFS is empty.

(II) In case $m \geq 1$. We assume the claim for $(m - 1)$, and assume that T_m and F_m are defined. Then we see that T_{m-1} and F_{m-1} are also defined, and by the induction hypothesis, $\langle T_{m-1}, F_{m-1} \rangle$ is inconsistent with $\mathcal{M}(\Gamma)|_{\Pi_0}$ for any EFS $\Gamma \in MLB^{[m-1]}[\![\Pi_0]\!]$. Therefore by Lemma 28, $\langle T_{m-1}, F_{m-1} \rangle$ is inconsistent with $\mathcal{M}(\Gamma)|_{\Pi_0}$ for any EFS $\Gamma \in MLB^{[m]}[\![\Pi_0]\!] \setminus MLB^{[m]}[\![\Pi_0]\!](l_m)$. Thus $\langle T_m, F_m \rangle$ is inconsistent with $\mathcal{M}(\Gamma)|_{\Pi_0}$ for any EFS $\Gamma \in MLB^{[m]}[\![\Pi_0]\!] \setminus MLB^{[m]}[\![\Pi_0]\!](l_m)$, because $T_{m-1} \subseteq T_m$ and $F_{m-1} \subseteq F_m$ hold.

Procedure RIIM(n, Π_0);
begin
  $T = \phi; F = \phi$;
  read_store(T, F);
  if $T = \phi$ **then begin**
    output the empty EFS;
    while $T = \phi$ **do** read_store(T, F);   .... (1)
  end;
  $T_0 = T; F_0 = F$;
  for $m = 1$ **to** n **do begin**
    $l_m = \max\{\|A\| \mid A \in T_{m-1}\}$;
    recursively generate $MLB^{[m]}[\![\Pi_0]\!](l_m)$, and set it to S;
    for each $\Gamma \in S$ **do**   .... (2)
      if $\langle T, F \rangle$ is consistent with $\mathcal{M}(\Gamma)|_{\Pi_0}$ **then begin**
        output Γ;
        while $\langle T, F \rangle$ is consistent with $\mathcal{M}(\Gamma)|_{\Pi_0}$ **do**   .... (3)
          read_store(T, F);
      end;
    $T_m = T; F_m = F$;
  end;
  refute the class $LB^{[\le n]}$ and stop;
end;

Procedure read_store(T, F);
begin
  read the next fact (w, t);
  if $t = $'+' **then** $T = T \cup \{w\}$ **else** $F = F \cup \{w\}$;
end.

Fig. 2. An RIIM for the class $LB^{[\le n]}$

Furthermore, since the for-loop (2) terminates, we see that $\langle T_m, F_m \rangle$ is also inconsistent with $\mathcal{M}(\Gamma)|_{\Pi_0}$ for any EFS $\Gamma \in MLB^{[m]}[\![\Pi_0]\!](l_m)$.

Hence we have the claim for m. Q.E.D. of the claim.

(A) Let $\Gamma_{\text{base}} \in LB^{[\le n]}$. Then assume that we present a predicate-restricted presentation σ of $\mathcal{M}(\Gamma_{\text{base}})$ w.r.t. Π_0.

By Lemma 27, there is an EFS $\Gamma \in MLB^{[\le n]}[\![\Pi_0]\!]$ such that $\mathcal{M}(\Gamma)|_{\Pi_0} = \mathcal{M}(\Gamma_{\text{base}})|_{\Pi_0}$. Therefore we see by the above claim that T_n and F_n are never defined. By Lemma 30, this means that the procedure outputs an EFS Γ with $\mathcal{M}(\Gamma)|_{\Pi_0} = \mathcal{M}(\Gamma_{\text{base}})|_{\Pi_0}$ and never terminates the while-loop (1) or (3).

(B) Let I be a subset of $HB[\Pi_0]$ with $I \ne \mathcal{M}(\Gamma)|_{\Pi_0}$ for any EFS $\Gamma \in LB^{[\le n]}$. When we present a predicate-restricted presentation σ of I w.r.t. Π_0, we see by using Lemma 30 n times that the procedure RIIM refutes the class $LB^{[\le n]}$.

Furthermore, it is easy to see that the procedure RIIM works consistently.   □

By Corollary 10, we see that if the number of axioms is not bounded by a finite number, then this class is not refutably inferable, because this class contains all finite concepts on $HB[\Pi_0]$.

Corollary 32. *For any $n \ge 0$, the class of languages definable by EFS's in $LB^{[\le n]}$ is refutably and consistently inferable from complete data.*

Note that Shinohara[26] showed that the classes in Theorem 31 or Corollary 32 are inferable from positive data. Moreover Moriyama&Sato[16] investigated some conditions for a class of EFS's to be inferable from positive data.

7 Concluding Remarks

We have pointed out that the essence of the computational logic of scientific discovery or the logic of machine discovery should be the refutability of the whole space of hypotheses by observed data or given facts. Then we have shown a series of such sufficiently large hypothesis spaces in terms of the elementary formal systems. More exactly, for any n the class of length-bounded EFS's with at most n axioms are refutably identifiable. The argument to prove this is also valid for the classes of weakly reducing EFS's, and linear or weakly reducing logic programs[7] with at most n axioms.

The refutability we proposed here forms an interesting contrast to the original one: In the logic of scientific discovery for scientists *each theory* in a hypothesis space is to be *refutable*, while in the computational version of the logic for machines or the logic of machine discovery the *space* itself is to be *refutable*.

The inductive inference machine that can refute the hypothesis space itself works as an automatic system for scientific discovery. If the machine for scientific discovery can not refute the whole space of hypotheses, it can just work for *computer aided* scientific discovery. That is, we need to check from time to time whether the machine is still searching for a possible hypothesis. In fact, in our previous works, we obtained very practical results in the area of Molecular Biology based on our theoretical studies on learnability[5, 6, 24]. Our learning system succeeded in discovering some simple and accurate knowledge, i.e. hypotheses, in a very short time. However the hypotheses the learning system found out happened to be in the spaces we initially gave. In fact, the spaces are natural subclasses of the refutably inferable classes we have shown in the last section.

References

1. Angluin, D.: *Finding patterns common to a set of strings*, Proc. 11th Annual Symposium on Theory of Computing (1979) 130–141.
2. Angluin, D.: *Inductive inference of formal languages from positive data*, Information and Control **45** (1980) 117–135.
3. Arikawa, S.: *Elementary formal systems and formal languages – simple formal systems*, Memoirs of Fac. Sci., Kyushu Univ. Ser. A, Math. **24** (1970) 47–75.
4. Arikawa, S., Shinohara, T. and Yamamoto, A.: *Learning elementary formal systems*, Theoretical Computer Science **95** (1992) 97–113.
5. Arikawa, S., Kuhara, S., Miyano S., Shinohara, A. and Shinohara, T.: *A learning algorithm for elementary formal systems and its experiments on identification of transmembrane domains*, Proc. 25th Hawaii Inter. Conf. System Sciences (1992) 675–684.
6. Arikawa, S., Kuhara, S., Miyano S., Mukouchi, Y., Shinohara, A. and Shinohara, T.: *A machine discovery from amino acid sequences by decision trees over regular patterns*, Proc. FGCS'92 (1992) 618–625.
7. Arimura, H.: *Completeness of depth-bounded resolution for weakly reducing programs*, Proc. Software Science and Engineering (1991) (World Scientific Series in Computer Science Vol. 31) 227–245.

8. Blum, L. and Blum, M.: *Toward a mathematical theory of inductive inference*, Information and Control **28** (1975) 125–155.

9. Gold, E.M.: *Language identification in the limit*, Information and Control **10** (1967) 447–474.

10. Kapur, S.: *Monotonic language learning*, Proc. 3rd Workshop on Algorithmic Learning Theory (1992) 147–158.

11. Lange, S. and Zeugmann, T.: *Monotonic versus non-monotonic language learning*, Proc. 2nd Inter. Workshop on Nonmonotonic and Inductive Logic (1991) (Lecture Notes in Artificial Intelligence, Vol. 659, 254–269).

12. Lange, S. and Zeugmann, T.: *Types of monotonic language learning and their characterization*, Proc. 5th Workshop on Comput. Learning Theory (1992) 377–390.

13. Lange, S. and Zeugmann, T.: *Learning recursive languages with bounded mind changes*, GOSLER-Report 16/92, FB Mathematik und Informatik, TH Leipzig, Sept. 1992.

14. Lloyd, J.W.: *Foundations of logic programming*, Springer-Verlag, 1984.

15. Minicozzi, E.: *Some natural properties of strong-identification in inductive inference*, Theoretical Computer Science **2** (1976) 345–360.

16. Moriyama, T. and Sato, M.: *Properties of language classes with finite elasticity*, to appear in Proc. 4th Workshop on Algorithmic Learning Theory (1993).

17. Mukouchi, Y.: *Characterization of pattern languages*, Proc. 2nd Workshop on Algorithmic Learning Theory (1991) 93–104 (also in IEICE Trans. Inf. &Syst., Vol. E75–D, No. 4 1992, 420–425).

18. Mukouchi, Y.: *Characterization of finite identification*, Proc. 3rd Inter. Workshop on Analogical and Inductive Inference (1992) (Lecture Notes in Artificial Intelligence Vol. 642) 260–267.

19. Mukouchi, Y. and Arikawa, S.: *Inductive Inference machines that can refute hypothesis spaces*, RIFIS-TR-CS-67, Research Institute of Fundamental Information Science, Kyushu University, March 1993.

20. Popper, R. K.: "Conjectures and refutations," Routledge and Kegan Paul, 1963.

21. Popper, R. K.: "The logic of scientific discovery," Harper&Row, 1965.

22. Sakurai, A.: *Inductive inference of formal languages from positive data enumerated primitive-recursively*, Proc. 2nd Workshop on Algorithmic Learning Theory (1991) 73–83.

23. Shapiro, E.Y.: *Inductive inference of theories from facts*, Technical Report 192, Yale University Computer Science Dept., 1981.

24. Shimozono, S., Shinohara, A., Shinohara, T., Miyano S., Kuhara S. and Arikawa S.: *Finding alphabet indexing for decision trees over regular patterns: an approach to bioinformatical knowledge acquisition*, Proc. 26th Hawaii Inter. Conf. System Sciences (1993) 763–772.

25. Shinohara, T.: *Inferring unions of two pattern languages*, Bull. of Informatics and Cybernetics **20** (1983) 83–87.

26. Shinohara, T.: *Inductive inference from positive data is powerful*, Proc. 3rd Workshop on Comput. Learning Theory (1990) 97–110.

27. Shrager, J. and Langley, P. (Editors): "Computational models of scientific discovery and theory formation," Morgan Kaufmann, 1990.

28. Smullyan, R.M.: "Theory of formal systems," Princeton Univ. Press, 1961.

29. Wright, K.: *Identification of unions of languages drawn from an identifiable class*, Proc. 2nd Workshop on Comput. Learning Theory (1989) 328–333.

30. Yamamoto, A.: *Procedural semantics and negative information of elementary formal system*, J. Logic Programming **13** No. 4 (1992) 89–98.

On the duality between mechanistic learners and what it is they learn*

Rūsiņš Freivalds[1] and Carl H. Smith[2],**

[1] Institute of Mathematics and Computer Science
University of Latvia
Raiņa bulvāris 29
LV-1459, Riga, Latvia
[2] Department of Computer Science
University of Maryland
College Park, MD 20912 USA

Abstract. All previous work in inductive inference and theoretical machine learning has taken the perspective of looking for a learning algorithm that successfully learns a collection of functions. In this work, we consider the perspective of starting with a set of functions, and considering the collection of learning algorithms that are successful at learning the given functions. Some strong dualities are revealed.

1 Introduction

One characteristic that is shared by all current theoretical approaches to machine learning is that the investigations are focused on the capabilities of a machine, or a class of machines. Typical results constructively exhibit a machine that can learn a collection of concepts, subject to some constraints. Sometimes these constraints concern complexity issues and sometimes they determine a criterion of successful learning [7, 1, 2, 3, 4, 5]. In this paper we investigate the dual notion that arises by asking the question: Given some set of concepts, which algorithms can learn all those concepts.

---

*This work was facilitated by an international agreement under NSF Grant 9119540.
†The second author was supported in part by NSF Grant 9020079.

From [6] we know that in the theory of inductive inference sometimes one concept must be mastered before an attempt to learn some other concept can be initiated. This observation is consistent with commonly observed human learning behavior. The intuition that we start from is that there are some things that are clearly easier to learn than others. For example, learning how to walk is something that all but a few of us humans masters very early in life. A much smaller number of us learn how to drive a car, while an even fewer number of us learn how to pilot an airplane. Although there may be some counterexamples, it seems safe to assert that anyone who can pilot an aircraft, also has learned how to operate an automobile. Based on this example, one would expect that for some concepts, there are other concepts such that any learning algorithm that can learn the second concept can also learn the first, and, there are some algorithms that can learn the first concept but not the second. Our results indicate that this is indeed the case, but only if "concept" is replaced by a suitable infinite set of concepts.

Below we present several instances where there is a strong duality between the notions of learning machine and set of functions that are learnable. For example, suppose S_1 and S_2 are sets of recursive functions and A_1 is the collection of all algorithms that can learn all the functions in S_1. Similarly, define A_2 to be the collection of all algorithms that can learn all the functions in S_2. One of our results is that the collection of algorithms that can learn all of the functions in $S_1 \cup S_2$ is precisely $A_1 \cap A_2$. By delving a little deeper, some cases where the duality fails are revealed. We continue with some notation and definitions.

2 Technical Preliminaries

As in traditional studies of inductive inference [7, 8], we model concepts as a set of stimulus and response pairs. Assuming that any concept associates only one response with each possible stimulus, it can be viewed as a *function* from stimuli to responses. It is possible to encode every string of ASCII symbols in the natural numbers. These strings include arbitrarily long texts and are certainly sufficient to express both stimuli and responses. For the purposes of a mathematical treatment of learning, it suffices to consider only the functions from natural numbers to natural numbers. Via suitable encodings, these functions can represent a wide range of phenomena. Since we will be concerned only with effective learning procedures, the recursive functions are used to model concepts. The examples then are ordered pairs $(x, f(x))$ from the graph of some function f. A program for computing the function f will be considered an explanation of f. By using standard encoding techniques, the natural numbers ($\mathbb{N}$) will serve as names for programs [14]. The function computed by program p will be denoted by φ_p. The collection of functions $\varphi_0, \varphi_1, \varphi_2, \cdots$, forms an *acceptable programming system*. Subset will be denoted by $\subseteq$ and $\subset$ will denote proper subset.

Inductive inference, the abstract study of generalization, has been well studied. We proceed by reviewing a few of the basic notions. Gold, in a seminal paper [13], defined the notion called *identification in the limit*. This definition concerned learning by algorithmic devices now called *inductive inference machines* (IIMs). An IIM inputs the range of a partial recursive function, an ordered pair at a time, and, while doing so, outputs computer programs. When discussing the inference of (total) recursive functions, we will assume, without loss of generality, that the input is received by an IIM in its natural domain increasing order, $f(0)$, $f(1)$, $\cdots$. An IIM, on input from a function f will output a potentially infinite sequence of programs p_0, p_1, $\cdots$. The IIM *converges* if either the sequence is finite, say of length $n+1$, or there is program p such that for all but finitely many i, $p_i = p$. In the former case we say the IIM converges to p_n, and in the latter case, to p. In general, there is no effective way to tell when, and if, an IIM has converged. The fact that an IIM M converged to p on input from the graph of f will be denoted by $M(f) \downarrow p$.

Following Gold, we say that an IIM M *identifies* a function f (written: $f \in EX(M)$), if, when M is given the range of f as input, it converges to a program p that computes f ($\varphi_p = f$). The "EX" stands for "explain" as the program p is an explanation of the function f since it can be used to predict any value in the range of f. If an IIM identifies some function f, then some form of learning must have taken place, since, by the properties of convergence, only finitely much of the range of f was known by the IIM at the (unknown) point of convergence. The terms *infer* and *learn* are commonly used as synonyms for identify. Each IIM will learn some set of partial recursive functions. The collection of all such sets, over the universe of effective algorithms viewed as IIMs, serves as a characterization of the learning power inherent in the Gold model. This collection is symbolically denoted by EX (for explanation) and is defined rigorously by $EX = \{S \mid \exists M(S \subseteq EX(M))\}$. From [9] we may assume that the *order* in which data is presented does not affect the composition of the class EX.

As an example, consider learning all the polynomial functions of a single variable with integer coefficients. The first step is to build an enumeration of all the polynomials in stages. At stage s (for $s \geq 0$) begin by listing all the $(s+1)$ tuples of integers ranging from 0 to s. Each tuple is then used as the coefficients of the terms 1, n, n^2, $\cdots$, n^s, making a polynomial. An inference machine starts by conjecturing the first polynomial on the list. The IIM then starts reading data. Every time a data item is read which contradicts the current hypothesis, the IIM abandons this conjecture and simply guesses the next polynomial in the list that is consistent with the data seen so far. Every incorrect hypothesis is eventually rejected, and a correct one is never abandoned.

A well studied modification of the above definition is to restrict the IIM to make a single conjecture. This has the effect of rendering the IIMs nonlimiting, they observe data and then produce a single output. The resulting class is called FIN in [12] and EX_0 in [10]. Here we will say $f \in FEX(M)$ if M, when given the range of f as input, outputs a single program that computes f. The class FEX is defined analogously to the class EX. Both the previously mentioned papers show that $FEX \subset EX$. If $f \in FEX(M)$ then we will say that M finitely identifies f.

We now proceed to give the definitions that are new to this paper. Given a set S of recursive functions, $L_{EX}(S)$ denotes the collection of all IIMs that can learn, in the EX sense, all the functions in S. In symbols, $L_{EX}(S) = \{M \mid S \subseteq EX(M)\}$. In the discussion that follows, we will have cause to consider *singleton* sets S, sets with a single member. In an analogous fashion, the classes $L_{FEX}(S)$ can be defined. For ease of notation, we will abbreviate $L_{EX}(S)$ simply as $L(S)$.

Another useful concept to define is the notion of a *focus* of a set of IIMs. For $\mathcal{M}$ a set of IIMs, the focus of $\mathcal{M}$ (written $F_{EX}(\mathcal{M})$) is the set of all recursive functions that can be learned by all the IIMs in $\mathcal{M}$. Again, for each set of criteria for successful inference, there is a notion of focus and we will omit the subscript "EX" when we are discussing the focus with respect to the EX criteria of successful learning. Hence,

$$F(\mathcal{M}) = \left\{ \bigcap_{M \in \mathcal{M}} EX(M) \right\}.$$

By definition, the focus of the empty set of IIMs is all of the recursive functions, as every machine in the empty set vacuously learns all the recursive functions. Notice that for any set of functions S, $F(L(S)) = S$.

Some notational conventions will be followed for the rest of this paper. f and g will be used to represent recursive functions and φ and ψ will be used to represent partial recursive functions. M will denote an algorithm that is viewed as an IIM. The notation "$M(\sigma)$" stands for the last hypothesis produced by the IIM M after processing all of the finite function σ as input data, but prior to requesting additional data.

3   On the Learning of Single Functions

To inaugurate our investigations we will consider the collection of IIMs that can learn single functions. From this, some of the basic properties of our new definitions will be revealed. It will also come out that we must consider learning infinite classes of functions in order to make statements like "learning X is harder than learning Y." Mathematically, the reason this is so is because of the *idiot savants*, IIMs that are oblivious to their input and only output a predetermined program. In fact, for every program p there is an IIM, that we will call M_p, that ignores its input and outputs only p. Clearly, $EX(M_p) = \{\varphi_p\}$. The first results in this section concern the structure of the classes $L(\{f\})$ and $L_{FEX}(\{f\})$ as f various. Then we consider how the relationship between $L(\{f\})$ and $L_{FEX}(\{f\})$ compares with the relationship between EX and FEX.

3.1   Basic Structural Properties

Theorem 1 *Suppose φ and ψ are partial recursive functions. Then either $[L(\{\varphi\}) = L(\{\psi\})]$ or $[L(\{\varphi\}) \cap L(\{\psi\}), L(\{\varphi\}) - L(\{\psi\})$ and $L(\{\psi\}) - L(\{\varphi\})$ are all infinite].*

Notice that in Case 1 of the proof of Theorem 1 the IIM used outputs only a single conjecture. In both Case 2 and Case 3, the IIM used may possibly output a second conjecture. This component of the proof is absolutely necessary, as indicated by following result. Our next result shows that no analogue of Theorem 1 holds for the criteria FIN

Theorem 2 *There exist partial recursive functions φ and ψ such that $L_{FEX}(\{\varphi\})$ $\cap L_{FEX}(\{\psi\}) = \emptyset$.*

For the notion of the inference of partial functions, we demand that, for success, the final program discovered by an IIM compute the input function exactly. In [9], an alternative definition was considered, where for successful inference the IIM was required to converge to a program computing an *extension* of the function serving as input. We say that an IIM M *at least learns* a partial recursive function ψ (written: $\psi \in \widehat{EX}(M)$) if $M(\psi) \downarrow p$ and $\varphi_p(x) = \psi(x)$, for all x in the domain of ψ. The class $\widehat{EX}$ is defined in the usual manner, $\widehat{EX} = \{S \mid \exists M(S \subseteq \widehat{EX}(M))\}$. The class $\widehat{FEX}$ is defined in an analogous manner. Clearly, $FEX \subseteq \widehat{FEX}$ and $EX \subseteq \widehat{EX}$.

By Theorem 1, the analogue of Theorem 2 does not hold for the EX identification type. Below we show that the analogue of Theorem 2 fails for identification types $\widehat{FEX}$ and $\widehat{EX}$ as well. Both of these results use the *amalgamation technique* from [10]. The amalgamation of programs $p_1, p_2, \cdots, p_n$ and a finite function σ is a program, denoted by $A_\sigma(p_1, \cdots, p_n)$, that executes the following algorithm.

Begin Program $A_\sigma(p_1, \cdots, p_n)$.

On input x, simulate the executions of each program p_j $(1 \leq j \leq n)$ on all arguments z in the domain of σ for t steps, where t is the length of σ.

If it is discovered that program p_j, for some j, converges on some input z from the domain of σ and $\varphi_{p_j}(z) \neq \sigma(z)$, then eliminate program p_j from further consideration.

Dovetail the computations, on argument x, of all programs from p_1, $\cdots$, p_n that were not eliminated by the previous step. If program p_j is the first program that is discovered to converge on input x, then output the value $\varphi_{p_j}(x)$. If no program in the remaining set converges on input x, then program $A_\sigma(p_1, \cdots, p_n)$ also diverges on input x.

End Program $A_\sigma(p_1, \cdots, p_n)$.

The basic property of amalgamation that is used in our arguments is encapsulated in the following Lemma.

Lemma 3 *[11]. For any finite set of programs P, and any sequence of finite functions $\sigma_0 \subseteq \sigma_1 \subseteq \sigma_2 \subseteq \cdots$, the sequence of programs $A_{\sigma_0}(P)$, $A_{\sigma_1}(P)$, $A_{\sigma_2}(P)$, $\cdots$ converges to a program that computes a function that agrees with σ_i, for all i, provided such a program is in P.*

Finally, we get to the results we are after. Below we prove analogues of Theorem 1 for the identification types just defined.

Theorem 4 *Suppose φ and ψ are partial recursive functions. Then either $[L_{\widehat{EX}}(\{\varphi\}) = L_{\widehat{EX}}(\{\psi\})]$ or $[L_{\widehat{EX}}(\{\varphi\}) \cap L_{\widehat{EX}}(\{\psi\})$, $L_{\widehat{EX}}(\{\varphi\}) - L_{\widehat{EX}}(\{\psi\})$ and $L_{\widehat{EX}}(\{\psi\}) - L_{\widehat{EX}}(\{\varphi\})$ are all infinite].*

Theorem 5 *Suppose φ and ψ are partial recursive functions. Then either $[L_{\widehat{FEX}}(\{\varphi\}) = L_{\widehat{FEX}}(\{\psi\})]$ or $[L_{\widehat{FEX}}(\{\varphi\}) \cap L_{\widehat{FEX}}(\{\psi\})$, $L_{\widehat{FEX}}(\{\varphi\}) - L_{\widehat{FEX}}(\{\psi\})$ and $L_{\widehat{FEX}}(\{\psi\}) - L_{\widehat{FEX}}(\{\varphi\})$ are all infinite].*

As a consequence of the above results, we have the following interesting observation.

Corollary 6 *FEX is properly contained in $\widehat{FEX}$.*

The initial motivation for this paper was that some tasks are harder to learn than others, and that this should be reflected in the theory of inductive inference. It was also speculated that learning some tasks should necessarily entail learning some other task. In our formalism this amounts to showing that $L_I(\{\varphi\}) \subset L_I(\{\psi\})$ for φ and ψ partial recursive functions and I some criteria of successful inference. From Theorems 1,4 and 5 we know that this cannot be the case for $I \in \{EX, \widehat{EX}, \widehat{FEX}\}$. Theorem 2 leaves open the possibility that perhaps $L_{FEX}(\{\varphi\}) \subset L_{FEX}(\{\psi\})$. However, this cannot be true either, since if p is a program for φ, then $M_p \in L_{FEX}(\varphi)$ and $M_p \notin L_{FEX}(\psi)$. So it appears that the idiot savants make it impossible to discuss the relative difficulty of learning specific functions. We will return to this topic later, but first we examine learning versus learnability of single functions from another perspective.

3.2 An Isomorphism Result

In this subsection we develop an isomorphism result between certain collections of IIMs.

Theorem 7 *Suppose $f_1 \neq f_2$ are recursive functions. Then $L(\{f_1\})$ and $L(\{f_2\})$ are isomorphic.*

3.3 Relationships between Notions

As noted earlier, $FEX \subset EX$. This is a statement concerning learnablity as a relation between identification types. Intuitively, it says that any set of functions that can be learned by some IIM according to the FEX criteria of successful inference, then there is some (possibly different) IIM that can learn the same set of functions when judged according to the EX criteria of successful learning. To prove this result formally, one must show that for any set U of partial recursive functions, if there exists an IIM M such that $U \subseteq FEX(M)$, then there is an IIM M' such that $U \subseteq EX(M')$. An analogous notion is to consider the "learners of" as a relation between identification types. Suppose $\mathcal{M}$ is an arbitrary collection of IIMs. If the existence of a $\varphi \in F_{FEX}(\mathcal{M})$ implies the existence of a $\varphi' \in F_{EX}(\mathcal{M})$, then we write $FEX \sqsubseteq EX$. If the inclusion in proper, we will write $FEX \sqsubset EX$. Intuitively, this means that all the FEX learners of some function also EX learn some other function. The same relation may be defined between any pair of identification types.

Proposition 8 $FEX \sqsubset EX$.

A question that comes to mind is, given any two identification criteria I_1 and I_2, if $I_1 \subseteq I_2$, is it the case that $I_1 \sqsubseteq I_2$? We proceed to answer this question negatively. To do so, we must define a new identification criteria. Suppose that M is an IIM. Say that $\psi \in AEX(M)$, if M, on input from the graph of ψ, outputs only finitely many different programs $p_0, p_1, \cdots, p_n$, and for some $i \leq n$, $\varphi_{p_i} = \psi$. To keep the notation clear, by writting $I_1 = I_2$ we mean that $I_1 \subseteq I_2$ and $I_2 \subseteq I_1$. If $I_1 \sqsubseteq I_2$ and $I_2 \sqsubseteq I_1$, then we will write $I_1 \simeq I_2$.

Proposition 9 $AEX = EX$.

Proposition 10 $EX \sqsubseteq AEX$.

Theorem 11 $EX \not\sqsubseteq AEX$.

Another analogous question that one could ask is, given any two identification criteria I_1 and I_2, if $I_1 \sqsubseteq I_2$, is it the case that $I_1 \subseteq I_2$? Again the answer is negative. Two new identification types must be defined in order to present the proof.

Suppose M is an IIM. A function ψ is in $BEX(M)$ if M, on input from ψ, outputs a convergent sequence of programs $p_0, p_1, \cdots, p_n$ such that $\varphi_{p_0} = \lambda x[0]$ and $\varphi_{p_n} = \psi$. A function ψ is in $CEX(M)$ if M on input from ψ. outputs a sequence of programs $p_0, p_1, p_2 \cdots$, such that $\varphi_{p_0} = \psi$. The classes BEX and CEX are defined in the usual manner.

Proposition 12 $BEX \sqsubseteq CEX$.

Proposition 13 $BEX \not\subseteq CEX$.

4　On Learning Collections of Functions

In section 3.1 we discovered that realizing our intuition that some learning some tasks must entail the learning of some other, isolatable, subtask would be impossible by considering the learning of single functions. In this section, we consider the learning of collections of functions. First we briefly consider the learning of finite collections of concepts. The brevity is due to the following result.

Theorem 14 *For any finite set C of partial recursive functions, there is an IIM, M_C, such that $EX(M_C) = C$.*

Theorem 1 relied on the existence of idiot savants that learned precisely a single function. Theorem 14 will enable us to replace a single function with a finite set of functions in those proofs. The amalgamation technique is already concerned with collections of programs, so the proofs of Section 3.1 will carry over to finite sets of functions. Hence, we must consider the learners of infinite collections of functions in order to verify the intuition of the introduction.

Before proceeding to present some duality results we show that there are cascadingly smaller and smaller sets of IIMs with larger and larger foci.

Theorem 15 *The is a sequence of sets of recursive functions $S_0 \subset S_1 \subset S_2 \cdots$ and a sequence of sets of IIMs $M_0 \supset M_1 \supset M_2 \cdots$ such that $M_i = L(S_i)$, for all i.*

4.1 Some Near Duality Results

Considering the learning of infinitely many functions, and correspondingly, the focus of an infinite set of inductive inference machines gives rise to simple presentation of when duality occurs, and when it does not.

Theorem 16 *Suppose M_1 and M_2 are collections of IIMs. Then*

$$F(M_1 \cup M_2) = (F(M_1) \cap F(M_2)).$$

Furthermore, if $(M_1 \cup M_2) \neq \emptyset$, then

$$F(M_1 \cap M_2) = (F(M_1) \cup F(M_2)).$$

Proof: Notice, that as a consequence of the definition of the focus of a collection of IIMs, that if we are given two collections of IIMs, M_1 and M_2 then $F(M_1 \cup M_2) = (F(M_1) \cap F(M_2))$. When $M_1 \cup M_2 \neq \emptyset$, $F(M_1 \cup M_2) \neq \emptyset$ and the theorem follows from the definitions. ⊠

Theorem 17 *Suppose S_1 and S_2 are sets of partial recursive functions. Then*

$$\begin{aligned} L(S_1) \cap L(S_2) &= L(S_1 \cup S_2) \\ L(S_1) \cup L(S_2) &\subseteq L(S_1 \cap S_2). \end{aligned}$$

Furthermore, for any sets S_1 and S_2 in EX, if S_1 and S_2 are incomparable, then

$$L(S_1 \cap S_2) \not\subseteq L(S_1) \cup L(S_2).$$

Finally, for any sets S_1 and S_2 in EX, if $S_1 \subseteq S_2$, then

$$L(S_1 \cap S_2) \subseteq L(S_1) \cup L(S_2).$$

Proof: Suppose S_1 and S_2 are sets of partial recursive functions. Suppose that M is an IIM such that $M \in L(S_1) \cap L(S_2)$. Hence, $S_1 \subseteq EX(M)$ and $S_2 \subseteq EX(M)$. Consequently, $(S_1 \cup S_2) \subseteq EX(M)$ and $M \in L(S_1 \cup S_2)$.

Suppose $M \in L(S_1 \cup S_2)$. Then $S_1 \subseteq EX(M)$ and $S_2 \subseteq EX(M)$. So, $M \in L(S_1) \cap L(S_2)$. Hence, $L(S_1) \cap L(S_2) = L(S_1 \cup S_2)$.

Suppose now that M_1 and M_2 are IIMs such that $M_1 \in L(S_1)$ and $M_2 \in L(S_2)$. In other words, $S_1 \subseteq EX(M_1)$ and $S_2 \subseteq EX(M_2)$. Since $S_1 \cap S_2 \subseteq S_1$, $M_1 \in L(S_1 \cap S_2)$. Similarly, $S_1 \cap S_2 \subseteq S_2$ and $M_2 \in L(S_1 \cap S_2)$. Hence, $L(S_1) \cup L(S_2) \subseteq L(S_1 \cap S_2)$.

Suppose S_1 and S_2 are incomparable and both sets are in EX. Choose witnesses M_1 and M_2 such that $S_1 \subseteq EX(M_1)$ and $S_2 \subseteq EX(M_2)$. Define an IIM M as follows. On input σ, find via simulation $p_1 = M_1(\sigma)$ and $p_2 = M_2(\sigma)$. M then outputs program p where

$$\varphi_p(x) = \begin{cases} \varphi_{p_1}(x) & \text{if } \varphi_{p_1}(x) = \varphi_{p_2}(x), \\ \uparrow & \text{otherwise.} \end{cases}$$

If $f \in S_1 \cap S_2$, then both M_1 and M_2 converge to correct programs. Consequently, $f \in EX(M)$. Suppose $f \in (S_1 - S_2)$. If M_2 fails to converge, then M will also fail to converge and $f \notin EX(M)$. Furthermore, if M_2 converges to an incorrect program, then this program will compute a different function than the program that M_1 converges to. In this case, again, $f \notin EX(M)$. Similarly, if $f \in (S_2 - S_1)$, then $f \notin EX(M)$. Hence, $M \notin (L(S_1) \cup L(S_2))$.

Finally, suppose $S_1 \subseteq S_2$ and both sets are in EX. Then $S_1 \cap S_2 = S_1$, so $L(S_1 \cap S_2) = L(S_1) \subset L(S_1) \cup L(S_2)$. ⊠

Essentially the same proof techniques can be used to show to following.

Theorem 18 *Suppose S_1 and S_2 are sets of partial recursive functions. Then*

$$L_{FEX}(S_1) \cap L_{FEX}(S_2) = L_{FEX}(S_1 \cup S_2)$$
$$L_{FEX}(S_1) \cup L_{FEX}(S_2) \subseteq L_{FEX}(S_1 \cap S_2)$$

Furthermore, for any sets S_1 and S_2 in FEX, if S_1 and S_2 are incomparable, then

$$L_{FEX}(S_1 \cap S_2) \not\subseteq L_{FEX}(S_1) \cup L_{FEX}(S_2).$$

Finally, for any sets S_1 and S_2 in FEX, if $S_1 \subseteq S_2$, then

$$L_{FEX}(S_1 \cap S_2) \subseteq L_{FEX}(S_1) \cup L_{FEX}(S_2).$$

Proof: Notice that the first three paragraphs of the proof of Theorem 17 do not involve simulations of the inference of specific sets. Hence, these arguments serve to prove the first two clauses of this corollary. The simulation of the final part of the proof of Theorem 17, adapted to the FEX case, results in an IIM M that produces a single conjecture. ⊠

4.2 Collections of Learners

The Blum's showed the the union of two inferrible sets of recursive functions was not necessarily inferrible, in symbols, there are $S_1 \in EX$ and $S_2 \in EX$ such that $(S_1 \cup S_2) \notin EX$ [9]. More general non union theorems were discovered in [16]. Multiple machine inference was shown to be equivalent to probabilistic learning [15] Subsequently, multiple machine inference has been studied extensively. In this subsection we prove some basic properties of collections of IIMs described as $L(\{f\})$ for some recursive functions f.

Theorem 19 *There are sets of IIMs $\mathcal{M}_1$ and $\mathcal{M}_2$ such that*

1. $\exists f_1 \in F(\mathcal{M}_1)$,

2. $\exists f_2 \in F(\mathcal{M}_2)$, and

3. $\not\exists f_3 \in F(\mathcal{M}_1 \cup \mathcal{M}_2)$.

Proof: Choose recursive functions $f_1 \neq f_2$. Let $\mathcal{M}_1 = L(f_1)$ and $\mathcal{M}_2 = L(f_2)$. Suppose programs p and q are such that $\varphi_p = f_1$ and $\varphi_q = f_2$. Then $M_p \in \mathcal{M}_1$ and $M_q \in \mathcal{M}_2$. Suppose by way of contradiction that there is a recursive function $f \in F(\mathcal{M}_1 \cup \mathcal{M}_2)$. Since $M_p \in \mathcal{M}_1$, it must be that $f = \varphi_p = f_1$. Similarly, since $M_q \in \mathcal{M}_2$, it must be that $f = \varphi_q = f_2$. This contradicts the choice of f_1 and f_2. ☒

Corollary 20 *Suppose $\mathcal{M}_1 = L(\{f_1\})$ and $\mathcal{M}_2 = L(\{f_2\})$ for recursive functions $f_1 \neq f_2$. Then, for any recursive function f, $(\mathcal{M}_1 \cup \mathcal{M}_2) \not\subseteq L(\{f_3\})$.*

Proof: Notice that since $f_1 \neq f_2$, $F(\mathcal{M}_1 \cup \mathcal{M}_2) = \emptyset$. ☒

Theorem 19 can be generalized to larger collections of sets of IIMs. However, a slightly different proof technique must be employed. From the proof of Theorem 21 below, it will be clear how to generalize the result to any finite number of collections of IIMs.

Theorem 21 *Given distinct recursive functions f_1, f_2 and f_3 there are distinct collections of IIMs $\mathcal{M}_1$, $\mathcal{M}_2$ and $\mathcal{M}_3$ such that $f_1 \in F(\mathcal{M}_1) \cap F(\mathcal{M}_2)$, $f_2 \in F(\mathcal{M}_1) \cap F(\mathcal{M}_3)$, and $f_3 \in F(\mathcal{M}_2) \cap F(\mathcal{M}_3)$, and there does not exist a function $f_4 \in F(\mathcal{M}_1) \cap F(\mathcal{M}_2) \cap F(\mathcal{M}_3)$.*

Proof: Suppose f_1, f_2 and f_3 are distinct recursive functions. By Theorem 14 there exist IIMs M_1, M_2 and M_3 such that $EX(M_1) = \{f_1, f_2\}$, $EX(M_2) = \{f_1, f_3\}$ and $EX(M_3) = \{f_2, f_3\}$. Then, $f_1 \in F(\mathcal{M}_1) \cap F(\mathcal{M}_2)$, $f_2 \in F(\mathcal{M}_1) \cap F(\mathcal{M}_3)$ and $f_3 \in F(\mathcal{M}_2) \cap F(\mathcal{M}_3)$.

By Theorem 14, choose $M_1 \in \mathcal{M}_1$ such that $EX(M_1) = \{f_1, f_2\}$ $M_2 \in \mathcal{M}_2$ such that $EX(M_2) = \{f_1, f_3\}$ and $M_1 \in \mathcal{M}_3$ such that $EX(M_3) = \{f_2, f_3\}$. Suppose by way of contradiction that $f_4 \in F(\mathcal{M}_1) \cap F(\mathcal{M}_2) \cap F(\mathcal{M}_3)$. Then it must be that $f_4 \in \{f_1, f_2, f_3\}$, as otherwise one of M_1, M_2 and M_3 would not be able to learn f_4, contradicting the selection of f_4.

If $f_4 = f_1$, then $f_4 \notin EX(M_3)$. Hence, $f_4 \notin F(M_3)$, a contradiction. If $f_4 = f_2$, then $f_4 \notin EX(M_2)$. Hence, $f_4 \notin F(M_2)$, a contradiction. If $f_4 = f_3$, then $f_4 \notin EX(M_1)$. Hence, $f_4 \notin F(M_1)$, a contradiction. $\boxtimes$

The general theorem can now be stated.

Theorem 22 *For any $n > 0$, given distinct recursive functions f_1, f_2, $\cdots$, f_n there are distinct collections of IIMs M_1, M_2, $\cdots$, M_n such that for all $1 \leq i \leq n$,*

$$f_i \in \left(\bigcap_{j \in \{\{1,\cdots,n\}-i\}} F(M_j) \right).$$

and there does not exist a recursive function f_{n+1} such that

$$f_{n+1} \in \left(\bigcup_{1 \leq j \leq n} F(M_j) \right).$$

4.3 Closure

When considering the learners of some set of functions S, it may be that any IIM that can learn all the functions in S, may also be able to learn some other functions as well. In symbols, it may be that there is an $S' \supset S$ such that if $S \subseteq EX(M)$, then $S' \subseteq EX(M)$. From another perspective, it may happen that $L(S) \subseteq L(S')$. To study this situation we say that a class of functions S is *closed* (with respect to EX) if there does not exist an $S' \supset S$ such that for any $M \in L(S)$, $S' \subseteq EX(M)$. Note that the collection of all recursive functions is vacuously closed. Our first result shows that there are several closed classes.

Theorem 23 *Every subset of any set in EX is closed.*

There are two interesting corollaries of Theorem 23

Corollary 24 *S is closed if and only if $S \in EX$.*

Corollary 25 *There is a continuum of closed classes.*

4.4 Noncompactness

Theorem 26 *There is a set of IIMs M such that $F(M) = \emptyset$ and for all $M' \subset M$, if M' is finite, then $F(M') \neq \emptyset$.*

5 Conclusions

A new perspective on learning algorithms, and what they learn, was introduced. Some strong dualities with conventional notions were discovered. The dualities however are not complete. Theorems 16 and 17 give an algebra of sorts for manipulating inferrible sets. For example, for S_1 and S_2 in EX, $F(L(S_1) \cup L(S_2)) = F(L(S_1)) \cap F(L(S_2)) = S_1 \cap S_2$.

References

[1] *Proceedings of the 1988 Workshop on Computational Learning Theory*, Palo Alto, CA., 1988. Morgan Kaufmann Publishers.

[2] *Proceedings of the Second Annual Workshop on Computational Learning Theory*, Palo Alto, CA., 1989. Morgan Kaufmann Publishers.

[3] *Proceedings of the Third Annual Workshop on Computational Learning Theory*, Palo Alto, CA., 1990. Morgan Kaufmann Publishers.

[4] *Proceedings of the 1991 Workshop on Computational Learning Theory*, Palo Alto, CA., 1991. Morgan Kaufmann Publishers.

[5] Proceedings of the 1992 workshop on computational learning theory, 1992.

[6] D. Angluin, W. I. Gasarch, and C. H. Smith. Training sequences. *Theoretical Computer Science*, 66:255–272, 1989.

[7] D. Angluin and C. H. Smith. Inductive inference: Theory and methods. *Computing Surveys*, 15:237–269, 1983.

[8] D. Angluin and C. H. Smith. Inductive inference. In S. Shapiro, editor, *Encyclopedia of Artificial Intelligence*, pages 409–418. John Wiley and Sons Inc., 1987.

[9] L. Blum and M. Blum. Toward a mathematical theory of inductive inference. *Information and Control*, 28:125–155, 1975.

[10] J. Case and C. Smith. Anomaly hierarchies of mechanized inductive inference. In *Proceedings of the 10th Symposium on the Theory of Computing*, pages 314–319, San Diego, CA, 1978.

[11] J. Case and C. Smith. Comparison of identification criteria for machine inductive inference. *Theoretical Computer Science*, 25(2):193–220, 1983.

[12] R. V. Freivalds and R. Wiehagen. Inductive inference with additional information. *Elektronische Informationsverabeitung und Kybernetik*, 15(4):179–184, 1979.

[13] E. M. Gold. Language identification in the limit. *Information and Control*, 10:447–474, 1967.

[14] M. Machtey and P. Young. *An Introduction to the General Theory of Algorithms*. North-Holland, New York, 1978.

[15] L. Pitt. A characterization of probabilistic inference. *Journal of the ACM*, 36(2):383–433, 1989.

[16] C. H. Smith. The power of pluralism for automatic program synthesis. *Journal of the ACM*, 29(4):1144–1165, 1982.

On Aggregating Teams of Learning Machines

Sanjay Jain[1] and Arun Sharma[2]

[1] Institute of Systems Science
National University of Singapore
Singapore 0511, Republic of Singapore
Email: sanjay@iss.nus.sg
[2] School of Computer Science and Engineering
The University of New South Wales
Sydney, NSW 2033, Australia
Email: arun@cse.unsw.edu.au

Abstract. The present paper studies the problem of when a team of learning machines can be aggregated into a single learning machine without any loss in learning power. The main results concern aggregation ratios for vacillatory identification of languages from texts. For a positive integer n, a machine is said to **TxtFex$_n$**-identify a language L just in case the machine converges to upto n grammars for L on any text for L. For such identification criteria, the aggregation ratio is derived for the $n = 2$ case. It is shown that the collection of languages that can be **TxtFex$_2$** identified by teams with success ratio greater than $5/6$ are the same as those collections of languages that can be **TxtFex$_2$**-identified by a single machine. It is also established that $5/6$ is indeed the cut-off point by showing that there are collections of languages that can be **TxtFex$_2$**-identified by a team employing 6 machines, at least 5 of which are required to be successful, but cannnot be **TxtFex$_2$**-identified by any single machine. Additionally, aggregation ratios are also derived for finite identification of languages from positive data and for numerous criteria involving language learning from both positive and negative data.

1 Introduction

The present paper investigates the problem of aggregating a team of learning machines into a single learning machine. In other words, we are interested in finding when a team of learning machines can be replaced by a single machine without any loss in learning power.

A team of learning machines is essentially a multiset of learning machines. A team is said to successfully learn a concept just in case each member of some nonempty subset of the team learns the concept. If the size of a team is n and if at least m machines in the team are required to be successful for the team to be successful, then the ratio m/n is referred to as the *success ratio* of the team. The present paper addresses the problem, "For what success ratios can a team be replaced by a single machine without any loss in learning power?" The answer to this question depends on the kind of concepts being learned and the the type of success criteria employed. For the problem of learning recursive functions from graphs, the answer is known for the three popularly investigated criteria of success, namely, **Fin** (finite identification), **Ex** (identification in the limit) and **Bc** (behaviorally correct identification). For **Ex** and **Bc**, Pitt and Smith [21] showed that a team can be aggregated into a single machine if the success ratio of the team is greater than 1/2. For finite function identification, **Fin**, it was reported in [12] that a team can be aggregated if the success ratio of the team is greater than 2/3 (this result can also be argued from a result of Freivalds [9] about probabilistic finite function identification).

The present paper describes aggregation results about language identification from positive data. The main results are in the context of vacillatory identification. To facilitate discussion of these results, we informally present some preliminaries from theory of language learning next.

Languages are sets of sentences and a sentence is a finite object; the set of all possible sentences can be coded into N — the set of natural numbers. Hence, languages may be construed as subsets of N. A grammar for a language is a set of rules that accepts (or equivalently, generates [11]) the language. Essentially, any computer program may be viewed as a grammar. Languages for which a grammar exists are called *recursively enumerable*.

A *text* for a language L is any infinite sequence that lists all and only the elements of L; repetitions are permitted. A learning machine is an algorithmic device that outputs grammars on finite initial sequences of texts. Two well studied criteria for a machine to successfully learn a language are *identification in the limit* and *behaviorlly correct identification*. We next give an informal definition of these criteria.

A learning machine **M** is said to **TxtEx** identify a language L just in case **M**, fed any text for L, converges to a correct grammar for L. This is essentially the seminal notion of identification in the limit introduced by Gold [10] (see also Case and Lynes [7] and Osherson and Weinstein [19]).

A learning machine **M** is said to **TxtBc**-identify L just in case **M**, fed any text for L, outputs an infinite sequence of grammars such that after a finite number of incorrect guesses, **M** outputs only grammars for L. This criterion was first studied by Case and Lynes [7] and Osherson and Weinstein [19], and is also referred to as "extensional" identification.

Osherson, Stob and Weinstein [17] first observed that for **TxtEx**-identification, a team can be aggregated if its success ratio is greater than 2/3. Hence, in matters of aggregation, identification in the limit of languages from positive data turns out to be similar to finite function identification. On the other hand, for **TxtBc**-identification, a result from Pitt [20] can easily be used to show that a team can be aggregated if its success ratio is greater than 1/2. Thus, **TxtEx** and **TxtBc** exhibit different behavior with respect to aggregation.

We now present two more criteria of successful language learning, namely, finite identification and vacillatory identification.

A machine **M** is said to **TxtFin**-identify a language L just in case **M**, fed any text for L, outputs only one grammar and that grammar is for L.

We show that for **TxtFin**-identification, a team can be aggregated only if its success ratio is greater than 2/3. Thus, **TxtFin**-identification shows similar behavior as **TxtEx**-identification and finite function identification so far as aggregation is concerned.

We next consider vacillatory identification of languages from texts in which a machine is required to converge to a finite set of grammars. This notion was studied by Osherson and Weinstein [19] and by Case [5]. It should be noted that in the context of function learning, vacillatory identification turns out to be the same as identification in the limit. This was first shown by Barzdin and Podnieks [2] (see also Case and Smith [8]).

Let n be a positive integer. A learning machine **M** is said to **TxtFex$_n$**-identify a language L just in case **M**, fed any text for L, converges in the limit to a finite set, with cardinality $\leq n$, of grammars for L. In other words, for any text T for L, there exists a set D of grammars of L, cardinality of $D \leq n$, such that **M**, fed T, outputs, after a finite number of incorrect guesses, only grammars from D.

If the upper bound n in **TxtFex$_n$**-identification is not specified and the only requirement is that the machine converge to some finite set of grammars for the language, then the criteria is referred to as **TxtFex$_*$**-identification.

We show that for **TxtFex$_*$**-identification, a team can be aggregated if its success ratio is greater than 1/2. It is interesting to note that in matters of aggregation **TxtFex$_*$**-identification behaves more like **TxtBc**-identification than like **TxtEx**-identification. The problem of aggregation for **TxtFex$_n$**, however, turns out to be more difficult. We are able to answer this question for the $n = 2$ case, by showing that for **TxtFex$_2$**-identification, a team can be aggregated only if its success ratio is greater than 5/6. We establish this by showing that the collections of languages that can be **TxtFex$_2$**-identified by teams with success ratios greater than 5/6 are exactly the same as those collections of languages that can be **TxtFex$_2$**-identified by a single machine. Our proof of this result involves a fairly complicated simulation argument. We also establish that 5/6 is indeed the cut-off point for **TxtFex$_2$** aggregation by employing a diagonalization argument to show that there are collections of languages that can be **TxtFex$_2$**-identified by a team of 6 machines, at least 5 of which are required to be successful, but cannot be **TxtFex$_2$**-identified by any single machine.

The problem of aggregation becomes somewhat more manageable if we are prepared to allow the aggregated machine to converge to extra number of grammars. In fact we are able to show that aggregation can be achieved at success ratios just above 1/2 if the aggregated machine is allowed to converge to extra number of grammars. For example, for any positive integer i, all the collections of languages that can be **TxtEx**-identified by teams of $2i + 1$ machines, at least $i + 1$ of which are required to be successful, can also be **TxtFex$_{i+1}$**-identified by a single machine. More generally, using a fairly straight simulation argument, it can be shown that all the collections of languages that can be **TxtFex$_j$**-identified by teams of $2i + 1$ machines, at least $i + 1$ of which are required to be successful, can also be **TxtFex$_{(i+1) \cdot j}$**-identified by a single machine.

In Section 3.4, we show that aggregation issues in the context of language identification from both positive and negative data follow a pattern similar to function learning.

We now proceed formally. Section 2 records the notation and describes preliminary notions and definitions from inductive inference literature. Our results are presented in Section 3.

2 Preliminaries

2.1 Notation

Any unexplained recursion theoretic notation is from [23]. The symbol N denotes the set of natural numbers, $\{0, 1, 2, 3, \ldots\}$. The symbol N^+ denotes the set of positive natural numbers, $\{1, 2, 3, \ldots\}$. Unless otherwise specified, i, j, m, n, s, t, x, y, with or without decorations[3], range over N. Symbols $\emptyset, \subseteq, \subset, \supseteq$, and $\supset$ denote empty set, subset, proper subset, superset, and proper superset, respectively. Symbols P and S, with or without decorations, range over finite sets. Cardinality of a set S is denoted by $\text{card}(S)$. We say that $\text{card}(A) \leq *$ to mean that $\text{card}(A)$ is finite. Intuitively, the symbol, $*$, denotes 'finite without any prespecified bound.' The maximum and minimum of a set are denoted by $\max(\cdot), \min(\cdot)$, respectively, where $\max(\emptyset) = 0$ and $\min(\emptyset) =\uparrow$.

Letters f, g, and h, with or without decorations, range over *total* functions with arguments and values from N. Symbol $\mathcal{R}$ denotes the set of all total computable functions. A pair $\langle i, j \rangle$ stands for an arbitrary, computable, one-to-one encoding of all pairs of natural numbers onto N [23]. Similarly, we can define $\langle \cdot, \ldots, \cdot \rangle$ for encoding multiple tuples of natural numbers onto N. By φ we denote a fixed *acceptable* programming system for the partial computable functions: $N \to N$ [22, 23, 16]. By φ_i we denote the partial computable function computed by program i in the φ-system. By Φ we denote an arbitrary fixed Blum complexity measure [3, 11] for the φ-system. By W_i we denote domain(φ_i). W_i is, then, the r.e. set/language ($\subseteq N$) accepted (or equivalently, generated) by the φ-program i. Symbol $\mathcal{E}$ will denote the set of all r.e. languages. Symbol L, with or without decorations, ranges over $\mathcal{E}$. Symbol $\mathcal{L}$, with or without decorations, ranges over subsets of $\mathcal{E}$. We denote by $W_{i,s}$ the set $\{x \leq s \mid \Phi_i(x) \leq s\}$. The quantifiers '$\overset{\infty}{\forall}$' and '$\overset{\infty}{\exists}$' mean 'for all but finitely many' and 'there exist infinitely many', respectively.

2.2 Learning Machines

We first consider function learning machines.

We assume, without loss of generality, that the graph of a function is fed to a machine in canonical order. For $f \in \mathcal{R}$ and $n \in N$, we let $f[n]$ denote the finite initial segment $\{(x, f(x)) \mid x < n\}$. Clearly, $f[0]$ denotes the empty segment. SEG denotes the set of all finite initial segments, $\{f[n] \mid f \in \mathcal{R} \wedge n \in N\}$.

---

[3] Decorations are subscripts, superscripts and the like.

Definition 1. [10] A *function learning machine* is an algorithmic device which computes a mapping, possibly partial, from SEG into N.

We now consider language learning machines. A *sequence* σ is a mapping from an initial segment of N into $(N \cup \{\#\})$. The *content* of a sequence σ, denoted content(σ), is the set of natural numbers in the range of σ. The *length* of σ, denoted by $|\sigma|$, is the number of elements in σ. For $n \leq |\sigma|$, the initial segment of σ of length n is denoted by $\sigma[n]$. Intuitively, $\#$'s represent pauses in the presentation of data. We let σ, τ, and γ, with or without decorations, range over finite sequences. SEQ denotes the set of all finite sequences.

Definition 2. A *language learning machine* is an algorithmic device which computes a mapping, possibly partial, from SEQ into N.

The set of all finite initial segments, SEG, can be coded onto N. Also, the set of all finite sequences of natural numbers and $\#$'s, SEQ, can be coded onto N. Thus, in both Definitions 1 and 2, we can view these machines as taking natural numbers as input and emitting natural numbers as output. Henceforth, we will refer to both function-learning machines and language-learning machines as just learning machines, or simply as machines. We let $\mathbf{M}$, with or without decorations, range over learning machines.

2.3 Criteria of Learning

FINITE FUNCTION IDENTIFICATION

For finite function identification only, we assume our learning machines to compute a mapping from SEG into $N \cup \{\perp\}$. The output of machine $\mathbf{M}$ on evidential state σ will be denoted by $\mathbf{M}(\sigma)$, where '$\mathbf{M}(\sigma) = \perp$' denotes that $\mathbf{M}$ does not issue any hypothesis on σ.

Definition 3. $\mathbf{M}$ **Fin**-*identifies* f (read: $f \in \mathbf{Fin}(\mathbf{M})$) $\iff$ $(\exists i \mid \varphi_i = f)$ $(\exists n_0)[(\forall n \geq n_0)[\mathbf{M}(f[n]) = i] \wedge (\forall n < n_0)[\mathbf{M}(f[n]) = \perp]]$. We define the class $\mathbf{Fin} = \{\mathcal{S} \subseteq \mathcal{R} \mid (\exists \mathbf{M})[\mathcal{S} \subseteq \mathbf{Fin}(\mathbf{M})]\}$.

FUNCTION IDENTIFICATION IN THE LIMIT

Definition 4. [10] $\mathbf{M}$ **Ex**-identifies f (read: $f \in \mathbf{Ex}(\mathbf{M})$) $\iff$ $(\exists i \mid \varphi_i = f)$ $(\overset{\infty}{\forall} n)[\mathbf{M}(f[n]) = i]$. We define the class $\mathbf{Ex} = \{\mathcal{S} \subseteq \mathcal{R} \mid (\exists \mathbf{M})[\mathcal{S} \subseteq \mathbf{Ex}(\mathbf{M})]\}$.

BEHAVIORALLY CORRECT FUNCTION IDENTIFICATION

Definition 5. [8] $\mathbf{M}$ **Bc**-identifies f (read: $f \in \mathbf{Bc}(\mathbf{M})$) $\iff$ $(\overset{\infty}{\forall} n)[\varphi_{\mathbf{M}(f[n])} = f]$. We define the class $\mathbf{Bc} = \{\mathcal{S} \subseteq \mathcal{R} \mid (\exists \mathbf{M})[\mathcal{S} \subseteq \mathbf{Bc}(\mathbf{M})]\}$.

The following proposition summarizes the relationship between the various function learning criteria.

Proposition 6. [8, 1] $\mathbf{Fin} \subset \mathbf{Ex} \subset \mathbf{Bc}$.

2.4   Language Learning

A *text* T for a language L is a mapping from N into $(N \cup \{\#\})$ such that L is the set of natural numbers in the range of T. The *content* of a text T, denoted content(T), is the set of natural numbers in the range of T.

FINITE LANGUAGE IDENTIFICATION

Again as in the case of finite function identification, we assume our learning machines to compute a mapping from SEQ into $N \cup \{\perp\}$. This assumption is for this definition only.

Definition 7. M **TxtFin**-*identifies* L (read: $L \in$ **TxtFin**(M)) $\iff$ ($\forall$ texts T for L) $(\exists i \mid W_i = L)$ $(\exists n_0)[(\forall n \geq n_0)[M(T[n]) = i] \land (\forall n < n_0)[M(T[n]) = \perp]]$. We define the class **TxtFin** $= \{\mathcal{L} \subseteq \mathcal{E} \mid (\exists M)[\mathcal{L} \subseteq$ **TxtFin**(M)$]\}$.

2.5   Language Identification in the Limit

Definition 8. [10] M **TxtEx**-identifies L (read: $L \in$ **TxtEx**(M)) $\iff$ ($\forall$ texts T for L) $(\exists i \mid W_i = L)$ $(\overset{\infty}{\forall} n)[M(T[n]) = i]$. We define the class **TxtEx** $= \{\mathcal{L} \subseteq \mathcal{E} \mid (\exists M)[\mathcal{L} \subseteq$ **TxtEx**(M)$]\}$.

BEHAVIORALLY CORRECT LANGUAGE IDENTIFICATION

Definition 9. [19, 7] M **TxtBc**-identifies L (read: $L \in$ **TxtBc**(M)) $\iff$ ($\forall$ texts T for L) $(\overset{\infty}{\forall} n)[W_{M(T[n])} = L]$. We define the class **TxtBc** $= \{\mathcal{L} \subseteq \mathcal{E} \mid (\exists M)[\mathcal{L} \subseteq$ **TxtBc**(M)$]\}$.

VACILLATORY LANGUAGE IDENTIFICATION

We now introduce the notion of a learning machine finitely converging on a text [5]. Let M be a learning machine and T be a text. $M(T)$ *finitely-converges* (written: $M(T)\Downarrow$) $\iff$ $\{M(\sigma) \mid \sigma \subset T\}$ is finite, otherwise we say that $M(T)$ *finitely-diverges* (written: $M(T)\Uparrow$). If $M(T)\Downarrow$, then $M(T)$ is defined $= \{i \mid (\overset{\infty}{\exists} \sigma \subset T)[M(\sigma) = i]\}$.

Definition 10. [19, 5] Let $b \in N^+ \cup \{*\}$.

(a) M **TxtFex**$_b$-*identifies* L (read: $L \in$ **TxtFex**$_b$(M)) $\iff$ ($\forall$ texts T for L)$(\exists P \mid \text{card}(P) \leq b \land (\forall i \in P)[W_i = L])[M(T)\Downarrow \land M(T) = P]$.
(b) **TxtFex**$_b = \{\mathcal{L} \subseteq \mathcal{E} \mid (\exists M)[\mathcal{L} \subseteq$ **TxtFex**$_b$(M)$]\}$.

The following proposition summarizes the relationship between the various language learning criteria.

Proposition 11. [19, 7, 5] **TxtFin** $\subset$ **TxtEx** $=$ **TxtFex**$_1 \subset$ **TxtFex**$_2 \subset \cdots \subset$ **TxtFex**$_* \subset$ **TxtBc**.

2.6   Team Learning

A team of learning machines is essentially a multiset of learning machines. Definition 12 introduces team learning of functions and Definition 13 introduces team learning of languages.

Definition 12. [24, 18] Let $\mathbf{I} \in \{\mathbf{Fin}, \mathbf{Ex}, \mathbf{Bc}\}$ and let $m, n \in N^+$.

(a) A team of n machines $\{\mathbf{M}_1, \mathbf{M}_2, \ldots, \mathbf{M}_n\}$ is said to $\mathbf{Team}_n^m\mathbf{I}$-*identify* f (written: $f \in \mathbf{Team}_n^m\mathbf{I}(\{\mathbf{M}_1, \mathbf{M}_2, \ldots, \mathbf{M}_n\})$) just in case there exist m distinct numbers $i_1, i_2, \ldots, i_m, 1 \leq i_1 < i_2 < \cdots < i_m \leq n$, such that each of $\mathbf{M}_{i_1}, \mathbf{M}_{i_2}, \ldots, \mathbf{M}_{i_m}$ I-identifies f.
(b) $\mathbf{Team}_n^m\mathbf{I} = \{\mathcal{S} \subseteq \mathcal{R} \mid (\exists \mathbf{M}_1, \mathbf{M}_2, \ldots, \mathbf{M}_n)[\mathcal{S} \subseteq \mathbf{Team}_n^m\mathbf{I}(\{\mathbf{M}_1, \mathbf{M}_2, \ldots, \mathbf{M}_n\})]\}$.

Definition 13. Let $b \in N^+ \cup \{*\}$. Let $\mathbf{I} \in \{\mathbf{TxtFin}, \mathbf{TxtEx}, \mathbf{TxtFex}_b, \mathbf{TxtBc}\}$. Let $m, n \in N^+$.

(a) A team of n machines $\{\mathbf{M}_1, \mathbf{M}_2, \ldots, \mathbf{M}_n\}$ is said to $\mathbf{Team}_n^m\mathbf{I}$-*identify* L (written: $L \in \mathbf{Team}_n^m\mathbf{I}(\{\mathbf{M}_1, \mathbf{M}_2, \ldots, \mathbf{M}_n\})$) just in case there exist m distinct numbers $i_1, i_2, \ldots, i_m, 1 \leq i_1 < i_2 < \cdots < i_m \leq n$, such that each of $\mathbf{M}_{i_1}, \mathbf{M}_{i_2}, \ldots, \mathbf{M}_{i_m}$ I-identifies L.
(b) $\mathbf{Team}_n^m\mathbf{I} = \{\mathcal{L} \subseteq \mathcal{E} \mid (\exists \mathbf{M}_1, \mathbf{M}_2, \ldots, \mathbf{M}_n)[\mathcal{L} \subseteq \mathbf{Team}_n^m\mathbf{I}(\{\mathbf{M}_1, \mathbf{M}_2, \ldots, \mathbf{M}_n\})]\}$.

For $\mathbf{Team}_n^m\mathbf{I}$-identification criteria, we refer to the fraction m/n as the *success ratio* of the criteria.

Definition 14. The fraction m/n is referred to as the *aggregation ratio* for the success criteria I-identification just in case

(a) $(\forall i, j \mid i/j > m/n)[\mathbf{Team}_j^i\mathbf{I} = \mathbf{I}]$, and
(b) $\mathbf{I} \subset \mathbf{Team}_n^m\mathbf{I}$.

In the following, for $i > j$, we take $\mathbf{Team}_j^i\mathbf{I} = \{\emptyset\}$.

3   Results

3.1   Previously Known Results

Aggregation results are known for all the function learning criteria defined in the previous section. For finite function identification, aggregation takes place at success ratios greater than $2/3$. This result, Theorem 15(a) below, appeared in [12] and can also easily be argued from a related result of Freivalds [9] about probabilistic finite identification. Theorem 15(b) shows that $2/3$ is the cut-off point for aggregation of **Fin**-identification; a diagonalization argument using the operator recursion theorem [4] suffices to establish this latter result.

Theorem 15. *(a)* $(\forall m, n \mid m/n > 2/3)[\mathbf{Team}_n^m \mathbf{Fin} = \mathbf{Fin}]$.
(b) $\mathbf{Fin} \subset \mathbf{Team}_3^2 \mathbf{Fin}$.

Pitt and Smith [21] settled the question for function identification in the limit and behaviorally correct function identification by showing the following Theorem 16(a) which implies that for both these criteria aggregation takes place at success ratios greater than 1/2. Theorem 16(b), due to Smith [24], shows that 1/2 is indeed the cut-off point.

Theorem 16. *Let* $\mathbf{I} \in \{\mathbf{Ex}, \mathbf{Bc}\}$.

(a) $(\forall m, n \mid m/n > 1/2)[\mathbf{Team}_n^m \mathbf{I} = \mathbf{I}]$
(b) $\mathbf{I} \subset \mathbf{Team}_2^1 \mathbf{I}$.

For language learning, the result is known for **TxtEx**-identification and **TxtBc**-identification. It was shown by Osherson, Stob, and Weinstein [17] that aggregation for **TxtEx** takes place at success ratios greater than 2/3, and 2/3 is also the cut-off point for aggregation of **TxtEx**-identification (see also [13, 15, 14] for extension of this result to anomalies in the final grammar).

Theorem 17. *(a)* $(\forall m, n \mid m/n > 2/3)[\mathbf{Team}_n^m \mathbf{TxtEx} = \mathbf{TxtEx}]$
(b) $\mathbf{TxtEx} \subset \mathbf{Team}_3^2 \mathbf{TxtEx}$.

Using a result from Pitt [20], it can be shown that aggregation for **TxtBc** takes place at success ratios greater than 1/2. This is Theorem 18(a) below. Part (b) of Theorem 18 implies that 1/2 is indeed the cut-off point for aggregation of **TxtBc** and a proof of this latter fact can easily be be obtained by considering a collection of recursive languages derived from the corresponding function learning result of Smith (Theorem 16(b)).

Theorem 18. *(a)* $(\forall m, n \mid m/n > 1/2)[\mathbf{Team}_n^m \mathbf{TxtBc} = \mathbf{TxtBc}]$
(b) $\mathbf{TxtBc} \subset \mathbf{Team}_2^1 \mathbf{TxtBc}$.

We now consider aggregation for **TxtFin**-identification and **TxtFex**$_b$-identification, $b \in N^+ \cup \{*\}$.

3.2 Aggregation for Finite Identification of Languages

It turns out that aggregation for finite identification of languages is no different than aggregation for limit identification of languages. Theorem 19(a) below shows that aggregation for **TxtFin**-identification takes place at success ratios greater than 2/3. A proof of this result can be obtained on the lines of a proof of Theorem 15(a). Part (b) of the following result implies that 2/3 is indeed the cut-off point for aggregation of **TxtFin**-identification; a proof follows by considering a collection of recursive languages derived from the proof of Theorem 15(b).

Theorem 19. *(a)* $(\forall m, n \mid m/n > 2/3)[\mathbf{Team}_n^m \mathbf{TxtFin} = \mathbf{TxtFin}]$
(b) $\mathbf{TxtFin} \subset \mathbf{Team}_3^2 \mathbf{TxtFin}$.

3.3 Aggregation for Vacillatory Identification of Languages

Our first result for team aggregation in the context of vacillatory identification is for **TxtFex$_*$**-identification. Theorem 20(a) below says that team aggregation for **TxtFex$_*$**-identification takes place at success ratios greater than 1/2. This result can be proved using a simulation argument. Theorem 20(b) confirms that 1/2 is indeed the cut-off point for aggregation of **TxtFex$_*$**-identification by implying that there are collections of languages which can be **TxtFex$_*$**-identified by a team employing 2 machines at least one of which is required to be successful, but cannot be **TxtFex$_*$**-identified by any single machine. It is interesting to observe that in matters of aggregation, **TxtFex$_*$**-identification behaves more like **TxtBc**-identification than like **TxtEx**-identification.

Theorem 20. *(a)* $(\forall m, n \mid m/n > 1/2)[\mathbf{Team}_n^m \mathbf{TxtFex}_* = \mathbf{TxtFex}_*]$
(b) $\mathbf{TxtFex}_* \subset \mathbf{Team}_2^1 \mathbf{TxtFex}_*$.

However, the problem of finding aggregation ratios for **TxtFex$_b$**-identification when $b \neq *$ turns out to be far more difficult. The difficulty arises in requiring the aggregated machine to also converge to upto b grammars. In the light of these difficulties, it is worth considering cases where the bound on the number of converged grammars for the aggregated machine is more than the bound allowed for the team. In this connection we can show that $\mathbf{Team}_5^3 \mathbf{TxtEx} - \mathbf{TxtFex}_2 \neq \emptyset$, but $\mathbf{Team}_5^3 \mathbf{TxtEx} \subseteq \mathbf{TxtFex}_3$. Hence, allowing more grammars in the limit can sometimes help achieve "pseudo aggregation." This result can be generalized to show the following.

Theorem 21. *Let $i \in N^+$.*

(a) $\mathbf{Team}_{2i+1}^{i+1} \mathbf{TxtEx} - \mathbf{TxtFex}_i \neq \emptyset$.
(b) $\mathbf{Team}_{2i+1}^{i+1} \mathbf{TxtEx} \subseteq \mathbf{TxtFex}_{i+1}$.

The next result generalizes Theorem 21(b).

Theorem 22. $\mathbf{Team}_{2i+1}^{i+1} \mathbf{TxtFex}_j \subseteq \mathbf{TxtFex}_{(i+1) \cdot j}$.

A number of results like the above can be worked out; we leave them for the full paper.

The above results however do not say anything about aggregation in the context of **TxtFex$_b$**-identification, when $b \neq *$. The following result shows that aggregation for **TxtFex$_2$**-identification does not take place at success ratio 2/3 and aggregation for **TxtFex$_3$**-identification does not take place at success ratio 3/4.

Theorem 23. *Let $i \in N^+$.* $\mathbf{Team}_{i+1}^i \mathbf{TxtFex}_i - \mathbf{TxtFex}_i \neq \emptyset$.

However, we are able to show that **TxtFex$_2$** aggregation does takes place for success ratios greater than 5/6 as implied by the following central result of this paper.

Theorem 24. *(a)* $(\forall m, n \mid m/n > 5/6)[\text{Team}_n^m\text{TxtFex}_2 = \text{TxtFex}_2]$
(b) $\text{TxtFex}_2 \subset \text{Team}_6^5\text{TxtFex}_2$.

Our proof of Theorem 24(a) involves a complicated simulation argument which will be given in the full paper. This proof uses the crucial Lemma 25 presented below which shows that there exist recursive functions G_1 and G_2, such that for any set S of r grammars, $(\forall L \mid \text{card}(\{i \in S \mid W_i = L\})) > 2r/5)(\exists i \in \{1,2\})[W_{G_i(S)} = L]$. Let m, n be as described in the hypothesis of the theorem. Suppose a team of n machines $\{\mathbf{M}_1, \ldots, \mathbf{M}_n\}$ are given. We now sketch a machine $\mathbf{M}$ that TxtFex_2-identifies any language which is $\text{Team}_n^m\text{TxtFex}_2$-identified by $\{\mathbf{M}_1, \ldots, \mathbf{M}_n\}$. Suppose $\{\mathbf{M}_1, \ldots, \mathbf{M}_n\}$, $\text{Team}_n^m\text{TxtFex}_2$-identify L. Let T be any text for L. Let $\mathbf{M}_{i_1}, \mathbf{M}_{i_2}, \ldots, \mathbf{M}_{i_m}$ be m of the machines in $\{\mathbf{M}_1, \ldots, \mathbf{M}_n\}$ which converge to at most two grammars on T. Let S denote the set formed by taking the last two grammars output by $\mathbf{M}_{i_1}, \ldots, \mathbf{M}_{i_m}$ on T. Now at least $m - (n - m)$ of the machines in $\{\mathbf{M}_{i_1}, \ldots, \mathbf{M}_{i_m}\}$ TxtFex_2-identify L. Thus at least $m - (n - m) = 2m - n$ of the grammars in S are grammars for L. Thus at least one of $G_1(S)$ and $G_2(S)$ is a grammar for L. Note that both $G_1(S)$ and $G_2(S)$ can be obtained from the text T in the limit. Now it is easy to construct a machine $\mathbf{M}$ such that (1) $\mathbf{M}$ in the limit (on T) converges to a subset of $\{G_1(S), G_2(S)\}$, and (2) only grammars of L are output infinitely often by $\mathbf{M}$. This completes the proof of the theorem. We sketch our proof of the crucial lemma below.

Lemma 25. *Suppose $r, w \in N$ are given such that $r > w > 2r/5$. There exist recursive functions G_1 and G_2 such that, $(\forall p_1, p_2, \ldots, p_r)(\forall L)[\text{card}(\{i \mid 1 \leq i \leq r \wedge W_{p_i} = L\}) \geq w \Rightarrow W_{G_1(p_1, \ldots, p_r)} = L \vee W_{G_2(p_1, \ldots, p_r)} = L].$*

Proof. (Sketch) We assume without loss of generality that $w \leq r/2$ (otherwise the lemma can be easily proved by just taking majority). Suppose $p_1, \ldots, p_r$ are given. Below we give a procedure to enumerate two languages L_1 and L_2 (the procedure depends on $p_1, \ldots, p_r$). We will then argue that

$$(\forall L)[\text{card}(\{i \mid 1 \leq i \leq r \wedge W_{p_i} = L\}) \geq w \Rightarrow L = L_1 \vee L = L_2]$$

It will be easy to see that grammars for L_1 and L_2 can be obtained effectively from $p_1, \ldots, p_r$. This will prove the lemma.

Let $\text{similar}(i, j, n) = \max(\{n_1 \leq n \mid W_{i,n_1} \subseteq W_{j,n} \wedge W_{j,n_1} \subseteq W_{i,n}\})$. Intuitively, similar computes the closeness between two grammars. It denotes the point where it appears that the languages accepted by the two grammars differ.

Let $n_0 = 0$, $m1_0 = m2_0 = 0$. Let $e1_0 = 1$ and $e2_0 = 2$. Let $P1_0 = \{1, \ldots, w\}$ and $P2_0 = \{w + 1, \ldots, 2w\}$. We will enumerate elements in L_1, L_2 in stages. $e1_s, e2_s$ will be a permutation of $1, 2$. $P1_s$ and $P2_s$ will be disjoint subsets of $\{1, \ldots, r\}$ of size w each. Let L_1^s and L_2^s denote L_1 and L_2 enumerated before stage s. The following invariants will be maintained by the construction.

Invariants (assuming that stage s is executed)
1. $L_{e1_s}^s = \bigcup_{i \in P1_s}[W_{p_i, m1_s}] \subseteq \bigcap_{i \in P1_s}[W_{p_i, n_s}]$.

2. $L_{e2_s}^s \supseteq \bigcup_{i \in P2_s} [W_{p_i, m2_s}]$.
3. $\bigcup_{i \in P2_s} [W_{p_i, m2_s}] \subseteq \bigcap_{i \in P2_s} [W_{p_i, n_s}]$.
4. $L_{e2_s}^s - \bigcup_{i \in P2_s} [W_{p_i, m2_s}] \subseteq L_{e1_s}^s$.
5. $(\forall x \in L_{e2_s}^s)[\text{card}(\{j \in \{1, 2, \ldots, r\} - P1_s \mid x \in W_{p_i, n_s}\}) \geq w/2]$.
6. $m1_{s+1} > n_s \geq m1_s \geq m2_s$.

Begin {stage s}

    Search for $n > n_s$ such that there exist a set $P \subseteq \{1, \ldots, r\}$ of cardinality w such that $\text{similar}(p_i, p_j, n) > n_s$, for $i, j \in P$.

    If and when such an n is found, let $n_{s+1} = n$.

    $P1_{s+1} \subseteq \{1, \ldots, r\}$ be of cardinality w such that $m1_{s+1} = \min(\{\text{similar}(p_i, p_j, n) \mid i, j \in P1_{s+1}\})$ is maximized.

    Let $P2_{s+1} \subseteq \{1, \ldots, r\} - P1_{s+1}$ be of cardinality w such that $m2_{s+1} = \min(\{\text{similar}(p_i, p_j, n) \mid i, j \in P2_{s+1}\})$ is maximized.

    If $\text{card}(P1_{s+1} \cap P1_s) > \text{card}(P1_{s+1} \cap P2_s)$, then let $e1_{s+1} = e1_s$ and $e2_{s+1} = e2_s$, else let $e1_{s+1} = e2_s$ and $e2_{s+1} = e1_s$.

    Enumerate $\bigcup_{i \in P1_{s+1}} [W_{p_i, m1_s}]$ in $L_{e1_{s+1}}$.

    Enumerate $\bigcup_{i \in P2_{s+1}} [W_{p_i, m2_s}]$ in $L_{e2_{s+1}}$.

    Go to stage $s + 1$.

End {stage s}

It is easy to see by induction that the invariants claimed earlier are satisfied. Moreover if there exists a language L such that $\text{card}(\{i \mid 1 \leq i \leq r \wedge L = W_{p_i}\}) \geq w$, then $m1_s$ is unbounded. If there exist two distinct languages L such that $\text{card}(\{i \mid 1 \leq i \leq r \wedge L = W_{p_i}\}) \geq w$, then both $m1_s$ and $m2_s$ are unbounded. It follows now from the invariants that

$$(\forall L)[\text{card}(\{i \mid 1 \leq i \leq r \wedge W_{p_i} = L\}) \geq w \Rightarrow L = L_1 \vee L = L_2]$$

                                                              ■

Our proof of Theorem 24(b) involves a diagnolization argument using the operator recursion theorem [4]. This proof will be presented in the full paper.

3.4 Aggregation for Language Identification from Informants

Results presented in the previous section were for language learning criteria in which learning takes place from positive data only. In the present section, we record similar results for learning criteria in which learning takes place from both positive and negative data. It should be noted that the proof techniques for language learning from informants and function learning from graphs are very similar, although identification of r. e. languages from informants differs from identification of recursive functions, at least ostensibly, on following two counts.

- Graphs of recursive functions are computable but informants for r. e. languages may not be computable.
- Identification of recursive functions requires conjecturing a total program eventually but this is not the case for identification of r. e. languages from informants because computable decision procedures do not always exist for r. e. languages.

Identification from texts is an abstraction of learning from positive data. Similarly, learning from both positive and negative data can be abstracted as identification from informants. The notion of informants, defined below, was first considered by Gold [10].

Definition 26. A text I is called an *informant* for a language L just in case content$(I) = \{\langle x, 1 \rangle \mid x \in L\} \cup \{\langle x, 0 \rangle \mid x \notin L\}$.

The next definition formalizes identification in the limit from informants.

Definition 27. (a) **M** **InfEx**-*identifies* L (read: $L \in \mathbf{InfEx}(M)$) $\iff$ ($\forall$ informants I for L)$(\exists i \mid W_i = L)(\overset{\infty}{\forall} n)[\mathbf{M}(I[n]) = i]$.
(b) $\mathbf{InfEx} = \{\mathcal{L} \subseteq \mathcal{E} \mid (\exists \mathbf{M})[\mathcal{L} \subseteq \mathbf{InfEx}(M)]\}$.

We leave it to the reader to similarly define **InfFin**, **InfBc**, and for each $b \in N^+ \cup \{*\}$, **InfFex**$_b$. Also, for $m, n \in N^+$ and for each $\mathbf{I} \in \{\mathbf{InfFin}, \mathbf{InfEx}, \mathbf{InfFex}_b, \mathbf{InfBc}\}$, we can define **Team**$_n^m$ **I**-identification. We now present aggregation results for these new criteria.

For finite identification from informants, team aggregation takes place at success ratios greater than 2/3 as implied by the following results. This is not unexpected given results about finite function identification and finite language identification from texts.

Theorem 28. *(a)* $(\forall m, n \mid m/n > 2/3)[\mathbf{Team}_n^m \mathbf{InfFin} = \mathbf{InfFin}]$.
(b) $\mathbf{InfFin} \subset \mathbf{Team}_3^2 \mathbf{InfFin}$.

For identification in the limit, however, aggregation turns out to be different for informants and texts. In fact language identification from informants behaves very much like function learning, as aggregation for **InfEx** takes place at success ratios greater than 1/2. Aggregation for **InfBc** also takes place at success ratios greater than 1/2. These observations are summarized in the following result.

Theorem 29. *Let* $\mathbf{I} \in \{\mathbf{InfEx}, \mathbf{InfBc}\}$.

(a) $(\forall m, n \mid m/n > 1/2)[\mathbf{Team}_n^m \mathbf{I} = \mathbf{I}]$.
(b) $\mathbf{I} \subset \mathbf{Team}_2^1 \mathbf{I}$.

So, we are left with aggregation for **InfFex**$_b$, for $b \in N^+ \cup \{*\}$. However, using techniques from Case and Smith [8], it is easy to show the following result which implies that identification in the limit from informants is the same as vacillatory identification from informants.

Theorem 30. $(\forall b \in N^+ \cup \{*\})[\mathbf{InfFex}_b = \mathbf{InfEx}]$.

Hence, Theorem 29 holds for vacillatory identification from informants, too.

4 Conclusion

Clearly, aggregation issues for for $\mathbf{TxtFex}_b$, where $b \neq * \wedge b > 2$, are open. Only partial results can be shown at this stage, as the combinatorial complexity of the simulation arguments become difficult to handle. We summarize the state of art about aggregation in the following table; the symbol '?' denotes open questions.

Type of Identification	Finite	Limit	Vacillatory				Behaviorally Correct
			2	3	$\cdots$	$*$	
Function (Graph)	$> \frac{2}{3}$	$> \frac{1}{2}$	$> \frac{1}{2}$	$> \frac{1}{2}$	$> \frac{1}{2}$	$> \frac{1}{2}$	$> \frac{1}{2}$
Language (Text)	$> \frac{2}{3}$	$> \frac{2}{3}$	$> \frac{5}{6}$	?	?	$> \frac{1}{2}$	$> \frac{1}{2}$
Language (Informant)	$> \frac{2}{3}$	$> \frac{1}{2}$	$> \frac{1}{2}$	$> \frac{1}{2}$	$> \frac{1}{2}$	$> \frac{1}{2}$	$> \frac{1}{2}$

5 Acknowledgement

We wish to thank the referees for many valuable comments.

References

1. J. M. Barzdin. Two theorems on the limiting synthesis of functions. *In Theory of Algorithms and Programs, Latvian State University, Riga*, 210:82–88, 1974. In Russian.
2. J. M. Barzdin and K. Podnieks. The theory of inductive inference. In *Mathematical Foundations of Computer Science*, 1973.
3. M. Blum. A machine independent theory of the complexity of recursive functions. *Journal of the ACM*, 14:322–336, 1967.
4. J. Case. Periodicity in generations of automata. *Mathematical Systems Theory*, 8:15–32, 1974.
5. J. Case. The power of vacillation. In D. Haussler and L. Pitt, editors, *Proceedings of the Workshop on Computational Learning Theory*, pages 133–142. Morgan Kaufmann Publishers, Inc., 1988. Expanded in [6].
6. J. Case. The power of vacillation in language learning. Technical Report 93-08, University of Delaware, 1992. Expands on [5]; journal article under review.
7. J. Case and C. Lynes. Machine inductive inference and language identification. In M. Nielsen and E. M. Schmidt, editors, *Proceedings of the 9th International Colloquium on Automata, Languages and Programming*, volume 140, pages 107–115. Springer-Verlag, Berlin, 1982.
8. J. Case and C. Smith. Comparison of identification criteria for machine inductive inference. *Theoretical Computer Science*, 25:193–220, 1983.

9. R. Freivalds. Functions computable in the limit by probabilistic machines. *Mathematical Foundations of Computer Science*, 1975.

10. E. M. Gold. Language identification in the limit.. *Information and Control*, 10:447–474, 1967.

11. J. Hopcroft and J. Ullman. *Introduction to Automata Theory Languages and Computation*. Addison-Wesley Publishing Company, 1979.

12. S. Jain and A. Sharma. Finite learning by a team. In M. Fulk and J. Case, editors, *Proceedings of the Third Annual Workshop on Computational Learning Theory, Rochester, New York*, pages 163–177. Morgan Kaufmann Publishers, Inc., August 1990.

13. S. Jain and A. Sharma. Language learning by a team. In M. S. Paterson, editor, *Proceedings of the 17th International Colloquium on Automata, Languages and Programming*, pages 153–166. Springer-Verlag, July 1990.

14. S. Jain and A. Sharma. Computational limits on team identification of languages. Technical Report 9301, School of Computer Science and Engineering; University of New S outh Wales, 1993.

15. S. Jain and A. Sharma. Probability is more powerful than team for language identification. In *Proceedings of the Sixth Annual Workshop on Computational Learning Theory, Santa Cruz, California*. ACM Press, July 1993.

16. M. Machtey and P. Young. *An Introduction to the General Theory of Algorithms*. North Holland, New York, 1978.

17. D. Osherson, M. Stob, and S. Weinstein. Aggregating inductive expertise. *Information and Control*, 70:69–95, 1986.

18. D. Osherson, M. Stob, and S. Weinstein. *Systems that Learn, An Introduction to Learning Theory for Cognitive and Computer Scientists*. MIT Press, Cambridge, Mass., 1986.

19. D. Osherson and S. Weinstein. Criteria of language learning. *Information and Control*, 52:123–138, 1982.

20. L. Pitt. *A characterization of probabilistic inference*. PhD thesis, Yale University, 1984.

21. L. Pitt and C. Smith. Probability and plurality for aggregations of learning machines. *Information and Computation*, 77:77–92, 1988.

22. H. Rogers. Gödel numberings of partial recursive functions. *Journal of Symbolic Logic*, 23:331–341, 1958.

23. H. Rogers. *Theory of Recursive Functions and Effective Computability*. McGraw Hill, New York, 1967. Reprinted, MIT Press 1987.

24. C. Smith. The power of pluralism for automatic program synthesis. *Journal of the ACM*, 29:1144–1165, 1982.

LEARNING WITH GROWING QUALITY

Juris Viksna
Department of Computer Science
University of Delaware
Newark, Delaware 19716
USA
[e-mail: viksna@cis.udel.edu]

Abstract

Usually "quality" of learning grows with experience. Here is given a formalization of that phenomenon within a recursion theoretic framework. We consider the learning of total recursive functions by some algorithmic device (inductive inference machine) and describe the "quality" of learning in two different ways: as probability with which machine identifies the given function correctly, and as density of a set of arguments for which the hypothesis given by machine coincides with the identifiable function. We prove that in both cases there exist classes of sets of total recursive functions, such that for each of these sets the "quality" with which a learning device can identify an arbitrary function from the set grows with the number of other functions, which learning device are trying to identify at the same time, i.e., these classes are identifiable only with learning devices that show some improvement of learning capabilities with practice.

Introduction

In real life situations human capacity to learn usually grows with experience. In particular, practice in learning allows more effectively acquire new information. Here we are trying to study the question, whether such improvement of learning capacity is simply a property of some learning devices, and these devices in principle can also be constructed in a such way, that they do not satisfy this property and still can perform the same learning task; or there exist learning tasks that can be performed only by such algorithmic learning devices which with growing experience shows some improvement of learning capacity.

We give a possible formalization of that phenomenon within a recursion theoretic framework. The obtained results in some sense show that there exist sets of objects that are learnable only in the way defined above, i.e., in order to learn any of these objects with required correctness, we need also learn sufficiently large number of other objects from the same set.

As objects for learning we will consider total recursive functions, and for the task of learning we will use inductive inference machines, a special type of Turing machines, which as input receive the graph of some total recursive function f, and which can, time by time, output some natural number as hypothesis about the input function f.

Inductive inference of total recursive functions is widely studied (see [2] and [4] for survey). At the same time, usually are considered only situations, when the task of learning does not depend from the previous experience of the learning device. It seems that the only results how to formalize the "learning how to learn" are due to D.Angluin, W.Gasarch and C.Smith (see [1] and [5]). In particular, they have shown that there exist a set of finite sequences of total recursive functions such that any sequence can be identified only in previously defined order, i.e., the first function in sequence can be identified without additional information, the second function can be identified only if we previously had identified the first, etc. This formalization could be considered as learning with the help of teacher, which presents for us information in some "good order", so that we can acquire it.

Here we obtain results of the similar type for another learning paradigm. We prove that there exist classes of sets of functions, such that the "quality" of learning grows with the number of functions which inductive inference machine is trying to identify simultaneously.

As a measure of "quality" of learning we will use two different notions:

probability with which inductive inference machine for the given function outputs correct hypothesis, and
density of a set of arguments, for which the hypothesis of inductive inference machine coincides with the given function.

Definitions and notations

Most of our definitions and notations are standard and can be found in [4] or [7]. We consider some fixed Gödel numbering $\phi_1, \phi_2, \ldots$ of all partial recursive functions, and the task of inductive inference machine, when given the graph of some total recursive function f, i.e., the sequence of values $f(0), f(1), \ldots$, is to find a natural number n, such that $f = \phi_n$.

Definition.

The subset A of total recursive functions is identifiable in the limit with probability $p \in (0,1]$, if there exists an inductive inference machine M, such that for every function $f \in A$, the machine M, when given the graph of f, with probability p outputs a finite sequence of natural numbers $i_1, i_2, \ldots, i_k$, such that $f = \phi_{i_k}$. ♦

We denote the class of all sets of functions identifiable in such sense by EX.

L.Pitt has shown, that the classes EX form a discrete hierarchy, i.e., $EX_p = EX_q$ if and only if p and q are both from one of the intervals: (1/2,1],(1/3,1/2],(1/4,1/3], etc. (See [6] and [9]).

For every set of natural numbers A we define the density of A as value

$$d(A) = \lim_{n \to \infty} \text{card } \{x \in A \mid x \leq n\} / n.$$

Intuitively, such definition of density for given set A describes "how many" of all natural numbers belong to A. It is easy to see that $d(A) \in [0,1]$.

Definition.

Recursive functions f and g are called equal up to a set with density $a \in [0,1]$, if $d(\{ x \mid f(x)=g(x) \}) \geq 1-a$. ♦

We denote such equality by $f =^{*a} g$.

Definition.

The subset A of total recursive functions is identifiable in the limit up to a set with density $a \in [0,1]$, if there exists an inductive inference machine M, such that for every function $f \in A$, the machine M, when given the graph of f, outputs a finite sequence of natural numbers $i_1, i_2, ..., i_k$, such that $f =^{*a} \phi_{i_k}$. ♦

We denote the class of all sets of functions identifiable in such sense by EX^{*a}.

For the classes EX^{*a} it is known that they form a dense hierarchy, i.e., for $a_1 > a_2$ $EX^{*a_2} \subset EX^{*a_1}$ and $EX^{*a_1} - EX^{*a_2} \neq \emptyset$. (see [8] and [10]).

To describe the situation where inductive inference machine can work with different amounts of experience, we consider IIM, which in parallel can identify several functions. Then the "experience" of IIM can be characterized by number of functions which IIM are trying to identify simultaneously.

Let Δ be a sequence (finite or infinite) of real numbers $\langle a_1, ..., a_n, ... \rangle$, such that $1 \geq a_1 \geq ... \geq a_n \geq ... \geq 0$. Cardinality of Δ we denote by ν (if sequence is finite, $\nu \in N$, otherwise $\nu = \omega$).

Definition.

The class A of sets with cardinality ν of total recursive functions is D-identifiable with respect to sequence of densities Δ, if there exists an inductive inference machine M, such that for arbitrary set $\{f_1, ..., f_n, ...\} \in A$ and for arbitrary natural number i, with $1 \leq i \leq \nu$, the machine M, when given graphs of arbitrary i functions $f_{k1}, ..., f_{ki}$ from this set, outputs i finite sequences of natural numbers $h(1,1), ..., h(1,b_1); ...; h(i,1), ..., h(i,b_i)$, such that for every j, with $1 \leq j \leq i$, $\phi_{h(j,b_j)} =^{*a_1} f_{kj}$. ♦

We denote the class of all D-identifiable (with sequence of densities Δ) classes of sets of functions by $D(\Delta)EX$.

Let Π be a finite sequence of real numbers $\langle p_1, ..., p_n \rangle$, such that $0 \leq p_1 \leq ... \leq p_n \leq 1$.

Definition.

The class A of sets with cardinality n of total recursive functions is P-identifiable with respect to sequence of probabilities Π, if there exists an inductive inference machine M, such that for arbitrary set $\{f_1, ..., f_n\} \in A$ and for arbitrary natural number i, with $1 \leq i \leq n$, the machine M, when given graphs of arbitrary i functions $f_{k1}, ..., f_{ki}$ from this set, outputs i sequences of natural numbers $h(1,1), ..., h(1,b_1), ...; ...; h(i,1), ..., h(i,b_i), ...$, such that for every j, with $1 \leq j \leq i$, the sequence $h(j,1), ..., h(j,b_j)$ with probability p_i is finite and $\phi h(j,b_j) = f_{kj}$. ♦

We denote the class of all P-identifiable (with sequence of probabilities Π) classes of sets of functions by $P(\Pi)EX$.

Here in the case of probabilistic identification, we consider only the finite sequences of probabilities Π (i.e., with $card(\Pi) < \omega$). The case when $card(\Pi) = \omega$ is not interesting, so that due to the results from [6], there does not exist infinite sequence of probabilities, such that for all of them, classes of functions, identifiable with these probabilities, are different.

At first, it is easy to observe, that for arbitrary Δ and Π, the corresponding classes $D(\Delta)EX$ and $P(\Pi)EX$ are not empty.

Let $\Delta = <a_1, ..., a_n, ...>$, $card(\Delta) = v$. Then for every set $A \in EX^{*a}$, where $a = \lim_{n \to \infty} a_n$, $A^v \in D(\Delta)EX$.

Similarly, for $\Pi = <p_1, ..., p_n>$ and for every set $A \in EX_p$, where $p=p_n$, $A^n \in P(\Pi)EX$.

However, these are not very interesting members of classes $D(\Delta)EX$ and $P(\Pi)EX$, because they do not show that quality of learning increases with number of considered functions. Further we are going to prove that actually there exist also more interesting elements in classes $D(\Delta)EX$ and $P(\Pi)EX$, which shows, that capacity of learning is growing, when we are trying to learn simultaneously more functions. We will say that such elements are the proper elements of classes $D(\Delta)EX$ and $P(\Pi)EX$.

Let $\Delta = <a_1, ..., a_n, ...>$ and $card(\Delta) = v$, $v > 1$.

Definition.

Class of sets of functions $A \in D(\Delta)EX$ is a proper element of $D(\Delta)EX$, if for every natural number i, with $2 \leq i \leq v$, there does not exist inductive inference machine M, such that for arbitrary set $\{f_1, ..., f_n, ...\} \in A$, the machine M, when given graphs of some i-1 functions $f_{k1}, ..., f_{ki-1}$ from this set, outputs i-1 finite sequences of natural numbers $h(1,1), ..., h(1,b_1); ...; h(i-1,1), ..., h(i-1,b_{i-1})$, such that there exist j, with $1 \leq j \leq i-1$, such that $\phi h(j,b_j) = {}^{*ai} f_{kj}$. ♦

Proper elements of $P(\Pi)EX$ we define similarly.

Let $\Pi = <p_1, ..., p_n>$, $n > 1$.

Definition.

Class of sets of functions $A \in P(\Pi)EX$ is a proper element of $P(\Pi)EX$, if for every natural number i, with $2 \leq i \leq n$, there does not exist inductive inference machine M, such that for arbitrary set $\{f_1, ..., f_n\} \in A$, the machine M, when given graphs of some i-1 functions $f_{k1}, ..., f_{ki-1}$ from this set, outputs i-1 finite sequences of natural numbers $h(1,1), ..., h(1,b_1); ...; ...; h(i-1,1), ..., h(i-1,b_{i-1}), ...$ such that for every j, with $1 \leq j \leq i-1$, the sequence $h(j,1), ..., h(j,b_j), ...$ with probability p_i is finite and $\phi h(j,b_j) = f_{kj}$. ♦

Main results

In this section we will show that there exist proper elements in classes $D(\Delta)EX$ and $P(\Pi)EX$.

At beginning we will consider classes $D(\Delta)EX$ with finite sequences of densities Δ.

Theorem 1.

For every $\Delta = <h_1, ..., h_n>$, $n > 1$, if $1 \geq h_1 > ... > h_n \geq 0$, then there exists a proper element A in $D(\Delta)EX$. ♦

Proof.

For reasons of simplicity here we will consider the case when $n=2$. Proof of the general case is in principle similar, but technically more complicated.

First, it is not hard to show that for arbitrary numbers h_1, $h_2 \in \{0,1\}$, the set of all natural numbers N can be split into 4 disjoint infinite recursive subsets $S_0, ..., S_3$, such that $d(S_0)=0$, $d(S_1)=1-h_1$, $d(S_2)=h_1-h_2$, $d(S_3)=h_2$.

For every subset of natural numbers S and for every natural number k, by S(k) we denote the k- th smallest element from S.

For every subset of natural numbers S and for every total recursive function f, by f(S) we denote a recursive function, such that, for every $k \in N$, $f(S)(k)=f(S(k))$.

Let $<.,.,.,.>: N \times N \times N \times N \to N$ be a recursive one to one mapping between the set of all 4 -tuples of natural numbers and the set of all natural numbers.

Class of pairs of total recursive functions A we define in the following way:
$A = \{\{f_1, f_2\}$ | the sequences $f(S_0)(n)$ and $f(S_0)(n)$ converges,

$$\lim_{n \to \infty} f_1(S_0)(n) = <a_1, 1, b_1, c_1>,$$

$$\lim_{n \to \infty} f_2(S_0)(n) = <a_2, 2, b_2, c_2>,$$

$$\phi_{a1} = f_1(S_1), \phi_{a2} = f_2(S_2),$$
$$\phi_{b1+b2} = f_1(S_2), \phi_{c1+c2} = f_2(S_2)\}.$$

It is easy to see that $A \in D(<h_1, h_2>)EX$, so that IIM which receives graph of one of the functions f_i from arbitrary pair $\{f_1, f_2\}$, can find in the limit value of a_i, and hence it can identify function f_i up to a set with density $1 - d(S_1)=h_1$; IIM which receives graphs of both of the functions f_1 and f_2 from arbitrary pair $\{f_1, f_2\}$, can find in the limit values a_1, a_2, b_1, b_2, c_1, c_2, and hence it can identify both functions f_1 and f_2 up to a set with density $1 - d(S_1) - d(S_2)=h_2$.

It remains to show that A is a proper element of $D(<h_1, h_2>)EX$, i.e., we have to show that for arbitrary inductive inference machine M there exists a pair of functions $\{f_1, f_2\} \in A$, such that M can identify none of the functions f_1 and f_2 only from its graph.

For each inductive inference machine M and for each total recursive function ϕ_g, such that M does not EX- identify ϕ_g with probability greater or equal to 1/2, we will construct a pair of functions $\{f_1, f_2\}$ in the following way.

First, it is not hard to show that we can algorithmically construct a sequence of natural numbers $m_0, m_1, ..., m_k, ...$, which can be either finite or infinite, and which satisfies the following properties:

if the sequence is infinite, then ϕ_{m0} is a total recursive function, such that

$\phi_{m0}(x)=< t, 1, 0, g >$, if $x \in S_0$ (where t is a natural number, such that $\phi_t(x)=1$ for all $x \in N$),

$\phi_{m0}(x)=1$, if $x \in S_1$, and

M does not EX- identify ϕ_{m0} up to a set with density h_2;

if the sequence is finite with last element m_k, then ϕ_{mk} is a total recursive function, such that

$\phi_{mk}(x)=< t, 1, 0, g >$, if $x \in S_0$ (where t is a natural number, such that $\phi_t(x)=1$ for all $x \in N$),

$\phi_{mk}(x)=1$, if $x \in S_1$, and

M does not EX- identify ϕ_{mk} up to a set with density h_2.

Now we define $f_1=\phi_{m0}$, $m=m_0$, if the sequence $m_0, ..., m_k, ...$ is infinite, and $f_1=\phi_{mk}$, $m=m_k$, if the sequence $m_0, ..., m_k$ is finite with last element m_k.
We define f_2 to be a total recursive function, such that

$f_2(S_0)(x)=< t,2,m,0 >$, for all $x \in N$,

$f_2(S_1)(x)=1$, for all $x \in N$, and

$f_2(S_2)(x)=\phi_g(x)$, for all $x \in N$.

Let A' be a class that consists of all such pairs $\{f_1, f_2\}$, i.e., for each IIM M and for each total recursive function ϕ_g, which M does not EX- identify with probability greater or equal to 1/2, we place into class A' a pair of functions, constructed in a way described above.

From the construction it follows that $A' \subset A$. If we assume that there exists an inductive inference machine M, such that for each pair $\{f_1, f_2\} \in A'$ machine M EX- identifies at least one of the functions f_1 and f_2 up to a set with density h_2 only from the graph of that function, then we can also construct another inductive inference machine M', such that M' can identify all the functions identifiable by M (up to a set with density h_2), and such that for each pair of functions $\{f_1, f_2\} \in A'$, in the case when M can identify function f_2, M' can also EX- identify function ϕ_g (which corresponds to the pair of functions f_1 and f_2) with probability 1/2 and up to a set with density h_2. Thus, for each pair of functions $\{f_1, f_2\} \in A'$ machine M' either EX- identifies function f_1, or EX- identifies the corresponding function ϕ_g with probability 1/2 (up to a set with density h_2), which contradicts our construction, so that for each IIM and for each $h \in [0,1)$ there exists a total recursive function that is not EX- identifiable with probability 1/2 up to a set with density h_2.

Therefore for each inductive inference machine M there exists a pair of functions $\{f_1, f_2\} \in A$, such that M can identify none of the functions f_1 and f_2 up to a set with density h_2 only from its graph, i.e., A is a proper element of $D(<h_1, h_2 >)EX$.

In general situation we can split the set of natural numbers N into $n+2$ disjoint subsets $S_0, ..., S_{n+2}$, such that $d(S_0)=0$, $d(S_1)=1-h_1$, ..., $d(S_{n+2})=h_n$, and define class A similarly, using a one to one mapping between N and all 2^n -tuples of natural numbers. ♦

Similar theorem we also can prove for the classes $P(\Pi)EX$. So that each function, identifiable in the limit with probability $p \in (0,1]$, is identifiable also with probability $1/k$, where $k \in N-\{0\}$ and $1/(k+1) < p \leq 1/k$, we will consider only sequences of probabilities Π in the form $\Pi = <1/k_1, ..., 1/k_n>$, $k_i \in N-\{0\}$.

Theorem 2.

For every $\Pi = <1/k_1, ..., 1/k_n>$, where $k_i \in N-\{0\}$, if $k_1 > ... > k_n \geq 1$, then there exist a proper element A in $P(\Pi)EX$. ♦

Sketch of proof.

We will here consider the case when $n=2$. For each of the functions f from the class A and for each $x \in N$, we will consider the value $f(x)$ as a 5- tuple of natural numbers $<t_1, t_2, t_3, t_4, t_5>$. The i- th component of $x=<x_1, x_2, x_3, x_4, x_5>$ we denote by $x[i]$ (where $i \in \{1, ...,5\}$).

Let Z_1, Z_2, Z_3 be subsets of N, such that $x \in Z_i$ if and only if $x=<x_1, x_2, i, x_4, x_5>$, where $i \in \{1, 2, 3\}$ and $x_1, x_2, x_4, x_5 \in N$.

Class of pairs of total recursive functions A we define in the following way:
$A=\{\{f_1, f_2\}|$ for all $n \in N$: $f_1(n)[1]=1$, $f_2(n)=2$,

$$\lim_{\substack{n \to \infty \\ f_1(n) \in Z_3}} f_1(n)[4] = a_1, \quad \lim_{\substack{n \to \infty \\ f_1(n) \in Z_3}} f_1(n)[5] = b_1,$$

$$\lim_{\substack{n \to \infty \\ f_2(n) \in Z_3}} f_2(n)[4] = a_2, \quad \lim_{\substack{n \to \infty \\ f_2(n) \in Z_3}} f_2(n)[5] = b_2,$$

$\phi_{a1+a2} = f_1$, $\phi_{b1+b2} = f_2$, and, either

$$\lim_{\substack{n \to \infty \\ f_1(n) \in Z_1}} f_1(n)[4] \quad \text{or} \quad \lim_{\substack{n \to \infty \\ f_1(n) \in Z_2}} f_1(n)[4]$$

converges to a, such that $\phi_a = f_1$, and, either

$$\lim_{\substack{n \to \infty \\ f_2(n) \in Z_1}} f_2(n)[4] \quad \text{or} \quad \lim_{\substack{n \to \infty \\ f_2(n) \in Z_2}} f_2(n)[4]$$

converges to b, such that $\phi_b = f_2 \}$. ♦

Analog of theorem 1 also holds for infinite sequences of densities Δ.

Theorem 3.

For every infinite $\Delta=<a_1, ..., a_n, ...>$, if $1 \geq a_1 > ... > a_n ... > 0$, then there exist a proper element A in $D(\Delta)EX$. ♦

This theorem shows that there exist class A of infinite sets of total recursive functions, such that for each infinite set S of recursive functions from A, in order to learn with sufficient correctness any function from set S, we must learn simultaneously sufficiently large number of other functions from the same set. However here we require that IIM that we use for learning must be the same for all sets from A, i.e., probably there still exist an IIM which can identify up to a set with density b any function f from some set $I \in A$ only from the graph of f, where $a_i > b \geq 0$ for all a_i from Δ (it is clear that such IIM will always exist, if we will require only that $a_i > b \geq 0$ for some a_i from Δ). For class A that we use for proof of theorem 3 really exist such IIM that can identify for some set $I \in A$ any function from I up to a set with density 0; however we expect that in described sense some stronger variant of theorem 3 still holds. Formulation of such potential result is given by the following conjecture.

Conjecture 1.

For every infinite $\Delta = <a_1, ..., a_n, ...>$, if $1 \geq a_1 > ... > a_n ... > 0$, then there exist a class $A \in D(\Delta)EX$, such that for every inductive inference machine M there exist a natural number $k \in N$, such that M cannot identify arbitrary function $f \in A$ up to a set with density a_i, with $i \geq k$, only from the graphs of f and any i-2 other functions $f_1, f_2, ...,f_{i-2}$. ♦

Some conclusions

Here we will try shortly discuss, what could be the connections between our theorems and actual learning processes.

Theorems 1 and 2 in some sense show that there exist sets of objects (here these objects are described by sets of recursive functions), such that, if we want sufficiently good "learn" some object from the set, we must use for learning available information about the whole object, even if we are interested only in part of it. The more we will restrict our attention to some part of the object, the worse will be the "quality" of our learning, where by quality we understand either probability, that answer will be correct, or the size of the set of inputs, for which our answer is correct.

It also seems, that for learning paradigm, discussed here, that is all what we can say about probabilistic learning.

For second type of learning, we can say something more.

Theorem 3 shows that there also exist sets of "infinite" objects with similar property, i.e., we never will be able to "learn" completely any of the objects from such set, but, the more we will try, the more correct our answer will be. At the same time, for any particular object there remains a possibility that some "smart" learner simply could "guess" the whole information about that object. Clearly that will always be the case, if the objects for learning are "finite" (i.e., such that can be described by finite number of recursive functions); for "infinite" objects we made a conjecture that in some cases even such "smart" learners do not exist. Existence of class A that satisfies conjecture 1 would also imply that there exist objects which is learnable with arbitrary correctness (after learning device had examined sufficiently large part of them), and which, at the same time, cannot be completely identified with any finite number of "right guesses" about them.

References

1. D.Angluin, W.Gasarch and C.Smith. Training sequences. Theoretical Computer Science 66 (1989), p.255-272.
2. D.Angluin and C.Smith. Inductive inference: theory and methods. ACM Computing Surveys 15 (1983), p.237-269.
3. L.Blum and M.Blum. Toward a mathematical theory of inductive inference. Information and Control 28 (1975), p.125-155.
4. J.Case and C.Smith. Comparison of identification criteria for machine inductive inference. Theoretical Computer Science 25 (1983), p.193-220.
5. W.Gasarch and C.Smith. On the inference of sequences of functions. Lecture Notes in Computer Science 265 (1987), p.23-41.
6. L.Pitt. Probabilistic inductive inference. Journal of the ACM 36 (1989), p.383-433.
7. H.Rogers. Theory of Recursive Functions and   Effective Computability. MIT Press, 1987.
8. J.Royer. Inductive inference of approximations.   Information and Computation 70 (1986), p.156-178.
9. C.Smith. The power of pluralism for automatic   program synthesis. Journal of the ACM 29 (1982), p.1144-1165.
10.C.Smith and M.Velauthapillai. On the inference of approximate programs. Theoretical Computer Science 77 (1990), p.249-266.

Use of Reduction Arguments in Determining Popperian FIN-Type Learning Capabilities

Robert Daley   Bala Kalyanasundaram

Computer Science Department, University of Pittsburgh
Pittsburgh, PA 15260, USA

Abstract

The main contribution of this paper is the development of analytical tools which permit the determination of team learning capabilities as well as the corresponding types of redundancy for Popperian FINite learning. The basis of our analytical framework is a reduction technique used previously by us.

Using our analytical tools we determine the redundancy types for all ratios of the form $\frac{4n}{9n-2}$, where $n \geq 2$, which defines the sequence of capabilities for probabilistic $PFIN$-type learners in the interval $(\frac{4}{9}, \frac{1}{2}]$. We also show that there is unbounded redundancy at the ratio $\frac{2}{5}$, i.e., quadrupling the team size and keeping the success ratio will always increase learning power. We also extend, using our tools, the region of known $PFIN$-type learning capabilities, by presenting an ω^2 sequence of learning capabilities beginning at $\frac{1}{2}$ which converges to $\frac{2}{5}$. We believe that this sequence forms the *backbone* of the learning capabilities in this interval, but that the actual capability structure is very much more complex.

1   Introduction

The research described in this paper deals fundamentally with the question of how teams of learners (i.e., learning algorithms) can work cooperatively to learn (from examples) classes of functions which individually they would be unable to learn. We refer to a team of learners as a *pluralistic learner*.

The objects to be learned in this paper are arbitrary total recursive functions, although all of the functions which we actually construct to separate learning capabilities will be functions which are eventually constant. Each individual learner will be a *FIN*-type learner, whereby it receives examples of the graph of the function to be learned and can hypothesize at most a single program for the function after having only seen a finite number of examples. In other words, a *FIN*-learner is not permitted to change its mind. In this paper we will further restrict our attention to *PFIN*-type (Popperian *FIN*-type) learners, which are permitted to hypothesize only programs which compute total functions (i.e., which halt on every input) [1]. We do this as a first step, because as evidenced by previous work the combinatorial complexity of the analysis is much simpler than that for FIN-type learners. As we shall presently see the situation for $PFIN$-type learners is so complex that the possibility for completely determining learning capabilities and redundancy types for FIN-type learners seems very remote.

Given a team of s learners we say that it can learn a class of functions $\mathcal{F}$ with plurality "r out of s" (denoted r/s) if given any function f from $\mathcal{F}$ at least r of the s members can correctly learn f. It is then natural to ask how the learning capability of such a team of learners with plurality r/s compares to that of a single probabilistic learner which can successfully learn any function from the class $\mathcal{F}$ with probability at least $\frac{r}{s}$. While in many situations the capabilities of these two types of learners coincide, there are other situations where they are quite different and this gives rise to the phenomenon called *redundancy*. Understanding the complex nature of redundancy is of central interest in the present work. Notice that one can also define a much simpler learning model where *every* learner in the team is required to output a program that computes a total function (Strong Popperian).

Work on these kinds of questions was begun by Freivalds [6] who showed that the learning capabilities of probabilistic learners beginning with probability 1 and ending with probability $\frac{1}{2}$ form discrete intervals (whose endpoints are given by the sequence $\frac{n}{2n-1}$ for $n \geq 2$) such that any two probabilities in an interval yield equivalent learning power. This was extended to teams of learners by Daley et. al in [5]. The phenomenon of redundancy was first observed for such learners by Velauthapillai [8] who showed that a team of 2 with plurality $1/2$ was less capable than a team of 4 with plurality $2/4$. This showed that for the ratio $\frac{1}{2}$ the capabilities of probabilistic and team learners are different.

Our work has continued the investigation of these questions (see [2, 3, 4]) for a variety of different types of learners and has resulted in what appear to be general properties associated with the learning capabilities of teams of learners as well as probabilistic learners. In particular, it was shown in [4] that the learning capabilities of $PFIN$-type learners in the interval $(\frac{4}{9}, \frac{1}{2}]$ is broken into intervals whose end-points are given by the sequence $\frac{4n}{9n-2}$ for $n \geq 2$. Perhaps, the most interesting and important thing revealed by these investigations is the intimate nature of the mutual information which team members must possess about one another and the extremely complex nature of their interaction required in order for them to fully exploit their joint learning capability.

The redundancy type of a ratio r/s, where r and s are relatively prime, is the partially ordered set formed by the relative learning capabilities of teams of learners with pluralities kr/ks, for $k \geq 1$. For example, from [4] we have that the $PFIN$ redundancy type at the ratio $\frac{4}{9}$ is $\langle \{1, 2\} \prec \{3m+1, 3m+2 : m \geq 1\} \prec \{3m : m \geq 1\} \rangle$, where we represent the equivalence classes as multiples of the base ratio $\frac{4}{9}$ and where $X \prec Y$ denotes that the class Y is *strictly* more capable than the class X. Thus, determining the redundancy type of a ratio is a special case of the general problem of determining the relative learning capabilities of two different teams of learners.

The process of determining the learning capability of a particular probabilistic or team *PFIN*-type learner involves showing that this learner (the pursuing learner) can simulate any probabilistic or team learner (the target learner) with a success ratio within its capability interval. The pursuing learner's simulation in turn involves the simulation of the programs produced by the target learner. Of course, the incorrect programs produced by the target learner can compute functions different from the (true) function to be learned, so that we see the target learner's programs split into groups (each group computing a different function), where only one group can compute the true function. Thus, the task of the pursuer's hypothesized programs will be to cover all these different functions (i.e., different groups of programs) in such a way that the correct plurality is maintained, where the correct plurality is roughly proportional to the size of the group following a particular function. This is the genesis of our reduction technique. This covering task is considerably simplified if the distribution of the values computed by the target learner's programs on any given input is known, and is the reason why we consider only *PFIN*-type learners in this paper. The amount of simplification can be seen by comparing the strategy given for the *PFIN*-type learner with plurality 2/4 in [4] with that for the analogous *FIN*-type learner given in [3].

Using the analytical tools developed in this paper, we are able to determine the *PFIN* redundancy type for each ratio in the interval $(\frac{4}{9}, \frac{1}{2}]$ of the form $\frac{4n}{9n-2}$ for $n \geq 2$. Generally speaking we show that there are two basic kinds of redundancy types: (1) no redundancy (for values of $n \not\equiv 2 \bmod 8$); and (2) odd-even redundancy (for values of $n \equiv 2 \bmod 8$), which is the redundancy type which exists for the ratio $\frac{1}{2}$. However, there are a finite number of exceptions to this general pattern, particularly for the ratio $\frac{17}{38}$ ($n = 34$), where the redundancy type $\langle \{1\} \prec \{7\} \prec \begin{smallmatrix} \{2,4\} \\ \{3,5,2i-1:i\geq 5\} \end{smallmatrix} \prec \{2i : i \geq 3\} \rangle$ forms a proper partially ordered (i.e., non-total ordered) set.

To obtain these results on redundancy types, we make extensive use of the reduction technique developed in [4] whereby the learning capability of a particular ratio is reduced to combinations of learning capabilities of previously determined sequences, in particular the sequence $\frac{n}{2n-1}$ giving the capabilities above $\frac{1}{2}$. In fact, this approach sheds considerable light on the origin of the sequences of intervals of learning capabilities. In particular, the sequence $\frac{4n}{9n-2}$ is generated by the combination of reductions $R_2^1 + R_n^1$, where R_i^1 is the reduction to the learning capability of the ratio $\frac{i}{2i-1}$ (from the sequence of capabilities above $\frac{1}{2}$). Furthermore, using these techniques it is possible to construct an ω^2 sequence of learning capabilities beginning at $\frac{1}{2}$ and converging to $\frac{2}{5}$. This sequence has the form $\frac{2mn}{(5m-1)n-m}$ for $m \geq 2$ and $n \geq m \cdot (m-1)$, and is based on the combination of reductions $R_m^1 + R_n^1$. For example, for $m = 2$ we obtain the sequence $\frac{4n}{9n-2}$, and letting $m = 3$ we obtain the next sequence $\frac{6n}{14n-3}$.

It is tempting to conjecture that this simple mechanism allows one to generate the complete set of capabilities in the interval $(\frac{2}{5}, \frac{1}{2}]$, but in fact the situation is considerably more complex than this, and we show that the capabilities in the interval $(\frac{3}{7}, \frac{4}{9}]$ is generated by three interleaved and overlapping sequences:

$\frac{6n}{14n-3}$ (generated by $R_3^1 + R_n^1$); $\frac{3n}{7n-1}$ (generated by $R_2^1 + R_2^1 + R_n^1$); and $\frac{8n}{19n-4}$ (generated by $R_4^1 + R_n^1$). The last sequence is actually the "backbone" sequence for the $m = 4$ case of our ω^2 sequence, and so only a finite number ($4 \leq n \leq 11$) of its points lie in the interval $(\frac{3}{7}, \frac{4}{9}]$. Turning to redundancy types for this interval we point out that our initial results (not reported here) demonstrate a general pattern of redundancy types plus infinitely many exceptions to this pattern, where all but finitely many of the exceptions (seem to) form a pattern also. Thus, the prospects of determining all the learning capabilities and all the redundancy types for even the interval $(\frac{2}{5}, \frac{1}{2}]$ appear to be bleak indeed.

The ratio $\frac{2}{5}$ is of particular interest because we show in this paper that there is unbounded redundancy for this ratio (i.e., the redundancy type contains an infinite ascending chain of learning capabilities). This is the highest ratio for which such unbounded redundancy has been observed (see [5]) and we conjecture that there are no larger ratios with unbounded redundancy.

2 Definitions and Notation

Consider a learning strategy L that receives values $(0, f(0)), (1, f(1)), (2, f(2)), \ldots$ of some (total recursive) function f from a class $\mathcal{F}$. In general, while observing this sequence of functional values, L will issue a sequence of hypotheses (i.e., programs) $h_1, h_2, h_3, \ldots$ not necessarily one for each new functional value L sees. Without loss of generality we may assume that $h_i \neq h_{i+1}$, so that each new conjecture represents a mind change from the previous hypothesis. We say that the algorithm L correctly learns the function f with respect to *FIN*-type learning (denoted $f \in FIN[L]$) if after a finite amount of computation (and after seeing a finite number of values of f), L produces a single hypothesis for f and this hypothesis is a correct program for f. Thus, L is not allowed any changes of mind. We say that a class $\mathcal{F}$ of functions is *FIN*-learnable (denoted by $\mathcal{F} \in FIN$) if there is a strategy L that can correctly identify every function f in $\mathcal{F}$ with respect to *FIN*-type learning.

As with any probabilistic algorithm, a probabilistic learning strategy L consults the outcome of a sequence of flips of a fair coin during the course of its computation. We say that the strategy L learns (with respect to *FIN*-type learning) a function f with probability p, if the probability, taken over all possible coin flips of L, that the strategy outputs a single correct program for f is at least p. We denote the fact that L so learns f by $f \in \textbf{FIN}\langle \textbf{P}: p\rangle[L]$, and we denote by $\textbf{FIN}\langle \textbf{P}: p\rangle$ the family of all classes of functions that are *FIN* learnable by some probabilistic strategy with probability at least p. The modifer $\langle \textbf{P}: p\rangle$ to *FIN* denotes that the **Probability** of success is at least p.

Analogous to pluralistic scientific communities, one can consider learning by a team of strategies. Here, the strategies in the team can cooperate to learn functions. Let $L = \{L_1, L_2, \ldots, L_s\}$ be the team that cooperates to learn a class $\mathcal{F}$ of functions. Each member of the team receives the values of a function f in $\mathcal{F}$, and the team L correctly learns f (with respect to *FIN*-type learning) with plurality r/s, if at least r out of s members output a single program which is correct for f. Notice that the remaining $s - r$ learners need not even produce any programs. The team L learns the class $\mathcal{F}$ of functions (with respect to *FIN*-type learning) (with plurality r/s) if the team L learns every f in $\mathcal{F}$ with respect to *FIN*-type learning. We denote the fact that the team L so learns f by $f \in \mathbf{FIN}\langle \mathbf{T} : r/s \rangle[L]$, and we denote by $\mathbf{FIN}\langle \mathbf{T} : r/s \rangle$ the collection of such classes of functions that are *FIN* learnable by some team of s learners with plurality at least r out of s. The modifer $\langle \mathbf{T} : r/s \rangle$ to *FIN* denotes that the learner consists of a **Team** of s learners with success ratio at least r/s. We are now ready to define *PFIN*-type learning.

For probabilistic or team *FIN*-type learning it is possible that many programs are produced out of which some of them (depending on the success ratio) compute f correctly. In this case, those programs that do not compute f correctly, need not always compute a total function. Further, some learners (in team learning) may not even participate in the learning process, i.e., they may not produce any programs at all. In this paper we impose the additional condition on our learners that all of the programs produced by any participating learner must compute a total function. A *PFIN*-type learner is a *FIN*-type learner which hypothesizes only programs for total computable functions. One can then analogously define $\mathbf{PFIN}\langle \mathbf{T} : r/s \rangle$ and $\mathbf{PFIN}\langle \mathbf{P} : p \rangle$.

We will examine the capabilities of probabilistic and pluralistic *PFIN*-type learning. Throughout this paper we distinguish between the expressions $\frac{r}{s}$ and r/s: the former indicates the numerical fraction and the latter indicates "r out of s".

We will now define the *capability type* of probabilistic or pluralistic learners.

Definition 1 *Suppose* $\{\alpha_n : n \geq 0\}$ *is a monotonically strictly decreasing sequence of rational numbers in the range* $(0, 1]$ *with* α_∞ *as the limiting value, called the accumulation point. We say that* $\{\alpha_n : n \geq 0\}$ *is the* capability type *of* PFIN-*type learners in the range* $(\alpha_\infty, \alpha_0]$, *if and only if*

1. *for integers* $n \geq 0$, *if* $\alpha_{n+1} < p \leq \alpha_n$, *then* $\mathbf{PFIN}\langle \mathbf{P} : p \rangle = \mathbf{PFIN}\langle \mathbf{P} : \alpha_n \rangle$,

2. *for integers* $n \geq 0$, $\mathbf{PFIN}\langle \mathbf{P} : \alpha_n \rangle \subset \mathbf{PFIN}\langle \mathbf{P} : \alpha_{n+1} \rangle$.

One can analogously define the notion of *capability type* for pluralistic learners, (where $\mathbf{PFIN}\langle \mathbf{T} : r/s \rangle$ replaces $\mathbf{PFIN}\langle \mathbf{P} : p \rangle$, and $\mathbf{PFIN}\langle \mathbf{T} : \alpha_n \rangle$ replaces $\mathbf{PFIN}\langle \mathbf{P} : \alpha_n \rangle$). Finally, we define the *redundancy type* of a ratio as follows.

Definition 2 *Let* r *and* s *be relatively prime, and let* $\preceq$ *denote the partial ordering defined by* $x \preceq y \Leftrightarrow \mathbf{PFIN}\langle \mathbf{T} : (xr)/(xs) \rangle \subseteq \mathbf{PFIN}\langle \mathbf{T} : (yr)/(ys) \rangle$. *Consider the equivalence classes of the natural numbers generated by* $\preceq$. *The* redundancy type *at ratio* $\frac{r}{s}$ *is the partially ordered set of these equivalence classes with the partial order induced by* $\preceq$.

3 Previous Results

The learning power of probabilistic *FIN*-type learners was first investigated by Freivalds [6].

Theorem 3 *(Freivalds) The capability type for FIN-type probabilistic learning in the range $(\frac{1}{2}, 1]$ is the sequence $\{\frac{n}{2n-1} : n \geq 1\}$.*

Velauthapillai [8] extended Freivalds' result to pluralistic learning.

Theorem 4 *(Velauthapillai) The capability type for FIN-type pluralistic learning in the range $(\frac{1}{2}, 1]$ is the sequence $\{\frac{n}{2n-1} : n \geq 1\}$.*

Finally, Daley et. al [5] showed that Freivalds' simulation construction also yields,

Theorem 5 *(Daley and Velauthapillai) For integers $n \geq 1$, if $\frac{n+1}{2n+1} < p \leq \frac{n}{2n-1}$ then $\mathbf{FIN}\langle \mathbf{P}: p \rangle = \mathbf{FIN}\langle \mathbf{T}: \frac{n}{2n-1} \rangle$.*

Thus for *FIN*-type learning probabilistic and pluralistic learners are equivalent with respect to their learning capabilities for ratios $> \frac{1}{2}$. In short, we say that the capability type for pluralistic learners coincides with that of probabilistic learners.

From Theorem 5 it follows that there is no redundancy for ratios $> \frac{1}{2}$. First evidence of the phenomenon of redundancy was given by Velauthapillai [8], by showing that at the ratio $\frac{1}{2}$ that a team with plurality 2/4 is a more capable learning team than one with plurality 1/2. Later, Jain and Sharma [7] showed that the amount of redundancy at $\frac{1}{2}$ is quite limited.

Theorem 6 *(Velauthapillai, Jain and Sharma) The redundancy type for pluralistic FIN-type learners at ratio $\frac{1}{2}$ is $\{\{2i - 1 : i \geq 1\} \prec \{2i : i \geq 1\}\}$.*

It was also shown in Daley et. al [5], however, that at the ratio $\frac{1}{3}$ there is an unlimited increase in learning power possible by (continually) doubling the team size and retaining the success ratio of $\frac{1}{3}$, i.e., there is infinitely much redundancy possible at $\frac{1}{3}$.

Recently in Daley et al. [3] we revealed much more about the relationship between probabilistic and pluralistic strategies and about the power of redundancy.

Theorem 7
a) For $\frac{24}{49} < p \leq \frac{1}{2}$, $\mathbf{FIN}\langle \mathbf{P}: p \rangle = \mathbf{FIN}\langle \mathbf{T}: 2/4 \rangle$.
b) $\mathbf{FIN}\langle \mathbf{T}: 2/4 \rangle \subset \mathbf{FIN}\langle \mathbf{T}: 24/49 \rangle$.

Observe that this theorem shows that $\mathbf{FIN}\langle \mathbf{P}: \frac{1}{2} \rangle = \mathbf{FIN}\langle \mathbf{T}: 2/4 \rangle$, thus completely characterizing the learning capability at probability $\frac{1}{2}$.

In [4] the capability types and redundancy types for $\hat{P}FIN$-type learning was investigated. The main results from that paper which relate to the present paper are the following.

Theorem 8 *(Daley, Kalyanasundaram and Veluathapillai)*
a) *The capability type for probabilistic as well as pluralistic PFIN-type learning in the interval $(\frac{1}{2}, 1]$ is the sequence $\{\frac{n}{2n-1} : n \geq 1\}$.*
b) *The redundancy type for pluralistic PFIN-type learning at $\frac{1}{2}$ is $\{\{2i + 1 : i \geq 0\} \prec \{2i : i \geq 1\}\}$.*

Theorem 9 *(Daley, Kalyanasundaram and Veluathapillai)*
a) The capability type for probabilistic PFIN-type learning in the interval $(\frac{4}{9}, \frac{1}{2}]$ is the sequence $\{\frac{4n}{9n-2} : n \geq 2\}$.
b) The redundancy type for pluralistic PFIN-type learning at $\frac{4}{9}$ is $\{\{1,2\} \prec \{3m+1, 3m+2 : m \geq 1\} \prec \{3m : m \geq 1\}\}$.

4 Results

We first show that the sequence $\{\frac{2mn}{(5m-1)n-m} : m, n \geq 1\}$ defines some of the critical probabilities in the probabilistic and pluralistic capability sequence of *PFIN*-type learning.

Theorem 10
For integers $m, n \geq 1$, $\mathbf{PFIN}\langle P: \frac{2mn}{(5m-1)n-m}\rangle \subset \mathbf{PFIN}\langle P: \frac{2m(n+1)}{(5m-1)(n+1)-m}\rangle$.

Notice that this sequence generalizes the diagonalization results in [4] where we have shown that the probabilistic capability sequence between $(\frac{4}{9}, \frac{1}{2}]$ is $\{\frac{4n}{9n-2} : n \geq 2\}$.

Our ultimate goal is to find necessary and sufficient conditions for a team with plurality r *out of* s to successfully simulate another team with plurality x *out of* y. At first, while looking at the simulation results for team pluralities in the interval $(\frac{1}{2}, 1]$, it appeared to be very promising to accomplish this goal. Unfortunately, the following results which deal with pluralities in the interval $(\frac{4}{9}, \frac{1}{2}]$ considerably weaken our hope.

The following four theorems deal with the redundancy type at $\{\frac{4n}{9n-2} : n \geq 2\}$. First we consider the case where n is odd.

Theorem 11 *For integers $n = 2k+1$ where $k \geq 1$, the redundancy type at $\frac{4n}{9n-2}$ is as follows:*
a) At $n \neq 31$, the redundancy type is $\langle\{2i - 1 : i \geq 1\} \prec \{2i : i \geq 1\}\rangle$.
b) At $n = 31$, the redundancy type is $\langle\{1,2\} \prec \{i : i \geq 3\}\rangle$.

We now consider the case where n is even. The redundancy type at $n = 2$ for the fraction $\frac{4n}{9n-2}$ has been shown in [4].

Theorem 12 *For integers $n = 8k+2$ where $k \geq 1$, the redundancy type at $\frac{4n}{9n-2}$ is as follows:*
a) At $n = 10$ and 26, the redundancy type is $\langle\{1\} \prec \{2i - 1 : i \geq 2\} \prec \{2i : i \geq 1\}\rangle$.
b) At $n = 18$, the redundancy type is $\langle\{2i - 1 : i \geq 1\} \prec \{2i : i \geq 1\}\rangle$.
c) At $n = 34$, the redundancy type is $\langle\{1\} \prec \{7\} \prec \begin{smallmatrix}\{2,4\}\\\{3,5,2i-1:i\geq5\}\end{smallmatrix} \prec \{2i : i \geq 3\}\rangle$.
(Notice that the redundancy type is a proper partial order.)

Theorem 13 *For integers $n = 4k + 2$ where k is odd, the redundancy type at $\frac{4n}{9n-2}$ is as follows:*
a) At $n = 14$ and 38, the redundancy type is $\langle\{1\} \prec \{i : i \geq 2\}\rangle$.
b) At all other values of n there is no redundancy.

Theorem 14 *For integers $n = 2k + 2$ where k is odd, the redundancy type at $\frac{4n}{9n-2}$ is as follows:*
a) At $n = 32$, the redundancy type is $\langle\{1,2\} \prec \{i : i \geq 3\}\rangle$.
b) At all other values of n there is no redundancy.

The following theorem deals with the capability sequence of probablistic *PFIN*-type learners in the interval $(\frac{3}{7}, \frac{4}{9}]$.

Theorem 15 *The capability type of PFIN-type learners in the range $(\frac{3}{7}, \frac{4}{9}]$ is $\{\frac{6n}{14n-3} : n \geq 6\} \cup \{\frac{3n}{7n-1} : n \geq 12\} \cup \{\frac{8n}{19n-4} : 4 \leq n \leq 11\}$.*

Notice that the capability type is the union of three sequences. It will become clear from the proofs of this theorem that the sequences are generated as a result of different combinations of cases that are defined by *PFIN*-type learning capabilities at higher success probabilities (reduction argument).

Finally, the following theorem shows infinite redundancy at $\frac{2}{5}$ for *PFIN*-type learning.

Theorem 16 *For $k \neq 0 \bmod 3$, $\mathbf{PFIN}\langle \mathbf{T}: 2k/5k \rangle \subset \mathbf{PFIN}\langle \mathbf{T}: 8k/20k \rangle$. In particular, for $i \geq 0$, $\mathbf{PFIN}\langle \mathbf{T}: (2 \cdot 4^i)/(5 \cdot 4^i) \rangle \subset \mathbf{PFIN}\langle \mathbf{T}: (2 \cdot 4^{i+1})/(5 \cdot 4^{i+1}) \rangle$.*

5 Reduction Technique

In this section we give a brief description of the reduction technique which is used to establish many of the results of this paper. As stated in the introduction the process of determining the learning capability of a particular probabilistic or team *PFIN*-type learner L involves showing that this learner (the pursuing learner) can simulate any probabilistic or team learner M (the target learner) with a success ratio within its capability interval, and the task of the programs produced by L is to cover the different functions computed by groups of programs produced by M. Suppose M is a $\mathbf{PFIN}\langle \mathbf{P}: p\rangle$-type learner, and let L be a $\mathbf{PFIN}\langle \mathbf{P}: \frac{r}{s}\rangle$ that successfully simulates M, i.e., learns every function that M can learn. Even though L is intended to be a probabilistic learner (with success ratio $\frac{r}{s}$), we will actually implement it as a team of learners (with plurality kr/ks, for an appropriate choice of k). Let $f \in \mathbf{PFIN}\langle \mathbf{P}: p\rangle[M]$ be an arbitrary function learnable by M. We will refer to M as the target learner and f as the target function, and we will refer to L as the pursuing team of learners. Observing the input graph of f, whenever M produces programs with a certain weight, L will also produce programs with a certain (but probably different) weight. Initially, L will wait until M produces programs with weight at least p. At this point kr of L's members output programs $\{S_1, \ldots, S_{kr}\}$.

Let g be one of the functions computed by M's programs and let q be the total weight of M's programs which compute g. Notice that M will produce $p - q$ new programs if it were the path of the target function. We call this a *breakaway* event. Then since weight $p - q$ of M's programs are wrong on g, the *reduced success ratio* of M on g is at most $\frac{p}{1-(p-q)}$. We call this the *reduced target ratio.* Similarly, L's programs must respond in a proportionate manner by having x of its programs compute g, where x is chosen so that its reduced success ratio $\frac{kr}{k(s-r)+x}$ (called the *reduced pursuer ratio*) on g is *equivalent* to the reduced target ratio. Since these reduced success ratios will be for less capable teams whose capabilities have already been established and known to form intervals, the weight q is partitioned into a number of cases, depending on the capability of the resulting reduced target ratio. In order for the reduced target ratio to have the capability of a team with plurality in the interval $(\frac{i+1}{2i+1}, \frac{i}{2i-1}]$ (the R_i^1 *Reduction*), it must be the case that $\frac{(3i-1)p}{i} - 1 \leq q < \frac{3(i+1)-1)p}{i+1}$. Similarly, in order for the reduced pursuer ratio to be of equivalent capability L must have $\frac{k((3i-1)r-is)}{n}$ of its programs compute g. If $q < Q_1 \equiv \frac{5p}{2} - 1$, then the function g can be safely ignored by L's programs since the reduced target ratio is equivalent to a deterministic learner, and L can issue an additional kr programs which will correctly compute g. If $q > T_1 \equiv p - Q_1 = 1 - \frac{3p}{2}$, then all other functions can be safely ignored, so that all of L's kr programs will correctly compute g.

For the reduction argument used in [4], which deals with the capability sequence $\frac{4n}{9n-2}$, and which will be needed to derive the results of Theorems 11 through 14, the value q is further decomposed into the five cases $(R_2^1, R_3^1, R_4^1, R_5^1$ and $R_{1,\infty}^1)$, defined by the following subintervals of $[Q_1, T_1]$: $[Q_1, \frac{8p}{3} - 1), [\frac{8p}{3} - 1, \frac{11p}{4} - 1), [\frac{11p}{4} - 1, \frac{14p}{5} - 1), [\frac{14p}{5} - 1, \frac{17p}{6} - 1), [\frac{17p}{6} - 1, T_1]$. The cases and the required number of L's programs which must follow the weight q programs of M is given in Table 1.

Table 1: Cases and Pursuer Strategy for interval $(\frac{4}{9}, \frac{1}{2}]$

Case	Target Team	Target Ratio	Pursuing Team	Pursuing Ratio
R_1^1	$0 \leq q < Q_1$	$> 2/3$	0	$4n/(5n-2)$
R_2^1	$Q_1 \leq q < \frac{8p}{3} - 1$	$> 3/5$	$k(n+2)$	$2/3$
R_3^1	$\frac{8p}{3} - 1 \leq q < \frac{11p}{4} - 1$	$> 4/7$	$k(\frac{7n}{3} - 2)$	$3/5$
R_4^1	$\frac{11p}{4} - 1 \leq q < \frac{14p}{5} - 1$	$> 5/9$	$k(2n+2)$	$4/7$
R_5^1	$\frac{14p}{5} - 1 \leq q < \frac{17p}{6} - 1$	$> 6/11$	$k(\frac{11n}{5} + 2)$	$5/9$
$R_{6,\infty}^1$	$\frac{17p}{6} - 1 \leq q \leq T_1$	$\geq \frac{2p}{4-5p}$	$k(3n-2)$	$2/4$

It is possible for multiple breakaways to occur, and so we must demonstrate that L has sufficient resources (i.e., programs) to handle all possibilities. For example, 2 Case R_2^1 breakaways (denoted $2R_2^1$) are possible, when $2Q_1 \leq p \equiv p \leq \frac{1}{2}$ or $n \geq 2$. Fortunately if $n \geq 2$, then (precisely) $2k(n+2) \leq k(4n)$, so that L will have enough initial programs to send along both breakaways. Table 2 summarizes the *critical* multiple events. We observe that the combination $R_2^1 + R_i^1$ is possible precisely when $\frac{5p}{2} - 1 + \frac{(3i-1)p}{i} - 1 \leq p$, and hence when $p \leq \frac{4i}{9i-2}$. Thus, this case combination accounts for the limitations on learning for ratios in the interval $(\frac{4}{9}, \frac{1}{2}]$, i.e., a team with ratio $\frac{4i}{9i-2}$ does *not* have sufficient team size to handle to combination $R_2^1 + R_{i+1}^1$. Moreover, the diagonalizing tree of functions for this case consists of a two-way branch at the root with two subtrees

Table 2: Multiple Events

Combination	Case Possible	L OK
$2R_2^1$	$n \geq 2$	$n \geq 2$
$R_2^1 + R_3^1$	$n \geq 3$	$n \geq 3$
$R_2^1 + R_4^1$	$n \geq 4$	$n \geq 4$
$R_2^1 + R_5^1$	$n \geq 5$	$n \geq 5$
$R_2^1 + R_6^1$	$n \geq 6$	$n \geq 6$
$3R_2^1$	$n \geq 6$	$n \geq 6$
$2R_3^1$	$n \geq 6$	$n \geq 6$
$R_3^1 + R_4^1$	$n \geq 12$	$n \geq 12$
$2R_2^1 + R_3^1$	$n \geq 18$	$n \geq 18$
$R_3^1 + R_5^1$	$n \geq 30$	$n \geq 30$
$4R_2^1$	$n \geq \infty$	NO
$2R_4^1$	$n \geq \infty$	NO
$2R_2^1 + R_4^1$	$n \geq \infty$	NO
$R_2^1 + R_{6,\infty}^1$	$n \geq \infty$	NO
$R_2^1 + 2R_3^1$	$n > \infty$	NO

which are the diagonalizing subtrees for the ratios $\frac{2}{3}$ and $\frac{i+1}{2i+1}$, respectively.

It is important to note that the success of the simulating team also depends on the fact that k is a sufficiently large integer (viz., 15) so that the fractions given in the table for the pursuing team are whole numbers. Unfortunately, while dealing with the redundancy issues required for Theorems 11 through 14, we should consider various values for k and the possibility that the fraction $\frac{4n}{9n-2}$ can be further simplified. To be more precise, given a ratio $\frac{4n}{9n-2}$, we reduce the fraction to base ratio $\frac{r}{s} = \frac{4n}{9n-2}$ such that r and s are relatively prime. Now the corresponding fractions for various cases for a team with plurality r out of s need not be whole numbers. So it is possible for some base teams to fail under certain case combinations. For each factor k, we identify the case combinations that a team with plurality $r \cdot k$ out of $s \cdot k$ can handle. As a consequence, the factors k are partitioned according to the case combination handling capability. It is then easy to show that the partitioning is the redundancy type for PFIN-type learning at the ratio r/s.

Table 3 gives the general forms for the simulation strategy required by Theorem 15 for capabilities in the interval $(\frac{3}{7}, \frac{4}{9}]$. We use the notation α_j to denote $\frac{(3j-1)p}{j} - 1$. Observe that infinitely many types of indivivual cases R_i^1 are possible, since the sequence $\frac{(3i-1)p}{i} - 1$ converges to $3p - 1 \leq T_1$. Thus, we have the case $R_{2,\infty}^2$ for $3p - 1 \leq q \leq T_1$ (analogous to $R_{6,\infty}^1$ in Table 1), which handles reductions to (a subset of the) capabilities in the interval $(\frac{4}{9}, \frac{1}{2}]$. The necessity of having three sequences to generate the capabilities in the interval $(\frac{3}{7}, \frac{4}{9}]$ can be seen as follows:

1. for all $n \geq 6$ a probabilistic learner with success ratio $\frac{6n}{14n-3}$ is unable to handle the combination $R_3^1 + R_{n+1}^1$, which is possible at the ratio $\frac{6(n+1)}{14(n+1)-3}$;

2. for all $n = 3j + 1$, where $j \geq 2$, a probabilistic learner with success ratio $\frac{6n}{14n-3}$ is unable to handle the combination $2R_2^1 + R_{2j+1}^1$, which is possible at the ratio $\frac{3(2j+1)}{7(2j+1)-1} < \frac{6(n+1)}{14(n+1)-3}$;

3. for all $n = 2j + 1$, where $j \geq 2$, a probabilistic learner with success ratio $\frac{3n}{7n-1}$ is unable to handle the combination $R_3^1 + R_{3j+2}^1$, which is possible at the ratio $\frac{6(3j+2)}{14(3j+2)-3} < \frac{3(n+1)}{7(n+1)-1}$;

4. for $n = 8$, a probabilistic learner with success ratio $\frac{6n}{14n-3}$ is unable to handle the combination $R_4^1 + R_5^1$, which is possible at the ratio $\frac{40}{91} < \frac{6(n+1)}{14(n+1)-3}$;

5. for $n = 11$, a probabilistic learner with success ratio $\frac{3n}{7n-1}$ is unable to handle the combination $R_4^1 + R_7^1$, which is possible at the ratio $\frac{56}{129} < \frac{3(n+1)}{7(n+1)-1}$;

6. for $n = 5$ (respectively, $n = 7$), a probabilistic learner with success ratio $\frac{8n}{19n-4}$ is unable to handle the combination $R_3^1 + R_j^1$, where $j = 9$ (respectively, $j = 17$), which is possible at the ratio $\frac{6j}{14j-3} < \frac{8(n+1)}{19(n+1)-1}$.

We observe that for even values of i the sequence $\frac{3i}{7i-1}$ is a subsequence of $\frac{6n}{14n-3}$, and for all $4 \leq i \leq 11$, *excepting $i = 5$ and $i = 7$*, the sequence $\frac{8i}{19i-4}$ is a subsequence of $\frac{6n}{14n-3}$.

Table 3: Cases and Pursuer Strategies for interval $(\frac{3}{7}, \frac{4}{9}]$

Case	Target Team	Pursuing Team $\frac{6n}{14n-3}$	Pursuing Team $\frac{3n}{7n-1}$	Pursuing Team $\frac{8n}{19n-4}$
R_1^1	$0 \leq q < Q_1$	0	0	0
R_2^1	$Q_1 \leq q < \frac{8p}{3} - 1$	$k(n+3)$	$k(n+2)$	$k(n+4)$
R_j^1	$\alpha_j \leq q < \alpha_{j+1}$	$k(\frac{4j-6}{j}n + 3)$	$k(\frac{4j-6}{j}n + 2)$	$k(\frac{5j-8}{j}n + 4)$
$R_{2,\infty}^2$	$3p - 1 \leq q \leq T_1$	$k(5n-3)$	$k(5n-2)$	$k(7n-4)$

6   Diagonalization Proofs

We give here the proof for Theorem 10. We begin by describing the class of functions $\mathcal{F}_n^m$ which can be learned by some $\mathbf{PFIN}\langle \mathbf{T}: 2mn/[(5m-1)n-m]\rangle$ learner L but not by any $\mathbf{PFIN}\langle \mathbf{T}: 2m(n-1)/[(5m-1)(n-1)-m]\rangle$ learner where $n > m(m-1)$. It is best to think of $\mathcal{F}_n^m$ as the set of all (groups of) functions which have a certain shape. The set $\mathcal{F}_n^m$ consists of all functions f satisfying

$$f(x) = \begin{cases} 0, & \text{if } 1 \leq x \leq a \\ c_1, & \text{if } a < x \leq b^{c_1} \\ c_2, & \text{if } x > b^{c_1} \end{cases}$$

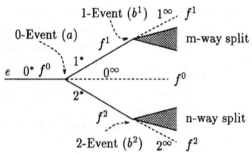

Figure 1.

where a and b^{c_1} are integers, and $c_1 \in \{1,2\}$, $c_2 \in \{3,4,\ldots,m+2\}$ if $c_1 = 1$ and $c_2 \in \{m+3, m+4, \ldots, m+n+2\}$ if $c_1 = 2$. The value $e = f(0)$ is decoded as a set of learners $M_1, \ldots, M_s$ and a number r such that $r \leq s$ (e.g., $r = 2m(n-1)$ and $s = [(5m-1)(n-1)-m])$. To simplify notation we will denote finite initial segments of functions by finite length sequences of their values. For example, $e0^m$ denotes the function which has value e for input 0 and value 0 for inputs 1, $\ldots$, m. We introduce the predicate $P(\sigma, k, t, m, y)$ which means that m is the point at which M_k produces its first (and only) hypothesis when fed the initial segments of the function σt^*, where σ is some finite initial segment, and y is an upper bound on the time it takes to do this. More formally, $P(\sigma, k, t, m, y)$ is

$$\forall j \leq m\ \Phi_{M_k}(\sigma t^j) \leq y \text{ and } \forall j < m\ \varphi_{M_k}(\sigma t^j) = 0 \text{ and } \varphi_{M_k}(\sigma t^m) > 0.$$

We use Φ_j to denote the computational complexity of the program with number j. In this learning model we have also adopted the convention that on each initial segment of a function a *PFIN* learner M will always produce *some* output, using the value 0 to indicate that it "doesn't know".

The values a and b^t (for $t = 1, 2$) are defined as follows:

$$a = \min y\ \exists \text{ distinct } i_1, \ldots, i_r\ \forall k \leq r\ \exists m \leq y\ P(e, i_k, 0, m, y)$$

$$b^t = \quad \min y\ \exists \text{ distinct } i_1, \ldots, i_r\ \forall k \leq r \text{ either } [\ \exists m \leq a\ P(e, i_k, 0, m, a)$$
$$\text{and } \Phi_{\varphi M_{i_k}}(e0^m)(a+1) \leq y \text{ and } \varphi_{\varphi M_{i_k}}(e0^m)(a+1) = t\ ]$$
$$\text{or } \exists m \leq y\ P(e0^a, i_k, t, m, y)$$

The expression $\varphi_{\varphi M_{i_k}}(e0^m)(a+1) = t$ indicates that the program produced by M_{i_k} (before a) is following the t-valued branch. It is possible for b^{c_1} to be undefined. In that case the function f will have value c_1 for all $x > a$. 5(see f^1 and f^2 in Figure 1). Similarly, it is possible for both b^{c_1} and a to be undefined and in that case f will have value 0 for all $x > 0$. However, if these values exist, then they can be computed (in real time) from the information encoded in the value e.

We now describe how the $2mn/[(5m-1)n-m]$ team can learn all the functions in the set $\mathcal{F}_n^m$. After seeing $e = f(0)$, $2mn$ members of L will each produce a program (say, $S_1, \ldots, S_{2mn}$). These **0-event** programs follow f^0, i.e., they produce output 0 for all inputs $1 \leq x \leq a$. If a is undefined, then these programs will forever follow f^0. If a is defined, then at a they split into two groups. The group $S_1, \ldots, S_{mn+m-n}$ follows f^1, i.e., they produce output 1 for all inputs $a < x \leq b^1$. If b^1 is undefined, then these programs will forever follow f^1. If b^1 exists, then at b^1 this group splits again into m subgroups. The number of followers and what they output is explained below.

Programs $S_{mn+m-n+1}, \ldots, S_{2mn}$ follow f^2, i.e., they produce output 2 for all inputs $a < x \leq b^2$. If b^2 is undefined, then these programs will forever follow f^2. If b^2 exists, then at b^2 this group splits again into n subgroups.

After seeing $f(a+1)$, the team of learners produce additional programs as follows: if $f(a+1) = 1$, then L will produce $mn + n - m$ additional programs called **1-event** programs. These programs follow the f^1-branch. Notice that at this point, the number of programs (**0** and **1-event**) following the f^1-branch is $(mn + m - n) + (mn + n - m) = 2mn$. Again if b^1 is undefined, then these programs will forever follow f^1. On the other hand, if b^1 exists, then at b^1 this group splits again into m subgroups where $2n$ programs follow each branch. If $f(a+1) = 2$, then L will produce $mn+m-n$ additional programs called **2-event** programs. These programs follow the f^2-branch. Notice that at this point, the number of programs (**0** and **2-event**) following the f^2-branch is $(mn+n-m) + (mn+m-n) = 2mn$. Again if b^2 is undefined, then these programs will forever follow f^2. On the other hand, if b^2 exists, then at b^2 this group splits again into n subgroups where $2m$ programs follow each branch.

After seeing $f(b^t + 1)$, the remaining members of L produce a program that outputs $f(b^t + 1)$ on any future input. It is easy to see that no matter which function (f^0, f^1, f^2, f_1^1, f_2^1, f_1^2, f_2^2, or f_3^2) is given as the target function, there are always $2mn$ programs produced by the team L, which correctly compute it.

We now show that the set $\mathcal{F}_n^m$ cannot be learned by any team of $2m(n-1)$ out of $[(5m-1)(n-1)-m]$ learners. Let $M = \{M_1, \ldots, M_{[(5m-1)(n-1)-m]}\}$ be such a team. We consider a function f from $\mathcal{F}_n^m$ where $M_1, \ldots, M_{[(5m-1)(n-1)-m]}$ and $r = 2m(n-1)$ are encoded as $f(0)$. We work from the top of the function tree to the bottom as follows. In the best case, just before the **1-event** exactly $2m(n-1)$ programs are following the f^1 branch. After the **1-event**, only $2(n-1)$ of the currently active programs can follow each branch, so $2(m-1)(n-1)$ new programs are required on each branch. Thus, the maximum remaining new programs which could be produced on the f^1 branch after the **0-event** is $m(n-1)+(n-1)-m$, so $m(n-1)-(n-1)+m$ of the original $2m(n-1)$ programs must follow the f^1 branch. This means that at most $m(n-1)+(n-1)-m$ of the original programs are available to follow the f^2 branch, and $m(n-1)-(n-1)+m$ new programs must be produced on the f^2 branch after the **0-event**. There is at least one branch (created after the **2-event**) which is followed by at most $2m-1$ of the currently active programs, so at least $2m(n-1) - 2m + 1$ new programs are required on this branch. However, this is not possible since already

$2m(n-1) + m(n-1) - (n-1) + m$ programs have been produced up to this point on the f^2 branch.

7 Acknowledgements

Both authors are supported in part by NSF Grant CCR-9202158. The first author also gratefully acknowledges the partial support provided by the Naval Center for Applied Research in Artificial Intelligence at the Naval Research Laboratory, Washington, D.C.

References

[1] J. Case, and C. Smith, Comparison of identification criteria for machine inductive inference, *Theoretical Computer Science*, **25**, 193-220.

[2] R. Daley, and B. Kalyanasundaram, Capabilities of Probabilistic Learners with Bounded Mind Changes, In *Proceedings of the 1993 Workshop on Computational Learning Theory*, 1993.

[3] R. Daley, B. Kalyanasundaram, and M. Velauthapillai, Breaking the probability $\frac{1}{2}$ barrier in FIN-type learning, In *Proceedings of the 1992 Workshop on Computational Learning Theory*, 1992.

[4] R. Daley, B. Kalyanasundaram, and M. Velauthapillai, The Power of Probabilism in Popperian FINite Learning, In *Proceedings of the 1992 Workshop on Analogical and Inductive Inference*, Lecture Notes in Computer Science **642**, 151-169.

[5] R. Daley, L. Pitt, M. Velauthapillai, and T. Will, Relations between probabilistic and team one-shot learners, In *Proceedings of the 1991 Workshop on Computational Learning Theory*, pages 228-239, 1991.

[6] R.V. Freivalds, Finite Identification of General Recursive Functions by Probabilistic Strategies, Akademie Verlag, Berlin, 1979.

[7] S. Jain, and A. Sharma, Finite learning by a team, In *Proceedings of the 1990 Workshop on Computational Learning Theory*, pages 163-177, 1990.

[8] M. Velauthapillai, Inductive inference with a bounded number of mind changes, In *Proceedings of the 1989 Workshop on Computational Learning Theory*, pages 200-213, 1989.

E-mail Addresses

daley@cs.pitt.edu
kalyan@cs.pitt.edu

Properties of Language Classes
with Finite Elasticity

Takashi Moriyama[1] and Masako Sato[2]

[1] 5th C&C System Operation Unit
NEC Corporation, Minato-ku, Tokyo 108-01, Japan
[2] Department of Mathematics, College of Arts and Integrated Sciences
University of Osaka Prefecture, Sakai, Osaka 593, Japan

Abstract. This paper considers properties of language classes with finite elasticity in the viewpoint of set theoretic operations. Finite elasticity was introduced by Wright as a sufficient condition for language classes to be inferable from positive data, and as a property preserved by (not usual) union operation for language classes. We show that the family of language classes with finite elasticity is closed under not only union but also various operations for language classes such as intersection, concatenation and so on.

As a framework defining languages, we introduce restricted elementary formal systems (EFS's for short), called max length-bounded by which any context-sensitive language is definable. We define various operations for EFS's corresponding to usual language operations and also for EFS classes, and investigate closure properties of the family G_e of max length-bounded EFS classes that define classes of languages with finite elasticity. Furthermore, we present theorems characterizing a max length-bounded EFS class in the family G_e, and that for the language class to be inferable from positive data, provided the class is closed under subset operation. From the former, it follows that for any n, a language class definable by max length-bounded EFS's with at most n axioms has finite elasticity. This means that G_e is sufficiently large.

1 Introduction

This paper follows a mathematical formulation of inductive inference due to Gold[4], and deals with inductive inference of an indexed family of recursive languages from positive data developed by Angluin[1].

There have been introduced various constraints for inductive inference machine's guess so far ([6, 7]), including conservativeness and monotonicity, and studied characterizations of inferable language classes under such constraints and relations among them.

On the other hand, various conditions (properties) for inferability from positive data have been presented ([1, 5, 10, 14]), which are set theoretic properties of a language class independent of indexing of the class. Among them, finite elasticity introduced by Wright[14] is a sufficient condition with *good* property in a sense that it is preserved by the operation to generate a *class of unions of pairs of languages*, and that the family L_e of language classes with finite elasticity includes interesting and important classes given by Shinohara[11, 12] using a framework of elementary formal systems described below.

In this paper, we investigate properties of the family L_e from the viewpoint of set theoretic operations. We show that the family L_e is closed under not only union but also various operations for language classes such as intersection, concatenation and so on.

Furthermore, we develop the above discussion on closure properties to elementary formal systems (EFS's for short) as a framework for defining languages. An elementary formal system introduced by Smullyan[13] works as a logic programming language and generates a language as a grammar. Arikawa and his colleagues[3] introduced some subclasses of EFS's which correspond to Chomsky hierarchy. Among them, the class of *length-bounded* EFS's that generates the class of context-sensitive languages was especially investigated as a framework for inductive inference of languages from positive data ([11, 12, 9]). In this paper, we introduce a *max length-bounded* EFS that is a slight modification of a *weakly reducing* EFS due to Yamamoto[15], and consider the family G_e of max length-bounded EFS classes that define classes of languages with finite elasticity. In order to develop the discussions on closure properties of L_e for the family G_e, we define various operations for EFS's corresponding to usual language operations, and also for EFS classes. It is shown that the class of max length-bounded EFS languages is closed under various language operations as well as that of context-sensitive languages. This implies closedness of the family G_e under various class operations.

Finally, we consider what classes of max length-bounded EFS's belong to G_e. We present a theorem characterizing a class in the family G_e, provided the class is closed under subset operation. From the characterization, we derive a corollary that for any n, a language class definable by max length-bounded EFS's with at most n axioms has finite elasticity, an extension of a result in [12]. Thus the family G_e (thus L_e) is sufficiently large, including such interesting classes. Moreover, we characterize a max length-bounded EFS class for the language class to be inferable from positive data, under the same condition mentioned above.

In §2, we consider closure properties of L_e. In §3, we deal with closure properties of G_e and present characterizing theorems of max length-bounded EFS classes mentioned above.

2 Closure Properties of Language Classes with Finite Elasticity

In this section, we introduce various operations for language classes, and discuss closure properties of classes with finite elasticity under such operations. First we give basic definitions on inductive inference from positive data. Since the empty language may be generated by intersection operation, we modify the definition to permit the empty language as a member of language class.

Let Σ be a fixed finite *alphabet* of symbols. The set of all finite strings over Σ is denoted by Σ^*. A *language* is any subset of Σ^*. Let N be the set of positive integers. A sequence of languages $\mathcal{L} = L_1, L_2, \cdots$ is called *an indexed family of recursive languages* if there is a computable function $f : N \times \Sigma^* \to \{0, 1\}$ such that

$$f(i, w) = \begin{cases} 1 & \text{if } w \in L_i, \\ 0 & \text{if } w \notin L_i. \end{cases}$$

In what follows, we assume that a class of languages is an indexed family of recursive languages without any notice. A *positive presentation* of a nonempty language L is an infinite sequence w_1, w_2, $\cdots$ such that $\{w_i \mid i \in N\} = L$. A class of languages $\mathcal{L} = L_1, L_2, \cdots$ is *inferable from positive data* if there is an inductive inference machine M such that the sequence of outputs produced by M converges to j with $L_j = L_i$ for any index $i \in N$, where $L_i \neq \phi$, and any positive presentation of L_i. For more details about inference machine, we refer to [1].

Let L be the collection of classes inferable from positive data. Angluin[1] presented a theorem characterizing L and interesting classes in L such as a pattern language class. Various sufficient conditions for inferability can be found in [1, 5, 10, 14]. Among them, we are concerned with finite elasticity[14] defined below.

Definition 1. A class of languages $\mathcal{L}$ has *infinite elasticity* if there are an infinite sequence of strings $w_0, w_1, \cdots$ and an infinite sequence of languages $L_1, L_2, \cdots$, each in $\mathcal{L}$ such that

$$(1) \quad \{w_0, w_1, \cdots, w_{k-1}\} \subseteq L_k \quad \text{and} \quad (2) \quad w_k \notin L_k$$

for any $k \in N$. A class $\mathcal{L}$ has *finite elasticity* if $\mathcal{L}$ does not have infinite elasticity.

Note that Motoki, Shinohara and Wright[8] showed that finite elasticity defined by Wright[14] is not a sufficient condition for inferability from positive data, and corrected the definition as the above. Wright's results given below are valid for the above correct definition of finite elasticity.

Given classes $\mathcal{L}_1$ and $\mathcal{L}_2$, let us define the following *union*, *intersection* and *concatenation* operations

$$\mathcal{L}_1 \,\tilde{\cup}\, \mathcal{L}_2 = \{L^{(1)} \cup L^{(2)} \mid L^{(1)} \in \mathcal{L}_1, L^{(2)} \in \mathcal{L}_2\},$$
$$\mathcal{L}_1 \,\tilde{\cap}\, \mathcal{L}_2 = \{L^{(1)} \cap L^{(2)} \mid L^{(1)} \in \mathcal{L}_1, L^{(2)} \in \mathcal{L}_2\},$$
$$\mathcal{L}_1 \,\tilde{\cdot}\, \mathcal{L}_2 = \{L^{(1)} \cdot L^{(2)} \mid L^{(1)} \in \mathcal{L}_1, L^{(2)} \in \mathcal{L}_2\}.$$

Throughout this paper, the set theoretic union operation is called *usual union* in order to distinguish *union* defined above, and denoted by $\cup$. As easily seen, if both $\mathcal{L}_1$ and $\mathcal{L}_2$ are indexed families of recursive languages, then so are all of the classes defined above.

For a given class of languages $\mathcal{L}$, let us define the following operations

$$\mathcal{L}^{\tilde{m}} = \{L^m \mid L \in \mathcal{L}\} \quad (m \geq 1), \qquad \mathcal{L}^{\tilde{+}} = \{L^+ \mid L \in \mathcal{L}\},$$
$$\mathcal{L}^{\tilde{*}} = \{L^* \mid L \in \mathcal{L}\}, \qquad\qquad \mathcal{L}^C = \{L^C \mid L \in \mathcal{L}\}.$$

Note that $\mathcal{L}\,\tilde{\cdot}\,\mathcal{L}$ is different from $\mathcal{L}^{\tilde{2}}$. Let L_e be the collection of classes with finite elasticity.

Theorem 2 [14]. $L_e \subsetneqq L$.

Theorem 3 [14]. *The family L_e is closed under union operation $\tilde{\cup}$, but L is not.*

Finite elasticity is preserved by not only union but also the other operations defined above.

Theorem 4. *The family L_e is closed under the operations* $\cup$, $\tilde{\cap}$, $\tilde{\div}$, $\tilde{\cdot}^m$, $\tilde{+}$ *and* $\tilde{*}$.

Proof. We only give the proof for the operation $\tilde{+}$. The proof for the other operations can be done analogously. We proceed by contradiction.

Assume that $\mathcal{L}^+$ has infinite elasticity for some $\mathcal{L} \in L_e$. Then there are two infinite sequences $(w_n)_{n\geq 0}$ and $(L_n^+)_{n\in N}$, where each w_n is a string and $L_n \in \mathcal{L}$ for $n \in N$ such that $\{w_0, w_1, \cdots, w_{k-1}\} \subseteq L_k^+$ and $w_k \notin L_k^+$ for any $k \in N$. Here we construct an infinite sequence of non-negative integers $(k_n)_{n\geq 0}$ and a tuple of strings $(w_{k_n 1}, w_{k_n 2}, \cdots, w_{k_n l_n})$ for $n \geq 0$ by the following stages: Let $k_0 = 0$ and $N_{-1} = N$. For a set S, $\#S$ means the cardinality of the set S.

Stage $n (\geq 0)$: Let $(w_{k_n 1}, w_{k_n 2}, \cdots, w_{k_n l_n})$ be a tuple of strings satisfying the following conditions

    (A) $w_{k_n} = w_{k_n 1} w_{k_n 2} \cdots w_{k_n l_n}$ and

$$\begin{cases} w_{k_n i} \in \Sigma^+ \ (1 \leq i \leq l_n), & \text{if } w_{k_n} \neq \epsilon, \\ l_n = 1, \ w_{k_n 1} = \epsilon, & \text{otherwise.} \end{cases}$$

    (B) $\#\{k \in N_{n-1} \mid \{w_{k_n 1}, \cdots, w_{k_n l_n}\} \subseteq L_k\} = \infty$.

For the tuple of $(w_{k_n 1}, w_{k_n 2}, \cdots, w_{k_n l_n})$, let us put

$$N_n = \{k \in N_{n-1} \mid \{w_{k_n 1}, w_{k_n 2}, \cdots, w_{k_n l_n}\} \subseteq L_k\} \quad \text{and}$$
$$k_{n+1} = \min\{k \in N_n\}.$$

    Go to *stage* $n + 1$.

Claim: For any $n \geq 0$, there is a tuple of $(w_{k_n 1}, w_{k_n 2}, \cdots, w_{k_n l_n})$ satisfying both (A) and (B), and $k_n < k_{n+1}$.

The proof of this claim proceeds by induction on n. Note that if $w_{k_n} = \epsilon$, it suffices to show (B) and $k_n < k_{n+1}$.

(I) In the case $n = 0$. If $w_0 = \epsilon$, then it is obvious that $w_{k_0 1}(= \epsilon) \in L_k$ for all $k \in N$, which implies that $N_0 = N$ and $k_1 = 1(> k_0)$. Otherwise, there are only finitely many tuples of strings satisfying (A). Among such tuples, there is at least one tuple satisfying (B) because of $w_0 \in L_k^+$ for *any* $k \in N$. For such a tuple, it follows from $N_0 \subseteq N$ that $0 = k_0 < k_1$.

(II) In the case $n \geq 1$. We assume that the claim is valid for $n - 1$. As easily seen, if $w_{k_n} = \epsilon$, then $w_{k_n 1} \in L_k$ for *any* $k > k_n$ and $w_{k_n 1} \notin L_{k_n}$. Since $N_{n-1} \supseteq N_n$, these lead that $N_n = N_{n-1} - \{k_n\}$, which implies $k_n < k_{n+1}$. From the induction hypothesis, clearly $\#N_n = \infty$.

If $w_{k_n} \neq \epsilon$, it can be seen as in (I) that there is a tuple of $(w_{k_n 1}, w_{k_n 2}, \cdots, w_{k_n l_n})$ satisfying both (A) and (B). Since $w_{k_n} \notin L_{k_n}^+$, it follows that $\{w_{k_n 1}, w_{k_n 2}, \cdots, w_{k_n l_n}\} \not\subseteq L_{k_n}$, thus $k_n \notin N_n$. Therefore we have $k_n < k_{n+1}$. By (I) and (II), the claim is valid.

In the above proof, we see that $N_0 \supsetneq N_1 \supsetneq \cdots \supsetneq N_{n-1}$ and $\{w_{k_n 1}, w_{k_n 2}, \cdots, w_{k_n l_n}\} \not\subseteq L_{k_n}$ for any $n \in N$. Hence for any $n \in N$ we can take a string $w'_{k_n} \in \{w_{k_n 1}, w_{k_n 2}, \cdots, w_{k_n l_n}\} - L_{k_n}$. So $w'_{k_n} \notin L_{k_n}$. By $k_n \in N_{n-1}$, clearly $k_n \in N_i$ for $i = 0, 1, \cdots, n - 1$. Therefore $\{w'_{k_0}, w'_{k_1}, \cdots, w'_{k_{n-1}}\} \subseteq L_{k_n}$. Consequently two infinite sequences of $(w'_{k_n})_{n\geq 0}$ and $(L_{k_n})_{n\in N}$ satisfy the conditions (1) and (2) in Definition 1. That is, $\mathcal{L} \notin L_e$, which contradicts our assumption. $\square$

Theorem 5. *The family L_e is not closed under complement operation ^{C}.*

Proof. Let a sequence of $w_0, w_1, \cdots$ be a recursive enumeration of Σ^*. Let $\mathcal{L} = L_1, L_2, \cdots$ be a class defined as follows: $L_k = \Sigma^* - \{w_0, \cdots, w_{k-1}\}$ for all $k \in N$. As easily seen, $\mathcal{L}$ has finite elasticity. However two infinite sequences of $(w_n)_{n \geq 0}$ and $(L_n^C)_{n \in N}$ satisfy the conditions (1) and (2) in Definition 1. Therefore $\mathcal{L}^C \notin \bar{L}_e$. $\square$

By Theorem 2 to 4, it follows immediately that:

Theorem 6. *Given language classes with finite elasticity, a class obtained by finitely applying operations $\cup$, $\tilde{\cup}$, $\tilde{\cap}$, $\tilde{\cdot}$, $\tilde{m}$, $\tilde{+}$, * to them has also finite elasticity, thus the class is inferable from positive data.*

3 Inductive Inference of EFS and Closure Properties

3.1 Max Length-bounded EFS's

In this section, we deal with a special type of EFS's, called max length-bounded to define target languages.

Let Σ, Π and X be mutually disjoint nonempty sets. We assume that Σ is finite. Elements in Σ, Π and X are called constant symbols, predicate symbols and variable symbols, respectively. A predicate symbol is associated with a positive integer called *arity*. For detailed definitions and results on EFS's, we refer to [2, 3, 13, 15].

Definition 7. A clause $A \leftarrow B_1, \cdots, B_n$ is *max length-bounded* if $B_1, \cdots, B_n$ are distinct atoms and $|A\theta| \geq |B_i\theta|$ for any substitution θ and $1 \leq i \leq n$, where $|A|$ means the length of A, i.e., the number of constant or variable symbols. An EFS Γ is *max length-bounded* if each clause $C \in \Gamma$ is max length-bounded.

Note that the restriction of $B_i \neq B_j$ for $i \neq j$ imposed on max length-bounded is essentially needed in our problem as seen later, particularly in the proof of Theorem 12 in §3.3.

In the above definition, the clause is called *length-bounded* (resp. *weakly reducing*) if $|A\theta| \geq \sum_{i=1}^{n} |B_i\theta|$ (resp. $|A\theta| \geq |B_i\theta|$ for $i = 1, \cdots, n$) for any substitution θ, and a length-bounded EFS and a weakly reducing EFS are defined in analogous manner as above ([3, 15]). For an EFS Γ, we denote by $M(\Gamma)$ the least Herbrand model. For a given weakly reducing EFS Γ we can get a max length-bounded EFS Γ' such that $M(\Gamma) = M(\Gamma')$, by taking off atoms of duplicated occurrences in the body of every clause of Γ. This implies that the following result is valid for a max length-bounded EFS as well as a weakly reducing EFS(cf. [15]).

Theorem 8. *For a max length-bounded EFS Γ, the least Herbrand model $M(\Gamma)$ is recursive.*

For an EFS Γ and a unary predicate symbol $p \in \Pi$, we denote by $L(\Gamma, p)$ the language defined by Γ and p. Without loss of generality, we can fix predicate symbols defining languages to one predicate symbol, say p. Since length-bounded EFS's correspond to context-sensitive grammars in Chomsky hierarchy[3], the class of languages definable by max length-bounded EFS's includes that of context-sensitive languages. But it is unknown whether the converse inclusion holds.

Given a class $\mathcal{G}$ of EFS's, we define $L(\mathcal{G}, p) = \{L(\Gamma, p) \mid \Gamma \in \mathcal{G}\}$. As easily seen, it follows from Theorem 8 that for any recursive enumeration $\mathcal{G}$ of max length-bounded EFS's the class $L(\mathcal{G}, p)$ is an indexed family of recursive languages. Hereafter, we assume that a class of EFS's is a recursive enumeration of max length-bounded EFS's.

3.2   Closure Properties of Max Length-bounded EFS Classes

In this section, we introduce operations for max length-bounded EFS's corresponding to the usual operations of languages. It is well known that the class of context-sensitive languages is closed under various language operations such as union, intersection, concatenation and so on. We show that the class of max length-bounded EFS languages is also closed under such operations.

Let $pred(\Gamma)$ be the set of predicate symbols appearing in EFS Γ.

First we introduce various operations for EFS's. Given EFS's Γ_1 and Γ_2, by renaming predicate symbols included in them we can get EFS's Γ_1' and Γ_2' such that

$$L(\Gamma_1, p) = L(\Gamma_1', p_1), \qquad L(\Gamma_2, p) = L(\Gamma_2', p_2),$$
$$pred(\Gamma_1') \cap pred(\Gamma_2') = \phi, \qquad p \notin pred(\Gamma_1') \cup pred(\Gamma_2').$$

Then we define the following operations for the EFS's:

$$\Gamma_1 \tilde{\cup} \Gamma_2 = \Gamma_1' \cup \Gamma_2' \cup \left\{ \begin{array}{l} p(x) \leftarrow p_1(x) \\ p(x) \leftarrow p_2(x) \end{array} \right\},$$
$$\Gamma_1 \tilde{\cap} \Gamma_2 = \Gamma_1' \cup \Gamma_2' \cup \{p(x) \leftarrow p_1(x), p_2(x)\},$$
$$\Gamma_1 \tilde{\ } \Gamma_2 = \Gamma_1' \cup \Gamma_2' \cup \{p(xy) \leftarrow p_1(x), p_2(y)\}.$$

Note that a clause $p(x) \leftarrow p_1(x), p_2(x)$ is not length-bounded but max length-bounded.

Similarly, given an EFS Γ, we can get an EFS Γ' satisfying $L(\Gamma, p) = L(\Gamma', p')$ and $p \notin pred(\Gamma')$. Then we define the following operations for the EFS :

$$\Gamma^{\tilde{m}} = \Gamma' \cup \{p(x_1 x_2 \cdots x_m) \leftarrow p'(x_1), \cdots, p'(x_m)\} \quad (m \geq 1),$$
$$\Gamma^{\tilde{+}} = \Gamma' \cup \left\{ \begin{array}{l} p(x) \leftarrow p'(x) \\ p(xy) \leftarrow p(x), p(y) \end{array} \right\}.$$

As easily seen, it follows from these definitions that:

$$L(\Gamma_1 \tilde{\cup} \Gamma_2, p) = L(\Gamma_1, p) \cup L(\Gamma_2, p), \qquad L(\Gamma^{\tilde{m}}, p) = L(\Gamma, p)^m,$$
$$L(\Gamma_1 \tilde{\cap} \Gamma_2, p) = L(\Gamma_1, p) \cap L(\Gamma_2, p), \qquad L(\Gamma^{\tilde{+}}, p) = L(\Gamma, p)^+,$$
$$L(\Gamma_1 \tilde{\ } \Gamma_2, p) = L(\Gamma_1, p) \cdot L(\Gamma_2, p),$$

where operations in right hand sides of equality symbol are usual language operations. Hence the following result is valid as well as context-sensitive languages:

Theorem 9. *The class of languages definable by max length-bounded EFS's is closed under the above usual language operations.*

Now we define operations for EFS classes as follows:

$$\mathcal{G}_1 \, \tilde{\cup} \, \mathcal{G}_2 = \{\Gamma_1 \, \tilde{\cup} \, \Gamma_2 \mid \Gamma_1 \in \mathcal{G}_1, \; \Gamma_2 \in \mathcal{G}_2\}, \quad \mathcal{G}^{\,\tilde{m}} = \{\Gamma^{\,\tilde{m}} \mid \Gamma \in \mathcal{G}\},$$
$$\mathcal{G}_1 \, \tilde{\cap} \, \mathcal{G}_2 = \{\Gamma_1 \, \tilde{\cap} \, \Gamma_2 \mid \Gamma_1 \in \mathcal{G}_1, \; \Gamma_2 \in \mathcal{G}_2\}, \quad \mathcal{G}^{\,\tilde{+}} = \{\Gamma^{\,\tilde{+}} \mid \Gamma \in \mathcal{G}\},$$
$$\mathcal{G}_1 \, \tilde{\cdot} \, \mathcal{G}_2 = \{\Gamma_1 \, \tilde{\cdot} \, \Gamma_2 \mid \Gamma_1 \in \mathcal{G}_1, \; \Gamma_2 \in \mathcal{G}_2\},$$

and $\mathcal{G}_1 \cup \mathcal{G}_2$ is the usual union set. It is easy to see that the above operations for EFS classes correspond to those for language classes defined in §2 ; for instance, $L(\mathcal{G}_1 \, \tilde{\cup} \, \mathcal{G}_2, p) = L(\mathcal{G}_1, p) \, \tilde{\cup} \, L(\mathcal{G}_2, p)$ and so on.

Let be $\mathbf{G}_e = \{\mathcal{G} \mid \mathcal{G} \text{ is a class of max length-bounded EFS's}, L(\mathcal{G}, p) \in \mathbf{L}_e\}$. Then it follows immediately from Theorem 6 and 9 that:

Theorem 10. $\mathbf{G}_e$ *is closed under the above operations for EFS classes. Therefore the class obtained by finitely applying such operations to classes in $\mathbf{G}_e$ defines a language class that is inferable from positive data.*

3.3 Characterizing Two Classes of Max Length-bounded EFS's

In this section, we characterize a max length-bounded EFS class in the family $\mathbf{G}_e$, and that for the language class to be inferable from positive data. In what follows, we assume that Π is finite.

Definition 11. An EFS Γ is *reduced with respect to* an Herbrand interpretation I if $I \subseteq M(\Gamma)$ but $I \not\subseteq M(\Gamma')$ for any $\Gamma' \subsetneq \Gamma$.

For EFS's Γ and Γ', if we can identify them by renaming variable symbols, then we identify Γ with Γ' because their least Herbrand models are equal. For a set $S \subseteq \Sigma^+$ and a unary predicate symbol $p \in \Pi$, let $p(S) = \{p(w) \mid w \in S\}$.

Theorem 12. *For any nonempty finite set S of strings, there exist only finitely many max length-bounded EFS's that are reduced with respect to $p(S)$.*

Proof. Let $S = \{w_1, \cdots, w_k\}$ be a nonempty set of strings and let Γ a max length-bounded EFS reduced w.r.t. $p(S)$. As easily seen, for any axiom $A \leftarrow B_1, \cdots, B_n$ of Γ, $|A| \leq \max\{|w_1|, \cdots, |w_k|\} (= l, \text{say})$. By Definition 7, clearly $|B_i| \leq |A|$ for any i. Since Π is finite, there exist only finitely many atoms whose lengths are at most l. Thus our Theorem is valid. □

Definition 13. A class $\mathcal{G}$ is said to be *closed under subset operation* if for any $\Gamma \in \mathcal{G}$, all subsets of Γ belong to $\mathcal{G}$.

Lemma 14. *Let $\mathcal{G}$ be a class of EFS's closed under subset operation. For any $\Gamma \in \mathcal{G}$ and any nonempty set $S \subseteq L(\Gamma)$, if Γ is not reduced with respect to $p(S)$, then there is an EFS $\Gamma' \in \mathcal{G}$ such that $\Gamma' \subsetneq \Gamma$ and Γ' is reduced with respect to $p(S)$.*

Proof. It is clear since $\mathcal{G}$ is closed under subset operation. □

Definition 15. A class $\mathcal{G}$ of EFS's is said to be *of infinite hierarchy with respect to* p if there are an infinite sequence of nonempty finite sets $(T_n)_{n \in N}$ of strings and an infinite sequence of EFS's $(\Gamma_n)_{n \in N}$, each in $\mathcal{G}$ such that

$$(1) \quad T_1 \subsetneq T_2 \subsetneq \cdots, \qquad (2) \quad \Gamma_1 \subsetneq \Gamma_2 \subsetneq \cdots, \qquad \text{and}$$
$$(3) \quad \Gamma_n \text{ is reduced with respect to } p(T_n) \text{ for any } n \in N.$$

A class $\mathcal{G}$ is said to be *of finite hierarchy with respect to* p if it is not of infinite hierarchy with respect to p.

Theorem 16. *Let $\mathcal{G}$ be a class of max length-bounded EFS's closed under subset operation and let $p \in \Pi$ a unary predicate symbol. Then $L(\mathcal{G}, p)$ has finite elasticity if and only if $\mathcal{G}$ is of finite hierarchy with respect to p.*

Proof. Only if part. Assume that $\mathcal{G}$ is not of finite hierarchy w.r.t. p and two infinite sequences of $(T_n)_{n \in N}$ and $(\Gamma_n)_{n \in N}$ satisfy conditions (1), (2) and (3) in Definition 15. As easily seen, $L(\Gamma_n, p) \subsetneq L(\Gamma_{n+1}, p)$ for any $n \in N$. Let $w_0, w_1, \cdots$ be a sequence such that $w_0 \in L(\Gamma_1, p)$ and $w_n \in L(\Gamma_{n+1}, p) - L(\Gamma_n, p)$ for $n \in N$. Then two infinite sequences of $(w_n)_{n \geq 0}$ and $(L(\Gamma_n, p))_{n \in N}$ satisfy the conditions (1) and (2) in Definition 1. Hence $L(\mathcal{G}, p)$ has infinite elasticity.

If part. Assume that $L(\mathcal{G}, p)$ has infinite elasticity and we have two infinite sequences $(w_n)_{n \geq 0}$ and $(\Gamma_n)_{n \in N}$, where $\Gamma_n \in \mathcal{G}$ such that $\{w_0, w_1, \cdots, w_{k-1}\} \subseteq L(\Gamma_k, p)$ and $w_k \notin L(\Gamma_k, p)$ for any $k \in N$. We define an infinite sequence of finite sets $(T_n)_{n \in N}$ of strings and an infinite sequence of finite sets $(\mathcal{F}_n)_{n \in N}$ of EFS's recursively by the following stages: Let $k_1 = 0$ and $S = \{w_i \mid i \geq 0\}$.

Stage $n \, (\in N)$: Let be

$$T_n = \{w_0, w_1, \cdots, w_{k_n}\},$$
$$\mathcal{F}_n = \{\Gamma \in \mathcal{G} \mid S \not\subseteq L(\Gamma, p) \text{ and } \Gamma \text{ is reduced w.r.t. } p(T_n)\} \quad \text{and}$$
$$k_{n+1} = \min\{k \mid \{w_0, w_1, \cdots, w_k\} \not\subseteq L(\Gamma, p) \text{ for any } \Gamma \in \mathcal{F}_n\}.$$

Go to stage $n + 1$.

The following claims are valid;

Claim A: For any $n \in N$, $\mathcal{F}_n$ is nonempty finite set, $\mathcal{F}_n \cap \mathcal{F}_{n+1} = \phi$ and $T_n \subsetneq T_{n+1}$.

The proof of the claim A. From the definition of k_1, $T_1 = \{w_0\}$. Below we shall prove that for any $n \in N$ if T_n is nonempty finite subset of S, then so is T_{n+1} and the claim A is valid for n.

We assume that T_n is nonempty finite subset of S. By Theorem 12, there exist only finitely many max length-bounded EFS's reduced w.r.t. $p(T_n)$. Thus $\mathcal{F}_n$ is *finite*. On the other hand, it follows from the definition of T_n and $w_{k_n+1} \notin L(\Gamma_{k_n+1}, p)$ that $T_n \subseteq L(\Gamma_{k_n+1}, p)$ and $S \not\subseteq L(\Gamma_{k_n+1}, p)$. If Γ_{k_n+1} is reduced w.r.t. $p(T_n)$, then $\Gamma_{k_n+1} \in \mathcal{F}_n$. Otherwise, from Lemma 14 there is an EFS $\Gamma' \in \mathcal{G}$ such that $\Gamma' \subsetneq \Gamma_{k_n+1}$ and Γ' is reduced w.r.t. $p(T_n)$. Clearly $L(\Gamma', p) \subseteq L(\Gamma, p)$, which implies $S \not\subseteq L(\Gamma', p)$. Hence $\Gamma' \in \mathcal{F}_n$. In any case, $\mathcal{F}_n$ is *nonempty finite set*.

For any $\Gamma \in \mathcal{F}_n$, we put $k_\Gamma = \min\{k \in N \mid \{w_0, \cdots, w_k\} \not\subseteq L(\Gamma, p)\}$. Then k_Γ is well-defined and $k_\Gamma > k_n$, because of $T_n \subseteq L(\Gamma, p)$ but $S \not\subseteq L(\Gamma, p)$. Since $\mathcal{F}_n$ is nonempty finite set, $k_{n+1} = \max\{k_\Gamma \mid \Gamma \in \mathcal{F}_n\}$ is well-defined and larger than k_n. These mean that T_{n+1} is *nonempty finite* subset of S and $T_n \subsetneq T_{n+1}$. Moreover, since $T_{k_{n+1}} \not\subseteq L(\Gamma, p)$ for any $\Gamma \in \mathcal{F}_n$, it turns out that $\mathcal{F}_n \cap \mathcal{F}_{n+1} = \phi$.

Claim B: For any $n \in N$ and any $\Gamma \in \mathcal{F}_{n+1}$, there is an EFS $\Gamma' \in \mathcal{F}_n$ such that $\Gamma' \subsetneq \Gamma$.

The proof of claim B.   Let Γ be any EFS of $\mathcal{F}_{n+1}$. From the claim A, $T_n \subsetneq T_{n+1}$ and $\Gamma \notin \mathcal{F}_n$. Hence Γ is not reduced w.r.t. $p(T_n)$ although $T_n \subseteq L(\Gamma, p)$. Appealing to Lemma 14 there is an EFS $\Gamma' \in \mathcal{G}$ such that $\Gamma' \subsetneq \Gamma$ and Γ' is reduced w.r.t. $p(T_n)$. Clearly $S \not\subseteq L(\Gamma', p)$, because of $S \not\subseteq L(\Gamma, p)$. This implies that $\Gamma' \in \mathcal{F}_n$.

Claim C: There is an infinite sequence of EFS's $(\Gamma'_n)_{n \in N}$ such that $\Gamma'_n \in \mathcal{F}_n$ for all $n \in N$ and $\Gamma'_1 \subsetneq \Gamma'_2 \subsetneq \cdots$.

The proof of claim C.   From the claim B, for any $n \in N$ and any $\Gamma \in \mathcal{F}_n$ there is a sequence of EFS's $\Gamma'_1, \Gamma'_2, \cdots, \Gamma'_n(= \Gamma)$ such that $\Gamma'_i \in \mathcal{F}_i$ for $i = 1, 2, \cdots, n$ and $\Gamma'_1 \subsetneq \Gamma'_2 \subsetneq \cdots \subsetneq \Gamma'_n$. From the claim A, $\mathcal{F}_1$ is finite. Hence there should be an infinite sequence of $(\Gamma'_n)_{n \in N}$ such that $\Gamma'_n \in \mathcal{F}_n$ for all n and $\Gamma'_1 \subsetneq \Gamma'_2 \subsetneq \cdots$.

From these claims, it turns out that infinite sequences of $(T_n)_{n \in N}$ and $(\Gamma'_n)_{n \in N}$ satisfy the conditions (1)-(3) of Definition 15. Therefore the class $\mathcal{G}$ is of infinite hierarchy w.r.t. p. □

For a class $\mathcal{G}$, let us denote by $\overline{\mathcal{G}}$ the closure under subset operation. That is, $\overline{\mathcal{G}} = \{\Gamma' \mid \exists \Gamma \in \mathcal{G} \text{ s.t. } \Gamma' \subseteq \Gamma\}$. As easily seen, if $L(\overline{\mathcal{G}})$ has finite elasticity, then so do $L(\mathcal{G})$. Therefore it follows immediately from Theorem 16 that:

Corollary 17. *Let $\mathcal{G}$ be a class of max length-bounded EFS's and let $p \in \Pi$ a unary predicate symbol. If $\overline{\mathcal{G}}$ is of finite hierarchy with respect to p, then $L(\mathcal{G}, p)$ has finite elasticity.*

The following result is obtained from Theorem 16 (cf. [12] for length-bounded EFS's):

Corollary 18. *For any $n \in N$, the class of languages definable by max length-bounded EFS's with at most n axioms has finite elasticity. Therefore the class is inferable from positive data.*

Finally we consider a characterization of a max length-bounded EFS class for the language class to be inferable from positive data. The present authors[9] dealt with length-bounded EFS's as a framework for inductive inference of languages from positive data, and gave characterizations of length-bounded EFS classes for inferability of their language (model) classes, without any restriction on Π. Results obtained in [9] are also valid for max length-bounded EFS's considered, because Theorem 8 and Theorem 12 obtained in the present paper play an essential role in derivations of them. Thus we only state a result without proof(cf. [9]).

Definition 19 [9]. An EFS $\Gamma \in \mathcal{G}$ is said to be *of finite hierarchy with respect to p within $\mathcal{G}$* if there is no pair of an infinite sequence of nonempty finite sets $(T_n)_{n \in N}$

of strings and an infinite sequence of EFS's $(\Gamma_n)_{n\in N}$ each in $\mathcal{G}$ such that

(1) $T_1 \subsetneq T_2 \subsetneq \cdots$,   (2) $\bigcup_{n=1}^{\infty} T_n = L(\Gamma, p)$,

(3) $\Gamma_1 \subsetneq \Gamma_2 \subsetneq \cdots$,   (4) Γ_n is reduced w.r.t. $p(T_n)$ for all n,

(5) $L(\Gamma_n, p) \subsetneq L(\Gamma, p)$ for all n.

Theorem 20. *Let $\mathcal{G}$ be a max length-bounded EFS class closed under subset operation and let $p \in \Pi$ a unary predicate symbol. Then the class $L(\mathcal{G}, p)$ is inferable from positive data if and only if any $\Gamma \in \mathcal{G}$ is of finite hierarchy with respect to p within $\mathcal{G}$.*

Acknowledgements

The authors wish to thank the anonymous referee for many suggestions, valuable comments and finding errors in earlier version of this paper.

References

1. D. Angluin: *Inductive inference of formal languages from positive data.* Information and Control **45** (1980) 117–135.
2. S. Arikawa: *Elementary formal systems and formal languages - simple formal systems.* Memoirs of Fac. Sci., Kyushu Univ. Ser. A, Math. **24** (1970) 47–75.
3. S. Arikawa, T. Shinohara, and A. Yamamoto: *Elementary formal system as a unifying framework for language learning.* Proc. 2nd Workshop on Comput. Learning Theory (1989) 312–327.
4. E.M. Gold: *Language identification in the limit.* Information and Control **10** (1967) 447–474.
5. S. Kapur: *Computational learning of languages.* PhD thesis, Technical Report 91-1234, Cornell University, September 1991.
6. S. Kapur: *Monotonic language learning.* Proc. 3rd Workshop on Algorithmic Learning Theory (1992) 147–158.
7. S. Lange and T. Zeugmann: *Learning recursive languages with bounded mind changes.* GOSLER-Report 16/92, FB Mathematik und Informatik, TH Leipzig, September 1992.
8. T. Motoki, T. Shinohara and K. Wright: *The correct definition of finite elasticity: corrigendum to identification of unions.* Proc. 4th Workshop on Comput. Learning Theory (1991) 375–375.
9. M. Sato and T. Moriyama: *Inductive inference of length-bounded EFS's from positive data.* In preparation, 1993.
10. M. Sato and K. Umayahara: *Inductive inferability for formal languages from positive data.* IEICE Trans. Inf. & Syst. **E75-D(4)** (1992) 84–92.
11. T. Shinohara: *Inductive inference from positive data is powerful.* Proc. 3rd Workshop on Comput. Learning Theory (1990) 97–110.
12. T. Shinohara: *Inductive inference of monotonic formal systems from positive data.* New Generation Computing **8** (1991) 371–384.
13. R.M. Smullyan: *Theory of Formal Systems.* Princeton Univ. Press, 1961.
14. K. Wright: *Identification of unions of languages drawn from an identifiable class.* Proc. 2nd Workshop on Comput. Learning Theory (1989) 328–333.
15. A. Yamamoto: *Procedual semantics and negative information of elementary formal system.* J. Logic Programming **13** (1991) 89–97.
16. A. Yamamoto: *Elementary formal system as a logic programming language.* Proc. Logic Program. Conf. '89, ICOT (1989) 123–132; also in Logic Programming'89, Lecture Notes in Artificial Intelligence **485** (1991).

Uniform Characterizations of Various Kinds of Language Learning

Shyam Kapur*

Institute for Research in Cognitive Science, University of Pennsylvania
3401 Walnut Street Rm 412C, Philadelphia, PA 19104 (USA)

Abstract. Learnability of families of recursive languages from positive data is studied in the Gold paradigm of inductive inference. A large amount of work has focused on trying to understand how the language learning ability of an inductive inference machine is affected when it is constrained. For example, derived from work in inductive logic, notions of monotonicity have been studied which variously reflect the requirement that the learner's guess must monotonically 'improve' with regard to the target language. A unique characterization theorem is obtained which uniformly characterizes all classes learnable under a number of different constraints specified via a parametric description. It is also shown how many known characterizations can be obtained by straightforward applications of this theorem. It is argued that the new parameterization scheme for specifying constraints works for a wide variety of constraints.

1 Introduction

In one version of the Gold paradigm for inductive inference [5], the language learner is presented with the *text* of a language, i.e., an infinite sequence of strings made up of all and only strings from the language. (For surveys of work in this model, see [21, 2].) The learner is said to learn a language if, on any text for it, the learner's guess *converges* to the same language, i.e., from some point onwards, the guess coincides with the language being presented. The learner is said to learn a family of languages if it learns each language in the family. This model is motivated by the well-established hypothesis that the child learns natural language from positive evidence alone. (For a discussion, see [3, 17].)

One of the central questions studied so far is whether or not various restrictions on the behavior of an inference machine limit its learning capability. In particular, the question whether it is possible to infer a language in such a way that the intermediate hypotheses are all monotonically better generalizations and/or specializations has been extensively investigated [15, 25]. Part of the motivation for this study can be derived from inductive logic and nonmonotonic reasoning [22, 6, 24] as well as natural language acquisition. It has sometimes been claimed that children are conservative (*non-overgeneralizing*) learners, that

---

* The author would like to thank Gianfranco Bilardi for a useful suggestion and Steffen Lange and Thomas Zeugmann for interesting discussions. The author was supported in part by ARO grant DAAL 03-89-C-0031, DARPA grant N00014-90-J-1863, NSF grant IRI 90-16592 and Ben Franklin grant 91S.3078C-1.

is, they never guess a subset language of a previous guess. (See [3, 18, 16, 7, 11] for relevant discussion.)

Characterizations of learnable classes of language families play an important role in the determination of the exact nature of various inference strategies. In this way, characterizations facilitate development of "uniform" learning procedures. In this paper, we show that uniformity can be achieved at yet another level for various characterizations can themselves be obtained in a uniform fashion. In particular, we show that the various monotonicity classes can be uniformly characterized. Our uniform characterization theorem for the classes of language families that can be learned under constraints specified in a particular parametric fashion is shown to subsume alternative characterizations. Our parameterization scheme is general enough to allow the possibility that while learning different languages in a family, the learner does not satisfy the same constraint.

2 Background

2.1 Model

Let Σ^* be a free monoid over Σ, a finite alphabet of symbols. Let Z_+ be the set of positive integers. Let $D_1, D_2, D_3, \ldots$ be a canonical enumeration of finite subsets of Σ^*; $M_1, M_2, M_3, \ldots$ a standard enumeration of all Turing machines over Σ. For any *index* $I \in Z_+$, let W_I denote the *language* (subset of Σ^*) accepted by the machine M_I. The complement of W_I in Σ^* is denoted as $\overline{W_I}$. An index I is *total* if the corresponding machine M_I is total and accepts a non-empty language. If $I_1, I_2, \ldots$ is a recursive enumeration of total indices, then $\mathcal{F} = W_{I_1}, W_{I_2}, \ldots$ is called an *indexed family of non-empty recursive languages* (hereafter, simply an *indexed family*)[2]. We denote by $\Delta_{\mathcal{F}}$ the set of all non-empty finite subsets of Σ^* that are contained in some language in the family $\mathcal{F}$.

A *text* is an infinite sequence of strings from Σ^*; t ranges over texts, t_n is the nth string in text t, $\bar{t}_n$ is the initial prefix of length n of text t, and $content(\bar{t}_n)$ is the set of strings in the prefix $\bar{t}_n$; t is *for* a language L if and only if the set of strings in t equals L.

An *inductive inference machine (IIM)* M is an algorithmic device whose input is a text $t_1, t_2, \ldots$ and whose output is a sequence of nonnegative integers $M(\bar{t}_1), M(\bar{t}_2), \ldots$ constrained to be either 0 or total indices. The procedure works in stages, but it can happen that a stage never gets completed. At the nth stage, t_n is input and $M(\bar{t}_n)$ is output. The interpretation is as follows: If $M(\bar{t}_n) = 0$, then the IIM makes no guess; otherwise, it guesses the language $W_{M(\bar{t}_n)}$.

An IIM M is said to learn the language L if and only if, for each text t for L, there is a k such that $W_{M(\bar{t}_k)} = L$ and, for all $n > k$, $M(\bar{t}_n) = M(\bar{t}_k)$. Intuitively, the guess (hypothesis) converges to a total index for the input language. (This is similar to the TxtEx-identification criterion [5].) We say that M learns a family $\mathcal{F}$ if M learns each language in $\mathcal{F}$.

---

[2] This is also referred to as a *uniformly recursive sequence of recursive sets* [23].

We consider *class preserving* language learning [15], i.e., where all the hypotheses ever produced describe languages that are contained in the family. Notice, however, that only the extensions are required to be identical and the particular description of the both the hypotheses themselves and the hypothesis space as a whole may be chosen as necessary. Keeping in mind the purposes of this study, such as natural language acquisition and other artificial intelligence applications, the restriction of the hypothesis space to an indexed family of recursive languages is quite natural. It ensures that a hypothesis is refutable by a string observed in the text and that the set of possible hypotheses can be generated automatically.

2.2 Variants of Monotonic Learning

In this section, we recall various definitions of monotonicity.

Definition 1. [12] An IIM M is said to learn a language L

(SM) strong-monotonically,
(M) monotonically, or
(WM) weak-monotonically

if and only if M learns L and on any text t for L and any two consecutive non-zero hypotheses, say $M(\bar{t}_n)$ and $M(\bar{t}_{n+k})$, $k \geq 1$, the corresponding condition is satisfied:

(SM) $W_{M(\bar{t}_n)} \subseteq W_{M(\bar{t}_{n+k})}$,
(M) $W_{M(\bar{t}_n)} \cap L \subseteq W_{M(\bar{t}_{n+k})} \cap L$, or
(WM) if $content(\bar{t}_{n+k}) \subseteq W_{M(\bar{t}_n)}$, then $W_{M(\bar{t}_n)} \subseteq W_{M(\bar{t}_{n+k})}$.

Let $SMON$, MON and $WMON$ denote the class of families for which there is an IIM M which learns each language strong-monotonically, monotonically and weak-monotonically, respectively.

Definition 2. [8] An IIM M is said to learn a language L

(DSM) dual strong-monotonically,
(DM) dual monotonically, or
(DWM) dual weak-monotonically

if and only if M learns L and on any text t for L and any two consecutive non-zero hypotheses, say $M(\bar{t}_n)$ and $M(\bar{t}_{n+k})$, $k \geq 1$, the corresponding condition is satisfied:

(DSM) $W_{M(\bar{t}_n)} \supseteq W_{M(\bar{t}_{n+k})}$,
(DM) $\overline{W_{M(\bar{t}_n)}} \cap \bar{L} \subseteq \overline{W_{M(\bar{t}_{n+k})}} \cap \bar{L}$, or
(DWM) if $content(\bar{t}_{n+k}) \subseteq W_{M(\bar{t}_n)}$, then $W_{M(\bar{t}_n)} \supseteq W_{M(\bar{t}_{n+k})}$.

Let $SMON^d$, MON^d and $WMON^d$ denote the class of families for which there is an IIM which learns each language dual strong-monotonically, dual monotonically and dual weak-monotonically, respectively.

Some well-known classes can be obtained by combining the constraints in Definitions 1 and 2. For example, if we insist that a machine learn a family strong-monotonically as well as dual strong-monotonically, it turns out to be equivalent to requiring that the machine makes only one guess which must be correct. Likewise, a machine that learns a family weak-monotonically and dual weak-monotonically may as well learn conservatively, i.e., exclusively perform justified mind changes.

Definition 3. [8] Let $SMON^{\&}$ represent the class of families learnable by some IIM that behaves strong-monotonically and dual strong-monotonically. Let $MON^{\&}$ and $WMON^{\&}$ be analogously defined.

Figure 1 below (reproduced from [15]) summarizes the known relationships between the various monotonicity classes defined above. The lines between the identification types indicate set inclusion, i.e., the lower type is properly contained in the upper one. Missing lines indicate incomparability of classes of families. (Note: LIM: class of learnable families; FIN: class of families learnable by IIMs that on any text for any language in the family produce only a single guess; $CONSERVATIVE$: class of families learnable by conservative IIMs.)

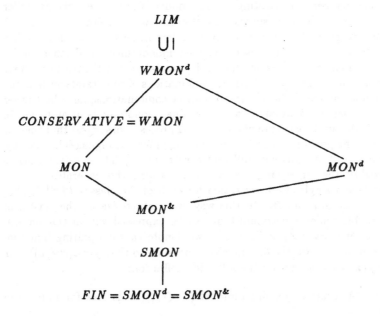

Fig. 1. Relationships between the classes generated under monotonicity constraints

3 Uniform Characterization Theorem

We next focus on the issue of uniform characterization of classes of families learnable under various monotonicity constraints. In this way, we characterize a number of known classes of language families; some of them for the first time.

Where R is any total index and S any finite set, let $\mathcal{A}_R$ and $\mathcal{B}_S$ be a binary and a unary relation on total indices, respectively. The notation $LIM[\mathcal{A}_R, \mathcal{B}_S]$ denotes classes of language families. $\mathcal{A}_R$ captures the constraint that must hold between consecutive non-zero guesses of the IIM given that the target language is R, while $\mathcal{B}_S$ stipulates whether this constraint can be gratuitously violated when the *content* of the input evidence is the set S. $\mathcal{B}_S$ is required to be a decidable relation with the *upward closure* property, i.e., if $I \in \mathcal{B}_S$ and, for some index R of a language in $\mathcal{F}$ and any finite set S', $W_R \supseteq S' \supseteq S$, then $I \in \mathcal{B}_{S'}$. There are a broad variety of constraints which have been investigated that can be captured within this parameterization scheme. Recall, for example, the conservativeness constraint that requires that the guess must not change as long as the input evidence is consistent with it. So here, for any total indices I, J "two adjoining guesses", and, for any R "target language", $(I, J) \in \mathcal{A}_R \iff I = J$. The relation $\mathcal{B}_S$ states the requirement that $\mathcal{A}_R$ has to be met as long as there is consistency. Therefore, for any I "current guess" and finite set S in R "target language", $I \in \mathcal{B}_S \iff S \not\subseteq W_I$.

There is an additional reason why parameterization of the constraints in this particular manner is interesting. In the model of language learning under study, the learner is said to be successful if it eventually converges to the target language and, in general, it is not possible to determine when the learner has converged. On the other hand, conservativeness, monotonicity, and other similar constraints have an absolute nature, i.e., it is mandated that the learner always meet that particular constraint. Further, combinations of constraints such as the ones that define the class $MON^{\&}$ are also strict combinations, i.e., the learner is required to observe monotonicity as well as dual monotonicity at all times. In harmony with the limiting nature of the learning process, we suggest that there is a need to allow the possibility that, while learning different languages in a family, the learner may end up satisfying different constraints [9]. The constraint $\mathcal{A}_R$ is parameterized on the target language and thus captures this generality.

We next give a single characterization for each of the classes $LIM[\mathcal{A}_R, \mathcal{B}_S]$ obtained by fixing $\mathcal{A}_R$ and $\mathcal{B}_S$. In this way, we characterize all the classes of families learnable under constraints that can be expressed within this parameterization scheme. By $c: Z_+ \times Z_+ \mapsto Z_+$ we denote cantor's pairing function. Also, *first* and *second* are the two inverse functions, so that $first(c(x, y)) = x$ and $second(c(x, y)) = y$. We use the following definition:

Definition 4. A recursively enumerable set $\zeta = \zeta_1, \zeta_2, \ldots$ is *good* for an indexed family $\mathcal{F}$ if

1. for all $k > 0$, $first(\zeta_k)$ is a total index of a language in $\mathcal{F}$, and
2. for each L in $\mathcal{F}$, there is some $k \in Z_+$ such that $D_{second(\zeta_k)} \subseteq W_{first(\zeta_k)} = L$.

Theorem 5. *An indexed family $\mathcal{F} \in LIM[\mathcal{A}_R, \mathcal{B}_S]$ if and only if a set ζ good for $\mathcal{F}$ with the following property is recursively enumerable: Given j and k such that $D_{second(\zeta_k)} \subseteq W_{first(\zeta_j)}$, a finite, non-empty sequence $\psi = \psi_1, \ldots \psi_m \subseteq \zeta$ is recursively enumerable such that*

1. $\psi_1 = \zeta_k$,
2. *for $1 \leq i \leq m$, $D_{second(\psi_i)}$ is a subset of $W_{first(\zeta_j)}$,*
3. *for $1 \leq i < m$, $first(\psi_i) \neq first(\psi_{i+1})$,*
4. *for $1 \leq i < m$, and any total index R of a language in $\mathcal{F}$,*

$$D_{second(\psi_{i+1})} \subseteq W_R \Rightarrow (first(\psi_i), first(\psi_{i+1})) \in \mathcal{A}_R, \text{ and}$$

5. *either*
   (a) *for some finite S in $W_{first(\zeta_j)}$), $first(\psi_m) \in \mathcal{B}_S$, or*
   (b) *$W_{first(\psi_m)} = W_{first(\zeta_j)}$ and, for any $L' \in \mathcal{F}$,*

$$D_{second(\psi_m)} \subseteq L' \subseteq W_{first(\zeta_j)} \Leftrightarrow L' = W_{first(\zeta_j)}.$$

Moreover, for any j, there is at least one k such that the condition 5(b) is satisfied.

Proof. ($\Leftarrow$) This requires a somewhat intricate construction. The essential idea is to exploit the behavior of a machine that is only incompletely constrained (the procedure IC below) in order to make the constrained machine M converge. The procedure IC below is a subroutine used by the IIM M we construct. ζ is the given recursively enumerable set.

PROCEDURE IC
Initialization: $IC(\bar{t}_0) = 0$;
Stage $n(n \geq 1)$:
Enumerate ζ for n steps. Find the least ζ_j (if any) such that both $D_{second(\zeta_j)} \subseteq content(\bar{t}_n) \subseteq W_{first(\zeta_j)}$ and ζ_j is equal to ψ_k, the last element of ψ given j and j enumerated in n steps. If such a j is found, then let $IC(\bar{t}_n) = \zeta_j$. Otherwise, let $IC(\bar{t}_n) = 0$.
END OF PROCEDURE IC
Two variables $GUESS$ and $GUESS\text{-}PREV$ are used by the IIM M defined below to keep track of the ζ_js which prompted some of its past guesses.
IIM M
Initialization: $M(\bar{t}_0) = 0$; (Note: We assume that $W_0 = \emptyset$ and, for any S, $0 \in \mathcal{B}_S$.)
Stage $n(n \geq 1)$:
If $M(\bar{t}_{n-1}) \in \mathcal{B}_{content(\bar{t}_n)}$, then (Case 1) if $IC(\bar{t}_n) \neq 0$, then let $M(\bar{t}_n) = first(IC(\bar{t}_n))$, $GUESS = IC(\bar{t}_n)$ and $GUESS\text{-}PREV = IC(\bar{t}_n)$; otherwise, let $M(\bar{t}_n) = 0$.
Otherwise, (Case 2) if $IC(\bar{t}_{n-1}) \neq IC(\bar{t}_n)$, let $GUESS = GUESS\text{-}PREV$ and $length = 1$. If $IC(\bar{t}_n) = 0$, then let $M(\bar{t}_n) = M(\bar{t}_{n-1})$. Otherwise, check to see if $W_{first(IC(\bar{t}_n))}$ and $W_{M(\bar{t}_{n-1})}$ agree on the first n strings in a canonical enumeration of Σ^*. If they do, let $M(\bar{t}_n) = M(\bar{t}_{n-1})$. Otherwise, check

to see if within n steps with respect to $IC(\bar{t}_n)$ and *GUESS*, the element ψ_{length} is enumerated and $D_{second(\psi_{length})} \subseteq content(\bar{t}_n)$. If so, increment *length* by 1, let $M(\bar{t}_n) = first(\psi_{length})$ and *GUESS-PREV*$= \psi_{length}$. Otherwise, let $M(\bar{t}_n) = M(\bar{t}_{n-1})$.

END OF IIM M

Let t be a text for any $L \in \mathcal{F}$. It is easy to see that the machine M is well-defined and that it will execute an infinite number of stages on t. Furthermore, beyond some stage n^*, for all $n > n^*$, the variable $IC(\bar{t}_n)$ will stabilize and $first(IC(\bar{t}_n))$ will be an index for L, say J. We first argue that the machine M necessarily converges to an index for L as well.

There are two cases to consider. First, suppose that for some $S \subseteq W_J$, $J \in \mathcal{B}_S$. If M enters Case 1 anytime after stage n^* when $content(\bar{t}_n) \supseteq S$, it will stay there and convergence to J is guaranteed. The other case. i.e., where M does not enter Case 1 an infinite number of times, we consider below. Suppose, on the contrary, that for no $S \subseteq W_J$, $J \in \mathcal{B}_S$. Then, if the machine enters Case 1 anytime after stage n^* the machine M will converge to J. Thus the remaining case to consider again is when the machine does not enter Case 1 an infinite number of times. We consider this next. Let us focus our attention on the stage beyond which the machine always enters Case 2 and the variable $IC(\bar{t}_n)$ has reached its final value. At the first such stage, the value of *GUESS* and *GUESS-PREV* will be the same, say ζ_j. Therefore, the machine's guess at such a stage is $first(\zeta_j)$. Thereafter, the machine M can only guess a finite number of different languages before it must guess an index for L and this guess will not be changed.

M behaves in a fashion consistent with the requirements $\mathcal{A}_R$ and $\mathcal{B}_S$; whenever the machine changes its guess and the condition $\mathcal{B}_S$ is not met, it is, by construction, a change to a language such that the appropriate $\mathcal{A}_R$s are all met. We omit the details.

($\Rightarrow$) We show how given an IIM M that learns a family $\mathcal{F}$ in such a way as to establish that $\mathcal{F} \in LIM[\mathcal{A}_R, \mathcal{B}_S]$, we can obtain an enumeration for the sequence ζ with all the requisite properties. Consider an enumeration of all finite sequences σ_i such that for all i, $content(\sigma_i) \in \Delta_{\mathcal{F}}$. For each sequence σ_i in this enumeration, in turn, determine if $M(\sigma_i)$ is non-zero, and if so, output $c(M(\sigma_i), J)$, where $D_J = content(\sigma_i)$. It is easy to argue that the set ζ enumerated in this way is good for $\mathcal{F}$.

We next show that, given j and k such that $D_{second(\zeta_k)} \subseteq W_{first(\zeta_j)}$, a finite, non-empty sequence $\psi = \psi_1, \ldots \psi_m \subseteq \zeta$ is recursively enumerable with the properties 1 through 5. By construction, there must be a sequence σ_i such that $D_{second(\zeta_k)} = content(\sigma_i)$ and $M(\sigma_i) = first(\zeta_k)$. Output ζ_k first. Let $x_1, x_2, x_3, \ldots$ be an enumeration of the language $W_{first(\zeta_j)}$. (Even for a finite language, by appropriate repetition of strings the enumeration can be made infinite.) Then, extend the sequence σ_i with first the string x_1 and then, in turn, all possible finite sequences of strings from within $W_{first(\zeta_k)}$. Recall that the machine M is stipulated to satisfy the condition $\mathcal{A}_R$ for each possible target language W_R at least as long as the condition $\mathcal{B}_S$ is not met. Consider the non-zero guesses that the machine makes on this text beyond the stage at which it had read σ_i.

Suppose there is a sequence τ_1 on which the machine made a guess different from $M(\sigma_i)$. There are only two possibilities. Either $first(\zeta_k) \in \mathcal{B}_{content(\tau_1)}$, so that the machine was free to make its new guess. In that case, stop the enumeration. Otherwise, output $c(M(\tau_1), J)$, where $D_J = content(\tau_1)$. Note that it must be the case that, for every R, $W_R \in \mathcal{F}$, such that $content(\tau_1) \subseteq W_R$, $(first(\zeta_k), M(\tau_1)) \in \mathcal{A}_R$. Now extend the sequence τ_1 with x_2 and then, in turn, all possible finite sequences of strings from within $W_{first(\zeta_j)}$. Continue to look at the sequence of guesses that the machine outputs and generate a ψ_3 (if necessary) likewise. Even if the enumeration is never stopped explicitly, it can be established by the standard *locking sequence* [4] argument that only finitely many numbers are enumerated. Furthermore, it can also be shown that all the properties required of ψ are indeed satisfied. We omit the details. $\qquad\qquad\square$

4 Applications

We next show how a number of classes of language families can be expressed in such a way that Theorem 5 directly applies to them. The specifications for the relations $\mathcal{A}_R$ and $\mathcal{B}_S$ in order to define a number of classes which correspond to the various monotonicity conditions in Section 2.2 are as follows:

Definition 6.
For any $R, S, I,$ and J,

1. LIM: $(I, J) \in \mathcal{A}_R$.
2. FIN: $(I, J) \in \mathcal{A}_R \iff I = J$.
3. $SMON$, $WMON$: $(I, J) \in \mathcal{A}_R \iff W_I \subseteq W_J$.
4. MON: $(I, J) \in \mathcal{A}_R \iff (W_I \setminus W_J) \cap W_R = \emptyset$.
5. $SMON^d$, $WMON^d$: $(I, J) \in \mathcal{A}_R \iff W_I \supseteq W_J$.
6. MON^d: $(I, J) \in \mathcal{A}_R \iff (W_J \setminus W_I) \subseteq W_R$.
7. $SMON^{\&}$, $WMON^{\&}$: $(I, J) \in \mathcal{A}_R \iff W_I = W_J$.
8. $MON^{\&}$: $(I, J) \in \mathcal{A}_R \iff ((W_I \setminus W_J) \cap W_R = \emptyset) \wedge ((W_J \setminus W_I) \subseteq W_R)$.

For $WMON$, $WMON^d$, and $WMON^{\&}$, $I \in \mathcal{B}_S \iff S \not\subseteq W_I$. For all other classes, $\mathcal{B}_S = \emptyset$.

Clearly, the relations $\mathcal{B}_S$s are decidable and upwardly closed. Many of the classes defined above have been characterized in previous work [1, 19, 7, 10, 12, 13, 20, 8, 25]. We demonstrate how some of those characterizations can be derived by straightforward applications of Theorem 5. We begin with the following important definitions.

Definition 7. [1] A finite set T is a *tell-tale* subset of L in $\mathcal{F}$ if $T \subseteq L$ and $(\forall L' \in \mathcal{F})(L' \subset L \Rightarrow T \not\subseteq L')$.

Definition 8. [13, 25] A family of finite sets $(T_j)_{j \in Z_+}$ is said to be *recursively generable* if and only if there is a total effective procedure which, on input j, generates all elements of T_j and stops.

Angluin [1] characterized *LIM* in the following way:

Theorem 9. *An indexed family $\mathcal{F} = I_1, I_2, \ldots$ is in LIM if and only if there is an effective procedure that, given as input any index I_j, recursively enumerates a tell-tale subset for W_{I_j}.*

Suppose first that a set ζ is given. We first show how a tell-tale subset for L can be recursively enumerated. Since ζ is good, we can find (in the limit) a k such that $D_{second(\zeta_k)} \subseteq W_{first(\zeta_k)} = L$. Given k and k, consider the enumeration of ψ. Notice that since *LIM* imposes no constraint on its guesses and $\mathcal{B}_S$ is empty, the condition 4 is trivially satisfied and condition 5(a) is never satisfied. It is easy to see that $\bigcup_{1 \leq i \leq m} D_{second(\psi_i)}$ can be enumerated and is a tell-tale subset for L. For the converse, the definition of ζ is built in an iterative fashion. Initially, let ζ include $c(I_j, J_j)$, where $D_{J_j} = \{x_j\}$, for each j and some $x_j \in W_{I_j}$. Given I_1, let the procedure that outputs a tell-tale subset for W_{I_1} output a sequence of strings $x_1, x_2, \ldots x_m$. Let there be a procedure to pad machines so as to produce different indices $I_{1,1}, I_{1,2}, \ldots, I_{1,m}$ for W_{I_1}. Then, given j and k such that $D_{second(\zeta_k)} \subseteq W_{first(\zeta_j)} = W_{I_1}$, for $1 \leq i \leq m$, let $\psi_i = c(I_{1,i}, J)$, where $D_J = D_{second(\zeta_k)} \cup \{x_1, x_2, \ldots, x_i\}$. Let ζ include all of ψ. Continue this process. It is easy to see that the resulting ζ is a recursively enumerable set and satisfies all the properties required of it.

It can easily be verified that the set *FIN* has the following characterization: (*FIN* has also been characterized by Mukouchi [20] and Kapur [8].)

Proposition 10. *[13, 25] An indexed family $\mathcal{F} = I_1, I_2, \ldots$ is in FIN if and only if there is a recursively generable family $(T_j)_{j \in N}$ of finite non–empty sets such that*

1. *$T_j \subseteq W_{I_j}$ for all $j \in N$.*
2. *For all $k, j \in Z_+$, if $T_k \subseteq W_{I_j}$, then $W_{I_j} = W_{I_k}$.*

Recall that $LIM[\mathcal{A}_R, \mathcal{B}_S] = FIN$ where, for any I, J, and R, $(I, J) \in \mathcal{A}_R \Longleftrightarrow I = J$, and, for any S, $\mathcal{B}_S = \emptyset$. Clearly, in this case, requirement 3 can only be met when $m = 1$. This also means that, for any k and j, if $D_{second(\zeta_k)} \subseteq W_{first(\zeta_j)}$, then $W_{first(\psi_1)} = W_{first(\zeta_k)} = W_{first(\zeta_j)}$. The equivalence of the two characterizations is now easy to establish.

By an argument very similar to that given above for *FIN*, it can be established that the characterization obtained from Theorem 5 is equivalent to the various characterizations of conservative learning [19, 7, 10, 13, 25]. We next consider a characterization of the class *SMON*.

Proposition 11. *[25] An indexed family $\mathcal{F} \in SMON$ if and only if there is a set of indices $I_1, I_2, \ldots$ that compose $\mathcal{F}$ and a recursively generable family T_j of finite and non–empty sets such that*

1. *For any $j \in Z_+$, $T_j \subseteq W_{I_j}$.*
2. *For any $j, k \in Z_+$, if $T_j \subseteq W_{I_k}$, then $W_{I_j} \subseteq W_{I_k}$.*

Recall that $LIM[\mathcal{A}_R, \mathcal{B}_S] = SMON$ where, for any $I, J,$ and $R,$ $(I, J) \in \mathcal{A}_R$ $\iff W_I \subseteq W_J,$ and, for any $S,$ $\mathcal{B}_S = \emptyset.$ In this case, for any k and $j,$ if $D_{second(\zeta_k)} \subseteq W_{first(\zeta_j)},$ then $W_{first(\zeta_k)} \subseteq W_{first(\zeta_j)}.$ The equivalence of the two characterizations is now easy to establish.

We omit the details showing the equivalence of characterizations for the other classes. We note that Theorem 5 provides the first known characterization of the class $WMON^d.$ The characterizations obtained from Theorem 5 can also be used to show that a particular family is not learnable by an IIM constrained in a certain way.

Proposition 12. *The following family is not in* MON^d: *Let* $\Sigma = \{a, b, c\}.$ *Let* $W_{I_1} = \{a\}^*.$ *For* $k > 1,$ *let* $W_{I_k} = \{a, a^2, \ldots a^{k-1}\} \cup \{c\}^*.$ *Further, for* $k, m \geq 1,$ *let*

$$W_{I_{k,m}} = \{a, a^2, \ldots a^k\} \cup \{c, c^2, \ldots c^m\} \cup \{b\}.$$

Proof. Suppose a suitable ζ is claimed to exist. Then, in particular, there must be a ζ_k such that $W_{first(\zeta_k)} = W_{I_1}.$ Suppose p is the largest number such that $a^p \in D_{second(\zeta_k)}.$ Likewise, there must be a ζ_j such that $W_{first(\zeta_j)} = W_{I_{p+1}}.$ Consider, the enumeration of $\psi = \psi_1, \ldots \psi_m$ corresponding to j and $k.$ Since $\mathcal{B}_S$ is empty, it must be the case that condition 5(b) is satisfied. Suppose q is the largest number such that $c^q \in D_{second(\psi_i)},$ where $1 \leq i \leq m.$ It is easy to see that condition (4) can not have been satisfied. $\square$

Theorem 5 can also be used to show that certain families are learnable under particular constraints.

Proposition 13. *The following families are in* $MON^\&$:

1. *The family of all finite sets over* $\Sigma = \{a\}$.
2. *(From [12].) Let* $\Sigma = \{a, b\}.$ *For any* $m \geq 1$ *and* $k_1, k_2, \ldots, k_m \geq 1,$ *let*

$$W_{I_{k_1, k_2, \ldots k_m}} = (\{a\}^* \setminus \{a^{k_1}, a^{k_2}, \ldots a^{k_m}\}) \cup \{b^{k_1}, b^{k_2}, \ldots b^{k_m}\}.$$

Proof. 1. Let $\zeta = c(I_1, 1), c(I_2, 2), \ldots$ where I_j is a total index for the finite set $D_j.$ It is easy to see that ζ satisfies all the necessary requirements.

2. For any $m \geq 1$ and $k_1, k_2, \ldots, k_m \geq 1,$ enumerate $c(I_{k_1, k_2, \ldots k_m}, I)$ in $\zeta,$ where $D_I = \{b^{k_1}, b^{k_2}, \ldots b^{k_m}\}.$ It is easy to check that ζ satisfies all the requirements.

$\square$

5 Conclusion

Motivated by studies in natural language acquisition and potential artificial intelligence applications, we only considered the relatively simple setting of indexed families of recursive languages. Further, in this paper, we restricted ourselves to the case of class-preserving learning, i.e., where the space of hypotheses is the same as the family to be learned. Clearly, the notions of parameterized constraints and uniform characterization theorems are very general and applicable to a wide array of possibilities. For example, one could consider these notions in the context of recursively enumerable languages, families that are not indexed, or *class comprising learning* [14].

References

1. Dana Angluin: Inductive inference of formal languages from positive data. Information and Control, 45:117–135, 1980.
2. Dana Angluin, Carl H. Smith: Formal inductive inference. In S. C. Shapiro (ed.) Encyclopedia of Artificial Intelligence, volume 1. Wiley-Interscience Publication, New York, 1987.
3. Robert Berwick: The Acquisition of Syntactic Knowledge. MIT press, Cambridge, MA, 1985.
4. L. Blum, M. Blum: Toward a mathematical theory of inductive inference. Information and Control, 28:125–155, 1975.
5. E. M. Gold: Language identification in the limit. Information and Control, 10:447–474, 1967.
6. Klaus P. Jantke: Monotonic and non-monotonic inductive inference. New Generation Computing, 8:349–360, 1991.
7. Shyam Kapur: Computational Learning of Languages. PhD thesis, Cornell University, September 1991. Computer Science Department Technical Report 91-1234.
8. Shyam Kapur: Monotonic language learning. In Proceedings of the third Workshop on Algorithmic Learning Theory, October 1992. Also in New Generation Computing (To appear).
9. Shyam Kapur: Language learning under teams of constraints. Manuscript, 1993.
10. Shyam Kapur, Gianfranco Bilardi: Language learning without overgeneralization. In Proceedings of the 9th Symposium on Theoretical Aspects of Computer Science (Lecture Notes in Computer Science 577), pages 245–256. Springer-Verlag, 1992.
11. Shyam Kapur, Barbara Lust, Wayne Harbert, Gita Martohardjono: Universal grammar and learnability theory: the case of binding domains and the subset principle. In Knowledge and Language: Issues in Representation and Acquisition. Kluwer Academic Publishers, 1993.
12. Steffen Lange, Thomas Zeugmann: Monotonic versus non-monotonic language learning. In Proceedings of the 2nd International Workshop on Nonmonotonic and Inductive Logic (Lecture Notes in Artificial Intelligence Series), 1991.
13. Steffen Lange, Thomas Zeugmann: Types of monotonic language learning and their characterization. In Proceedings of the 5th Conference on Computational Learning Theory. Morgan-Kaufman, 1992.
14. Steffen Lange, Thomas Zeugmann: Language learning in dependence on the space of hypotheses. In Proceedings of the 6th Conference on Computational Learning Theory. Morgan-Kaufman, 1993.
15. Steffen Lange, Thomas Zeugmann, Shyam Kapur: Class preserving monotonic and dual monotonic language learning. Technical Report GOSLER-14/92, FB Mathematik und Informatik, TH Leipzig, August 1992.
16. M. R. Manzini, Kenneth Wexler: Parameters, binding theory and learnability. Linguistic Inquiry, 18:413–444, 1987.

17. Gary Marcus: Negative evidence in language acquisition. Cognition, 46:53–85, 1993.
18. Irene Mazurkewich, Lydia White: The acquisition of dative-alternation: unlearning overgeneralizations. Cognition, 16(3):261–283, 1984.
19. Tatsuya Motoki: Consistent, responsive and conservative inference from positive data. In Proceedings of the LA Symposium, pages 55–60, 1990.
20. Yasuhito Mukouchi: Characterization of finite identification. In Proceedings of the International Workshop on Analogical and Inductive Inference, Lecture notes in Artificial Intelligence (Subseries of Lecture Notes in Computer Science), volume 642, 1992.
21. Daniel N. Osherson, Michael Stob, Scott Weinstein: Systems that Learn: An Introduction to Learning. MIT press, Cambridge, MA, 1986.
22. E. Y. Shapiro: Inductive inference of theories from facts. Technical Report 192, Yale University, 1981.
23. Robert Irving Soare: Recursively enumerable sets and degrees : a study of computable functions and computably generated sets. Springer-Verlag, Berlin; New York, 1987.
24. Rolf Wiehagen: A thesis in inductive inference: In Proceedings of the 1st International Workshop on Nonmonotonic and Inductive Logic. Springer-Verlag, 1991. Lecture Notes in Artificial Intelligence Vol. 543.
25. Thomas Zeugmann, Steffen Lange, and Shyam Kapur: Characterizations of class preserving monotonic and dual monotonic language learning. Technical Report IRCS-92-24, Institute for Research in Cognitive Science, University of Pennsylvania, September 1992.

How to invent characterizable inference methods for regular languages

Timo Knuutila

Department of Computer Science, University of Turku,
Lemminkäisenkatu 14 A, SF-20520 Turku, Finland.
E-mail: knuutila@euroni.cs.utu.fi

Abstract. We present a general framework for constructing characterizable inference algorithms for regular languages. This general approach
is based on the fact, that certain families of non-trivial and infinite regular languages can be described with finite tuples of finite sets of strings
(or trees). It follows from this property, that the methods are able to
inductively learn in the limit from positive samples only. It is also shown
that if the mapping from these presentations to languages is monotonic,
and can be approximated in an acceptable way, then this approximation
can be used to construct the presentation of the smallest language in
the family containing a given sample. We show the applicability of this
framework for both the familiar regular string languages and the regular
tree languages.

1   Introduction

Grammatical inference, the inference of a formal language from a given sample,
is a central topic in syntactic pattern recognition. It was shown already in 1967
that this approach has a fundamental limitation: the problem of identifying
even arbitrary regular languages from a positive sample (set of members of the
language) is undecidable [Go1]. Fortunately, there do exist non-trivial families
of regular languages that are shown to be identifiable in the limit from positive
samples only [An2, RN1, Mu1, GV1]. The inference methods for these families
are even *characterizable* in the sense that the result of the inference is known to
be a language in this family; moreover, this language is often the smallest one
containing the given sample. The characterizable methods are naturally to be
preferred to the *heuristic* ones, because the results of the latter (see [BF1, GT1,
Fu1, Mi1], for example) cannot be formally described. The reader is referred to
Knuutila and Steinby[KS1] for a formal treatment covering some of the heuristic
inference methods.

We generalize in this paper our earlier results on the inference of k-testable
tree languages [Kn1] and obtain a general framework for constructing characterizable inference methods. These methods can be used (under certain general
conditions) to identify the target language inductively in the limit from a positive
sample.

2 Preliminaries

In this section we define formally the concepts of *strings, string language recognizers* and *trees*, and review some basic universal algebra which is needed in the algebraic formulation of the theory of *tree language recognizers* [GS1]. It should be noted that the presentation of the tree case alone suffices to formulate the string case, too, since strings can be viewed as unary trees. Finally, we define some concepts on strings and trees which are needed in the formulation of the examples of language families used in the sequel.

2.1 Strings and string language recognizers

An *alphabet* is a finite set of symbols, and in what follows, X always denotes an alphabet. The set of all finite *strings* (sequences of symbols) over X is denoted by X^* and the *empty string* by e. Any subset of X^* is an *X-language*.

Let $w = x_1 x_2 \cdots x_n \in X^*$ ($x_i \in X, 0 \le i \le n$). Then the *length* of w, denoted by $\lg(w)$, is n. Especially, $\lg(e) = 0$. The set of *prefixes*, *suffixes* and *subwords* of an individual string $w \in X^*$, $\operatorname{pref}(w)$, $\operatorname{suff}(w)$, and $\operatorname{subw}(w)$ are defined as follows:

$$\begin{aligned}
\operatorname{pref}(w) &= \{\, u \in X^* \mid (\exists v \in X^*)\ uv = w \,\}; \\
\operatorname{suff}(w) &= \{\, u \in X^* \mid (\exists v \in X^*)\ vu = w \,\}; \\
\operatorname{subw}(w) &= \{\, u \in X^* \mid (\exists v, v' \in X^*)\ vuv' = w \,\} \ .
\end{aligned}$$

The concepts just defined are generalized to sets of strings as follows. Suppose $L \subseteq X^*$ and $\phi \in \{\operatorname{pref}, \operatorname{suff}, \operatorname{subw}\}$. Then $\lg(L) = \max\{\, \lg(w) \mid w \in L \,\}$ and $\phi(L) = \bigcup_{w \in L} \phi(w)$.

A *finite X-recognizer* $\mathfrak{A} = (A, X, \delta, a_0, A')$ consists of a finite and nonempty state set A, the *input alphabet* X, a *transition function* $\delta \colon A \times X \to A$, an *initial state* $a_0 \in A$ and a set $A' \subseteq A$ of *final states*. When δ is extended in the natural way to a function $\delta^* \colon A \times X^* \to A$, the *language recognized* by $\mathfrak{A}$, $L(\mathfrak{A})$, can be defined by $L(\mathfrak{A}) = \{\, w \in X^* \mid \delta^*(a_0, w) \in A' \,\}$. A language $L \subseteq X^*$ is *regular* if it is recognized by a finite X-recognizer.

Two states a and b of an X-recognizer $\mathfrak{A}$ are said to be *equivalent*, if

$$(\forall w \in X^*)\ (\delta^*(a, w) \in A' \Leftrightarrow \delta^*(b, w) \in A') \ .$$

We denote the equivalence relation (actually a *congruence* of $\mathfrak{A}$) defined by the condition above by $\sim_{\mathfrak{A}}$ in the sequel. In a similar manner, two words u and v of X^* are considered equivalent wrt. a regular language L, if $(\forall w \in X^*)\ (uw \in L \Leftrightarrow vw \in L)$. The equivalence relation (the *Nerode congruence* of L) defined by the condition above is denoted by $\sim_L$.

2.2 Trees, universal algebra and tree language recognizers

A *ranked alphabet* is a finite operator domain, *i.e.* it is a finite set of symbols each of which has been assigned a nonnegative integer *arity*. In what follows, Σ always denotes a ranked alphabet. For any $m \ge 0$, Σ_m is the set of m-ary symbols in

Σ, and rank$(\sigma) = m$ for each $\sigma \in \Sigma_m$. The set of Σ-*terms* or Σ-*trees* $T(\Sigma)$ is defined as the smallest set of strings (over the alphabet Σ augmented with the parentheses and the comma) T such that (1) $\Sigma_0 \subseteq T$; and (2) $\sigma(t_1, \ldots, t_m) \in T$ whenever $m > 0$, $\sigma \in \Sigma_m$ and $t_1, \ldots, t_m \in T$. It is always assumed that $\Sigma_0 \neq \varnothing$; this guarantees that there are Σ-trees. A Σ-*tree language* is any subset of $T(\Sigma)$.

In a Σ-*algebra* $\mathcal{A} = (A, \Sigma)$, A is a nonempty set of *elements* and each σ in Σ is realized as an m-ary operation $\sigma^{\mathcal{A}} : A^m \to A$ of A, where $m = \mathrm{rank}(\sigma)$. In particular, each nullary symbol γ fixes a constant $\gamma^{\mathcal{A}}$ in A. The algebra $\mathcal{A}$ is *finite*, if A is a finite set. A Σ-tree t has a unique value $t^{\mathcal{A}}$ in any given Σ-algebra $\mathcal{A} = (A, \Sigma)$ which is recursively defined as

$$t^{\mathcal{A}} = \begin{cases} \gamma^{\mathcal{A}}, & \text{if } t = \gamma \in \Sigma_0; \\ \sigma^{\mathcal{A}}(t_1^{\mathcal{A}}, \ldots, t_m^{\mathcal{A}}), & \text{if } t = \sigma(t_1, \ldots, t_m) \end{cases}.$$

The Σ-terms form a Σ-algebra $\mathcal{T}(\Sigma) = (T(\Sigma), \Sigma)$, called the Σ-*term algebra*, which is defined so that $\gamma^{\mathcal{T}(\Sigma)} = \gamma$ for any $\gamma \in \Sigma_0$, and $\sigma^{\mathcal{T}(\Sigma)}(t_1, \ldots, t_m) = \sigma(t_1, \ldots, t_m)$, for $m > 0$, $\sigma \in \Sigma_m$ and $t_1, \ldots, t_m \in T(\Sigma)$.

Definition 1. A Σ-*recognizer* $\mathbf{A} = (\mathcal{A}, A')$ consists of a Σ-algebra $\mathcal{A} = (A, \Sigma)$ and a set of *final states* A' $(\subseteq A)$. The elements of A are called *states* of $\mathbf{A}$. The Σ-recognizer $\mathbf{A}$ is *finite*, if A is finite. The *tree language recognized* by $\mathbf{A}$ is the Σ-tree language

$$T(\mathbf{A}) = \{\, t \in T(\Sigma) \mid t^{\mathcal{A}} \in A' \,\}.$$

A Σ-tree language T is called *recognizable* or *regular*, if $T = T(\mathbf{A})$ for some finite Σ-recognizer $\mathbf{A}$.

Let us recall how regular string languages can be interpreted as unary recognizable tree languages [Bü1, St1]. If we define the ranked alphabet $\Sigma = \Sigma_1 \cup \Sigma_0$ so that $\Sigma_0 = \{\epsilon\}$ and $\Sigma_1 = X$, the recursive conversion (1) $t_e = \epsilon$, (2) $t_{wx} = x(t_w)$ $(w \in X^*, x \in X)$ yields a natural bijection $X^* \to T(\Sigma)$, $w \mapsto t_w$.

From any finite X-recognizer $\mathfrak{A} = (A, X, \delta, a_0, A')$ we obtain a finite Σ-recognizer $\mathbf{A} = (\mathcal{A}, A')$ such that (*) $T(\mathbf{A}) = \{\, t_w \mid w \in L(\mathfrak{A}) \,\}$, if the Σ-algebra $\mathcal{A} = (A, \Sigma)$ is defined so that $\epsilon^{\mathcal{A}} = a_0$ and $x^{\mathcal{A}}(a) = \delta(a, x)$ for all $x \in \Sigma_1$ $(= X)$ and $a \in A$. The converse transformation of a finite Σ-recognizer $\mathbf{A}$ into a finite X-recognizer $\mathfrak{A}$ such that (*) holds, is equally natural.

2.3 Some properties of trees

Let $t \in T(\Sigma)$. We define the *height* $\mathrm{hg}(t)$, the *size* $\mathrm{sz}(t)$ and the set of *subtrees* $\mathrm{sub}(t)$ of t as follows:

1. if $t = \gamma$ with $\gamma \in \Sigma_0$, then $\mathrm{hg}(t) = 0$, $\mathrm{sz}(t) = 1$ and $\mathrm{sub}(t) = \{\gamma\}$;
2. if $t = \sigma(t_1, \ldots, t_m)$ with $m > 0$, then
$$\mathrm{hg}(t) = \max\{\, \mathrm{hg}(t_i) \mid 1 \le i \le m \,\} + 1,$$
$$\mathrm{sz}(t) = \mathrm{sz}(t_1) + \ldots + \mathrm{sz}(t_m) + 1, \text{ and}$$
$$\mathrm{sub}(t) = \mathrm{sub}(t_1) \cup \ldots \cup \mathrm{sub}(t_m) \cup \{t\}.$$

These concepts are generalized to sets of trees in the natural way. Suppose that $S \subseteq T(\Sigma)$ is finite. Then $\mathrm{hg}(S) = \max_{t \in S} \mathrm{hg}(t)$, $\mathrm{sz}(S) = \sum_{t \in S} \mathrm{sz}(t)$, and $\mathrm{sub}(S) = \bigcup_{t \in S} \mathrm{sub}(t)$.

The concatenation of trees is somewhat more involved than with strings. Let ξ be a symbol which does not appear in Σ, and let $\Sigma\xi$ be the ranked alphabet which results when ξ is added to Σ as a nullary symbol. Suppose that $t \in T(\Sigma\xi)$ and $S \subseteq T(\Sigma\xi)$. The ξ-product of t and S, denoted by $t \cdot_\xi S$, consists of all trees that can be constructed by replacing all ξ-labelled leaves of t by trees s in S, formally:

$$t \cdot_\xi S = \begin{cases} S, & \text{if } t = \xi; \\ \{\gamma\}, & \text{if } t = \gamma, \ \gamma \in \Sigma_0; \\ \{\sigma(s_1, \ldots, s_m) \mid s_1 \in t_1 \cdot_\xi S, \ldots, s_m \in t_m \cdot_\xi S\}, & \text{if } t = \sigma(t_1, \ldots, t_m). \end{cases}$$

This operation is extended to sets of trees in the natural way. Let $S, T \subseteq T(\Sigma\xi)$. Then the *forest product* of T and S, denoted by $T \cdot_\xi S$, is defined as $T \cdot_\xi S = \bigcup(t \cdot_\xi S \mid t \in T)$. Especially, $\varnothing \cdot_\xi S = \varnothing$.

A *special Σ-tree* is a $\Sigma\xi$-tree in which the symbol ξ appears exactly once. The set of special trees is denoted by $\mathrm{Sp}(\Sigma)$. As a tree a special Σ-tree differs from a Σ-tree only in that one of its leaves is labelled with the new symbol ξ.

The set of subtrees $\mathrm{sub}(t)$ can be understood as the set of 'prefixes' of a tree t. The tree counterparts of suffixes and subwords are formalized using the ξ-products of trees. We define the sets *root trees* $\mathrm{rt}(t)$ and the *forks*[1] $\mathrm{fork}(t)$ of a tree t as follows:

$$\mathrm{rt}(t) = \{s \in T(\Sigma\xi) \mid (\exists s_1, \ldots, s_n \in T(\Sigma)) \ t \in s \cdot_\xi \{s_1, \ldots, s_n\}\};$$
$$\mathrm{fork}(t) = \{s \in T(\Sigma\xi) \mid (\exists s' \in \mathrm{sub}(t)) \ s \in \mathrm{rt}(s')\} \ .$$

It should be noted that $t \in \mathrm{rt}(t)$ and $t \in \mathrm{fork}(t)$ for all $t \in T(\Sigma)$.

2.4 Strings: k-prefix, k-suffix and k-subwords

Suppose $k \geq 0$, $w \in X^*$ and let ϕ be one of the operations pref, suff and subw. We define the (sets of) *k-prefix*, *k-suffix* and *k-subwords* of w, $\mathrm{pref}(w, k)$, $\mathrm{suff}(w, k)$ and $\mathrm{subw}(w, k)$ respectively, as the intersection of $\phi(w)$ and X^k, the set of all X-words of length k. It should be noted that the sets $\mathrm{pref}(w, k)$ and $\mathrm{suff}(w, k)$ are either empty (if $\mathrm{lg}(w) < k$) or singletons. These concepts are generalized to sets of strings in the natural way. It is obvious from the definition that for any $L \subseteq X^*$, all the sets $\phi(L, k)$ are finite. Moreover, if $L \subseteq L' \subseteq X^*$ then $\phi(L, k) \subseteq \phi(L', k)$.

---

[1] The concept of a fork defined here is of a more general form than the one used in [GS1].

2.5 Trees: k-subtrees, k-root and k-forks

Suppose $k \geq 0$ and $t = \sigma(t_1, \ldots, t_m)$ $(m \geq 0)$. A straightforward extension of the definitions of k-bounded string sets to k-bounded trees by defining the sets of k-roots, k-subtrees and k-forks of t, $\mathrm{rt}(t, k)$, $\mathrm{sub}(t, k)$ and $\mathrm{fork}(t, k)$ respectively, as $\mathrm{rt}(t, k) = \{\, s \in \mathrm{rt}(t) \mid \mathrm{hg}(t) = k \,\}$ etc. leads to ambiguities with roots and forks, since a single tree may now have many k-roots. Thus, we define the k-root with the following recursive definition, which gives a unique k-root for each $t \in T(\Sigma)$ and $k \geq 0$:

$$
\mathrm{rt}(t, k) = \begin{cases} \xi, & \text{if } k = 0; \\ t, & \text{if } k \geq 1 \text{ and } \mathrm{hg}(t) \leq k; \\ \sigma(\mathrm{rt}(t_1, k-1), \ldots, \mathrm{rt}(t_m, k-1)) & \text{if } t = \sigma(t_1, \ldots, t_m) \\ & \text{and } \mathrm{hg}(t) > k \geq 1 \ . \end{cases}
$$

The sets $\mathrm{sub}(t, k)$ and $\mathrm{fork}(t, k)$ are defined as

$$
\mathrm{sub}(t, k) = \{\, s \in \mathrm{sub}(t) \mid \mathrm{hg}(s) = k \,\};
$$
$$
\mathrm{fork}(t, k) = \{\, s \mid (\exists s' \in \mathrm{sub}(t))\ s = \mathrm{rt}(s', k) \,\} \ .
$$

Obviously $\mathrm{rt}(t, k) \in T(\Sigma\xi)$, $\mathrm{sub}(t, k) \subseteq T(\Sigma)$, and $\mathrm{fork}(t, k) \subseteq T(\Sigma\xi)$ for any $t \in T(\Sigma), k \geq 0$. Note that $\mathrm{sub}(t, k) = \varnothing$ if $k > \mathrm{hg}(t)$. We write shortly σ_ξ instead of $\sigma(\xi, \ldots, \xi)$. For example, suppose $\sigma \in \Sigma_2$ and $\gamma \in \Sigma_0$. Then

$$
\begin{aligned}
\mathrm{rt}(\sigma(\gamma, \sigma(\gamma, \gamma)), 2) &= \sigma(\gamma, \sigma_\xi), \\
\mathrm{sub}(\sigma(\gamma, \sigma(\gamma, \gamma)), 1) &= \{\sigma(\gamma, \gamma)\}, \text{ and} \\
\mathrm{fork}(\sigma(\gamma, \sigma(\gamma, \gamma)), 2) &= \{\sigma(\gamma, \sigma_\xi), \sigma(\gamma, \gamma)\} \ .
\end{aligned}
$$

The concepts are extended to sets of trees as usual. It is to be noted that for any $k \geq 0$ and $T \subseteq T(\Sigma)$, the sets $\mathrm{rt}(T, k)$, $\mathrm{sub}(T, k)$ and $\mathrm{fork}(T, k)$ are finite. The reason for this is that the trees of these sets are of a bounded size, and that Σ is finite.

The remark that if $T \subseteq T' \subseteq T(\Sigma)$ then $\phi(T, k) \subseteq \phi(T', k)$, where $\phi \in \{\mathrm{rt}, \mathrm{sub}, \mathrm{fork}\}$, applies to tree languages, too.

2.6 The canonical recognizers of string and tree languages

Let $S \subseteq X^*$ be a finite set of strings, the *sample*. Then the *canonical recognizer* of S is $\mathbf{C}_S = (\mathrm{pref}(S) \cup \{0\}, X, \delta, \epsilon, S)$, where 0 is a new symbol and

$$
\delta(u, x) = \begin{cases} ux, & \text{if } ux \in \mathrm{pref}(S), \text{ and} \\ 0 & \text{otherwise} \end{cases}
$$

for all $u \in X^*$ and $x \in X$. It is obvious from the construction that $L(\mathbf{C}_S) = S$.

We define the canonical recognizer of a sample of a tree language in a similar manner. Let S be a finite set of Σ-trees. The canonical recognizer $\mathbf{C}_S$ of S is the Σ-recognizer $(\mathcal{A}_S, S)$, where $\mathcal{A}_S = (\text{sub}(S) \cup \{0\}, \Sigma)$ (0 is a new symbol) is the Σ-algebra defined as follows:

$$\text{for } \gamma \in \Sigma_0, \ \gamma^{\mathcal{A}_S} = \begin{cases} \gamma, \text{ if } \gamma \in \text{sub}(S), \text{ and} \\ 0 \text{ otherwise;} \end{cases}$$

$$\text{for } \sigma \in \Sigma_m, \ m > 0 \text{ and } a_1, \ldots, a_m \in \mathcal{A}_S,$$
$$\sigma^{\mathcal{A}_S}(a_1, \ldots, a_m) = \begin{cases} \sigma(a_1, \ldots, a_m), \text{ if } \sigma(a_1, \ldots, a_m) \in \text{sub}(S), \text{ and} \\ 0 \text{ otherwise.} \end{cases}$$

It is obvious that for any Σ-tree t, $t^{\mathcal{A}_S} = t$ if $t \in \text{sub}(S)$, and $t^{\mathcal{A}_S} = 0$ otherwise. This implies that for any finite set S of Σ-trees, $T(\mathbf{C}_S) = S$.

3 A general framework for inference methods

3.1 Some previous results

The following theorem [An1] gives a general description of the classes of languages that can be identified in the limit from positive data only.

Theorem 2 (Angluin, 1980). *Let X be an alphabet and $L_1, L_2, \ldots$ an indexed class of recursive languages such that for every nonempty finite set $S \subseteq X^*$, the cardinality of the set*

$$C(S) = \{ L \mid S \subseteq L, \ (\exists i \geq 0) \, L = L_i \}$$

is finite. Then this class is identifiable in the limit from positive samples only.

$\square$

Angluin's theorem gives the identifiability for a known class of languages — it does *not* characterize the languages for which this property holds. However, it was shown in [An2] that the class of *k-reversible* languages falls into the category above. Moreover, the inferred language was the smallest k-reversible language containing the given sample. The same properties were derived by Muggleton [Mu1] for a similar class, the *k-contextual* languages. These families of languages are defined as follows.

Definition 3. Language $L \subseteq X^*$ is *k-reversible* if

$$(\forall \, u, u', w \in X^*, \ v \in X^k) \quad (uvw \in L) \wedge (u'vw \in L) \Leftrightarrow (uv \sim_L u'v) .$$

L is *k-contextual* if

$$(\forall \, u, u', w, w' \in X^*, \ v \in X^k) \quad (uvw \in L) \wedge (u'vw' \in L) \Leftrightarrow (uv \sim_L u'v) .$$

Both the results of Angluin and Muggleton are based on the fact, that growing the sample will eventually lead to a *characterizing sample* containing all the strings needed to define the language accurately. These samples are, however, rather complex and unintuitive, and the algorithms based on the utilization of the restrictions (like k-reversibility) on these languages do not seem to be generalizable to other families of languages.

Let us consider the k-reversible languages as an example [An2]. Suppose that the language is recognized by an unknown recognizer $\mathbf{A} = (\mathcal{A}, A')$. We define the set of k-*leaders* of a state $a \in A$ as

$$\text{lead}(a, k) = \{ w \in X^k \mid (\exists b \in A)\ \delta^*(b, w) = a \} \ .$$

Let $u(a, v)$ denote the[2] shortest string u that leads to a from the initial state via v, one of the k-leaders of a, *i.e.* $\delta^*(a_0, u(a, v) \cdot v) = a$, and $w(a)$ the[3] shortest string w that leads to a final state from a. Then the characterizing sample of a k-reversible language $L = L(\mathfrak{A})$ is

$$\begin{aligned}
&\{ w \mid w \in L, \lg(w) < k \} \cup \\
&\{ u(a, v) \cdot v \cdot w(a) \mid v \in \text{lead}(a, k) \} \cup \\
&\{ u(a, v) \cdot v \cdot x \cdot w(b) \mid v \in \text{lead}(a, k), b = \delta(a, x) \},
\end{aligned}$$

where a and b range over the state set A. The inference of the k-reversible languages proceeds by constructing the canonical automaton of the sample and by transforming it into a k-reversible one.

The method proposed by García and Vidal [GV1] for the k-*testable* languages [McN1, BS1, Za1] has a more comprehensible basis: each language in this class is naturally associated with a 3-tuple of finite sets of strings. Let us denote by $X^{<k}$ the set $\cup_{i=0}^{k-1} X^i$. If $L \subseteq X^*$ is a k-testable language, then the L can be defined by three finite sets, $A, B \subseteq X^{<k}$ and $C \subseteq X^k$, as follows:

$$L = (AX^* \cap X^* B) - (X^* C X^*) \ .$$

Informally, the words of L begin with a string from A, end with a string of B and do not contain as a substring any member of the set C. This leads to a natural interpretation of the sample S: $\text{pref}(S) \cap X^{<k}$ gives us the set A, $\text{suff}(S) \cap X^{<k}$ gives the set B, and $\text{subw}(S, k)$ gives the set $X^k - C$. Thus, each sample generates a definition of a k-testable language; the inference algorithm has only to construct the corresponding recognizer. This can be done effectively, since each 3-tuple (A, B, C) used in the definition of a k-testable language L can effectively used as the basis of the construction of the *associated recognizer* recognizing L.

---

[2] If there are several alternatives to choose from, we take the alphabetically first.

[3] Again, the alphabetically first is taken.

3.2 Inferability in the limit

We define next more formally the connection between language families and their presentations.

Definition 4. Let $\mathcal{L}$ be a language family and $\mathbb{D} = D_1 \times \cdots \times D_n$ $(n > 0)$. $\mathbb{D}$ is a *base* of $\mathcal{L}$, and we write $\mathcal{L} = \lambda(\mathbb{D})$, if there exists a surjective mapping $\lambda : \mathbb{D} \to \mathcal{L}$. If $\mathcal{L} = \lambda(\mathbb{D})$, then each member $M \in \mathbb{D}$ is a *presentation* of a language $L = \lambda(M) \in \mathcal{L}$.

We have not fixed the domains of the sets D_i in order to keep Definition 4 as general as possible. It should be noted that λ is not necessarily injective; two different presentations may define the same language.

For example, the family of k-testable string languages, test(k), has a base $\mathbb{D}(k) = X^{<k} \times X^{<k} \times X^k$, since we can define a surjective mapping $\lambda : \mathbb{D}(k) \to$ test(k) by $(A, B, C) \mapsto (AX^* \cap X^* B) - (X^* C X^*)$ (for all $(A, B, C) \in \mathbb{D}(k)$).

The remarks made at the end of the previous section lead to the following informal framework for the basis of inference. We assume that our target language L belongs to some family $\mathcal{L}$ of regular languages.

1. Determine a set $\mathbb{D}$ and a surjective mapping λ such that $\mathcal{L} = \lambda(\mathbb{D})$.
2. Given a finite sample $S \subseteq L$, construct a presentation $\mu(S) \in \mathbb{D}$, as a guess of a possible presentation M of L. This proposal is usually derived from the properties of $\mathbb{D}$ and λ. It is required that $S \subseteq \lambda(\mu(S))$, *i.e.* our guess must at least contain the given sample.
3. Build the recognizer for the language $\lambda(\mu(S))$.

The result is a recognizer of a language in $\mathcal{L}$ — thus we can call the framework characterizable. We can already now state the following important property.

Proposition 5. *Let $\mathcal{L} = \lambda(\mathbb{D})$ be a class of regular languages such that $\mathbb{D}$ is finite and λ can be calculated effectively. If $L \in \mathcal{L}$ then L can be inductively inferred in the limit.*

Proof. Since $\mathbb{D}$ is finite, there exists only a finite number of different presentations and a finite number of languages defined by them. Consequently, the cardinality of the set $C(S) = \{ L \mid S \subseteq L, (\exists i \geq 0) L = L_i \}$ must be finite for each finite sample S. $\quad\square$

As an example, the k-testable string languages have a finite base $\mathbb{D}(k) = X^{<k} \times X^{<k} \times X^k$.

3.3 Characterizable inference

The practical value of our approach depends mainly on the 'goodness' of our proposal $\mu(S)$. Since λ is a mapping from the presentations M to the languages $\lambda(M)$, then a choice from the set $\lambda^{-1}(S)$ seems to be the most natural. However, λ must be generalized to a mapping η in such a way that η^{-1} can be applied to all samples — not only to the languages in the current family $\mathcal{L}$. In order to do this generalization in a formal manner, we need first to consider some relations between λ and $\mathcal{L}$.

Definition 6. Let $\mathcal{L} = \lambda(\mathbb{D})$ and $\mathbb{D} = D_1 \times \cdots \times D_n$. The function λ is *isotonic for component* i of $\mathbb{D}$ $(1 \leq i \leq n)$, if $\lambda(M) \subseteq \lambda(M')$ for all $M = (M_1, \ldots, M_i, \ldots, M_n)$ and $M' = (M_1, \ldots, N_i, \ldots, M_n)$ $(M, M' \in \mathbb{D})$, where $M_i \subseteq N_i$. Function λ is *antitonic for component* i of $\mathbb{D}$ if $\lambda(M') \subseteq \lambda(M)$ holds for all such pairs M and M'. We denote $M \leq N$ if $M_i \subseteq N_i$ for each component i λ is isotonic for and $M_i \supseteq N_i$ for each antitonic one.

Definition 7. Let $\mathcal{L} = \lambda(\mathbb{D})$ and $\mathbb{D} = D_1 \times \cdots \times D_n$. Function λ is *monotonic* wrt. $\mathcal{L}$, if it is either isotonic or antitonic for each component i $(1 \leq i \leq n)$ of $\mathbb{D}$. If λ is monotonic wrt. $\mathcal{L}$, then it follows from $M \leq N$ that $\lambda(M) \subseteq \lambda(N)$.

Let us consider the family $test(k) = \lambda((X^{<k}, X^{<k}, X^k))$. Here λ is isotonic with components 1 and 2 and antitonic with component 3. Thus, λ is monotonic wrt. $test(k)$.

Definition 8. Let $\mathcal{L} = \lambda(\mathbb{D}) \subseteq \mathfrak{p}X^*$ (or $\mathcal{L} \subseteq \mathfrak{p}T(\Sigma)$ for trees). A function $\mu : \mathfrak{p}X^* \to \mathbb{D}$ is an *acceptable guess for* λ^{-1}, if the following conditions hold.

1. The proposed language contains the sample: $S \subseteq \lambda(\mu(S))$.
2. The guesses are correct for the member languages of $\mathcal{L}$: $L = \lambda(\mu(L))$ for all $L \in \mathcal{L}$; or alternatively, $\mu(L) \in \lambda^{-1}(L)$.
3. The guesses are monotonic: if $S \subseteq T \subseteq \mathfrak{p}X^*$, then $\mu(S) \leq \mu(T)$.

Consider again the family $test(k) = \lambda(\mathbb{D}(k))$. Let $S \subseteq X^*$ and $\mu : \mathfrak{p}X^* \to \mathbb{D}(k)$ defined by $S \mapsto (A_k(S), B_k(S), C_k(S))$, where

$$A_k(S) = \operatorname{pref}(S) \cap X^{<k},$$
$$B_k(S) = \operatorname{suff}(S) \cap X^{<k}, \text{ and}$$
$$C_k(S) = X^k - \operatorname{subw}(S, k) .$$

It is easy to show that μ is an acceptable guess of λ^{-1}.

Proposition 9. Let $\mathcal{L} = \lambda(\mathbb{D}) \subseteq \mathfrak{p}X^*$ where λ is monotonic wrt. $\mathcal{L}$ and μ an acceptable guess of λ^{-1}. If $S \subseteq T \subseteq X^*$, then $\lambda(\mu(S)) \subseteq \lambda(\mu(T))$.

Proof. Since $S \subseteq T$ and μ is an acceptable guess, it holds that (see Definition 8) $\mu(S) \leq \mu(T)$. The claim follows now from the monotonicity of λ. □

Proposition 9 shows that the result of inference (the language $\lambda(\mu(S))$), grows as the sample S increases. However, we must avoid growing the guess larger than the target language. This over-generalization can be avoided, if μ has the property that $\lambda(\mu(S))$ is the *smallest* language in $\mathcal{L}$ containing S. This result (Proposition 10) also justifies the use of the word 'characterizable' with our methods.

Proposition 10. *Let $\mathcal{L} = \lambda(\mathbb{D})$ ($\subseteq \mathfrak{p}X^*$), where λ is monotonic wrt. $\mathcal{L}$ and let μ be an acceptable guess for λ^{-1}. For each $S \subseteq X^*$, the language $\lambda(\mu(S))$ is the smallest language of $\mathcal{L}$ containing S.*

Proof. It follows from the acceptability of μ that $\lambda(\mu(S))$ always exists, and $\lambda(\mu(S)) \in \mathcal{L}$. Suppose that there exists a language $L = \lambda(M) \in \mathcal{L}$ such that $S \subseteq L$ and $L \subseteq \lambda(\mu(S))$. Since μ is an acceptable guess and λ is monotonic wrt. $\mathcal{L}$, it must hold that $\mu(S) \leq M$ and $M \leq \mu(S)$. But this is possible only if $M = \mu(S)$ and thus $L = \lambda(\mu(S))$. $\square$

Together propositions 9 and 10 ensure, that our guesses of an unknown language L come more and more accurate as the sample $S \subseteq L$ increases. Let us denote $S_0 = \varnothing$ and $S_{i+1} = S_i \cup \{s\}$, where $s \in L, s \notin S_i$. Initially $\lambda(\mu(S_0)) \subseteq L$, and $\lambda(\mu(S_i)) \subseteq \lambda(\mu(S_{i+1}))$ for all $i \geq 0$. Now $\lambda(\mu(S_\infty)) = L$ — which is exactly what 'inductive inference in the limit' means.

3.4 A general framework

We are now ready to make our framework more specific. The process starts with an arbitrary family of languages $\mathcal{L}$ and results with an algorithm constructing the recognizer for the smallest language in $\mathcal{L}$ containing a given sample S. If the base of $\mathcal{L}$ is finite, then the algorithm is able to infer in the limit.

The steps to follow are:

1. Determine a base for $\mathcal{L}$. This requires the definition of the surjective mapping λ between the presentations and languages. This is usually already given in some form in the definition of $\mathcal{L}$.
2. Define an acceptable guess μ of λ^{-1}. The requirement $L = \lambda(\mu(L))$ (Definition 8), which can be transformed to the form $\mu(L) \in \lambda^{-1}(L)$ for each $L \in \mathcal{L}$, gives a natural basis for μ. Program the function μ.
3. Determine the connection between presentations and associated recognizers in such a way that given a presentation, the construction of a recognizer can be done efficiently. Program the function REC that constructs a recognizer from a presentation.
4. Your algorithm is $\text{REC}(\mu(S))$.

For practical reasons, the constructions of the presentation and the recognizer should be intertwined in such a way that the recognizer can be created directly from the sample (see [GV1, Kn1], for example). However, the sequential construction (1) $\mu(S)$, (2) $\text{REC}(\mu(S))$ serves well as a prototype for the final implementation. The resulting recognizer may be rather large and must thus be minimized. There exist asymptotically time-optimal ($O(n \log n)$) algorithms for the minimization of string recognizers [Ho1, Gr1], but the minimization of tree language recognizers [Br1, GS1] is much harder.

Consider again the family test(k) and the construction of the recognizer associated with a presentation $M = (A, B, C)$ of a sample S. The recognizer

$\mathfrak{A}_M = (A_M, X_M, \delta_M, a_0, A'_M)$ associated with M can be constructed as follows:

$$
\begin{aligned}
X_M &= \text{the set of letters occurring in the sample } S; \\
A_M &= X^{<k} \cup (X^k - C) \cup \{0\}; \\
A'_M &= (AX^* \cap X^* B) \cap X^{<k+1}; \\
a_0 &= e;
\end{aligned}
$$

$$
\delta(w, x) = \begin{cases}
wx & \text{if } \lg(w) < k; \\
w'x & \text{if } \lg(w) = k, w = yw' \text{ and } w'x \notin C; \\
0 & \text{if } \lg(w) = k, w = yw' \text{ and } w'x \in C \ (x, y \in X) \ .
\end{cases}
$$

It is shown in [GV1] that this construction can be done in time $O(n \log n)$, where n is the size of the sample.

4 Examples

We show next how our framework can be used with a few different families of string and tree languages.

4.1 Definite string languages

We define the families of *k-definite*, *reverse k-definite* [Pil] and *simple k, m-definite* string languages, def(k), rdef(k) and sdef(k, m) respectively. The families def(k) and rdef(k) have base $X^k \times X^{<k}$ and sdef(k, m) has base $X^k \times X^m \times X^{<k+m}$. The mappings λ are defined as follows:

- *k-definite*: $\lambda((A, B)) = X^* A \cup B$;
- *reverse k-definite*: $\lambda((A, B)) = AX^* \cup B$;
- *simple k, m-definite*: $\lambda((A, B, C)) = AX^* B \cup C$.

All the mappings λ just defined are monotonic wrt. their corresponding family. The acceptable guesses for the mappings λ^{-1} are constructed from the prefixes and suffixes of the language L:

- *k-definite*: $\mu(L) = (\text{suff}(L, k), \ L \cap X^{<k})$;
- *reverse k-definite*: $\mu(L) = (\text{pref}(L, k), \ L \cap X^{<k})$;
- *simple k, m-definite*: $\mu(L) = (\text{pref}(L, k), \ \text{suff}(L, m), \ L \cap X^{<k+m})$.

It is straightforward to show that each μ really is an acceptable guess. The construction of the associated recognizer can be done similarly as with the k-testable string languages.

4.2 Definite tree languages

Let us denote the set of trees of height k with $T(\Sigma, k) = \{ t \in T(\Sigma) \mid \text{hg}(t) = k \}$, and the set of trees of lower than k with $T(\Sigma, < k) = \bigcup_{i=0}^{k-1} T(\Sigma, i)$. The families of *k-definite*, *reverse k-definite* [He1] and *simple k, m-definite* tree languages have the following (finite) bases and mappings λ:

- k-definite:
  - $\mathbb{D} = \mathrm{rt}(T(\Sigma), k) \times T(\Sigma, < k)$,
  - $\lambda((A, B)) = A \cdot_\xi T(\Sigma) \cup B$; and
- reverse k-definite:
  - $\mathbb{D} = T(\Sigma, k) \times T(\Sigma, < k)$,
  - $\lambda((A, B)) = T((\Sigma - \Sigma_0) \cup \{\xi\}) \cdot_\xi A \cup B$;
- simple k, m-definite:
  - $\mathbb{D} = \mathrm{rt}(T(\Sigma), k) \times T(\Sigma, m) \times T(\Sigma, < k + m)$
  - $\lambda((A, B, C)) = A \cdot_\xi T(\Sigma\xi) \cdot_\xi B \cup C$.

For example, in the case of reverse k-definite tree languages, λ is monotonic, since it is isotonic with both components of the presentations $M = (A, B) \in T(\Sigma, k) \times T(\Sigma, < k)$. The acceptable guesses are defined as follows:

- k-definite: $\mu(L) = (\mathrm{rt}(L, k), \; T(\Sigma, < k) \cap L)$;
- reverse k-definite: $\mu(L) = (\mathrm{sub}(L, k), \; T(\Sigma, < k) \cap L)$;
- simple k, m-definite: $\mu(L) = (\mathrm{rt}(L, k), \; \mathrm{sub}(L, m), \; T(\Sigma, < k + m) \cap L)$.

We show that μ is an acceptable guess for λ^{-1} in the case of k-definite languages. The property that $\mu(L) \in \lambda^{-1}(L)$ is verified by showing that $\mu(\lambda(M)) = M$ for an arbitrary $M = (A, B) \in (\mathrm{rt}(T(\Sigma), k) \times T(\Sigma, < k))$.

$$
\begin{aligned}
\mu(\lambda((A, B))) &= \mu(A \cdot_\xi T(\Sigma) \cup B) \\
&= (\mathrm{rt}(A \cdot_\xi T(\Sigma) \cup B, k), T(\Sigma, < k) \cap (A \cdot_\xi T(\Sigma) \cup B)) \\
&= (\mathrm{rt}(A \cdot_\xi T(\Sigma), k), T(\Sigma, < k) \cap B) \\
&= (A, B)
\end{aligned}
$$

Suppose that $S \subseteq T$. Then $\mathrm{rt}(S, k) \subseteq \mathrm{rt}(T, k)$ and $(T(\Sigma, < k) \cap S) \subseteq (T(\Sigma, < k) \cap T)$ and thus $\mu(S) \leq \mu(T)$.

5 Conclusion and further work

We have presented a general framework for the inference in families of regular languages with monotonic mappings from their presentations to the languages themselves. If an acceptable guess for the (inverse of the) mapping can be found, then the framework gives a characterizable inference algorithm. If the base of the family is finite, we have an algorithm for the inference in the limit, too.

The restrictions stated on the language families considered raise the following important questions.

- Can we give a general description of families of languages for which the proposed framework can be applied?
- If we allow a more general definition for the base of a family, can we still state similar propositions as with the currently used concept? One obvious enlargement would be to define the base as a (finite) set of tuples of sets. For example, the *general k, m-definite* languages, which are of the form

$$
A_1 X^* B_1 \cup A_2 X^* B_2 \cup \cdots \cup A_n X^* B_n \cup C \quad (n \geq 0),
$$

where $A_i \subseteq X^k, B_i \subseteq X^m$ for $1 \leq i \leq n$ and $C \subseteq X^{<k+m}$, have a natural base $\mathfrak{p}(X^k, X_m) \times X^{<k+m}$.

References

[An1] D. Angluin. Inductive inference of formal languages from positive data. *Information and Control*, 45:117–135, 1980.

[An2] D. Angluin. Inference of reversible languages. *Journal of the ACM*, 29(3):741–765, July 1982.

[BF1] A. W. Biermann and J. A. Feldman. On the synthesis of finite state machines from samples of their behavior. *IEEE Transactions on Computers*, C-21:592–597, June 1972.

[Br1] W. S. Brainerd. The minimalization of tree automata. *Information and Control*, 13:484–491, 1968.

[BS1] J. A. Brzozowski and I. Simon. Characterizations of locally testable events. *Discrete Mathematics*, 4:243–271, 1973.

[Bü1] J. R. Büchi. *Finite Automata, Their Algebras and Grammars*. Springer-Verlag, 1989. Edited by D. Siefkes.

[Fu1] K-S. Fu. *Syntactic Pattern Recognition and Applications*. Prentice-Hall, Englewood Cliffs, 1982.

[Go1] E. M. Gold. Language identification in the limit. *Information and Control*, 10:447–474, 1967.

[Gr1] D. Gries. Describing an algorithm by Hopcroft. *Acta Informatica*, 2:97–109, 1973.

[GS1] F. Gécseg and M. Steinby. *Tree Automata*. Akadémiai Kiadó, Budapest, 1984.

[GT1] R. C. Gonzalez and M. G. Thomason. *Syntactic Pattern Recognition: An Introduction*. Addison-Wesley, Reading, Massachuttes, 1978.

[GV1] P. García and E. Vidal. Inference of k-testable languages in the strict sense and application to syntactic pattern recognition. *IEEE Transactions on Pattern Analysis and Machine Intelligence*, PAMI-12(9):920–925, September 1990.

[He1] U. Heuter. Definite tree languages. *Bulletin of the EATCS*, 35:137–143, 1988.

[Ho1] J. Hopcroft. *An $n \log n$ algorithm for minimizing states in a finite automaton*, pages 189–196. Academic Press, 1971.

[Kn1] T. Knuutila. Inference of k-testable tree languages. In H. Bunke, editor, *Advances in Structural and Syntactic Pattern Recognition*. World Scientific, 1993. Proceedings of the International Workshop on Structural and Syntactic Pattern Recognition, Bern, Switzerland, 1992.

[KS1] T. Knuutila and M. Steinby. Inference of tree languages from a finite samples: an algebraic approach. Technical Report R-92-2, University of Turku, May 1992. To appear in the Journal of Theoretical Computer Science.

[McN1] R. McNaughton. Algebraic decision procedures for local testability. *Math. Syst. Theor.*, 8:60–76, 1974.

[Mi1] L. Miclet. *Structural Methods in Pattern Recognition*. Springer-Verlag, 1986.

[Mu1] S. Muggleton. *Inductive Acquisition of Expert Knowledge*. Addison-Wesley, 1990.

[Pi1] J. E. Pin. *Varieties of Formal Languages*. North Oxford Academic, 1986.

[RN1] V. Radhakrishnan and G. Nagaraja. Inference of regular grammars via skeletons. *IEEE Transactions on Systems, Man, and Cybernetics*, SMC-17:982–992, 1987.

[St1]   M. Steinby. Some algebraic aspects of recognizability and rationality. In
        G. Goos and J. Hartmanis, editors, *Proceedings of 1981 International Confer-*
        ence on Fundamentals of Computation Theory, number 117 in Lecture Notes
        in Computer Science, pages 360–372. Springer-Verlag, 1981.
[Za1]   Y. Zalcstein. Locally testable languages. *Journal of Computer and System*
        Sciences, 6:151–167, 1972.

Neural Discriminant Analysis

Jorge Ricardo Cuellar
Siemens Corporate Research
Otto-Hahn-Ring 6
W-8000 München 83
jorge@inf21.zfe.siemens.de

Hans Ulrich Simon
Fachbereich Informatik
Universität Dortmund
W-4600 Dortmund 50
simon@nereus.informatik.uni-dortmund.de

Abstract

Statistical Discriminant Analysis is a classical technique in pattern matching with applications for classification problems and more general decision tasks. In this paper, we use a specific class of discriminant functions which we call product discriminant functions, or simply PDF's. Our main results for PDF's are the following:

- They are quite expressive, e.g., probability distributions defined by Chow-Expansions, Unique Probabilistic Automata or Unique Markov Models can also succinctly be written as PDF's.

- It is possible to obtain with high confidence almost optimal decisions for classification problems which can be modelled by PDF's. The number of training examples needed for that is bounded by a polynomial of low degree (in the relevant parameters).

- The evaluation of the training examples can be implemented on shallow neural nets.

1 Introduction

A classical problem in pattern matching is the problem of classifying objects on the base of a given feature vector. This paper presents analytical tools for designing confident neural classificators. For this purpose, we use statistical discriminant analysis and show that a simple learning rule (which is a variant of the well-known Hebbian learning rule) leads to a quite elegant and efficient neural implementation.

In order to discriminate between different *states* $\omega_1, \ldots, \omega_r$, we use *discriminant functions* $G_1, \ldots, G_r$. For a given *feature vector* x, the value of $G_i(x)$ should express how strongly we believe that ω_i is the underlying state. The state with a maximal value of its discriminant function 'wins', i.e., if $G_i(x) \geq G_j(x)$ for all j, then we decide that ω_i is the underlying state (ties are broken arbitrarily). Of course, the decision may be wrong. We will however try to keep the expected prediction error as small as possible. Discriminant functions leading to a smallest error are called *Bayes discriminant functions*, or *BDF's* simply. It is known from Bayes decision theory that, for instance,

$$\text{pr}(\omega_i) \cdot \text{pr}(x|\omega_i), i = 1, \ldots, r$$

is a system of such functions. Here, $\text{pr}(\omega_i)$ denotes the a-priori probability of ω_i and $\text{pr}(x|\omega_i)$ the conditional probability of feature vector x given that ω_i is the underlying state of nature. In practice, neither the a-priori probabilities nor the conditional probability functions are known in advance. We will therefore try to approximate the true conditional probabilities by a simpler *model of probability* with fewer unknown parameters. This model could be, for instance, a parametric form for a class of distributions, a neural net architecture, a probabilistic automaton or a hidden Markov model (concrete examples for such models are presented in the next section).

In our paper, we propose a particular kind of a parametric form, called product discriminant functions (PDF's), as a model of probability. Our main results for PDF's are the following:

- They are quite expressive, e.g., probability distributions defined by Chow-Expansions, Unique Probabilistic Automata or Unique Markov Models can also succinctly be written as PDF's.

- It is possible to obtain with high confidence almost optimal decisions for classification problems which can be modelled by PDF's. The number of training examples needed for that is bounded by a polynomial of low degree (in the relevant parameters).

- The evaluation of the training examples can be implemented on shallow neural nets.

The most critical assumption underlying these results is that the classification problem at hand can be modelled by PDF's. In its strongest form it would demand that the conditional probability functions $\text{pr}(x|\omega)$ can exactly be written in the specific form of the PDF which represents the model of probability. This strong assumption is however very unlikely to be satisfied in practice. Fortunately, a weaker version of it is sufficient for our purposes. We only require that the best probability distribution in our model has a small 'distance' to the true target distribution. In this sense, our results are robust against slight inaccuracies of our model.[1]

We restrict ourselves to the case of 2 different states of nature and binary feature vectors although generalizations to any finite number of states or finitely many values per feature are possible (more or less straightforward). The paper is structured as follows:
In section 2, product discriminant functions are defined and the aforementioned

---

[1] This robustness has recently been shown by Anoulova and Pölt in [2].

main results for them are proven. In section 3, we show that the neural implementation of PDF's bears inherent parallelism: it turns out that it can be performed on quite shallow neural nets for a wide subclass of PDF's. Section 4 mentions some open problems.

2 Product Discriminant Functions

We start this section with a formal definition of product discriminant functions. Let P be a probability distribution on $X = \{0,1\}^n$ and $E \subseteq X$ an event. The *characteristic function* for E is given by

$$\chi_E(x) = \begin{cases} 1 & \text{if } x \in E, \\ 0 & \text{if } x \notin E. \end{cases}$$

The *elementary function* for E is given by

$$\varphi_E(x) = (P(E))^{\chi_E(x)} = \begin{cases} P(E) & \text{if } x \in E, \\ 1 & \text{if } x \notin E. \end{cases}$$

Let $S, S' \subseteq 2^X$ be two systems of events. The *product discriminant function (PDF)* for S, S' is given by

$$\prod_{E \in S} \varphi_E(x) / \prod_{E' \in S'} \varphi_{E'}(x).$$

Although the definition of PDF's is extremely technical, it is possible to show that several natural classes of distributions such as

1. (Conditional) Product Distributions on Multidimensional Discrete Feature Spaces

2. Distributions defined by Unique Probabilistic Automata (UPA's) or Unique Markov-Models (UMM's).

can succinctly be written as PDF's. We sketch this here roughly for UPA's. It is discussed in greater detail in the full paper.

Quite flexible representations of distribution classes are *Probabilistic Automata (PA)* which we are going to define now. We will then show that subclasses, called *Unique Probabilistic Automata (UPA)* can be succinctly written as PDF's. A PA is given by (definition from [1]):

- a finite set Q of *states*,

- an alphabet A of *letters*,

- an *initial distribution* $I : Q \to [0, 1]$ satisfying $\sum_{q \in Q} I(q) = 1$,

- a *transition distribution* $T : Q \times Q \times A \to [0,1]$ satisfying

$$\sum_{q' \in Q, a \in A} T(q, q', a) = 1 \text{ for all } q \in Q.$$

Each PA represents a probability distribution on A^n, the strings of length n over A (for arbitrary, but fixed, n). A random string $w \in A^n$ is generated as follows:

1. An initial state is selected at random according to I.

2. A sequence of n random transitions is selected as follows: Given that q is the current state, we select transition (q, q', a) with probability $T(q, q', a)$. As a result, we generate letter a and transit from state q to state q'.

The resulting probability distribution is then formally written as

$$P_{I,T}(w) = \sum_{q_0, \ldots, q_n \in Q} I(q_0) \cdot \prod_{i=1}^{n} T(q_{i-1}, q_i, w_i).$$

A-priori knowledge about the distribution class, modelled by the PA, is incorporated by *constraints* $C = (Q_0, \mathcal{T})$, where $Q_0 \subseteq Q$ is the set of *possible initial states*, and $\mathcal{T} \subseteq Q \times Q \times A$ the set of *possible transitions*. We say that a PA satisfies constraint C if $I(q) = 0$ for all $q \notin Q_0$, and $T(q, q', a) = 0$ for all $(q, q', a) \notin \mathcal{T}$.

A PA with unique sequences of n transitions for all generated words $w \in A^n$ is called *Unique Probabilistic Automaton (UPA)*. We denote the sequence of states on the unique generating path for w by $q(w, 0), \ldots, q(w, n)$, and the unique sequence of transitions by $t(w, 1), \ldots, t(w, n)$. The probability distribution for UPA's has the following simple form:

$$P_{I,T}(w) = I(q(w, 0)) \cdot \prod_{i=1}^{n} T(t(w, i)).$$

Let $E(t, i)$ denote the event that $t = t(w, i)$, i.e., t is the i-th transition on the generating path of w. Similarly, $E(q, i)$ denotes the event that $q = q(w, i)$. The probabilities of these events are denoted by $p_{t,i}$ and $p_{q,i}$, respectively. We then arrive at the form of a PDF:

$$P_{I,t}(w) = \prod_{q \in Q_0} p_{q,0}^{\chi_{E(q,0)}(w)} \cdot \prod_{i=1}^{n} \left(\prod_{t \in \mathcal{T}} p_{t,i}^{\chi_{E(t,i)}(w)} \right) \Big/ \left(\prod_{q \in Q} p_{q,i}^{\chi_{E(q,i)}(w)} \right).$$

The full paper contains a more detailed derivation of this formula. The following subsections discuss important properties of PDF's.

2.1 Statistical Analysis

We start this section by presenting the formal ingredients and an outline of the analysis to follow. For $i = 0, 1$, let $p_i = \mathrm{pr}(\omega_i)$ denote the a-priori probability of ω_i, and $P_i(x) = \mathrm{pr}(x|\omega_i)$ the conditional probability function given that ω_i is the state of nature. $H_i(x) = p_i \cdot P_i(x)$ are Bayes discriminant functions, and

$$h(x) = \begin{cases} 1 & \text{if } H_1(x) \geq H_0(x), \\ 0 & \text{if } H_1(x) < H_0(x), \end{cases}$$

is a decision function with minimal prediction error. Let $\tilde{h}$ denote the decision function which results from substituting approximations $\tilde{p}_i, \tilde{P}_i(x)$ for the exact quantities. The resulting additional prediction error can be bounded, according to the lemma of Duin (see [5, 2]), in terms of the Manhattan distance

$$\|F - G\| = \sum_{x \in X} |F(x) - G(x)|$$

between two functions F, G:

Lemma 2.1

$$error(\tilde{h}) - error(h) \leq \sum_{i=0,1} |\tilde{p}_i - p_i| + \max_{i=0,1} \|\tilde{P}_i - P_i\|.$$

Finding good approximations for the a-priori probabilities is the minor problem. We will therefore focus on the problem of finding an approximation $\tilde{P}$ of a target distribution P on X ($P = P_0, P_1$ in our application). The following definition introduces a central concept for our analysis:
Let $0 < \mu, \beta \leq 1$. $\tilde{P}$ is called (μ, β)-*estimation* of P if

$$P\left(P(x)/(1 + \mu) \leq \tilde{P}(x) \leq P(x) \cdot (1 + \mu)\right) \geq 1 - \beta.$$

In other words, $\tilde{P}(x)$ approximates $P(x)$ up to a factor of $1 + \mu$ for 'most' $x \in X$. The total probability of the set $X' \subseteq X$ of *exceptions* is bounded by β.
We are now going to show that (μ, β)-estimations of P have two important properties:

- Their Manhattan distance to P is bounded by $2\mu + 2\beta$.

- They are 'efficiently' computable from an empirical sample for P, provided that P is a PDF.

These properties, combined with the lemma of Duin, will then lead to a quite satisfactory solution for classification problems which 'can succinctly be modelled' by PDF's. In the remainder of this section, we follow this outline and fill in the details.

Lemma 2.2 *If $\tilde{P}$ is a (μ, β)-estimation of P, then $\|\tilde{P} - P\| \leq 2\mu + 2\beta$.*

Proof The proof uses a partition of X into

$$Y = \{x|\ \tilde{P}(x) \leq P(x)\} \text{ and } Z = \{x|\ \tilde{P}(x) > P(x)\},$$

and a partition of Y into $Y' = Y \cap X'$ (the exceptions in Y) and $Y'' = Y \setminus Y'$. From the definition of (μ, β)-estimations, the following equalities are easily derived:

$$
\begin{aligned}
\|\tilde{P} - P\| &= \sum_{x \in X} |\tilde{P}(x) - P(x)| \\
&= \sum_{y \in Y} (P(x) - \tilde{P}(x)) + \sum_{z \in Z} (\tilde{P}(z) - P(z)) \\
&= (P(Y) - \tilde{P}(Y)) + (\tilde{P}(Z) - P(Z)) \\
&= 2(P(Y) - \tilde{P}(Y)).
\end{aligned}
$$

The last equality holds because the probabilities of Y and Z sum up to 1. We may now proceed by using the partition of Y into Y', Y'':

$$
\begin{aligned}
P(Y) - \tilde{P}(Y) &= \sum_{y' \in Y'} (P(y') - \tilde{P}(y')) + \sum_{y'' \in Y''} (P(y'') - \tilde{P}(y'')) \\
&\leq \sum_{y' \in Y'} ((1+\mu)\tilde{P}(y') - \tilde{P}(y')) + \sum_{y'' \in Y''} P(y'') \\
&\leq \mu + \beta
\end{aligned}
$$

This completes the proof of the lemma. ●

Before stating the next result, we need to capture formally the 'complexity' of a PDF. Let P be a PDF for two systems S, S' of events. An event $E \in S \cup S'$ is called *active* on $x \in X$ if $x \in E$. Note that the evaluation of $P(x)$ depends only on the probabilities of active events (the other events contribute factor 1 to $P(x)$). We define $size(P)$ as the total number $|S| + |S'|$ of events, and $depth(P)$ as the maximal number of *active* events on any $x \in X$.

For instance, the depth of a unique probabilistic automaton (written as a PDF) corresponds to the length of the generated strings and the size to the number of edges and nodes in its state diagram.

Let $\bar{x}_1, \ldots, \bar{x}_m$ be an empirical sample of m points drawn independently according to P. A straightforward idea is to estimate the probabilities of all events in $S \cup S'$ empirically by their relative frequencies, i.e, $\tilde{P}(E)$ is set to $1/m$ times the number of $i \in [1 : m]$ with $\bar{x}_i \in E$. The resulting approximation $\tilde{P}(x)$ is called *empirical estimation* for $P(x)$. We note without proof that these empirical estimations coincide with the Maximum Likelihood Estimations for all distributions defined by UPA's.[2] The next result gives a bound on the sample size m, required for getting (μ, β)-estimations of P 'with high confidence'.

---

[2] A proof for a slightly less general class of PA's is contained in [1].

Lemma 2.3 *Let P be a PDF and $\tilde{P}$ the empirical estimation for P resulting from a sample of size m. Let $s = size(P)$, $d = depth(P)$ and*

$$K(\mu, \beta, \delta) = \frac{3\ln(2/\delta)}{\mu^2 \beta}.$$

$\tilde{P}$ is a (μ, β)-estimation of P with a confidence of at least $1 - \delta$ if

$$m \geq K\left(\frac{\mu}{2d}, \frac{\beta}{s}, \frac{\delta}{s}\right) = O\left(\frac{d^2 \cdot s \cdot \ln(s/\delta)}{\mu^2 \cdot \beta}\right).$$

Proof The proof requires the following fact from statistics. Let p be the probability of an event and p_m an empirical estimation based on a sample of size m. According to Chernoffs inequalities [3]:

$$\mathrm{pr}\left(p_m \geq (1+\mu)p\right) \leq e^{-\mu^2 mp/3}, \mathrm{pr}\left(p_m \leq (1-\mu)p\right) \leq e^{-\mu^2 mp/2}.$$

Chernoffs inequalities yield huge sample bounds if p is close to zero. A straightforward computation shows however that $p/(1+\mu) \leq p_m \leq p \cdot (1+\mu)$ with a confidence of at least $1 - \delta$ if $p \geq \beta$ and $m \geq K(\mu, \beta, \delta)$.

A PDF is composed of elementary functions $\varphi_E(x) = (P(E))^{\chi_E(x)}$. We claim that $\tilde{\varphi}_E(x) = (\tilde{P}(E))^{\chi_E(x)}$ is a (μ, β)-estimation of φ_E with a confidence of at least $1 - \delta$ if $m \geq K(\mu, \beta, \delta)$. Since $\varphi_E(x) = \tilde{\varphi}_E(x) = 1$ for all $x \notin E$, the estimation is only potentially bad for $x \in E$. If $P(E) \geq \beta$, we get the desired result by applying the aforementioned statistical fact. If, on the other hand, $P(E) \leq \beta$, we may consider all $x \in E$ as exceptions. In both cases, the assertion of the claim is valid.

Due to the multiplicative structure of PDF's, it is easy to see how the sample bound scales with increased complexity of the PDF. The basic technique is 'distributing δ, β, μ over the components of the PDF'. More formally, we must substitute δ/s for δ, since the incofidencies of all s parameter estimations should sum up to δ. For the same reason, β/s is substituted for β. The treatment of μ is a little bit more involved. Since there are only d active events on any x, the inaccuracies μ' of the elementary functions accumulate to a value of at most $(1 + \mu')^d$. Using the formula $1 + x \leq e^x \leq 1 + x/2$, which is valid for all $0 \leq x \leq 1/2$, and setting $\mu' = \mu/(2d)$, we obtain

$$(1 + \mu/(2d))^d \leq e^{\mu/2} \leq 1 + \mu.$$

Therefore the sample size scales with s, d, μ, β, δ as described in the assertion of the lemma. ●

We want to apply these results to classification problems modelled by PDF's. For simplicity, we assume that the a-priori probabilities are equal. The conditional probability distributions $P_i(x), i = 0, 1$, are assumed to be PDF's with size s and depth d. Let h denote the Bayes decision function, and $\tilde{h}$ its *empirical estimation*, i.e, the decision function resulting from empirical estimations $\tilde{P}_i$ of P_i. We want to bound with confidence at least $1 - \delta$ the additional error of $\tilde{h}$ by ϵ. According to the lemma of Duin, it is sufficient to bound $||\tilde{P}_i - P_i||$ by ϵ for $i = 0, 1$. According to Lemma 2.2, it is sufficient that $\tilde{P}_i$ is a $(\epsilon/4, \epsilon/4)$-estimation of P_i. An application of Lemma 2.3 leads now directly to the main result of this subsection:

Theorem 2.4 *Let h be the Bayes decision for a 2-state classification problem modelled by PDF's of size s and depth d. Let $\tilde{h}$ denote the empirical estimation of h resulting from two samples (one for ω_0, the other for ω_1) of size m. Then*

$$error(\tilde{h}) - error(h) \leq \epsilon$$

with a confidence of at least $1 - \delta$ if

$$m \geq K\left(\frac{\epsilon}{8d}, \frac{\epsilon}{4s}, \frac{\delta}{2s}\right) = O\left(\frac{d^2 \cdot s \cdot \ln(s/\delta)}{\epsilon^3}\right).$$

Proof In the bound for m in Lemma 2.3, we substituted $\epsilon/4$ for μ, β, respectively, and $\delta/2$ for δ because there are now two target distributions which contribute to the total inconfidence. $\bullet$

We state without proof that the bound on the sample size does not change its order of magnitude if the a-priori probabilities are unknown. The formal treatment for the general case of unknown a-priori probabilities can be looked up in [7].

2.2 Neural Implementation

A *linear unit (or linear neuron)* is a computational unit which computes a linear function of the form

$$L_w(y_1, \ldots, y_r) = w \cdot y = \sum_{i=1}^{r} w_i \cdot y_i.$$

A *threshold unit (or threshold neuron)* is a computational unit which computes a function of the form

$$T_w(y_1, \ldots, y_r) = \begin{cases} 1 & \text{if } w \cdot y \geq 0, \\ 0 & \text{if } w \cdot y < 0. \end{cases}$$

The reals $w_1, \ldots, w_r$ are called *weights*. They are either hardwired or adjustable. We restrict ourselves to neural networks which are computational circuits with linear and threshold units as base. If the network performs a binary decision, we use only threshold units. If the network approximates a BDF, we use a linear unit as output unit (the other units are again threshold units). If the network is in *performance mode*, all weights are fixed. Thus the network computes a fixed real (or boolean) function (which is defined in the obvious way). In *learning or training mode* a subset of the weights is adjustable (some weights may be hard-wired all the time). In this mode, the network passes on-line through a sample of labelled training examples and adjusts its weights according to a predefined learning rule.

Let's assume that networks N_0, N_1 compute (appoximations of) BDF's P_0, P_1 for two states ω_0, ω_1 of nature, respectively. Then we (approximately) obtain a Bayes classifier N by 'amalgamating' their linear output units z_0, z_1 to a single threshold output unit z (as shown in Figure 1). Network N just compares $P_1(x)$ and $P_0(x)$ and determines the winner (i.e., outputs 1 iff $P_1(x) - P_0(x) \geq 0$). We may therefore concentrate on the design of neural approximators for BDF's.

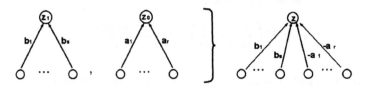

Figure 1: The construction of a neural classifier N from neural BDF-approximators N_0, N_1.

The BDF-property is not destroyed by monotone transformations, i.e., monotone transformations of BDF's are again BDF's. If the BDF has the particular form of a PDF, we will make use of the logarithmic transformation. The logarithm of a PDF for systems $S, S' \subseteq 2^X$ of events and probability distribution P, has the following form:

$$Q(x) = \sum_{E \in S} \log(P(E)) \cdot \chi_E(x) - \sum_{E' \in S'} \log(P(E')) \cdot \chi_{E'}(x).$$

Note that Q is a linear function in $\chi_E, \chi_{E'}$. We show in Section 3 how to implement the characteristic functions by completely hardwired and quite shallow neural nets for a variety of PDF's (UPA's and UMM's). Thus the only remaining task is the training problem for the linear output neuron z which should compute a linear function with weights of the form $\log P(E), \log P(E')$. The true probabilities of the events are unknown. Assumed that the units can perform exact real arithmetic, we could substitute $\log \tilde{P}(E), \log \tilde{P}(E')$ for the ideal weights $\log P(E), \log P(E')$, respectively. Existing machines operate with finite precision and are therefore not able to perform exact real arithmetic. In the following, we present a simple learning rule which

- uses only finite precision,

- allows a control of the resulting round-up error,

- operates on-line and allows therefore a flexible switching from learning to performance mode and vice versa,

- can be interpreted as a variant of the well-known Hebbian learning rule.

We may assume w.l.o.g. that S and S' contain the same number of events (this can always be forced by adding redundant events of probability 1 to the smaller system of events). This has the advantage that we can substitute absolute frequencies $K(E), K(E')$ for the corresponding relative frequencies without loosing the BDF-property. A first rough approximation of $\log K$ is the binary coding length of K, which is denoted by $L(K)$ in the sequel. The relation between these two numbers is the following:

$$L(K) - 1 \leq \log K < L(K).$$

If we are satisfied by this relation, there is a very simple learning rule:
Unit z counts on-line the number K of occurences of event E and uses $L(K)$ as the approximation for the ideal weight $\log K$.
In Subsection 2.1, we argued that (μ, β)-estimations $\tilde{P}(E)$ for $P(E)$ are desirable. It is reasonable to keep a sort of balance between the statistical estimation error and the round-up error. Thus, we need a tool to dynamically decrease the round-up error. Such a tool is provided by the following relation which holds for all positive integers t:

$$\frac{L(K^t)}{t} - \frac{1}{t} \leq \log K < \frac{L(K^t)}{t}.$$

In the *learning rule with precision* $1/t$, K is updated as before and $L(K^t)/t$ is used as approximation for $\log K$.[3] It is now easy to derive the 'neural analog' to Theorem 2.4.

Corollary 2.5 *Let h be the Bayes decision for a 2-state classification problem modelled by PDF's of size s and depth d. Let N be the corresponding neural classifier which is trained on two samples (one for ω_0, the other for ω_1) of size m, and updates its adjustable weights according to the learning rule with precision $1/t$. Then*

$$error(\tilde{h}) - error(h) \leq \epsilon$$

with a confidence of at least $1 - \delta$ if

$$t \geq \frac{48d}{\epsilon \log e} = O\left(\frac{d}{\epsilon}\right) \quad \text{and} \quad m \geq K\left(\frac{\epsilon}{16d}, \frac{\epsilon}{4s}, \frac{\delta}{2s}\right) = O\left(\frac{d^2 \cdot s \cdot \ln(s/\delta)}{\epsilon^3}\right).$$

---

[3] It is convenient to choose $t = 2^\tau$ for some τ. In this case, we can compute K^t by iteratively squaring a variable with initial value K τ-times. Then $L' = L(K^t)$ is computed by counting the number of bits in the binary representation of K^t. Finally, L' is transformed into the rational $L = L' \cdot 2^{-\tau}$ by 'shifting the point' τ positions to the left. All digits until τ positions behind the point then coincide with the corresponding digits of $\log K$.

The proof is omitted in this abstract.

A second possible learning rule is obtained as follows. Let E be one of the events of the systems S, S' constituting a PDF. Let $K = K(E)$ be the number of occurences of E within a random sample. Like for the preceding rule, we are going to design an on-line learning rule which adjusts a weight parameter to $\ln K$ (this time, we prefer the natural logarithm). Let $H(K) = 1+1/2+1/3+\cdots+1/K$ and $\gamma = 0.5772156649\ldots$ Euler's constant. With these notations the following relation holds (see [6]):

$$H(K) - \gamma - \frac{1}{2K} < \ln K < H(K) - \gamma.$$

This relation suggests the following rule. Initialize a counter to $1-\gamma$ when event E occurs the first time (forget for a second that γ is not finitely presentable). For all $i \geq 2$, we increase the counter by $1/i$ when we observe the i-th occurence of E. At any time, the current value $H(K) - \gamma$ of the counter is an approximation of $\ln K$ with precision $1/(2K)$. In order to get a result similar to Corollary 2.5, we have to overcome two difficulties:

1. A precision of $1/(2K)$ is not quite good for small K. It seems therefore that the estimations of $\ln K$ are too bad if $K = K(E)$ is small. However, it can be shown by a rigorous analysis that this problem does not hurd us. The reason is that probability parameters $P(E)$ which are 'very small' need not be well approximated (compare with the similar reasoning in Subsection 2.1). For parameters $P(E)$ which are sufficiently bounded away from zero, $K = K(E)$ is 'big' and $1/(2K)$ is a fairly good precision.

2. We must use finite presentations for Euler's constant and terms like $1/i$. This can be done subject to the following rule. If $1/t$ is the desired precision and there are at most m training examples, we may use a precision of $1/(tm)$ per term.

Since the numbers $H(K)$ are called *harmonic numbers*, we refer to the learning rule just described as the *harmonic learning rule with precision $1/t$ and sample size m*. It is not hard to show the following

Corollary 2.6 *Let h be the Bayes decision for a 2-state classification problem modelled by PDF's of size s and depth d. Let N be the corresponding neural classifier which is trained on two samples (one for ω_0, the other for ω_1) of size m, and updates its adjustable weights according to the harmonic learning rule with precision $1/t$ and sample size m. Then*

$$error(\tilde{h}) - error(h) \leq \epsilon$$

with a confidence of at least $1 - \delta$ if

$$t \geq \frac{48d}{\epsilon} = O\left(\frac{d}{\epsilon}\right) \quad \text{and} \quad m \geq K\left(\frac{\epsilon}{16d}, \frac{\epsilon}{4s}, \frac{\delta}{2s}\right) = O\left(\frac{d^2 \cdot s \cdot \ln(s/\delta)}{\epsilon^3}\right).$$

The proof is omitted in this abstract.

In the full version of the paper (see [4]), it is shown that our learning rules can be interpreted as variants of the well-known Hebbian learning rule.

3 Construction of Shallow Nets

The *size* of a network N is the number of its units. Its *depth* is the length of the longest path in N. The size represents the amount of hardware within N. The depth represents the time required to perform a computation with N. Shallow networks (i.e., networks with a small depth) have a high degree of parallelism and perform their computations quite fast. The aim of this section is to 'compile' UPA's into shallow networks. In combination with the results in Subsection 2.2, we obtain quite time-efficient approximately Bayes-optimal neural classifiers.

Given a PDF for systems $S, S' \subseteq 2^X (X = \{0,1\}^n)$ of events, we want to construct a system of neural networks $N_E, N_{E'}$ computing the characteristic functions $\chi_E, \chi_{E'}$, respectively. We assume that there are $2n$ input nodes for inputs $x_1, \ldots, x_n$ and their negations $\bar{x}_1, \ldots, \bar{x}_n$. For convenience, we assume furthermore that there are two additional input nodes ZERO and ONE which are clamped to the constants 0 and 1, respectively. The main result of this section is the following

Theorem 3.1 *Let A be a UPA for words of length n with state set Q, alphabet $A = \{0,1\}$, and $C = (Q_0, T)$ a constraint which is satisfied by A. The probability distribution P on A^n induced by A can be computed by a neural network of depth $2\lceil \log n \rceil + 6$ and size $O\left(n \cdot (|Q|^3 + |T|)\right)$.*

Proof According to section 2, P can be written as a PDF whose event systems contain the events $E(q, i), E(t, j)$ for $q \in Q, t \in T, 0 \leq i \leq n, 1 \leq j \leq n$. Let $\chi_{q,i}$ and $\chi_{t,j}$ denote the corresponding characteristic functions. According to Subsection 2.2, it suffices to work out the neural implementation of these functions. We use 'dynamic programming' and, in addition, a 'path doubling technique' in order to keep the resulting network shallow. The basic idea is quite simple. Let $\mathcal{D}$ be the state diagram of A. For all $q_1, q_2 \in Q, 0 \leq l \leq n, w \in A^l$ let

$$P_{q_1, q_2}^l(w) = \begin{cases} 1 & \text{if } \mathcal{D} \text{ contains a path labelled } w \text{ from } q_1 \text{ to } q_2, \\ 0 & \text{otherwise.} \end{cases}$$

The proof proceeds by

- synthesizing the characteristic functions out of these auxiliary functions,
- presenting a neural implementation for the auxiliary functions.

The implementation will have the special feature that the layers of the network are alternating AND- and OR-layers.

The functions $\chi_{q,i}$ can be computed as follows:

$$\chi_{q,i}(w) = \bigvee_{q_0 \in Q_0, q_1 \in Q} \left(P^i_{q_0,q}(w_1 \ldots w_i) \wedge P^{n-i}_{q,q_1}(w_{i+1} \ldots w_n) \right).$$

For $a, b \in A$, let $b^a = b$ if $a = 1$, and $b^a = \bar{b}$ otherwise. Note that $b^a = 1$ if and only if $a = b$. The functions $\chi_{t,j}$ with $t = (q, q', a)$ can be computed as follows:

$$\chi_{t,j}(w) = \chi_{q,j-1}(w) \wedge \chi_{q',j}(w) \wedge w_i^a.$$

Given the auxiliary functions, three layers (AND,OR,AND) are sufficient to compute the characteristic functions.

We now turn to the implementation of the auxiliary functions. Let λ be the empty word. Note that $P^0_{q_1,q_2}(\lambda) = 1$ if $q_1 = q_2$, and 0 otherwise. The auxiliary functions with $l = 0$ are therefore available at the two input nodes ZERO and ONE. For $a \in A$, $P^1_{q_1,q_2}(a) = 1$ if T contains a transition of the form (q_1, q_2, a), and 0 otherwise. Thus:

$$P^1_{q_1,q_2}(a) = (a \wedge P^1_{q_1,q_2}(1)) \vee (\bar{a} \wedge P^1_{q_1,q_2}(0)),$$

where $P^1_{q_1,q_2}(1)$ and $P^1_{q_1,q_2}(0)$ are available at ZERO and ONE. The auxiliary functions with $l > 1$ can be computed by a combination of the dynamic programming and the path doubling technique. Let k be given by $2^k < l \leq 2^{k+1}$ and $K = 2^k$. We may split $w \in A^l$ into $w' = w_1 \ldots w_K$ and $w'' = w_{K+1} \ldots w_l$. Then the following relation holds:

$$P^l_{q_1,q_2}(w) = \bigvee_{q \in Q} \left(P^K_{q_1,q}(w') \wedge P^{l-K}_{q,q_2}(w'') \right).$$

Thus $2(1 + \lceil \log n \rceil)$ layers ($1 + \lceil \log n \rceil$ alternations of OR, AND) are sufficient to compute the auxiliary functions.

The computing scheme contains $2\lceil \log n \rceil + 5$ layers altogether. One additional layer results from the linear unit which combines the characteristic functions. Counting the number of units within the computing scheme, we see that the assertion of the theorem is valid. $\bullet$

This section is closed by the following remarks:

- The restriction to a binary alphabet within Theorem 3.1 is not essential (the letters of a larger alphabet can be encoded in binary).

- It is easy to prove an analogous theorem for Unique Markov Models.

4 Open Problems

Our paper presented some simple learning rules for neural networks which are closely related to statistical discriminant analysis. This relationship enabled us to present a massively parallel neural implementation for selfadjusting almost Bayes classifiers. We believe that there are more fruitful relationships of this kind.

Our neural implementation does not use the full power of threshold units (since all hidden layers are AND- or OR-layers). This raises the question whether the network can be made even more shallow.

Theorems 2.4 and Corollary 2.5 use the unrealistic assumption that the unknown conditional probability functions can succinctly be written as PDF's. Although a recent paper [2] shows that this 'perfect model assumption' can be considerably weakened, it would be important to get more and stronger robustness results.

Acknowledgements

The second author gratefully acknowledges the support of Bundesministerium für Forschung und Technologie grant 01IN102C/2. The author takes the responsibility for the contents.

References

[1] Naoki Abe and Manfred K. Warmuth. On the computational complexity of approximating distributions by probabilistic automata. In *Proceedings of the 3rd Annual Workshop on Computational Learning Theory*, pages 52–66, 1990.

[2] Svetlana Anoulova and Stefan Pölt. Using Kullback-Leibler divergence in Learning Theory. Research report in preparation.

[3] Herman Chernoff. A measure of asymptotic efficiency for tests of a hypothesis based on the sum of observations. *Ann. Math. Statist.*, 23:493–507, 1952.

[4] Jorge Ricardo Cuellar and Hans Ulrich Simon. Neural Discriminant Analysis. *Forschungsbericht Nr. 469 der Universität Dortmund*, 1993.

[5] R. Duin. A sample size dependent error bound. In *Proceedings of the 3rd International Conference in Pattern Recognition*, 1976.

[6] Donald E. Knuth. *The Art of Computer Programming: Fundamental Algorithms*, volume 1. Addison Wesley, second edition, 1973.

[7] Stefan Pölt. Extensions of the Pab-Decision Model. *Forschungsbericht Nr. 468 der Universität Dortmund*, 1993.

A New Algorithm for Automatic Configuration of Hidden Markov Models

Makoto Iwayama[1] and Nitin Indurkhya[2] and Hiroshi Motoda[1]

[1] Hitachi Advanced Research Laboratory, Hatoyama, Saitama 350-03, Japan
[2] Telecom Research Laboratories, 770 Blackburn Road, Clayton, Vic 3168, Australia

Abstract. Hidden Markov Models (HMM) (i.e. doubly stochastic probabilistic networks) have been widely used in analyzing time-series data such as those obtained from speech and molecular biology. A crucial issue in modeling time-series data using HMM, is the problem of determining the appropriate model architecture: the number of states and the links between the states. While current HMM training procedures iteratively optimize model parameters, they usually require the model configuration to be fixed. The task of model configuration is done manually by trained experts. In this paper we present a procedure that addresses the problem of automatically configuring HMM's. It starts with a large, possibly over-fitted HMM, and attempts to prune it down to the appropriate complexity fit. The procedure can be seen as a generalization of the well-known iterative Baum-Welch Algorithm. The parameter re-estimates in our procedure can be formally derived and its local convergence can be formally proved. Compared to existing methods, our procedure offers the following advantages: (1) better convergence characteristics than the standard Baum-Welch algorithm, (2) automatic reduction of model size to the right complexity fit, (3) better generalization, and (4) relative insensitivity to the initial model size. We demonstrate these features by presenting empirical results on the problem of recognizing DNA promoter sequences.

1   Introduction

Hidden Markov Models (HMM) (i.e., doubly stochastic probabilistic networks) have been widely used in analyzing time-series data such as those obtained from speech[15, 10] and sequence data in domains such as molecular biology[8, 4, 3]. A crucial issue in modeling time-series data using HMM, is the problem of determining the appropriate model architecture: the number of states and the links between the states. Model complexity and predictive performance are highly related. While increased complexity gives better fits on the training data, it usually results in poorer predictive performance on new cases. Determining the optimal model complexity is crucial for good performance on new cases. For HMM's, this task is done manually by trained experts. Once the configuration is decided, several HMM training procedures are available that optimize model parameters while keeping the model configuration fixed [11, 15]. In domains where human expertise for determining model configuration is unavailable, efficient use of HMM's is difficult. This *knowledge acquisition bottleneck* hinders the application of HMM technology. Automatic methods for determining model configuration would be of considerable assistance in applying HMM's in new domains.

In this paper we present a new procedure that addresses the problem of automatically configuring an HMM to the right complexity fit. It starts with a large, possibly over-fitted, HMM and attempts to iteratively prune it down to the appropriate complexity fit. The parameter re-estimates in our procedure can be formally derived and it can be shown that the procedure optimizes the maximum likelihood. Other features of our procedure include: (1) better convergence characteristics than the standard Baum-Welch algorithm, (2) automatic reduction of model size to the right complexity fit, (3) better generalization, and (4) relative insensitivity to the initial model size.

We consider only sequences that consist of discrete symbols. Our pruning strategy focuses on single link deletions. We always start with an ergodic (i.e. fully-connected network) model, although our procedure is not limited by this condition.

2   Hidden Markov Models

Hidden Markov Models (HMM) are a special class of Markov models that exhibit two levels of stochastic behavior, such that one level is not directly observable (hence the word "hidden") but can only be inferred through another stochastic process that produces a sequence of observable symbols. Several well-written introductions to HMM's, such as [10, 15], are available in existing literature, which emphasize their applications to speech recognition, a domain where HMM's have proven quite successful.

A HMM consists of a set of *states* which are inter-connected by *links*. In each state, it is possible to make a probabilistic transition to another state via a link between the two states. As a result of the transition, an output symbol is produced according to a probability distribution associated with the link. The sequence of output symbols is directly observed and represents the observable stochastic process. The underlying sequence of state transitions is not directly observed, but can only be inferred through the sequence of output symbols. Given a HMM and a sequence of symbols, a standard procedure based on dynamic programming concepts is available to determine the probability that the sequence was produced by the HMM. For details, the reader may refer to [15]. Our interest is in methods for training HMM's and this shall be the focus of this paper.

The best known training procedure for HMM's is the *Baum-Welch Algorithm*, an iterative algorithm that re-estimates model parameters so as to maximize the probability of generating the time-series [11]. The algorithm has a solid theoretical foundation and many analytical results as to its optimality are available. A major restriction is that it requires the model configuration to be fixed throughout the training process. This means that model configuration must be determined beforehand – usually based on human expertise. We use the procedure as a starting point of our method described in Sect. 3. Unlike the standard Baum-Welch, we allow the number of model parameters to be reduced in each iteration.

3 Modified Baum-Welch Algorithm: Link Deletion

In this section we describe our procedure for adjusting model size by link deletion. This procedure is a generalization of the Baum-Welch Algorithm. We first derive expressions for parameter re-estimation with link deletion and then show how it can be used within a HMM training procedure.

3.1 Parameter Re-estimation with Link Deletion

In this section we formally derive expressions for parameter re-estimation with link deletion. These expressions can be used in an iterative training procedure to be described in Sect. 3.2. In addition to deriving these expressions, we also show that the re-estimation process can reach a (local) optimum in the maximum likelihood sense. Most of the notation used in this section is summarized in Table 1.

Table 1. HMM-related notation

S:	Set of states $S = \{s_i\}$
Y:	Set of discrete output symbols
A:	State Transition Matrix; $A = \{a_{ij}\}$; a_{ij} is probability of transition from s_i to s_j, Note that $\sum_j a_{ij} = 1$.
B:	Output probability matrix; $B = \{b_{ij}(k)\}$; $b_{ij}(k)$ represents the probability that symbol k will be output when a transition is made from state s_i to state s_j. Again, $\sum_k b_{ij}(k) = 1$.
π:	Initial probability Matrix; $\pi = \{\pi_i\}$; π_i is the initial probability of being in state s_i. Here too, $\sum_i \pi_i = 1$.

An HMM model θ is the triple, $\theta = (A, B, \pi)$. These represent the parameters of the HMM. Given an HMM θ and a sequence of observed symbols $y = y_1, y_2, \ldots, y_T$, the probability that the HMM generates y is given by

$$P(y|\theta) = \sum_x P(x, y|\theta) \ , \tag{1}$$

where x is a (hidden unobserved) state sequence. However, because $P(x, y|\theta)$ can be expanded to

$$P(x, y|\theta) = \pi_{x_o} \prod_{t=0}^{T-1} a_{x_t x_{t+1}} \prod_{t=0}^{T-1} b_{x_t x_{t+1}}(y_{t+1}) \ , \tag{2}$$

substituting (2) into (1) yields

$$P(y|\theta) = \sum_x \pi_{x_o} \prod_{t=0}^{T-1} a_{x_t x_{t+1}} \prod_{t=0}^{T-1} b_{x_t x_{t+1}}(y_{t+1}) \ . \tag{3}$$

Rather than work with $P(y|\theta)$ directly, our re-estimation transformation is based on a function $Q(\theta, \widehat{\theta})$ of the current HMM θ (before link deletion) and the new HMM $\widehat{\theta}$ (after link deletion) defined by

$$Q(\theta, \widehat{\theta}) = \sum_x P(x, y|\theta) \log P(x, y|\widehat{\theta}) \ . \tag{4}$$

We assume that the link from s_p to s_q is to be deleted.

The utility of $Q(\theta, \widehat{\theta})$ stems from the following theorem.

Theorem A-1

If $Q(\theta, \widehat{\theta}) \geq Q(\theta, \theta)$, then $P(y|\widehat{\theta}) \geq P(y|\theta)$, with equality if and only if $\theta = \widehat{\theta}$.

Proof

Since $\log Z \leq Z - 1$ with equality if and only if $Z = 1$,

$$Q(\theta, \widehat{\theta}) - Q(\theta, \theta) = \sum_x P(x, y|\theta) \log[P(x, y|\widehat{\theta})/P(x, y|\theta)]$$

$$\leq \sum_x P(x, y|\theta)[P(x, y|\widehat{\theta})/P(x, y|\theta) - 1]$$

$$= \sum_x P(x, y|\widehat{\theta}) - \sum_x P(x, y|\theta)$$

$$= P(y|\widehat{\theta}) - P(y|\theta).$$

∎

The theorem says that if Q increases by re-estimation (i.e. link deletion), the probability of y also increases. So, maximizing $Q(\theta, \widehat{\theta})$ with respect to $\widehat{\theta}$, is equivalent to maximum likelihood estimation.

Parameter re-estimates for maximizing $Q(\theta, \widehat{\theta})$ under link deletion are given by the following theorem.

Theorem A-2

If deletion of link from s_p to s_q increases the Q-function, then the parameter re-estimates for the pruned HMM are given by:

$$\widehat{\pi}_i = \frac{\sum_{j'} \gamma(i, j', 1)}{\sum_{i', j'} \gamma(i', j', 1)} \tag{5}$$

$$\widehat{a}_{ij} = \begin{cases} 0 & i = s_p, j = s_q \\ \dfrac{\sum_t \gamma(i, j, t)}{\sum_{j': j' \neq s_q} \sum_t \gamma(i, j', t)} & i = s_p, j \neq s_q \\ \dfrac{\sum_t \gamma(i, j, t)}{\sum_{j'} \sum_t \gamma(i, j', t)} & \text{otherwise} \end{cases} \tag{6}$$

$$\widehat{b}_{ij}(k) = \frac{\sum_{t: y_t = k} \gamma(i, j, t)}{\sum_t \gamma(i, j, t)} \tag{7}$$

where $\gamma(i, j, t)$ is defined as the probability of making a transition from state i to state j at time t [3].

---

[3] $\gamma(i, j, t)$ is estimated from the training data by using the Forward-Backward Algorithm [15].

Proof

According to (4) and (2),

$$Q(\theta, \widehat{\theta}) = \sum_{\boldsymbol{x}} P(\boldsymbol{x}, \boldsymbol{y} | \theta) \{ \log \widehat{\pi}_{x_0} + \sum_{t} \log \widehat{a}_{x_t x_{t+1}} + \sum_{t} \log \widehat{b}_{x_t x_{t+1}}(y_{t+1}) \}. \quad (8)$$

In this expression, each term of $\widehat{\pi}_i$, $\widehat{a}_{ij}$ and $\widehat{b}_{ij}$ being independent each other, we can maximize Q by maximizing each term separately. We only show the derivation for the case of a_{ij} ($i = s_p$ and $j \neq s_q$), the other cases can be derived in a similar fashion.

By maximizing

$$\sum_{\boldsymbol{x}} P(\boldsymbol{x}, \boldsymbol{y} | \theta) \sum_{t} \log \widehat{a}_{x_t x_{t+1}}, \quad (9)$$

we can obtain the most likely $\widehat{a}_{ij}$. Recall that the link from s_p to s_q is to be deleted. This means that the transition probability $a_{s_p s_q}$ should be 0 in the re-estimated $\widehat{\theta}$. So, maximizing should be done under the constraint

$$\sum_{j : j \neq s_q} \widehat{a}_{ij} = 1 \quad (10)$$

Indexing (9) by i gives

$$\sum_{\boldsymbol{x}} P(\boldsymbol{x}, \boldsymbol{y} | \theta) \sum_{t} \log \widehat{a}_{x_t x_{t+1}} = \sum_{i} [\sum_{\boldsymbol{x}} P(\boldsymbol{x}, \boldsymbol{y} | \theta) \sum_{t : x_t = i} \log \widehat{a}_{i, x_{t+1}}]. \quad (11)$$

Here, we define $n_{ij}(\boldsymbol{x})$ as the number of transition t which satisfies $x_t = i$ and $x_{t+1} = j$ in the state sequence $\boldsymbol{x}$. Using $n_{ij}(\boldsymbol{x})$, the right hand side of (11) can be rewritten to

$$\sum_{i} \sum_{j} \sum_{\boldsymbol{x}} n_{ij}(\boldsymbol{x}) P(\boldsymbol{x}, \boldsymbol{y} | \theta) \log \widehat{a}_{ij} = \sum_{i} \sum_{j} A_{ij} \log \widehat{a}_{ij}, \quad (12)$$

where $A_{ij} = \sum_{\boldsymbol{x}} n_{ij}(\boldsymbol{x}) P(\boldsymbol{x}, \boldsymbol{y} | \theta)$. Since maximizing the right hand side of (12) is equivalent to maximizing each term with respect to i independently, we consider only i term of (12) from now on. By using the Lagrange multiplier method, $\widehat{a}_{ij}$ is the solution of

$$\frac{\partial}{\partial \widehat{a}_{ij}} \{ \sum_{j'} A_{ij'} \log \widehat{a}_{ij'} + \lambda (1 - \sum_{j' : j' \neq s_q} \widehat{a}_{ij'}) \} = 0, \quad (13)$$

where λ is the Lagrange multiplier. (13) becomes

$$A_{ij} - \lambda \widehat{a}_{ij} = 0. \quad (14)$$

By summing (14) on $j \neq s_q$, we obtain

$$\sum_{j' : j' \neq s_q} A_{ij'} - \lambda \sum_{j' : j' \neq s_q} \widehat{a}_{ij'} = 0. \quad (15)$$

Substituting (10) into (15) gives

$$\lambda = \sum_{j':j' \neq s_q} A_{ij'}. \tag{16}$$

Then substituting (16) into (14) yields

$$\widehat{a}_{ij} = \frac{A_{ij}}{\sum_{j':j' \neq s_q} A_{ij'}}. \tag{17}$$

Because $n_{ij}(x)$ is the number of transition from i to j in the sequence x and $P(x, y|\theta)$ is the probability of x (and y),

$$A_{ij} = \sum_{x} n_{ij}(x) P(x, y|\theta) = \sum_{t} \gamma(i, j, t). \tag{18}$$

Substituting (18) into (17) yields

$$\widehat{a}_{ij} = \frac{\sum_{t} \gamma(i, j, t)}{\sum_{j':j' \neq s_q} \sum_{t} \gamma(i, j', t)} \quad (i = s_p, j \neq s_q). \tag{19}$$

These parameter re-estimates can be used to determine new model parameters if pruning is considered necessary. In the next section we describe the overall training procedure in which these parameter re-estimates are used.

3.2 The Modified Baum-Welch Procedure

The standard Baum-Welch algorithm is an iterative procedure in which the model parameters are iteratively modified so as to increase the maximum likelihood estimate (or decrease the average *entropy*, the negative log probability) of the training data in each iteration. Our modifications to this procedure can be seen as increasing the search-space. Besides the usual re-estimates from the standard Baum-Welch procedure (these are described in [11]), we also consider re-estimates due to *single link deletion* in each iteration. These were described in Sect. 3.1. A decision is made as to whether or not to delete a link, model parameters are updated accordingly and this process is iterated. The complete modified Baum-Welch procedure is shown in Fig. 1. Because the decision to delete a link is made irrevocably, our algorithm can be seen as a hill-climbing search procedure. In deciding which link to consider for deletion, the procedure focuses on the *weakest link* – the link that has the lowest transition probability. The stopping criterion is the rate of improvement – if it is less than a user-specified threshold, the procedure terminates.

While the algorithm uses the resubstitution estimate (using training data for evaluation), it is possible to use unseen test cases for deciding whether to delete a link or not. The evaluation function used is the well-known entropy metric that measures the log probability. This can easily be replaced by alternative model evaluation functions such as those based on MDL principle [16, 17] or AIC [1, 2].

If link deletion is never done in any iteration, then our procedure reduces to the standard Baum-Welch algorithm. However, since link deletion is done only if

Input: y a set of N training data;
      y' a set of N' evaluation data $(= y)$;

Initialize: $\theta^1 :=$ randomized HMM;
        $k := 1$;
        $e' :=$ mean_entropy(y, θ_1);
        $d :=$ value of threshold;

repeat
  $e := e'$;
  $a^k_{s_p s_q} := \min_{i,j} a^k_{ij}$; (the weakest link)
  $\widehat{\theta}^{k+1}_{std} :=$ re-estimation by standard Baum-Welch;
  $\widehat{\theta}^{k+1}_{del} :=$ re-estimation by modified Baum-Welch(delete link from s_p to s_q);
  if $\left(\text{mean\_entropy}(y',\widehat{\theta}^{k+1}_{std}) \leq \text{mean\_entropy}(y',\widehat{\theta}^{k+1}_{del})\right)$
    then
      $\theta^{k+1} := \widehat{\theta}^{k+1}_{std}$;
    else
      $\theta^{k+1} := \widehat{\theta}^{k+1}_{del}$;
  $e' :=$ mean_entropy(y',θ_{k+1});
  $k := k + 1$;
until $e - e' \leq d$
output θ_k

function mean_entropy(y, θ)
  return $\dfrac{-\sum_{1 \leq i \leq N} \log P(y_i|\theta)}{N}$;

Fig. 1. The Modified Baum-Welch Algorithm

it gives better results, it can be expected that our procedure will perform at least as well as standard Baum-Welch (we provide empirical results in support of this claim in the next section). If no link deletion is performed, then the procedure will converge to a global minimum (in the entropy space) [11]. However, when links are being deleted, this cannot be guaranteed. But, we can show that a *locally optimal* model will be obtained in the following sense:

Consider a model trained by our procedure. Let the size of the final model be N. This model is obtained by iteratively deleting links from an initial larger model. Consider the behavior of the algorithm after the last such link has been deleted. Since no more links are deleted, the procedure behaves like the standard Baum-Welch algorithm. Hence, by the theorem in [11], it will converge on to the minimum entropy model of size N. Hence our procedure finds the minimum entropy model among the models having the same configuration as the final model. While there may be models having the same size but different configuration that have a lower entropy than the one obtained by our procedure, the hill-climbing search process cannot find such globally optimal models.

4 Empirical Results: DNA Sequence Analysis

We empirically analyzed the behavior of our procedure by setting it the task of learning HMM's to model DNA Promoter sequences [13]. DNA sequences consist of a string of nucleotides (there are 4 of them, we refer to them by the mnemonics A, G, T and C). A *promoter* is a particularly well-studied DNA sequence with various results on recognizing them from sequence information alone. The dataset we used is the same as described in [13]. From this we took 53 promoter sequences each of length 57 (i.e. consisting of 57 symbols, each of which was one of A, G, T or C) to learn HMM's. Model evaluation was done by computing the average entropy. We also had 53 non-promoter sequences for testing purposes. Some sample promoter sequences are shown in Fig. 2.

tactagcaatacgcttgcgttcggtggttaagtatgtataatgcgcgggcttgtcgt
tgctatcctgacagttgtcacgctgattggtgtcgttacaatctaacgcatcgccaa
gtactagagaactagtgcattagcttatttttttgttatcatgctaaccacccggcg

⋮

Fig. 2. DNA promoter sequences

Since little is known about what HMM configuration would be best for modeling promoter sequences, we began with the least structured models – ergodic HMM's in which each state was linked to every other state in the model thereby giving a fully-connected Markov network. While our procedure initializes model parameters with random values, we found that repeated trials with different starting values gave similar results[4]. Hence, in the experiments below, only the results on one training run are reported. After experimenting with various threshold values for the stopping criterion, we found best results were obtained with a threshold of 10^{-5} and this value was used in all the experiments. In estimating the true entropy of a model on unseen cases, we used the resampling technique of *leaving-one-out* [6].

The experiments can be divided into two groups. First, we show that varying the model size does indeed affect performance. Second, we illustrate several characteristics of the modified Baum-Welch algorithm and compare its behavior to that of the standard Baum-Welch procedure.

4.1 Effect of Model Size on System Performance

These experiments are intended to demonstrate that changing model size of an HMM does affect performance. We show that while larger models exhibit better discriminability (for example, promoter v/s non-promoter), they have a tendency to overfit the training data.

Model Size v/s Discriminability Model discriminability refers to a model's ability to distinguish objects that it is supposed to model from those it is not supposed to model. For example, a good probabilistic model for promoter sequences should return a very high probability for a promoter sequence and a very low probability for a non-promoter sequence. Model discriminability between two hypotheses can be measured by the *information index for mutual discrimination* [9]. High values of this index indicate good discrimination between the concepts. We investigated how HMM size affects model discriminability and the results are

---

[4] average entropy results differed in the fifth decimal place.

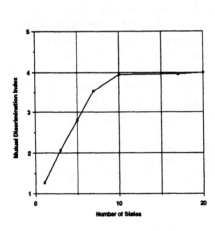

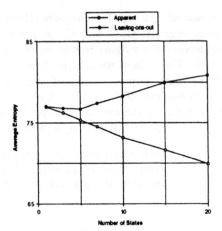

Fig. 3. Increased Discriminability with Model Size

Fig.4. Training and Test Results for DNA

summarized in Fig. 3. We trained a series of ergodic HMM's (fully connected) of varying sizes trained by the standard Baum-Welch algorithm to model promoter sequences. The models were trained using the 53 promoter sequences available. The performance on positive test cases was estimated by using the leaving-one-out resampling method. Performance on negative test cases was obtained by using the 53 non-promoter sequences available. We see that with increasing HMM size, model discriminability improves and then levels off. Thus, increase in the HMM model size results in better model discriminability although the gains for models with more than 10 states are not significant.

The Problem of Over-fitting Overfitting the training data is a well-known and common problem in machine learning and its occurrence has been noted in various representations such as decision trees, rules and neural networks [21]. In Fig. 4 we demonstrate its presence in HMM's. We trained a series of HMM's of varying sizes, as in the previous section, and obtained training and test curves as shown. The test curve was obtained by the leaving-one-out resampling method. With increasing model size, the average entropy of the training curve decreases monotonically giving better and better training results. However, the performance curve for the test data follows the classical pattern of improvement and subsequent deterioration with increasing model size. For the given dataset, optimal performance is obtained for HMM's with 5 to 7 states.

4.2 Characteristics of the Modified Baum-Welch Procedure

In Sect. 4.1 we illustrated the effect of model size on system performance for the HMM representation. In this section we show how our procedure adjusts model size. We examine various empirical characteristics such as convergence rate, training dynamics, test-set error-rates and effect of initial model size. In all cases, we compare it to the standard Baum-Welch algorithm.

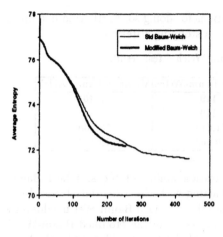 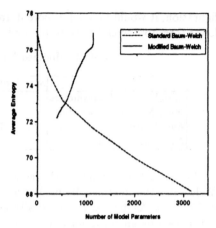

Fig. 5. Comparison of Convergence Characteristics

Fig. 6. Pruning a Model – Training Dynamics

Convergence Characteristics We have formally shown that our procedure does indeed converge to a locally optimal model in the maximum likelihood sense. In this section we examine how many iterations are necessary to reach the optimal model. In Fig. 5 we compare convergence characteristics of the modified Baum-Welch algorithm to that of the standard version. The initial model size was the same in both cases – a 15 state ergodic HMM. However, while the standard Baum-Welch keeps the model size fixed, our procedure deletes links if possible. The average entropy was measured only on the training data. Because the modified Baum-Welch procedure is able to consider link deletion if it lowers the entropy by a larger amount, hence the average entropy falls more sharply than it does for the standard Baum-Welch. However, because it can get caught in local minima, it can stop earlier than standard Baum-Welch, as illustrated in the figure. Also, as illustrated in the figure, it is possible for the standard Baum-Welch to train to a lower entropy although this is for a substantially larger model.

Effect of Model Pruning on Performance The modified Baum-Welch always checks to see if the current model can be made smaller while at the same time, driving down the entropy on the training set. The effect of this is that not only does the entropy decrease monotonically, the model size also decreases. The training dynamics of model size are shown in Fig. 6 in which we also show the standard Baum-Welch training curve for reference. The initial model was a 15 state ergodic HMM with 1140 parameters[5]. The final model has 395 parameters. Note that the final model obtained by the modified Baum-Welch has a lower average entropy than that of a same sized model trained by the standard Baum-Welch. While it is possible, in theory, for the standard procedure to reach the same performance level as the modified Baum-Welch for the same final con-

---

[5] A n-state ergodic HMM for the Promoter sequence problem has $5n^2 + n$ parameters.

figuration, it would take a lot more iterations for doing so. Furthermore, a much lower threshold would have to be used as a stopping criterion.

Table 2. Modified Baum-Welch – test results

	Standard Baum-Welch	Modified Baum-Welch
Initial Num. of parameters	1140	1140
Final Num. of parameters	1140	395
Ave. Entropy (Train)	71.613	72.200
Ave. Entropy (Test)	79.968	79.147

In Table 2 we compare the generalization ability of the standard Baum-Welch and the modified Baum-Welch starting with the same initial model size of 15 states. As discussed earlier, although the standard Baum-Welch achieves a lower training entropy, not only is the final model of the Modified Baum-Welch substantially smaller, it also generalizes slightly better as evidenced by the lower entropy on the test cases (measured by leaving-one-out). The final model with 395 parameters roughly corresponds to an 8 or 9 state HMM. As discussed in Sect. 4.1, for the given dataset, the optimal model size is 5 to 7 states. Hence, starting from an overly large 15 state HMM, the modified Baum-Welch comes quite close to reaching the optimal model size.

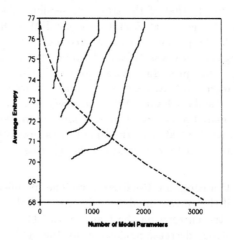

Fig. 7. Modified Baum-Welch: Different Starting Points

Effect of Initial Model Size Our procedure only considers model size changes that involve pruning. The central assumption behind this strategy is that the initial model most probably overfits the data. However, when little is known about the domain or the dataset, it is difficult to determine in advance the validity of this assumption. If this assumption is critical in obtaining the right-sized model, then without some way of assessing its validity, any procedure (such as ours) that relies only on pruning would be of limited applicability. On the other hand, if the pruning procedure were shown to be relatively insensitive to the initial model size in that it reaches approximately the same final model size no matter what the initial model size (i.e. no pruning done for models that are too small, and substantial pruning for very large models), it would demonstrate the procedure's usefulness even in situations where the user was totally ignorant of the approximate region of the appropriate complexity fit.

Figure 7 illustrates training results for our procedure with several different initial model sizes. We have also plotted the standard Baum-Welch curve for reference. While the performance on the training set varied a lot depending on the initial model size, note that the final model sizes are quite close to one another. In fact, for larger models the pruning was quite significant while the smaller models had almost minimal pruning done to them. This shows that the modified Baum-Welch is relatively insensitive to the initial model size.

5 Related Work

There have been several studies related to choosing the right-sized HMM based on performance results. The simplest approach is to find complete solutions for different model configurations, and compare their performance by testing on independent test cases (or by resampling methods). While this approach, used in [7], can potentially find the optimal solution, it is time-consuming and a more automated approach is often desirable.

A more efficient search strategy, called *model surgery*, is proposed in [8]. After standard training of a fixed-sized model, model surgery adjusts the architecture by deleting/inserting states, following which the training process is iterated until no more changes are made to the model. The effects of model surgery are confined to adjusting model length, and the basic configuration is kept fixed. While model surgery decouples parameter optimization and model size adjustment, in contrast, our procedure addresses both the problems in an integrated fashion and allows the configuration to change more flexibly.

An incremental model adjusting method is described in [18]. Their procedure starts with an initial large HMM which directly encodes each training sequence as a separate path. An iterative merging procedure then attempts to merge redundant states if possible. Model parameters of the new state are re-estimated based on the the parameters of the merged states. While our procedure shares some similarities, such as continuously decreasing model size, there are several significant differences. Our re-estimation process is more formally grounded. Also, we feel that for most realistic datasets, the initial model required by the state-merging method would be unacceptably large.

Another approach towards automating the process of model construction is discussed in [19], wherein a special kind of HMM, called HM-Net, is automatically configured and trained. The configuration process involves beginning with a single-state model and iteratively adding states (by splitting existing states) as necessary. This approach is particularly interesting from our perspective because it nicely complements our own method. We start with a large model and prune it down to the appropriate size whereas the HM-Net training procedure incrementally grows a model to the right size.

The use of Rissanen's MDL principle [16, 17] in evaluating different HMM configurations is discussed in [12]. However, it still requires obtaining complete solutions for a number of alternative model configurations. In this respect, the use of MDL is really restricted to evaluating a given model based on its configuration, performance, and training set size. Model adjustment is still done manually. The MDL principle can be integrated within our own algorithm by changing the evaluation function from entropy to the MDL criterion, and this is one of the directions along which our work could be extended.

Pruning methods have been used with great success in automatically adjusting model configuration in induction procedures involving other representations such as decision trees [5] and rules [14, 20]. In this paper we have shown how pruning methods can be used for adjusting configuration of markov models.

6 Discussion and Future Work

We have described a new iterative algorithm for training HMM's that automatically adjusts the model size by pruning. It starts with a large, possibly over-fitted model and attempts to prune it down the appropriate complexity fit. The pruning method operates by making single link deletions to the model in an iterative fashion with parameters being re-estimated after every iteration. The procedure can be seen as a generalization of the Baum-Welch algorithm. While the standard Baum-Welch procedure requires the model size to be fixed, we have generalized it by incorporating the notion of link deletions.

Current methods for learning HMM's require that the model configuration be determined manually with considerable reliance on the intuitions and expertise of human experts. Our goal is to develop a fully automated procedure for training HMM's that addresses the two issues of model configuration and parameter estimation issues in an integrated fashion. The pruning method described in this paper is a first step towards this goal. We have provided empirical results comparing our procedure with the current method on the task of modeling DNA promoter sequences. Our algorithm exhibits several attractive characteristics. It shows fast convergence, good generalization, automatic adjustment of model size to approximately the right complexity fit, and relative insensitivity to the initial model size.

There are several directions along which this work could be extended. Our procedure provides a semi-formal framework within which several heuristic approaches towards automatic HMM configuration can be applied: (1) incorporation of link, or even state, addition which would compensate for over-pruning, (2) use of more sophisticated evaluation functions, such as MDL or AIC, that might better reflect the interplay between model size and performance, (3) stronger pruning such as state deletion which might help preserve model structure characteristics under pruning, and (4) use of a more sophisticated search strategy, such as beam search, instead of the current hill-climbing control strategy.

References

1. H. Akaike. Information theory and an extension of the maximum likelihood principle. In B. Petrov and F. Csaki, editors, *Proc. of the 2nd International Symposium on Information Theory*, pages 267–281, Budapest, 1972. Akademiai Kaido.

2. H. Akaike. A new look at the statistical model identification. *IEEE Trans. Automatic Control*, 19(6):716–723, 1974.

3. K. Asai, S. Hayamizu, and K. Onizuka. HMM with protein structure grammar. In *HICSS-93*, 1993. to appear.

4. P. Baldi, Y. Chauvin, T. Hunkapiller, and McClure M. A. Hidden Markov Models in molecular biology: New algorithms and applications. In C. L. Giles, S. J. Hanson, and J. D. Cowan, editors, *Advances in Neural Information Processing Systems 5*. Mougan Kaufman, 1993. to appear.

5. L. Breiman, J. H. Friedman, Olshen R. A., and C. J. Stone. *Classification and Regression Trees*. The Wadsworth statistics/probability series. Wadsworth, Inc., 1984.

6. B. Efron. The jackknife, the bootstrap and other resampling plans. In *SIAM*, 1982.

7. T. Hanazawa, T. Kawabata, and K. Shikano. Recognition of japanese voiced stops using Hidden Markov Models. *Nihon-Onkyou-Gakkai-Shi*, 45(10):776–785, 1989. (in Japanese).

8. D. Haussler, A. Krogh, I. S. Mian, and K. Sjölander. Protein modeling using Hidden Markov Models: Analysis of globins. Technical Report UCSC-CRL-92-23, University of California, 1992.

9. S. Kullback and R. A. Leibler. On information and sufficiency. *Ann. Math. Stat.*, 22:79–86, 1951.

10. S. E. Levinson, L. R. Rabiner, and M. M. Sondhi. An introduction to the application of the theory of probabilitic functions of a Markov process to automatic speech recognition. *Bell Systems Technical Journal*, 62:1035–74, 1983.

11. L. A. Liporace. Maximum likelihood estimation for multivariate observations of Markov sources. *IEEE Trans. Info. Theory*, IT-28(5):729–734, 1982.

12. Morita. From information coding to MDL principle. *SUURI-KAGAKU*, 25(8):25–31, 1987. (in Japanese).

13. M. Noordewier, G. Towell, and J. Shavlik. Learning to recognize promoters in DNA sequences. In *Proceedings of 1990 AAAI Spring Symposium on Artificial Intelligence and Molecular Biology*, 1990.

14. J. R. Quinlan. Generating production rules from decision trees. In *Proc. of IJCAI-87*, pages 304–307, 1987.

15. L. R. Rabiner and B. H. Juang. An introduction to Hidden Markov Models. *IEEE ASSP MAGAZINE*, pages 4–16, January 1986.

16. J. Rissanen. Modelling by shortest data description. *Automatica*, 14:465–471, 1978.

17. J. Rissanen. Stochastic complexity. *Journal of Royal Statistical Society, B*, 49(3):223–239, 1987.

18. A. Stolcke and S. Omohundro. Hidden Markov Model induction by Bayesian model merging. In C. L. Giles, S. J. Hanson, and J. D. Cowan, editors, *Advances in Neural Information Processing Systems 5*. Morgan Kaufman, 1993. to appear.

19. J. Takami and S. Sagayama. A successive state splitting algorithm for efficient allophone modeling. In *Proceedings of ICASSP*, 1992.

20. S. M. Weiss and N. Indurkhya. Reduced complexity rule induction. In *Proc. of IJCAI-91*, pages 678–684, 1991.

21. S. M. Weiss and C. Kulikowski. *Computer Systems That Learn*. Morgan Kaufmann, 1991.

On the VC-dimension of Depth Four Threshold Circuits and the Complexity of Boolean-valued Functions

Akito Sakurai

Advanced Research Laboratory, Hitachi Ltd.,
Hatoyama-cho, Saitama 350-03, JAPAN

Abstract. We consider the problem of determining VC-dimension $\partial_3(h)$ of depth four n-input 1-output threshold circuits with h elements. Best known asymptotic lower bounds and upper bounds are proved, that is, when $h \to \infty$, $\partial_3(h)$ is upper bounded by $((h^2/3) + nh)(\log h)(1 + o(1))$ and lower bounded by $(1/2)((h^2/4) + nh)(\log h)(1 - o(1))$. We also consider the problem of determining complexity $c_3(N)$ of Boolean-valued functions defined on N-pointsets in $\mathcal{R}^n$, measured by the number of threshold elements, with which we can construct a depth four circuit to realize the functions. We also show the best known upper and lower bounds, that is, when $N \to \infty$, the complexity is upper bounded by $\sqrt{16(N/\log N)(1 + o(1)) + 4n^2} - 2n$ and lower bounded by $\sqrt{6(N/\log N)(1 - o(1)) + (9/4)n^2} - (3/2)n$.

1 Introduction

Many researchers in the neural network community want to know the generalization capabilities of the neural networks that they have built, or want to guarantee the capabilities from a theoretical grounding. There are a few methods, such as a cross validation to check the capabilities during a learning period or a correlation method to construct a network with a certain capability certified by statistics, that can provide the capability. However we still want some measure of the capabilities of the networks we happen to have, which can estimate the capabilities regardless of how the networks were built and in some situations can estimate the size of the networks before we build them.

One well-known result in this line was obtained by research in the PAC learning paradigm ([2],[3]). These results are stated in the form "if you are to estimate a dichotomy of $\mathcal{R}^n$ from randomly drawn samples in $\mathcal{R}^n$ and associated categories, you have to have $F(\epsilon, \delta, \text{VC-dimension})$ samples to make the probability of your risk having an error rate greater than ϵ less than δ." VC-dimension is a measure of the richness of your set of representations of the dichotomies.

Although its importance is now well-known, the VC-dimension of complex representations such as neural networks is not well understood. We have studied the VC-dimension of neural networks of threshold elements, and have already reported one of the results for the one-hidden layer case ([9]). In this paper we report results for the three-hidden layer case.

Let us look at the problem from a different point of view. The neural networks we consider in this paper are just threshold circuits which were under attack by many prominent researchers in the 1960's and 1970's. Among the themes they studied was a problem of determining complexity of Boolean functions (functions which map $\{0, 1\}^n$ into $\{0, 1\}$), which is the number of threshold elements necessary to realize any Boolean function. The problem of complexity is far more popular in terms of "and" and "or" elements. The latter problem is deeply connected with the complexity of algorithms and still many researchers are investigating it. In contrast, the complexity in terms of threshold elements had nothing to do with other research areas, and interests in it faded away after the research on threshold elements as a whole declined. Only recently the study started again from different points of view.

Let a function whose region is $\{0, 1\}$ be called a Boolean-valued function. The VC-dimension of a set of Boolean-valued functions $\mathcal{F}$ is:

the maximum of N such that there exists an N-pointset P, on which any Boolean-valued function defined is equal to $f|_P$ for some $f \in \mathcal{F}$.

Let $\mathfrak{d}_L(h)$ represent the VC-dimension of depth $L + 1$ threshold circuits with h elements, and $c_L(N)$ the complexity of Boolean valued functions on N-pointset:

the minimum number of threshold elements necessary to realize by a depth $L + 1$ circuit any Boolean-valued function on an N-pointset in $\mathcal{R}^n$,

where the minimum is taken for all possible arrangement of N points, and $\mathfrak{C}_L(N)$ the complexity of partially defined Boolean functions:

the minimum number of threshold elements necessary to realize by a depth $L + 1$ circuit any partially defined n-input Boolean functions (defined on N vertices in n dimensional hypercube).

Our results are roughly stated as follows:

$$\frac{1}{2}\left(\frac{h^2}{4} + nh\right)(\log h)(1 - o(1)) \leq \mathfrak{d}_3(h) \leq \left(\frac{h^2}{3} + nh\right)(\log h)(1 + o(1)),$$

$$\sqrt{\frac{6N}{\log N}(1 - o(1)) + \frac{9n^2}{4}} - \frac{3n}{2} \leq c_3(N) \leq \sqrt{\frac{16N}{\log N}(1 + o(1)) + 4n^2} - 2n,$$

$$2\sqrt{\frac{N}{\log N}(1 - o(1))} \leq \mathfrak{C}_3(N) \leq 4\sqrt{2}\sqrt{\frac{N}{\log N}(1 + o(1))},$$

For the upper bound of VC-dimension of threshold circuits, we have a looser upper bound shown by Baum and Haussler([2]) which corresponds to $2((h^2/3) + (n + 1)h)\log(eh)$ for the depth-four (i.e. three-hidden-layer) case, which is worse than ours by the factor 2. For the lower bound, Maass([6]) pointed out that

"For an arbitrary large n, there exists a set of n-input 1-output neural networks with at most $33n$ connections which have the VC-dimension $n \log n$."

referring to O.B. Lupanov's result([5]). However Lupanov's result would be better interpreted in our present framework as:

"The VC-dimension of depth four n-input 1-output threshold circuits with $h = 24\sqrt{2^n/n}(1 + o(1))$ threshold elements is, when $n \to \infty$, lower bounded by $(1/144)(h^2/2)(\log h)(1 - o(1))$. In particular, if n is a power of 2, the bound improves to $(1/36)(h^2/2)(\log h)(1 - o(1))$ for $h = 12\sqrt{2^n/n}(1 + o(1))$."

Note that we have done a little bit of work to reduce the depth by one and consequently reduced the element count.

Section 2 defines terminologies. Section 3 proves the upper bound of the VC-dimension and the lower bound of the complexity, and section 4 and 5 the lower bound of the VC-dimension and the upper bound of the complexity.

2 Terminology

A *linear threshold element*, abbreviated as a *threshold element* hereafter, is an element with k inputs $(x_1, x_2, \ldots, x_k)$, 1 output, and $k + 1$ parameters $(w_1, w_2, \ldots, w_k)$, θ which outputs $\text{sgn}(\sum_{i=1}^{k} w_i x_i - \theta)$, where $\text{sgn}(x)$ is the sign function. A *threshold circuit* is a feedforward circuit (or network) composed of threshold elements. It is just a combinatorial network, except that the input values to elements can be any real values. The threshold circuits we deal with in this paper have only one binary output. The inputs to the circuit or the output from the circuit are called *external inputs* or *external output* to distinguish them from intra-circuit inputs and outputs. The *depth* of a circuit is the length of the longest path from its external inputs to its external output, where the length is the number of the elements on the path. We can naturally assign the *depth* to each element in a circuit as the length of the longest path from the output of the element to the external output, where the length is the number of the elements on the path. The depth of the external output element is 0. A *hidden layer* is a set of elements with the same depth other than 0. Therefore a depth L circuit has $L - 1$ hidden layers. A point x in $\mathcal{R}^n$ is on the *output value boundary* of a threshold circuit when in any of its neighborhood there exist a point for which the circuit outputs 1 and another point 0.

A function whose range is $\{0, 1\}$ (a set of 0 and 1) is called a *Boolean-valued function*. In particular the Boolean-valued function whose domain is $\{0, 1\}^n$ is called a *Boolean function*. A *0-1vector* is a vector whose elements are 0 or 1. $\mathbf{a} \cdot \mathbf{b}$ is the inner product of two vectors $\mathbf{a}$ and $\mathbf{b}$. An *N-pointset* is a pointset whose cardinality is N. $\#(P)$ represents the cardinality of a set P. A function whose range is $\{0, 1, \ldots, 2^s - 1\}$ (a set of integers between 0 and $2^s - 1$) is called an *s-bit-integer-valued function*. When an s-bit-integer-valued function is realized by a threshold circuit, a linear element (which outputs $\sum_{i=1}^{k} w_i x_i$ for inputs $x_1, \ldots, x_k$) is used as its external output element. A point x in $\mathcal{R}^n$ is on the *output value boundary* of some threshold circuit which realizes an s-bit-integer-valued function when in any of its neighborhood there exist a point for which the circuit outputs one value and another point another different value.

Let $\mathcal{F}$ be a set of Boolean-valued functions, and S be an N-pointset in $\mathcal{R}^n$. Set $\Pi_{\mathcal{F}}(S) \stackrel{\text{def}}{=} \{D : D \subset S, \exists f \in \mathcal{F} \ [\forall p \in D \ f(p) = 1 \text{ and } \forall p \in S - D \ f(p) = 0] \}$.

If $\Pi_{\mathcal{F}}(S) = 2^S$, that is, if for any dichotomy $S_0 \cup S_1$ ($S_0 \cap S_1 = \{\}$) of S there exists some $f \in \mathcal{F}$ such that $f(S_0) = \{0\}$ and $f(S_1) = \{1\}$, S is *shattered by* $\mathcal{F}$. The *VC-dimension* $VCdim(\mathcal{F})$ of a class $\mathcal{F}$ of Boolean-valued functions is the maximum cardinality of the sets which are shattered by $\mathcal{F}$. If we define $\Pi_{\mathcal{F}}(m) = \max(\#(\Pi_{\mathcal{F}}(S)))$ for a positive integer m where S in the expression varies in all possible set in $\mathcal{R}^n$ with cardinality m, $VCdim(\mathcal{F})$ is the maximum of d which satisfies $\Pi_{\mathcal{F}}(d) = 2^d$. The VC-dimension of a set of circuits is the VC-dimension of the set of functions represented by the circuits.

When $m \geq d$, $\Phi_d(m) \overset{\text{def}}{=} \sum_{i=0}^{d} \binom{m}{i}$, otherwise $\Phi_d(m) \overset{\text{def}}{=} 2^m$, assuming $d \geq 0$, $m \geq 0$. Then we have ([3],[4]):

(i) If $VCdim(H) = d$, for any $m \geq 0$, $\Pi_H(m) \leq \Phi_d(m)$.

(ii) For any $m \geq d \geq 1$, $\Phi_d(m) \leq 2(m^d/d!) \leq (em/d)^d$.

(iii) In particular, if T_n is a set of Boolean-valued functions realizable by an n-input threshold element, its VC-dimension is $n + 1$ and for any $m \geq 0$, $\Pi_{T_n}(m) = 2\Phi_n(m - 1) \leq \Phi_{n+1}(m)$.

3 Binding by the Function Counting Argument

We prove the upper bound of the VC-dimension and the lower bound of the complexity by counting an upper bound of the number of all the Boolean-valued functions on an N-pointset realizable by a set of threshold circuits. The first half of the proof is identical to that of Theorem 1 in Baum and Haussler ([2]).

Theorem 3.1. *The VC-dimension of depth $L + 1$ n-input 1-output threshold circuits with $h = h_0 + h_1 + \cdots + h_L$ linear threshold elements, where h_i elements are in depth $L - i$ layer and $h_L = 1$, upper bounded by $(\sum_{i<j} h_i h_j + nh)(\log h + 2 \log \log h) + h$ where $\log$ is the logarithm base 2 and $h \geq 46$.*

(Proof) The number of Boolean-valued functions defined on an N-pointset in $\mathcal{R}^n$ realizable by the set of threshold circuits mentioned above, is upper bounded by

$$\prod_{l=0}^{L} \left(\Pi_{T_{n+\Sigma_{i=0}^{l-1} h_i}}(N) \right)^{h_l} \leq 2^h \prod_{l=0}^{L} \left(\frac{eN}{n + \Sigma_{i=0}^{l-1} h_i} \right)^{(n+\Sigma_{i=0}^{l-1} h_i)h_l}$$

Applying Shannon's inequality to the left hand side, we get

$$\log \prod_{l=0}^{L} \left(\Pi_{T_{n+\Sigma_{i=0}^{l-1} h_i}}(N) \right)^{h_l} \leq \left(\sum_{i<j} h_i h_j + nh \right) \log \frac{eNh}{\sum_{i<j} h_i h_j + nh} + h.$$

If we put $N = (\sum_{i<j} h_i h_j + nh)(\log h + 2 \log \log h) + h$, then we get

$$\log \prod_{l=0}^{L} \left(\Pi_{T_{n+\Sigma_{i=0}^{l-1} h_i}}(N) \right)^{h_l} - N$$

$$< \left(\sum_{i<j} h_i h_j + nh \right) \left(\log e + \log \left(1 + \frac{2 \log \log h}{\log h} \right) - \log \log h \right.$$

$$\left. + \frac{h}{(\sum_{i<j} h_i h_j + nh)(\log h + 2 \log \log h)} \right)$$

$$< 0 \qquad \text{(when } h \geq 46\text{)}.$$

$\square$

In the same way we can prove the following general theorem for the upper bound of VC-dimension, which gives a slightly looser bound for the linear threshold circuit case (Corollary 3.3.). The proof is a combination of that of Theorem 1 in [2] and a careful estimation of inequalities as in the above proof.

Theorem 3.2. *The VC-dimension of a set of circuits with h elements whose VC-dimensions are d_i's is upper bounded by $(\sum_{i=1}^{h} d_i)(\log h + 2 \log \log h)$ when $h \geq 37$, where log is the logarithm base 2.*

Corollary 3.3. *The VC-dimension of depth $L + 1$ n-input 1-output neural networks with $h = h_0 + h_1 + \cdots + h_L$ linear threshold elements, where h_i elements are in depth $L - i$ layer and $h_L = 1$, is upper bounded by $(\sum_{i<j} h_i h_j + (n + 1)h)(\log h + 2 \log \log h)$ when $h \geq 37$, where log is the logarithm base 2.*

If we look at the same property from the view point of capacity, we have the next theorem.

Theorem 3.4. *To realize any Boolean-valued function defined on any N-point set by n-input 1-output threshold circuits with $h = h_0 + h_1 + \cdots + h_L$ elements, where $h_i = r_i h$ elements are in the depth $L - i$ layer and $0 < r_i < 1$ are fixed to some value, h should be eqaul to or greater than $\sqrt{2rN/(\log(2rN) + 4 \log \log(2rN)) + r^2 n^2/4} - rn/2$, where $r = 1/(\sum_{i<j} r_i r_j)$ and log is the logarithm base 2.*

(Proof) According to the proof of Theorem 3.1 it is enough to show that

$$h < \sqrt{2rN/(\log(2rN) + 4 \log \log(2rN)) + r^2 n^2/4} - rn/2$$

$$\implies \left(\sum_{i<j} h_i h_j + nh \right)(\log h + 2 \log \log h) + h < N.$$

For such an h

$$\sum_{i<j} h_i h_j + nh = \frac{1}{r} h^2 + nh < \frac{2N}{\log(2rN) + 4 \log \log(2rN)}$$

holds. And, since $h < \sqrt{2rN/(\log(2rN) + 4 \log \log(2rN))}$, we have

$$\log h < \frac{1}{2} \log \frac{2rN}{\log(2rN) + 4 \log \log(2rN)} < \frac{1}{2} (\log(2rN) - \log \log(2rN)),$$

$$\log \log h < \log \log(2rN).$$

Therefore

$$\left(\sum_{i<j} h_i h_j + nh\right)(\log h + 2\log\log h) + h$$

$$= N\frac{\log(2rN) + 3\log\log(2rN) + \sqrt{\frac{2r}{N}}\,(\log(2rN) + 4\log\log(2rN))}{\log(2rN) + 4\log\log(2rN)}$$

$$< N \qquad\qquad (\text{when } N \to \infty). \qquad\qquad \square$$

The following corollaries are easily obtained from the above theorems.

Corollary 3.5. When $h \to \infty$,

$$\partial_L(h) \le \left(\left(1 - \frac{1}{L}\right)\frac{h^2}{2} + nh\right)(\log h)\left(1 + O\left(\frac{\log\log h}{\log h}\right)\right) \qquad (L \ge 1).$$

Corollary 3.6. When $N \to \infty$,

$$\frac{N}{n\log N} \le c_1(N) \quad (n \ge 4),$$

$$\sqrt{\frac{4L}{L-1}\frac{N}{\log N}\left(1 - O\left(\frac{\log\log N}{\log N}\right)\right) + \left(\frac{L}{L-1}\right)^2 n^2} - \frac{L}{L-1}n \le c_L(N) \quad (L > 1).$$

4 Binding by the Construction of Circuits

Since the proofs of the target theorems have many technical details, we divide the statements into two large parts and prove them separately. In this section, we prove the following statements, which are looser bounds than our final results.

Theorem 4.1. *The VC-dimension of depth four n-input 1-output threshold circuits with h linear threshold elements is lower bounded by $(1/2)(h^2/4)(\log h)(1 - O(\log\log h/\log h))$. Moreover the set of the N-pointset on which the above bound is attained has positive Lebesgue measure.*

Theorem 4.2. *For any N-pointset in $\mathcal{R}^n$, any Boolean-valued function defined on the pointset is realizable by a depth four circuit with at most $4\sqrt{N/\log N}(1 + O(\log\log N/\log N))$ linear threshold elements. Moreover the set of these N-pointsets have positive Lebesgue measure.*

Theorem 4.3. *For any N ($\le 2^n$) there exists N-pointset in $\{0, 1\}^n$ such that any Boolean-valued function defined on the pointset is realizable by a depth four circuit with at most $4\sqrt{2}\sqrt{N/\log N}(1 + o(1))$ linear threshold elements.*

These are easily proven from the following theorem.

Theorem 4.4. *Suppose that positive integers h_1, h_2, s (where h_2 is even) are given. Let P be any $N = (1/2)h_1 h_2$ pointset in $\mathcal{R}^1$ and Q be any s-pointset in*

$\mathcal{R}^1$. Any Boolean-valued function $f(x,y)$ defined on the product set $P \times Q$ can be realized by a depth four circuit with $h_1 + h_2 + 2 \cdot 2^s + s + 1$ threshold elements. Moreover any point in $P \times Q$ does not reside on the output value boundary of the circuit.

The proofs of Theorem 4.1 through Theorem 4.3 based on Theorem 4.4 are easily done as follows.

(Proof of Theorem 4.1 based on Theorem 4.4) For any N-pointset in $\mathcal{R}^n$, there exists an **a** for which every $\mathbf{a} \cdot \mathbf{x}$ differs for $\mathbf{x} \in P$. Then we can use $\mathbf{a} \cdot \mathbf{x}$ as the external input to the circuit constructed in the proof of Theorem 4.4. If we put in Theorem 4.4 as $h_1 \leftarrow \lceil h/2 \rceil - 2 \cdot 2^s - s - 4$, $h_2 \leftarrow \lceil h/4 \rceil$, $M \leftarrow s$ (where $s = \lfloor \log(h/\log h) \rfloor$), we get the desired results. □

(Proof of Theorem 4.2 based on Theorem 4.4) As in the previous proof, it suffices to put: $N \leftarrow \lceil N/s \rceil$, $h_1 \leftarrow \left\lceil \sqrt{2}\sqrt{\lceil N/s \rceil} \right\rceil$, $h_2 \leftarrow 2\left\lceil (1/\sqrt{2})\sqrt{\lceil N/s \rceil} \right\rceil$, and $M \leftarrow s$ where $s = \left\lceil \log \sqrt{N/(\log N)^2} \right\rceil$. □

(Proof of Theorem 4.3 based on Theorem 4.4) Let $r = \lfloor \log\log \sqrt{N/(\log N)^2} \rfloor$. Let us select arbitrarily at most $\lceil N/2^r \rceil$ points from vertices of $n - r$ dimensional hypercube. Let the set be called P. Let Q be the set of all vertices on r dimensional hypercube. Clearly the product set $P \times Q$ $(\#(P \times Q) \geq N)$ is a subset of the set of vertices on n dimensional hypercube. Again as in the previous proof, it suffices to put: $N \leftarrow \lceil N/s \rceil$, $h_1 \leftarrow \left\lceil \sqrt{2}\sqrt{\lceil N/s \rceil} \right\rceil$, $h_2 \leftarrow 2\left\lceil (1/\sqrt{2})\sqrt{\lceil N/s \rceil} \right\rceil$, and $M \leftarrow s$ where $s = 2^r$. □

It is easily seen that Theorem 4.4 is proven from the following two theorems.

Theorem 4.5. *We assume the same conditions as in Theorem 4.4. Let Q be $\{y_i : 1 \leq i \leq s, y_i < y_{i+1}\}$, and $f(x) \stackrel{\text{def}}{=} \sum_{i=1}^{s} f(x, y_i)2^{i-1}$. Any s-bit-integer-valued function defined on P is realized by a depth four circuit with $h_1 + h_2 + 2^s$ threshold elements and one linear element (for the external output). Moreover, in the circuit there are 2^s elements in the depth one layer which satisfy*
· *For any external input $x \in P$,*
$$\forall i \leq f(x)\ G_i^1(x) = 1 \quad \text{and} \quad \forall i > f(x)\ G_i^1(x) = 0,$$
where $G_i^1(x)$ $(0 \leq i \leq 2^s - 1)$ is the output value of the element g_i^1 when the external input x is fed to the circuit. Moreover any point in $P \times Q$ does not reside on the output value boundary of the circuit.

Theorem 4.6. *Let us consider a Boolean-valued function $f(x, y)$ defined on the product set $P \times Q$ for pointsets P and Q in $\mathcal{R}^1$. Let Q be $\{y_i : 1 \leq i \leq s, y_i < y_{i+1}\}$, and $f(x) \stackrel{\text{def}}{=} \sum_{i=1}^{s} f(x, y_i)2^{i-1}$. Suppose that $f(x)$ is realized by a depth $L \geq 3$ threshold circuit $\mathfrak{G}$ with H threshold elements and a linear external output element, which has 2^s elements in the depth one layer that satisfy*
· *For any external input $x \in P$,*
$$\forall i \leq f(x)\ G_i^1(x) = 1 \quad \text{and} \quad \forall i > f(x)\ G_i^1(x) = 0,$$

where $G_i^1(x)$ $(0 \leq i \leq 2^s - 1)$ *is the output value of the element g_i^1 when x is fed to the external input of the circuit. Then $f(x, y)$ is realized by a depth L circuit with $H + 2^s + s + 1$ threshold elements.*

(Proof of Theorem 4.5) We sort the points in P into h_1 groups in ascending order of their coordinate values, where each group has $h_2/2$ points. Let the subsets be $P_1, P_2, \ldots, P_{h_1}$. Set

$$t_i = \frac{1}{2} \left(\max\{x : x \in P_i\} + \min\{x : x \in P_{i+1}\} \right) \qquad (1 \leq i \leq h_1 - 1)$$

And we name the points x in each P_i as x_j^i $(1 \leq j \leq h_2/2)$ in the non-decreasing order of $f(x)$, so that we have $f(x_j^i) \leq f(x_{j+1}^i)$.

We define weights and thresholds of the elements in the circuit as follows, where only the external input x is shown as an argument to each element g_j^i for simplicity.

Depth 3 elements $(1 \leq i \leq h_1 - 1)$: $g_i^3(x) = \text{sgn}(x - t_i)$
Depth 2 elements. $(1 \leq j \leq h_2/2)$:

$$g_{2j-1}^2(x) = \text{sgn}\left(x - \left(\sum_{i=1}^{h_1-1}(x_j^{i+1} - x_j^i)g_i^3 + (x_j^1 - \epsilon) \right) \right)$$

$$g_{2j}^2(x) = \text{sgn}\left(x - \left(\sum_{i=1}^{h_1-1}(x_j^{i+1} - x_j^i)g_i^3 + (x_j^1 + \epsilon) \right) \right)$$

Depth 1 elements $(0 \leq k \leq 2^s - 1)$:

$$g_k^1(x) = \text{sgn}\left(\sum_{j=1}^{h_2/2} j(g_{2j-1}^2 - g_{2j}^2) \right.$$
$$- \left(\sum_{i=1}^{h_1-1} \left(\min\{j : k \leq f(x_j^{i+1})\} - \min\{j : k \leq f(x_j^i)\} \right) g_i^3 \right.$$
$$\left. \left. + \min\{j : k \leq f(x_j^1)\} + 0.5 \right) \right)$$

Depth 0 element : $g^0 = \sum_{k=0}^{2^s-1} (g_k^1 - g_{k+1}^1)k$

Next we show that the requirements in the theorem statement are fulfilled. Let $G_j^i(x)$ be the output value of the element g_j^i, when x is fed as the external input to the circuit G constructed above. Let us suppose now that $x_{j_0}^{i_0}$ such that $f(x_{j_0}^{i_0}) = k_0$ is input to G.

The $G_j^i(x)$'s are calculated as follows.

Depth 3 elements $(1 \leq i \leq h_1 - 1)$:
Since $x_{j_0}^{i_0} > t_i$ if $i < i_0$ and $x_{j_0}^{i_0} < t_i$ otherwise,

$$G_i^3(x_{j_0}^{i_0}) = \text{sgn}(x_{j_0}^{i_0} - t_i) = \begin{cases} 1, & \text{if } i < i_0; \\ 0, & \text{if } i_0 \leq i. \end{cases}$$

Depth 2 elements $(1 \leq j \leq h_2)$:

$$G_{2j-1}^2(x_{j_0}^{i_0}) = \text{sgn}\left(x_{j_0}^{i_0} - \left(\sum_{i=1}^{h_1-1}(x_j^{i+1} - x_j^i)G_i^3(x_{j_0}^{i_0}) + (x_j^1 - \epsilon) \right) \right)$$

$$= \text{sgn}\left(x_{j_0}^{i_0} - (x_j^{i_0} - \epsilon) \right)$$

With the same argument we get $G_{2j}^2(x_{j_0}^{i_0}) = \text{sgn}\left(x_{j_0}^{i_0} - (x_j^{i_0} + \epsilon) \right)$.
Therefore

$$G^2_{2j-1}(x_{j_0}^{i_0}) - G^2_{2j}(x_{j_0}^{i_0}) = \begin{cases} 1, & \text{if } j = j_0; \\ 0, & \text{otherwise.} \end{cases}$$

Depth 1 elements $(0 \le k \le 2^s - 1)$:

Since we have $\min\{j : k \le f(x_j^{i_0})\} \le j_0$ when $k \le k_0$ and $j_0 < \min\{j : k \le f(x_j^{i_0})\}$ when $k_0 < k$, we get

$$G_k^1(x_{j_0}^{i_0}) = \text{sgn}\left(j_0 - (\min\{j : k \le f(x_j^{i_0})\} + 0.5)\right) = \begin{cases} 1, & \text{if } k \le k_0; \\ 0, & \text{if } k_0 < k. \end{cases}$$

In other words, the elements in the depth 1 layer behave as expected. Accordingly, for $x \in P$, when $k = f(x)$, $G_k^1(x) - G_{k+1}^1(x) = 1$, otherwise $G_k^1(x) - G_{k+1}^1(x) = 0$, which implies $G^0(x) = f(x)$. The number of threshold elements we used is clearly $h_1 + h_2 + 2^s$. $\quad\square$

(Proof of Theorem 4.6) Let $G_j^i(x)$ represent the output value of the element g_j^i when x is fed as the external input to the circuit $\mathfrak{G}$, and $\mathcal{C}_1$ be $\mathfrak{G}$ with the output element deleted.

Let us prepare s linear threshold elements $u_1^2, u_2^2, \ldots, u_s^2$ connected in the following way, where $y_0 \in \mathcal{R}^1$ is any point such that $y_0 < y_1$.

$$u_i^2(y) = \text{sgn}(y - (1/2)(y_{i-1} + y_i)) \quad (1 \le i \le s),$$

and $u_{s+1}^2 \equiv 0$. Clearly for any $y \in Q$, when $y = y_j$, $u_j^2(y) - u_{j+1}^2(y) = 1$, otherwise $u_j^2(y) - u_{j+1}^2(y) = 0$. Let $org\_act\_g_i^1$ be the activation function of g_i^1 that appeared in the proof of Theorem 4.5, that is, $g_i^1(x) = \text{sgn}\left(org\_act\_g_i^1(x)\right)$. Let us duplicate g_i^1 and add connections from u_j^2 to the pair of g_i^1 as

$$g_i^{1+}(x, y) = \text{sgn}\left(org\_act\_g_i^1(x) + K\left(\sum_{j=1}^{s} a_{i,j}\left(u_j^2(y) - u_{j+1}^2(y)\right) - 1\right)\right)$$

$$g_i^{1-}(x, y) = \text{sgn}\left(org\_act\_g_{i+1}^1(x) + K\left(\sum_{j=1}^{s} a_{i,j}\left(u_j^2(y) - u_{j+1}^2(y)\right) - 1\right)\right)$$

where $\sum_{j=1}^{s} a_{i,j} 2^{j-1}$ is the binary expansion of i, and K is set large enough so that $\forall i \; \forall x \in P \; org\_act\_G_i^1(x) < K$ holds. Let the external output element (the depth 0 element) be as follows and the resultant network be called $\mathcal{C}$.

$$g^0(x, y) = \text{sgn}\left(\sum_{i=0}^{2^s-1} (g_i^{1+}(x, y) - g_i^{1-}(x, y)) - 0.5\right)$$

Let $\mathcal{G}_j^i(x, y), \mathcal{U}_j^i(x, y)$ represent the outputs of g_j^i, u_j^i in $\mathcal{C}$ when x and y are fed to circuit $\mathcal{C}$. Then we get

$$\mathcal{G}_i^{1+}(x, y_j) = \text{sgn}\left(org\_act\_\mathcal{G}_i^1(x) + K\left(a_{i,j} - 1\right)\right)$$

$$= \begin{cases} \text{sgn}\left(org\_act\_\mathcal{G}_i^1(x)\right), & \text{if } a_{i,j} = 1; \\ 0, & \text{otherwise (supposing } x \in P); \end{cases}$$

$$= \begin{cases} \mathcal{G}_i^1(x), & \text{if } a_{i,j} = 1; \\ 0, & \text{otherwise}; \end{cases}$$

Similarly, when $a_{i,j} = 1$, $\mathcal{G}_i^{1-}(x, y_j) = \mathcal{G}_{i+1}^1(x)$, otherwise $\mathcal{G}_i^{1-}(x, y_j) = 0$.

Since $i \leq f(x)$ holds if and only if $G_i^1(x) = 1$,

$$\mathcal{G}^0(x, y_j) = \operatorname{sgn}\left(\sum_{i=0}^{2^s-1}(\mathcal{G}_i^{1+}(x, y_j) - \mathcal{G}_i^{1-}(x, y_j)) - 0.5\right) = \begin{cases} 1, & \text{if } a_{f(x),j} = 1; \\ 0, & \text{otherwise.} \end{cases}$$

Since "$f(x, y_j) = 1$" if and only if "$a_{f(x),j} = 1$", we have $f(x, y_j) = \mathcal{G}^0(x, y_j)$. $\quad\square$

5   Improving the Bounds

We prove the desired lower bound of the VC-dimension and upper bound of the complexity based on the proofs in the previous section for looser bounds. The improvements are made by utilizing the higher dimensionality of the input space. Note that the theorem numbers in this section are given in accordance with the ones in the previous section, so that some numbers are missing.

Theorem 5.1. *The VC-dimension of depth four n-input 1-output threshold circuits with h linear threshold elements is lower bounded by $(1/2)\,(h^2/4 + nh)$ $(\log h)\,(1 - O(\log\log h/\log h))$. Moreover the set of these N-pointsets on which the above bound is attained has positive Lebesgue measure.*

Theorem 5.2. *For some N-pointset in $\mathcal{R}^n$, any Boolean-valued function defined on the pointset is realizable with a depth four circuit with at most $\sqrt{16(N/\log N)(1 + O(\log\log N/\log N))} + 4n^2 - 2n$ linear threshold elements. Moreover the set of these N-pointsets have positive Lebesgue measure.*

The above two theorems are easily proven from the following theorem.

Theorem 5.4. *Suppose that positive integers n, h_1, h_2, s (where h_1 and h_2 are even and $2^s < (1/2)(h_1 + h_2)$) are given. Let $N^- = (1/2)h_1 h_2$, $N^+ = (n-1)\,((1/2)(h_1 + h_2) - 2^s)$, $N = N^- + N^+$, and $H = h_1 + h_2$. Let P^+ be any N^+-pointset in $\mathcal{R}^n$ in general position, P^- be a suitably chosen N^--pointset in $\mathcal{R}^n$, and $P = P^- \cup P^+$. Any Boolean-valued function $f(x, y)$ defined on the product set $P \times Q$ for any s-pointset Q in $\mathcal{R}^1$ can be realized by a depth four circuit with at most $h_1 + h_2 + 4 \cdot 2^s + s + 2$ threshold elements. Moreover any point in $P \times Q$ does not reside on the output value boundary of the circuit.*

The proofs of Theorem 5.1 and Theorem 5.2 are done in the same way as those of Theorem 4.1 and Theorem 4.2 and are omitted. Theorem 5.4 is easily proven from the following two theorems as in the previous section.

Theorem 5.5. *We assume the same conditions as in Theorem 5.4. Let Q be $\{y_i : 1 \leq i \leq s, y_i < y_{i+1}\}$, and $f(x) \stackrel{\text{def}}{=} \sum_{i=1}^{s} f(x, y_i)2^{i-1}$. Any s-bit-integer-valued function defined on P is realized by a depth four circuit with $h_1 + h_2 + 2 \cdot 2^s + 1$ threshold elements and one linear element (for the external output). Moreover,* *in the circuit there are $2 \cdot 2^s$ elements in the depth one layer which satisfy the conditions that for any external input $\mathbf{x} \in P$, either*

$\cdot$ $\forall i \leq f(\mathbf{x})$ $G_i^{1-}(\mathbf{x}) = 1$, $\forall i > f(\mathbf{x})$ $G_i^{1-}(\mathbf{x}) = 0$, $\forall j$ $G_j^{1+}(\mathbf{x}) = 0$, *or*

$\cdot$ $\forall i \leq f(\mathbf{x})$ $G_i^{1+}(\mathbf{x}) = 1$, $\forall i > f(\mathbf{x})$ $G_i^{1+}(\mathbf{x}) = 0$, $\forall j$ $G_j^{1-}(\mathbf{x}) = 0$,

where $G_i^{1-}(\mathbf{x})$ and $G_i^{1-}(\mathbf{x})$ ($0 \leq i \leq 2^s - 1$) designate the output value of the elements g_i^{1-} and g_i^{1+} in the depth one layer when the external input $\mathbf{x}$ is fed to the circuit. Moreover any point in $P \times Q$ does not reside on the output value boundary of the circuit.

Theorem 5.6. *Let us consider a Boolean-valued function $f(\mathbf{x}, y)$ defined on the product set $P \times Q$ for a pointset P in $\mathcal{R}^n$ and a pointset Q in $\mathcal{R}^1$. Let Q be $\{y_i : 1 \leq i \leq s, y_i < y_{i+1}\}$, and $f(\mathbf{x}) \overset{def}{=} \sum_{i=1}^{s} f(\mathbf{x}, y_i) 2^{i-1}$. Suppose that $f(\mathbf{x})$ is realized by a depth $L \geq 3$ threshold circuit $\mathfrak{G}$ with H linear threshold elements and a linear element (for the external output), which has $2 \cdot 2^s$ elements in the depth one layer that satisfy the conditions that for any external input $\mathbf{x} \in P$, either*

$\cdot$ $\forall i \leq f(\mathbf{x})$ $G_i^{1-}(\mathbf{x}) = 1$, $\forall i > f(\mathbf{x})$ $G_i^{1-}(\mathbf{x}) = 0$, $\forall j$ $G_j^{1+}(\mathbf{x}) = 0$, *or*

$\cdot$ $\forall i \leq f(\mathbf{x})$ $G_i^{1+}(\mathbf{x}) = 1$, $\forall i > f(\mathbf{x})$ $G_i^{1+}(\mathbf{x}) = 0$, $\forall j$ $G_j^{1-}(\mathbf{x}) = 0$,

where $G_i^{1-}(\mathbf{x})$ and $G_i^{1-}(\mathbf{x})$ ($0 \leq i \leq 2^s - 1$) designate the output value of the elements g_i^{1-} and g_i^{1+} in the depth one layer when the external input $\mathbf{x}$ is fed to the circuit. Then $f(\mathbf{x}, y)$ is realized by a depth L circuit with $H + 2 \cdot 2^s + s + 1$ threshold elements.

The proof of Theorem 5.6 is similar to that of Theorem 4.6 and is omitted.

(Proof of Theorem 5.5) We construct P^- supposing that P^+ is given. First we obtain a point o in $\mathcal{R}^1$, a line $\mathcal{L}$ which contains o, and a hyperplane $\mathcal{H}$ which satisfy the following conditions (their existence is obvious).

(P1) o is in the negative region of $\mathcal{H}$, and P^+ the positive region, and

(P2) for any R ($\subset P^+$, $\#(R) \leq n-1$), there exists a hyperplane which passes through o and every point in R but does not include $\mathcal{L}$ and does not pass through any point in $P^+ - R$.

We take arbitrarily N^- points on $\mathcal{L}$ which are in $\mathcal{N}(\delta)$ for small enough δ, where $\mathcal{N}(r) \overset{def}{=} \{\mathbf{x} : |\mathbf{x} - \mathbf{o}| < r\}$. We will show how small δ should be in the below. Let P^- be the pointset thus we obtained. Let us define any orthogonal coordinate system whose first coordinate axis is $\mathcal{L}$. Hereafter all the points in $\mathcal{R}^n$ are represented according to the coordinate system. It is easily proven that the following property holds.

(P3) For any $\mathbf{x}_1, \mathbf{x}_2, \ldots, \mathbf{x}_{n-1} \in P^-$, letting $\mathbf{x}_i' = {}^t(x_{i,2}, \ldots, x_{i,n})$ for $\mathbf{x}_i = {}^t(x_{i,1}, x_{i,2}, \ldots, x_{i,n})$, $\det(\mathbf{x}_1', \mathbf{x}_2', \ldots, \mathbf{x}_{n-1}') \neq 0$ holds.

Let P^- correspond to P in Theorem 4.5, and for simplicity we use x to represent the first coordinate value of $\mathbf{x}$. We sort points in P^-, as in the proof of Theorem 4.5, into h_1 groups with $h_2/2$ points for each group, in ascending order of their first coordinate values. Let the subsets be $P_1, P_2, \ldots, P_{h_1}$. Set

$$t_i = (1/2)(\max\{x : \mathbf{x} \in P_i\} + \min\{x : \mathbf{x} \in P_{i+1}\}) \quad (1 \leq i \leq h_1 - 1).$$

And we name the points $\mathbf{x}$ in each P_i as $\mathbf{x}_j^i$ $(1 \leq j \leq h_2/2)$ in the non-decreasing order of $f(\mathbf{x})$, so that we have $f(\mathbf{x}_j^i) \leq f(\mathbf{x}_{j+1}^i)$. Note that in the following x_j^i represents the first coordinate value of $\mathbf{x}_j^i$.

We define weights and thresholds of the elements in the circuit as follows, where for $\mathbf{w}_j^l$ $(l = 2, 3)$ only the first coordinate values are 1 and others are undefined for the moment. Note that regardless of $\mathbf{w}_j^l$'s undefinedness the function of the circuit is well-defined for the points in P^-, which is identical to the function shown in the proof of Theorem 4.5. Note also that as usual only the external input $\mathbf{x}$ is shown as an argument to each element g_j^i for simplicity.

Depth 3 elements $(1 \leq i \leq h_1/2)$:
$$g_{2i-1}^3(\mathbf{x}) = \mathrm{sgn}(\mathbf{w}_i^3 \cdot \mathbf{x} - t_{2i-1})$$
$$g_{2i}^3(\mathbf{x}) = \mathrm{sgn}(\mathbf{w}_i^3 \cdot \mathbf{x} - t_{2i})$$
Depth 2 elements. $(1 \leq j \leq h_2/2)$:
$$g_{2j-1}^2(\mathbf{x}) = \mathrm{sgn}\left(\mathbf{w}_j^2 \cdot \mathbf{x} - \left(\textstyle\sum_{i=1}^{h_1-1}(x_j^{i+1} - x_j^i)g_i^3 + (x_j^1 - \epsilon)\right)\right)$$
$$g_{2j}^2(\mathbf{x}) = \mathrm{sgn}\left(\mathbf{w}_j^2 \cdot \mathbf{x} - \left(\textstyle\sum_{i=1}^{h_1-1}(x_j^{i+1} - x_j^i)g_i^3 + (x_j^1 + \epsilon)\right)\right)$$

In the following $G_j^l(\mathbf{x})$ represents the output value of the element g_j^l in the circuit thus constructed when $\mathbf{x}$ is fed to the external input to the circuit.

Next we are going to define $\mathbf{w}_j^l$ $(l = 2, 3)$. Let $T_i^+ \stackrel{\text{def}}{=} \{\mathbf{x} : \mathbf{x} \in P^+, f(\mathbf{x}) = i\}$, and ϵ be any constant such that $0 < \epsilon < (1/2)\min|x_j^i - x_{j'}^{i'}|$. Note that the arbitrariness of ϵ will not affect the function of the following circuit for $\mathbf{x} \in P$.

Step 1: $i \leftarrow -1$, $j \leftarrow 1$, $l \leftarrow 3$.

Step 2: If $i = 2^s - 1$ then stop. Otherwise, $i \leftarrow i + 1$.

Step 3: If $\#(T_i^+) = 0$ then go back to Step 2.

Step 4: If $\#(T_i^+) > n - 1$, take $n - 1$ points or otherwise take $\#(T_i^+)$ points from T_i^+ arbitrarily and delete them from T_i^+. Define $\mathbf{w}_j^l$ so that

$$\begin{cases} \mathbf{w}_j^3 \cdot \mathbf{x} - t_{2j-1} = (1/2)(-t_{2j-1} + t_{2j}), & \text{if } l = 3, \\ \mathbf{w}_j^2 \cdot \mathbf{x} - \left(\sum_{i=1}^{h_1-1}(x_j^{i+1} - x_j^i)G_i^3(\mathbf{x}) + (x_j^1 - \epsilon)\right) = \epsilon, & \text{if } l = 2, \end{cases}$$

hold for these points. By the property (P3), we can always define $\mathbf{w}_j^l$ as stated. Let the set of the extracted points be called T_j^l.

Step 5: $j \leftarrow j + 1$. If $j > h_{4-l}$ then $j \leftarrow 1$, $l \leftarrow l - 1$. Go back to Step 3.

It is obvious that $\forall \mathbf{x} \in T_j^l$ $[G_{2j-1}^l(\mathbf{x}) = 1 \wedge G_{2j}^l(\mathbf{x}) = 0]$ holds; however, its converse does not necessarily hold. Namely we have possibility that "for some $\mathbf{x} \in P^+ - T_j^l$ $[G_{2j-1}^l(\mathbf{x}) = 1 \wedge G_{2j}^l(\mathbf{x}) = 0]$." Nevertheless if $|(1/2)(t_{2j-1} - t_{2j})|$ is small enough the converse holds, because

(1) when $\delta = 0$ (that is when P^- is degenerated to $\mathbf{o}$) there exists a solution (let this be called $\mathbf{w}_j^l(0)$), and since the solution of simultaneous linear equations is a continuous function of constant column vector, for small enough $\delta > 0$, there exists a solution $\mathbf{w}_j^l(\delta)$ which is close to $\mathbf{w}_j^l(0)$, and

(2) since the hyperplane designated by $\mathbf{w}_j^l(0)$ passes through $\mathbf{o}$ and every point in T_j^l but no point in $P^+ - T_j^l$, there are no points of $P^+ - T_j^l$ in some neighborhood of the hyperplane, which implies

(3) "For small enough $\delta > 0$, we can define $\mathbf{w}_j^l(\delta)$ so that no point of $P^+ - T_j^l$ locates near the hyperplane designated by $\mathbf{w}_j^l(\delta)$."

The other part of the circuit is defined as follows. In the depth two layer we use one more threshold element. For the depth one layer we adopt elements whose connections are slightly modified from those in the proof of Theorem 4.5 (a term "$-Ku^2$" is added to get g^{1+}), and newly defined ones (g^{1-}). We assume that the hyperplane $\mathcal{H}$ is denoted by $\mathbf{w}^0 \cdot \mathbf{x} = \theta^0$. Note that, from the definition of $\mathcal{H}$, $\forall x \in P^+$ $\mathbf{w}^0 \cdot \mathbf{x} - \theta^0 > 0$ holds.

Depth 2 element : $u^2(\mathbf{x}) = \text{sgn}(\mathbf{w}^0 \cdot \mathbf{x} - \theta^0)$

Depth 1 element $(0 \le k \le 2^s - 1)$:

$$g_k^{1-}(\mathbf{x}) = \text{sgn}\Big(\sum_{j=1}^{h_2/2} j(g_{2j-1}^2 - g_{2j}^2)$$
$$- \Big(\sum_{i=1}^{h_1-1} \big(\min\{j : k \le f(\mathbf{x}_j^{i+1})\} - \min\{j : k \le f(\mathbf{x}_j^i)\} \big) g_i^3$$
$$+ \min\{j : k \le f(\mathbf{x}_j^1)\} + 0.5 \Big) - Ku^2 \Big)$$

$$g_k^{1+}(\mathbf{x}) = \text{sgn}\Big(\sum_{j \ s.t. \ \exists x[x \in T_j^3 \ \wedge \ f(x) \ge k]} (g_{2j-1}^3 - g_{2j}^3)$$
$$+ \sum_{j \ s.t. \ \exists x[x \in T_j^2 \ \wedge \ f(x) \ge k]} (g_{2j-1}^2 - g_{2j}^2)$$
$$- K(1 - u^2) \Big)$$

Depth 0 element : $g_l^0 = \sum_{k=0}^{2^s-1} \Big(\big(g_k^{1-} - g_{k+1}^{1-} \big) + \big(g_k^{1+} - g_{k+1}^{1+} \big) \Big) k$

where $g_{2^s}^{1+} \equiv 0$, $g_{2^s}^{1-} \equiv 0$, and K is defined large enough so that

$$\forall k \ \forall \mathbf{x} \in P^- \ G_k^{1+}(\mathbf{x}) = 0 \ \wedge \ \forall k \ \forall \mathbf{x} \in P^+ \ G_k^{1-}(\mathbf{x}) = 0$$

holds. Then that the circuit works properly for any $\mathbf{x} \in P^-$ is shown in the same way as the proof for Theorem 4.5, using

$$G_k^{1-}(\mathbf{x}) = \begin{cases} 1, & \text{if } k \le f(\mathbf{x}), \\ 0, & \text{if } f(\mathbf{x}) < k, \end{cases} \quad \text{and} \quad G_k^{1+}(\mathbf{x}) = 0.$$

And for the case when $\mathbf{x} \in P^+$, the facts that
· if $\mathbf{x} \in P^+$ then $G_k^{1-}(\mathbf{x}) = 0$,
that for $\mathbf{x} \in T_j^3$,
· $G_{2j-1}^3(\mathbf{x}) = 1$, $G_{2j}^3(\mathbf{x}) = 0$,
· $\forall k \ne j$ $[G_{2k-1}^3(\mathbf{x}) = G_{2k}^3(\mathbf{x})]$ and $\forall k$ $[G_{2k-1}^2(\mathbf{x}) = G_{2k}^2(\mathbf{x})]$
hold, and so

$$G_k^{1+}(\mathbf{x}) = \text{sgn}\Big(\sum_{j \ s.t. \ \exists x[x \in T_j^3 \wedge f(x) \ge k]} (G_{2j-1}^3 - G_{2j}^3) \Big) = \begin{cases} 1, & \text{if } k \le f(\mathbf{x}), \\ 0, & \text{if } f(\mathbf{x}) < k, \end{cases}$$

and that the similar property holds for $\mathbf{x} \in T_j^2$ imply the desired function of the circuit.

The elements we used are at most $h_1 + h_2 + 2 \cdot 2^s + 1$ threshold elements and one linear element (for the external output) □

6 Concluding Remarks

We have already shown ([9]) that the VC-dimension of depth two threshold circuits is lower bounded by about half of the upper bound obtained in Section 3. Therefore for the shallow network case there remains a problem of bridging a gap between bounds of the VC-dimension of the depth three case, since the best lower bound we have obtained so far is $h^2/4 + nh$ which exhibits significant gap to the upper bound $(h^2/4 + nh)(\log h + 2\log\log h)$ obtained in Section 3. The proof for the lower bound is a subset of the proof of Theorem 4.5. Inability to get the $\log h$ gain comes from the fact that the equivalent of the "output function" performed by the depth one element in the proof of Theorem 4.5 can not be squashed into one output threshold element nor into the hidden layers.

For the deep threshold circuits, roughly speaking, the best lower bounds of VC-dimension we obtained so far are again half of their upper bounds shown in Section 3. These results will be published elsewhere. Considering these results and results obtained by Lupanov ([5]) with their proofs, we expect that, at least for deep networks,

$$\partial_L(h) \approx ((1 - 1/L)(h^2/2) + nh)(1 \pm O(\log\log h/\log h)),$$

which is an open problem.

References

1. Baum, E. B. : On the capabilities of multilayer perceptrons, *Journal of Complexity*, vol. 4, 193-215 (1988).
2. Baum, E.B., and D. Haussler: What size net gives valid generalization?, *Neural Computation*, vol.1, 151-160 (1989).
3. Blumer, A., A. Ehrenfeucht, D. Haussler, and M. K. Warmuth : Learnability and the Vapnik-Chervonenkis Dimension, *Journal of the ACM*, vol.36, no.4, 929-965 (Oct. 1989).
4. Cover T. M. : Geometrical and statistical properties of systems of linear inequalities with applications in pattern recognition, *IEEE Transactions on Electronic Computers*, vol. 14, 326-334 (1965).
5. Lupanov, O. B. : Circuits using threshold elements, *Soviet Physics – Doklady*, vol.17, no.2, 91-93 (1972), (translated from *Doklady Akademii Nauk SSSR*, vol. 202, no.6, 1288-1291 (Feb. 1972)).
6. Maass, W. : Bounds for the computational power and learning complexity of analog neural nets, *preprint* (1992).
7. Nechiporuk, E.I. : The synthesis of networks from threshold elements, *Automation Express*, vol.7, no.1, 35-39; no.2, 27-32 (1964), (translated from *Probl. Kibern.*, no.11, 49-62 (April 1964)).
8. Sakurai, A. : n–h–1 networks store no less n·h+1 examples but sometimes no more, *Proceedings of IJCNN92*, III-936 - III-941 (June 1992).
9. Sakurai, A. : Tighter Bounds of the VC-Dimension of Three-layer Networks, to be presented at WCNN93 (1993).

On the Sample Complexity of Consistent Learning with One-Sided Error

Eiji Takimoto and Akira Maruoka

Graduate School of Information Sciences
Tohoku University, Sendai 980, Japan

Abstract. Although consistent learning is sufficient for PAC-learning, it has not been found what strategy makes learning more efficient, especially on the sample complexity, i.e., the number of examples required. For the first step towards this problem, only classes that have consistent learning algorithms with one-sided error are considered. A combinatorial quantity called maximal particle sets is introduced, and an upper bound of the sample complexity of consistent learning with one-sided error is obtained in terms of maximal particle sets. For the class of n-dimensional parallel axis rectangles, one of those classes that are consistently learnable with one-sided error, the cardinality of the maximal particle set is estimated and $O(\frac{d}{\varepsilon} + \frac{1}{\varepsilon}\log\frac{1}{\delta})$ upper bound of the learning algorithm for the class is obtained. This bound improves the bounds due to Blumer et al.[2] and meets the lower bound within a constant factor.

1 Introduction

In machine learning, it is one of the most fundamental and important problems to estimate the sample complexity, the number of examples required. When dealing with this problem, we may focus only on the sample size of learning processes without considering computational feasibility. From this point of view, [2] treated learning processes as functions and formalized the most basic learning model under the PAC framework. Employing the notion of VC dimension, they showed both upper and lower bounds of the sample complexity of learning functions. In fact, they showed that for any consistent learning function, sample of size $O(\frac{d}{\varepsilon}\log\frac{1}{\varepsilon} + \frac{1}{\varepsilon}\log\frac{1}{\delta})$ is sufficient and for any learning function, $\Omega(d + \frac{1}{\varepsilon}\log\frac{1}{\delta})$ is necessary, where d denotes the VC dimension of the concept class to be learned. [1] improved the constant factor of the upper bound, and [3] improved the lower bound to $\Omega(\frac{d}{\varepsilon} + \frac{1}{\varepsilon}\log\frac{1}{\delta})$. However, there is an $O(\log\frac{1}{\varepsilon})$ gap between the upper and lower bounds, and it is still open whether any concept class of VC dimension d has a learning function with sample complexity as small as the lower bound.

On the other hand, [4] introduced another learning scheme, called the prediction model, and showed that any concept class of VC dimension d has a prediction strategy with sample complexity $O(\frac{d}{\varepsilon})$ and that this size is optimal within a constant factor. So, it is expected that the learning function transformed from the optimal prediction strategy (this transformation is straightforward) has the optimal sample complexity, though the hypothesis class for the learning algorithm obtained in this way would be different from the target class in general.

It was observed that if a concept class C of VC dimension d is closed under intersections, then the optimal prediction strategy for C is transformed to the consistent learning function for C with one-sided error that, given any sample as its input, produces the minimum hypothesis among those which are consistent with the given sample [5]. We notice here the fact that since C is closed under intersections, the minimum consistent hypothesis is in C, that is, the hypothesis class is the same as the target class. So, consistent learning functions with one-sided error are expected to have optimal sample complexities.

In this paper, in order to take a first step toward this problem, we introduce a combinatorial quantity, which we call maximal particle sets, and show an upper bound of the probability that the learning function produces a bad hypothesis in terms of maximal particle sets. It should be noticed that once the cardinality of a maximal particle set is found, then an upper bound of the sample complexity of the learning function is obtained. Although a general upper bound of the cardinality has not been found yet, we find one for the class of parallel axis rectangles in n dimensional Euclidean space, which is one of the classes that is consistently learnable with one-sided error. As a result, we obtain the upper bound of the sample complexity of the learning function for the class. It turns out that this bound is better than the previously known ones and matches the lower bound within a constant factor.

In Section 2, we give the definition of learning functions that learn concept classes with one-sided error, and characterize the classes by some closure property for concept classes. In Section 3, we introduce a notion of maximal particle sets and compare it with another combinatorial quantity from which the sample complexity of consistent learning functions are obtained in [2]. In Section 4, by keeping track of the proof in [2] mostly, we show the upper bound of the probability that the learning function produces a bad hypothesis in terms of maximal particle sets. In Section 5, we show an upper bound of the cardinality of the maximal particle set for the class of n-dimensional parallel axis rectangles. We also apply this bound to the result in Section 4 to obtain an optimal upper bound of the sample complexity of the learning function for the class.

2 Preliminaries

Let X denote a set. A subset of X is called a concept. A concept r represents the characteristic function of r as well that, for any $x \in X$, takes value 1 if $x \in r$ and 0 otherwise. An example of r is a pair of x and its value $r(x)$ in $\{0,1\}$ for some $x \in X$. Let C and H be concept classes. In what follows, C is called the target class and H is called the hypothesis class. Throughout the paper, we will assume that, when the domain X is real-valued, C and H are well-behaved in the measure-theoretic sense defined in [2]. A sequence of m examples of $c \in C$, $\langle(x_1, c(x_1)), \ldots, (x_m, c(x_m))\rangle$, is called a sample of size m of c, or simply an m-sample of c. A sample of c, $\langle(x_1, c(x_1)), \ldots, (x_m, c(x_m))\rangle$, is also denoted $\mathrm{sam}_c(\bar{x})$, where $\bar{x} = (x_1, \ldots, x_m) \in X^m$. The sample space of C, denoted S_C, is the set of all samples of concepts in C. That is,

$$S_C = \bigcup_{c \in C} \bigcup_{m \geq 1} \bigcup_{\bar{x} \in X^m} \{\mathrm{sam}_c(\bar{x})\}.$$

A function $A : S_C \to H$ which maps any sample of C to a concept in H is called a learning function for C by H. If A uses C as its hypothesis class (i.e., A is a learning function for C by C), then we call A a learning function for C. If $h = A(\mathrm{sam}_c(\bar{x}))$, then we call c a target concept and h a hypothesis (produced by A for $\mathrm{sam}_c(\bar{x})$). A hypothesis h is consistent with $\mathrm{sam}_c(\bar{x})$ if $h(x_i) = c(x_i)$ for any $x_i \in \bar{x}$. A learning function A for C (by H) is consistent if, for any sample in S_C, A produces a hypothesis consistent with the given sample.

Definition 1. A learning function A for C (by H) learns C with one-sided error if A always produces a hypothesis contained in the target concept. That is, for any sample $\mathrm{sam}_c(\bar{x})$ in S_C, $A(\mathrm{sam}_c(\bar{x})) \subseteq c$.

In Proposition 4 below we shall characterize a concept class that is consistently learnable with one-sided error. A similar characterization was also made in [6]. Before proceeding to the proposition, we introduce a closure property for concept classes.

Definition 2. For $\mathrm{sam}_c(\bar{x})$ in S_C, the set of all concepts in C that are consistent with $\mathrm{sam}_c(\bar{x})$ is denoted $\mathrm{CONS}_c(\bar{x})$. If there exists the smallest concept in $\mathrm{CONS}_c(\bar{x})$, then the concept is denoted $\hat{h}_{c,\bar{x}}$. Note that $\hat{h}_{c,\bar{x}} = \bigcap_{h \in \mathrm{CONS}_c(\bar{x})} h$ and $\hat{h}_{c,\bar{x}} \in C$.

Definition 3. C is closed under intersections over CONS if, for any $\mathrm{sam}_c(\bar{x})$ in S_C, there exists $\hat{h}_{c,\bar{x}}$.

Note that if C is closed under (infinitely many) intersections, then C is closed over CONS, too. On the other hand, if C is only closed under finitely many intersections (i.e., for any $c_1, c_2 \in C$, $c_1 \cap c_2 \in C$), C is not necessarily closed over CONS. See the next example.

Example 1. Let X be a real line (the set of real numbers) and let C be the class of open intervals over X. That is, $C = \{(a,b) | a, b \in X\}$. Clearly, C is closed under finitely many intersections. Let $c = (0,1)$ and $\bar{x} = (0.2, 0.5, 0.8)$. Since $0.2, 0.5, 0.8 \in c$, $\mathrm{sam}_c(\bar{x}) = \langle (0.2, 1), (0.5, 1), (0.8, 1) \rangle$. Although $(0.2 - \varepsilon, 0.8 + \varepsilon) \in \mathrm{CONS}_c(\bar{x})$ for any $\varepsilon > 0$, $[0.2, 0.8] \notin \mathrm{CONS}_c(\bar{x})$. So, there does not exist the smallest concept in $\mathrm{CONS}_c(\bar{x})$.

Proposition 4. Let A be a learning function for C. Then, C is closed under intersections over CONS and for any $\mathrm{sam}_c(\bar{x}) \in S_C$, $A(\mathrm{sam}_c(\bar{x})) = \hat{h}_{c,\bar{x}}$ if and only if A is a consistent learning function with one-sided error for C.

Proof. Sufficiency is trivial. We show only the necessity part of the proposition.

Suppose that A is a consistent learning function with one-sided error for C. Fix a sample $\text{sam}_c(\bar{x})$ and put $h = A(\text{sam}_c(\bar{x}))$. Since the hypothesis class used by A is C ($h \in C$) and h is consistent with $\text{sam}_c(\bar{x})$, we have

$$h \in \text{CONS}_c(\bar{x}). \tag{1}$$

On the other hand, for any $h' \in \text{CONS}_c(\bar{x})$, since $\text{sam}_{h'}(\bar{x}) = \text{sam}_c(\bar{x})$, $A(\text{sam}_{h'}(\bar{x})) = h$. Moreover, since A learns C with one-sided error, $A(\text{sam}_{h'}(\bar{x})) \subseteq h'$. Thus, we have

$$h \subseteq h'. \tag{2}$$

From (1) and (2), h is the smallest concept in $\text{CONS}_c(\bar{x})$. □

We have observed in Example 1 that the set of open intervals is not closed under intersections over CONS. So, the above proposition says that this class is not learnable with one-sided error (by open intervals as hypotheses).

3 Maximal particle sets

In this section, we introduce a notion of maximal particle sets, which plays an important role in obtaining an upper bound of the sample complexity of a consistent learning function that learns target concepts with one-sided error.

Throughout the rest of the paper, $S \subseteq X$ is assumed to denote a multiset. The intersection and the difference between a multiset and a non-multiset are defined as follows. Let S be a multiset and T be a non-multiset. Then $S \cap T$ denotes the set of elements in T that also belong to S ($S \cap T = \{x \in T | x \in S\}$), and $S - T$ denotes the set of elements in S that don't belong to T ($S - T = \{x \in S | x \notin T\}$). For example, letting $S = \{1, 1, 2, 3, 3\}$ and $T = \{1, 2, 4\}$, we have $S \cap T = \{1, 2\}$ and $S - T = \{3, 3\}$. Note that $S \cap T$ is a non-multiset, whereas $S - T$ is a multiset.

Definition 5 (Particle sets). For $S \subseteq X$, the particle set of S by C, denoted $\Pi_C(S)$, is the set of all subsets of S that can be obtained by intersecting S with a concept in C, that is, $\Pi_C(S) = \{S \cap c \mid c \in C\}$. An element of $\Pi_C(S)$ is called a particle (of S by C). Furthermore, for $l \geq 0$, $\Pi_C(S, l)$ denotes the set of all particles T such that there exist at least l elements in S that do not belong to T. That is, $\Pi_C(S, l) = \{T \in \Pi_C(S) \mid |S - T| \geq l\}$.

Note that $\Pi_C(S) = \Pi_C(S, 0)$. Note also that since $c \in C$ is a non-multiset, any particle of S by C is a non-multiset as well. This implies that $\Pi_C(S)$ depends only on the distinct elements in S. So, particle sets of S can be defined to be ones with S being non-multiset. In this sense, in [2], $\Pi_C(S)$ was defined so that S is assumed to be a non-multiset. On the other hand, in defining $\Pi_C(S, l)$, we have to take S to be a multiset in general, because $\Pi_C(S, l)$ depends also on the occurrences of each element in S.

Definition 6 (Maximal particle sets). For $l \geq 0$, the l-maximal particle set of S by C, denoted $\Lambda_C(S, l)$, is the set of all particles in $\Pi_C(S, l)$ that are maximal. That is,

$$\Lambda_C(S, l) = \{T \in \Pi_C(S, l) \mid \forall T' \in \Pi_C(S, l), T \not\subset T'\}.$$

Note that for any T in $\Pi_C(S)$, $T \in \Lambda_C(S, |S - T|)$ holds.

Example 2. Let X be a real line and C be the class of closed intervals over X. Let $S = \{0.2, 0.2, 0.5, 0.8\}$. Then

$$\Pi_C(S) = \{\phi, \{0.2\}, \{0.5\}, \{0.8\}, \{0.2, 0.5\}, \{0.5, 0.8\}, \{0.2, 0.5, 0.8\}\},$$
$$\Pi_C(S, 2) = \{\phi, \{0.2\}, \{0.5\}, \{0.8\}, \{0.5, 0.8\}\},$$
$$\Lambda_C(S, 2) = \{\{0.2\}, \{0.5, 0.8\}\}.$$

Note that $S - \{0.5, 0.8\} = \{0.2, 0.2\}$, whose cardinality is greater than or equal 2. This is why $\{0.5, 0.8\} \in \Pi_C(S, 2)$.

We need some additional definitions for combinatorial quantities which will be used to show upper bounds of the sample complexities of learning functions.

Definition 7. For $m \geq 1$, $\Pi_C(m)$ denotes the cardinality of the largest particle set of S by C such that $|S| = m$. For $m \geq 1$ and $l \geq 0$, $\Lambda_C(m, l)$ denotes the cardinality of the largest l-maximal particle set of S by C such that $|S| = m$.

Example 3. Let X be a real line and C be the class of closed intervals over X. Let S be a distinct element set of size m. Suppose that $S = \{a_1, \ldots, a_m\}$, where $a_1 < \cdots < a_m$. Then, clearly $\Pi_C(S) = \{\phi\} \cup \{\{a_i, \ldots, a_j\} \mid 1 \leq i \leq j \leq m\}$ and $\Lambda_C(S, l) = \{\{a_{i+1}, \ldots, a_{m-l+i}\} \mid 0 \leq i \leq l\}$, and thus, $|\Pi_C(S)| = \frac{m(m+1)}{2} + 1$, and $|\Lambda_C(S, l)| = l + 1$ (if $l < m$). It is easily shown that whatever S is of size m, $|\Pi_C(S)|$ and $|\Lambda_C(S, l)|$ are bounded above by the values obtained in the right-hand sides in the equalities above, respectively. Therefore, $\Pi_C(m) = \frac{m(m+1)}{2} + 1$, and $\Lambda_C(m, l) \leq l + 1$. Note that the upper bound of $\Lambda_C(m, l)$ does not depend on m.

[2] showed for a concept class H an upper bound of $\Pi_H(m)$ in terms of the VC dimension of H.

Proposition 8 [2]. Let H be a concept class whose VC dimension is d. Then, for any $m \geq 1$,

$$\Pi_H(m) \leq \sum_{i=0}^{d} \binom{m}{i} \leq \left(\frac{em}{d}\right)^d.$$

The estimation for $\Lambda_H(m, l)$ corresponding to the above proposition has not been succeeded (of course, trivially $\Lambda_H(m, l) \leq \Pi_H(m)$ holds). We conjecture that if H is closed under intersections (over CONS), then $\Lambda_H(m, l)$ is significantly smaller than $\Pi_H(m)$ (unless $\Pi_H(m)$ is small).

4   Sample Complexity of a Consistent Learning Function with One-Sided Error

In this section, we discuss the sample complexity of a consistent learning function that learns C with one-sided error.

Let A be a learning function for C by H. Let D be a probability distribution over X and (X, D) denote the probability space. In what follows, we shall assume that for any $m \geq 1$, $\bar{x} \in X^m$ is a random variable on (X^m, D^m). For $c \in C$, $m \geq 1$ and $0 < \varepsilon \leq 1$, $\mathrm{err}_{A,c,D}(m)$ denotes the probability that the error of hypothesis $A(\mathrm{sam}_c(\bar{x}))$ is not smaller than ε. That is, $\mathrm{err}_{A,c,D}(m) = \mathrm{Pr}_{D^m}\left(D(c\Delta A(\mathrm{sam}_c(\bar{x}))) \geq \varepsilon\right)$, where Δ denotes the symmetric difference. Furthermore, $\mathrm{err}_{A,\varepsilon}(m)$ denotes the supremum of $\mathrm{err}_{A,c,\varepsilon,D}(m)$ over all $c \in C$ and D. The sample complexity of A, denoted $m_A(\varepsilon, \delta)$, is defined as the minimum sample size m such that $\mathrm{err}_{A,\varepsilon}(m)$ is at most δ. That is, $m_A(\varepsilon, \delta) = \min_{m \geq 1}\left\{m \mid \mathrm{err}_{A,\varepsilon}(m) \leq \delta\right\}$.

[2] gave an upper bound of $\mathrm{err}_{A,\varepsilon}(m)$ in terms of the cardinality of the largest particle sets.

Theorem 9 [2]. Let the VC dimension of H be d and A be a consistent learning function for C by H. Then, $\mathrm{err}_{A,\varepsilon}(m) \leq 2 \cdot \Pi_H(2m) \cdot 2^{-\frac{\varepsilon m}{2}}$.

This theorem together with Proposition 8 gives an upper bound of the sample complexity of a consistent learning function.

Theorem 10 [2]. Let the VC dimension of H be d and A be a consistent learning function for C by H(thus, the VC dimension of C must be at most d). Then,

$$m_A(\varepsilon, \delta) \leq \max\left\{\frac{4}{\varepsilon}\log\frac{2}{\delta}, \frac{8d}{\varepsilon}\log\frac{13}{\varepsilon}\right\}.$$

On the other hand, [3] gave an information-theoretic lower bound of the sample complexity of any learning function.

Theorem 11 [3]. Let the VC dimension of C be d. Then, for any hypothesis class H and learning function A for C by H, for sufficiently small ε and δ,

$$m_A(\varepsilon, \delta) \geq \max\left\{\frac{1-\varepsilon}{\varepsilon}\ln\frac{1}{\delta}, \frac{d-1}{32\varepsilon}\right\}.$$

Note that there is an $O(\frac{1}{\varepsilon})$ gap between the upper bound and the lower bound.

Now, we exhibit the main theorem of this paper which gives an upper bound of $\mathrm{err}_{A,\varepsilon}(m)$ for consistent learning function A that learns C with one-sided error.

Theorem 12. Let A be a consistent learning function with one-sided error for C by H. Then,

$$\mathrm{err}_{A,\varepsilon}(m) \leq 2 \cdot \max_{|S| \leq 2m} \sum_{T \in \Lambda_H(S, \frac{\varepsilon m}{2})} 2^{-(|S-T|)}.$$

Since $|S-T| \geq \frac{\varepsilon m}{2}$ for any $T \in \Lambda_H(S, \frac{\varepsilon m}{2})$, we immediately have the following corollary. This is the same form as Theorem 9 except that $\Pi_H(2m)$ is replaced by $\Lambda_H(2m, \frac{\varepsilon m}{2})$.

Corollary 13. Let A be a consistent learning function with one-sided error for C by H. Then,

$$\text{err}_{A,\varepsilon}(m) \leq 2 \cdot \Lambda_H(2m, \frac{\varepsilon m}{2}) \cdot 2^{-\frac{\varepsilon m}{2}}.$$

When C is closed under intersections over CONS, then this corollary yields an upper bound of the sample complexity of the consistent learning function that learns C with one-sided error. If $\Lambda_C(m, l)$ is sufficiently small, say at most $2^{\sigma l + O(d)}$ for some $\sigma < 1$, then it turns out that the learning function has an optimal sample complexity.

Below, we give the proof of Theorem 12.

Proof of Theorem 12. Let A be a consistent learning function with one-sided error for C by H. Take a target concept $c \in C$ and a probability distribution D on X arbitrarily. Let $r(\bar{x}) = c \Delta A(\text{sam}_c(\bar{x}))$. Note that $\text{err}_{A,c,\varepsilon,D}(m) = \text{Pr}_{D^m}(D(r(\bar{x})) \geq \varepsilon)$. For $r \subseteq X$ and $\bar{x} \in X^m$, the multiset of the components of $\bar{x}$ that belong to r is denoted $r \cap \bar{x}$, that is, if $\bar{x} = \{x_1, \ldots, x_m\}$, then $r \cap \bar{x} = \{x_i | x_i \in r, 1 \leq i \leq m\}$. For vector $\bar{z}$ in X^{2m} of length $2m$, the vectors composed of the first (last, resp.) half part of $\bar{z}$ are denoted $\bar{z}_l$ ($\bar{z}_r$, resp.). Note that these vectors are of length m and $\bar{z}$ is the concatenation of $\bar{z}_l$ and $\bar{z}_r$.

For $\bar{z} \in X^{2m}$, $J_\varepsilon^{2m}(\bar{z})$ denotes the event that A, when given m-sample of c induced by the first half part of $\bar{z}$, produces h such that the error region of h ($c \Delta h$) has probability at least ε and contains at least $\frac{\varepsilon m}{2}$ components of the last half part of $\bar{z}$. That is,

$$J_\varepsilon^{2m}(\bar{z}) \equiv D(r(\bar{z}_l)) \geq \varepsilon \text{ and } |r(\bar{z}_l) \cap \bar{z}_r| \geq \frac{\varepsilon m}{2}.$$

Assuming $\bar{z}$ to be a random variable according to D^{2m}, we have

$$\text{Pr}_{D^{2m}}(J_\varepsilon^{2m}(\bar{z})) = \text{Pr}_{D^m}(D(r(\bar{z}_l)) \geq \varepsilon) \cdot \text{Pr}_{D^m}\left(|r(\bar{z}_l) \cap \bar{z}_r| \geq \frac{\varepsilon m}{2} \;\middle|\; D(r(\bar{z}_l)) \geq \varepsilon\right)$$

$$\geq \frac{1}{2} \cdot \text{err}_{A,c,\varepsilon,D}(m),$$

because the conditional probability means, in m independent Bernoulli trials each with probability at least ε of success, the probability of more than $\varepsilon m/2$ successes occurring, which is well known to be more than $1/2$. Thus, $\text{err}_{A,c,\varepsilon,D}(m) \leq 2 \cdot \text{Pr}_{D^{2m}}(J_\varepsilon^{2m}(\bar{z}))$.

For each i, $1 \leq i \leq (2m)!$, let σ_i be a distinct permutation of the indices $1, \ldots, 2m$. It is clear that

$$\text{Pr}_{D^{2m}}(J_\varepsilon^{2m}(\sigma_i(\bar{z}))) = \text{Pr}_{D^{2m}}(J_\varepsilon^{2m}(\bar{z}))$$

for all σ_i, and thus,

$$\Pr{}_{D^{2m}}\left(J_\varepsilon^{2m}(\bar{z})\right) = \frac{1}{(2m)!} \sum_{i=1}^{(2m)!} \Pr{}_{D^{2m}}\left(J_\varepsilon^{2m}(\sigma_i(\bar{z}))\right).$$

Therefore,

$$\Pr{}_{D^{2m}}\left(J_\varepsilon^{2m}(\bar{z})\right) = \frac{1}{(2m)!} \sum_{i=1}^{(2m)!} \int_{X^{2m}} \mathbf{1}\left(J_\varepsilon^{2m}(\sigma_i(\bar{z}))\right) dD^{2m}$$

$$= \int_{X^{2m}} \frac{1}{(2m)!} \sum_{i=1}^{(2m)!} \mathbf{1}\left(J_\varepsilon^{2m}(\sigma_i(\bar{z}))\right) dD^{2m},$$

where $\mathbf{1}(E)$ is the characteristic function which takes value 1 if event E holds, and 0 otherwise. So, it suffices to show that for any (fixed) $\bar{z} \in X^{2m}$,

$$\frac{1}{(2m)!} \sum_{i=1}^{(2m)!} \mathbf{1}\left(J_\varepsilon^{2m}(\sigma_i(\bar{z}))\right) \le \sum_{T \in \Lambda_H(S; \frac{\varepsilon m}{2})} 2^{-(|S-T|)},$$

where S denotes the multiset of positive examples in $\bar{z}$. That is, $S = c \cap \bar{z}$. Note that since $|S| \le 2m$, we have

$$\sum_{T \in \Lambda_H(S, \frac{\varepsilon m}{2})} 2^{-(|S-T|)} \le \max_{|S| \le 2m} \sum_{T \in \Lambda_H(S, \frac{\varepsilon m}{2})} 2^{-(|S-T|)}.$$

Suppose that $J_\varepsilon^{2m}(\sigma_i(\bar{z}))$ holds for some i. In other words, for $\bar{x} = \sigma_i(\bar{z})_l$ and $\bar{y} = \sigma_i(\bar{z})_r$, we have $D(r(\bar{x})) \ge \varepsilon$ and $|r(\bar{x}) \cap \bar{y}| \ge \frac{\varepsilon m}{2}$. Since A is a consistent learning function with one-sided error, the event

$$\exists \hat{h} \in H, \; h \subseteq c, \; (c - h) \cap \bar{x} = \phi, \; |(c - h) \cap \bar{y}| \ge \frac{\varepsilon m}{2}$$

occurs(this actually occurs for $h = A(\mathrm{sam}_c(\bar{x})))$. Let $h \cap S$ be denoted $T \in \Pi_H(S)$. Then, since $S - T = (c - h) \cap \bar{z}$ can be partitioned into $(c - h) \cap \bar{x}$ and $(c - h) \cap \bar{y}$, the above event implies

$$\exists T \in \Pi_H(S), \; |S - T| \ge \frac{\varepsilon m}{2}, \; (S - T) \cap \bar{x} = \phi,$$

and this also implies

$$\exists T \in \Pi_H(S, \frac{\varepsilon m}{2}), \; (S - T) \cap \bar{x} = \phi.$$

Moreover, since $(S - T) \cap \bar{x} = \phi$ implies $(S - T') \cap \bar{x} = \phi$ for any $T' \supseteq T$, the above event implies

$$\exists T \in \Lambda_H(S, \frac{\varepsilon m}{2}), \; (S - T) \cap \bar{x} = \phi.$$

This result says that the following inequality holds:

$$\frac{1}{(2m)!} \sum_{i=1}^{(2m)!} \mathbf{1}\big(J_\varepsilon^{2m}(\sigma_i(\bar{z}))\big) \leq \frac{1}{(2m)!} \sum_{i=1}^{(2m)!} \mathbf{1}\left(\exists T \in \Lambda_H(S, \tfrac{\varepsilon m}{2}), \ (S-T) \cap \sigma_i(\bar{z})_l = \phi\right)$$

$$\leq \sum_{T \in \Lambda_H(S, \frac{\varepsilon m}{2})} \frac{1}{(2m)!} \sum_{i=1}^{(2m)!} \mathbf{1}\big((S-T) \cap \sigma_i(\bar{z})_l = \phi\big).$$

The summand of T means the fraction of σ_i such that all elements in $S - T$ (there exist $|S - T|$ elements) appear in the last half part of $\sigma_i(\bar{z})$. So, this is bounded above by

$$\frac{\binom{m}{|S-T|}}{\binom{2m}{|S-T|}} = \frac{m(m-1)\cdots(m-(|S-T|+1))}{2m(2m-1)\cdots(2m-(|S-T|+1))} \leq \underbrace{\frac{1}{2}\cdot\frac{1}{2}\cdots\frac{1}{2}}_{|S-T|} = \left(\frac{1}{2}\right)^{|S-T|}.$$

Thus, it follows that

$$\frac{1}{(2m)!} \sum_{i=1}^{(2m)!} \mathbf{1}\big(J_\varepsilon^{2m}(\sigma_i(\bar{z}))\big) \leq \sum_{T \in \Lambda_H(S, \frac{\varepsilon}{2})} 2^{-(|S-T|)}.$$

$\square$

5 Sample Complexity of the Learning Function for Parallel Axis Rectangles

When C is consistently learnable with one-sided error, Corollary 13 says that an upper bound of $\Lambda_C(m, l)$ yields an upper bound of the sample complexity of a consistent learning function that learns C with one-sided error. Though any non-trivial upper bound of $\Lambda_C(m, l)$ is unknown in general, we show one for the class of parallel axis rectangles, which is one of the classes that is consistently learnable with one-sided error. Then, from this result, we give the upper bound of the sample complexity of the learning function for the class, which turns out to meet a lower bound within a constant factor.

Definition 14. A parallel axis rectangle c in n Euclidean dimensional space is an direct product of n closed intervals over a real line X. That is, c is represented as $c = [l_1, r_1] \times \cdots \times [l_n, r_n]$ for some n closed intervals $[l_1, r_1], \ldots, [l_n, r_n]$. The class of parallel axis rectangles in n dimensional space is denoted C_n.

An upper bound of the sample complexity for C_n was given in [2] by providing a learning function A for C_n. A was defined as the learning function that, for any $\mathrm{sam}_c(\bar{x}) \in S_{C_n}$, produces hypothesis $\hat{h}_{c,\bar{x}}$, the minimum hypothesis among those which are consistent with $\mathrm{sam}_c(\bar{x})$. Note that since C_n is closed under intersections, such $\hat{h}_{c,\bar{x}}$ exists.

Theorem 15 [2]. Let A be the learning function defined above. Then,

$$m_A(\varepsilon, \delta) \leq \frac{2n}{\varepsilon} \ln \frac{2n}{\delta}.$$

Since the VC dimension of C_n is $2n$, Theorem 10 gives another upper bound of $m_A(\varepsilon, \delta)$ as follows:

$$m_A(\varepsilon, \delta) \leq \max \left\{ \frac{4}{\varepsilon} \log \frac{2}{\delta}, \frac{16n}{\varepsilon} \log \frac{13}{\varepsilon} \right\}.$$

Either of these two upper bounds on $m_A(\varepsilon, \delta)$ can be better than the other depending on the values of ε and δ.

Now we proceed to show our bound. We need some preliminaries beforehand. For vector $x \in X^n$, the i-th component of x is denoted x_i.

Definition 16. Let $S \subseteq X^n$ be a multiset and $V \subseteq X$ be a non-multiset. Define multisets $S_V \subseteq X^{n-1}$ and $S' \subseteq X$ as follows:

$$S_V = \bigcup_{x \in S} \{(x_1, \ldots, x_{n-1}) \mid x_n \in V\},$$

$$S' = \bigcup_{x \in S} \{x_n\}.$$

Note that since, based on the elements of S, S_V and S' become multisets, S is partitioned into two sets, $\{x \in S \mid x_n \notin V\}$ and $\{x \in S \mid x_n \in V\}$ whose cardinalities are $|S' - V|$ and $|S_V|$, respectively.

Lemma 17. Let $S \subseteq X^n$ be a multiset and $T_1 \subseteq S_{T_2}$ and $T_2 \subseteq S'$ be non-multisets. Let $T = \{x \mid x \in S; (x_1, \ldots, x_{n-1}) \in T_1, x_n \in T_2\}$ (this is re-written as $(T_1 \times T_2) \cap S$). Then,

$$|S - T| = |S_{T_2} - T_1| + |S' - T_2|.$$

Proof. S is partitioned into $\{x \in S \mid x_n \notin T_2\}$ and $\{x \in S \mid x_n \in T_2\}$. Clearly the first set does not contain T and of size $|S' - T_2|$. The difference set between the second set and T is $\{x \in S \mid (x_1, \ldots, x_{n-1}) \in S_{T_2} - T_1, x_n \in T_2\}$, which is of size $|S_{T_2} - T_1|$. $\square$

Lemma 18. Let $S \subseteq X^n$ be a multiset. Then, for any $T_1 \in \Pi_{C_{n-1}}(S_{T_2}, l_1)$ and $T_2 \in \Pi_{C_1}(S', l_2)$, $(T_1 \times T_2) \cap S \in \Pi_{C_n}(S, l_1 + l_2)$.

Proof. From the definition of particle sets, there are $c_1 \in C_{n-1}$ and $c_2 \in C_1$ such that $T_1 = c_1 \cap S_{T_2}$ and $T_2 = c_2 \cap S'$. Hence,

$$(T_1 \times T_2) \cap S = \{x \mid (x_1, \ldots, x_{n-1}) \in T_1, x_n \in T_2, x \in S\}$$
$$= \{x \mid (x_1, \ldots, x_{n-1}) \in c_1, x_n \in c_2, x \in S\}$$
$$= \{x \in c_1 \times c_2 \mid x \in S\}$$
$$= (c_1 \times c_2) \cap S.$$

Since $c_1 \times c_2 \in C_n$, $(T_1 \times T_2) \cap S \in \Pi_{C_n}(S)$. Furthermore, Lemma 17 says that $|S - T| = |S_{T_2} - T_1| + |S' - T_2| \geq l_1 + l_2$, completing the proof of lemma. $\square$

Theorem 19. For any $n, m \geq 1$ and any $l \geq 0$,

$$\Lambda_{C_n}(m, l) \leq \binom{l + 2n - 1}{2n - 1}.$$

Proof. We first show that the following recursively defined inequalities hold: For any $m \geq 1$ and $l \geq 0$,

$$\Lambda_{C_n}(m, l) \leq \begin{cases} l + 1 & \text{if } n = 1, \\ \sum_{k=0}^{l} \Lambda_{C_{n-1}}(m - k, l - k) \cdot \Lambda_{C_1}(m, k) & \text{if } n \geq 2. \end{cases}$$

In the case $n = 1$, since C_n is the class of closed intervals over a real line, Example 3 says that $\Lambda_{C_n}(m, l) \leq l + 1$.

Then, we verify the inequality in the case $n \geq 2$. Fix a multiset $S \subseteq X^n$ with $|S| = m$, and fix an l-maximal particle T in $\Lambda_{C_n}(S, l)$ arbitrarily. Let T be represented as $c \cap S$ for some $c = [l_1, r_1] \times \cdots \times [l_n, r_n]$. Let $c_1 = [l_1, r_1] \times \cdots \times [l_{n-1}, r_{n-1}]$ and $c_2 = [l_n, r_n]$, and let $T_1 = c_1 \cap S_{T_2}$ and $T_2 = c_2 \cap S'$. Note that $c = c_1 \times c_2$, where $c_1 \in C_{n-1}$ and $c_2 \in C_1$. Then we show the following three claims hold.

Claim 1.

$$T = (T_1 \times T_2) \cap S.$$

This is easily verified and the proof is omitted.

Claim 2. There exists a $0 \leq k \leq l$ such that

$$T_2 \in \Lambda_{C_1}(S', k).$$

$T_2 \in \Pi_{C_1}(S')$ since $T_2 = c_2 \cap S'$ and $c_2 \in C_1$. So, letting $k' = |S' - T_2|$, we have $T_2 \in \Lambda_{C_1}(S', k')$. Hence, if $k' \leq l$, the claim holds for $k = k'$. For the case $k' > l$, we show that $T_2 \in \Lambda_{C_1}(S', l)$ by contradiction. Suppose that $T_2 \notin \Lambda_{C_1}(S', l)$, i.e., there exists a $T_2' \in \Pi_{C_1}(S', l)$ such that $T_2' \supsetneq T_2$. Note that $S_{T_2'} \supsetneq S_{T_2}$. Let $c_1' \subseteq C_{n-1}$ be a sufficiently large parallel axis rectangle in $n - 1$ dimension space such that $c_1' \supset S_{T_2'}$. Let $T_1' = c_1' \cap S_{T_2'} \ (= S_{T_2'})$. Then, clearly $T_1' \supseteq T_1$ and $T_1' \in \Pi_{C_{n-1}}(S_{T_2'}, 0)$. Therefore, we have $T' = (T_1' \times T_2') \cap S \supsetneq T$, and by Lemma 18, $T' \in \Pi_{C_n}(S, l)$. This contradicts the fact that T is an l-maximal particle of S by C_n.

Claim 3.

$$T_1 \in \Lambda_{C_{n-1}}(S_{T_2}, l - k).$$

By Lemma 17, $|S_{T_2} - T_1| = |S - T| - |S' - T_2|$. So, if $|S' - T_2| = k(\leq l)$, then $|S_{T_2} - T_1| \geq l - k$. If $|S' - T_2| > l$, then the proof of Claim 2 says that $k = l$, which implies $|S_{T_2} - T_1| \geq l - k$ trivially. Thus, in either case we have $|S_{T_2} - T_1| \geq l - k$. Hence, $T_1 \in \Pi_{C_{n-1}}(S_{T_2}, l - k)$. Next we show that T_1 is a $(l-k)$-maximal particle of S_{T_2} by contradiction. Suppose that this is not the case, i.e., there exists a $T_1' \in \mathrm{II}_{C_{n-1}}(S_{T_2}, l - k)$ such that $T_1' \supsetneq T_1$. This implies that there exists an $x \in S$ such that $x_n \in T_2$, $(x_1, \ldots, x_{n-1}) \in T_1'$ and $(x_1, \ldots, x_{n-1}) \notin T_1$. In other words, $T' = (T_1' \times T_2) \cap S \supsetneq (T_1 \times T_2) \cap S = T$. Moreover, by Lemma 18, we have $T' \in \mathrm{II}_{C_n}(S, l)$. This contradicts the fact that T is an l-maximal particle of S by C_n.

From these claims, we have

$$T \in \left\{ (T_1 \times T_2) \cap S \mid T_1 \in \Lambda_{C_{n-1}}(S_{T_2}, l - k), T_2 \in \Lambda_{C_1}(S', k), 0 \le k \le l \right\}.$$

Therefore,

$$|\Lambda_{C_n}(S, l)| \le \sum_{k=0}^{l} \sum_{T_2 \in \Lambda_{C_1}(S', k)} |\Lambda_{C_{n-1}}(S_{T_2}, l - k)|$$

$$\le \sum_{k=0}^{l} \Lambda_{C_1}(m, k) \cdot \Lambda_{C_{n-1}}(m - k, l - k).$$

Since $S \subseteq X^n$ was arbitrarily chosen, we have

$$\Lambda_{C_n}(m, l) \le \sum_{k=0}^{l} \Lambda_{C_1}(m, k) \cdot \Lambda_{C_{n-1}}(m - k, l - k).$$

So, it suffices to show that for any $n, m \ge 1$ and $l \ge 0$, $\Lambda_{C_n}(m, l) \le a_{n,l}$, where $a_{n,l} = \binom{2n+l-1}{2n-1}$. We prove this by an induction on n.

The case $n = 1$ is trivial.

If $n \ge 2$, then we suppose that the claim holds for $n - 1$. Then, by the recursion obtained above, we have

$$\Lambda_{C_n}(m, l) \le \sum_{k=0}^{l} \Lambda_{C_{n-1}}(m - k, l - k) \cdot \Lambda_{C_1}(m, k)$$

$$= \sum_{k=0}^{l} \Lambda_{C_{n-1}}(m - l + k, k) \cdot \Lambda_{C_1}(m, l - k)$$

$$\le \sum_{k=0}^{l} (l - k + 1) \cdot \binom{k + 2n - 3}{2n - 3}.$$

Putting $\alpha = 2n - 3$ and using the formula $\binom{N-1}{M-1} + \binom{N-1}{M} = \binom{N}{M}$, we have

$$(l - k + 1) \binom{k + \alpha}{\alpha} = (l - k + 1) \left\{ \binom{k + \alpha + 1}{\alpha + 1} - \binom{k + \alpha}{\alpha + 1} \right\}$$

$$= (l - k + 1) \binom{k + \alpha + 1}{\alpha + 1} - (l - k + 2) \binom{k + \alpha}{\alpha + 1} + \binom{k + \alpha}{\alpha + 1}.$$

Applying the formula again to the last term, we have

$$\binom{k+\alpha}{\alpha+1} = \binom{k+\alpha+1}{\alpha+2} - \binom{k+\alpha}{\alpha+2}.$$

Thus, we finally have

$$\Lambda_{C_n}(m,l) \le \binom{l+\alpha+1}{\alpha+1} + \binom{l+\alpha+1}{\alpha+2} = \binom{l+\alpha+2}{\alpha+2} = a_{n,l}.$$

□

This theorem yields a better upper bound of the sample complexity of A than those given previously in literature. Before showing it, we need an estimation of the number of a combination.

Lemma 20 [7]. For any $N \ge 1$ and $0 < p \le 1$, if pN is an integer, then

$$\binom{N}{pN} \le 2^{H(p)N}.$$

Here, $H(p)$ denotes the entropy function $(H(p) = -p \log p - (1-p) \log(1-p))$.

Corollary 21. Let A be a consistent learning function with one-sided error for C_n. Then,

$$m_A(\varepsilon, \delta) \le \max\left\{ \frac{4}{\varepsilon} \log \frac{2}{\delta}, \frac{39n}{\varepsilon} \right\}.$$

Proof. We show that for any $0 < \varepsilon, \delta \le 1$, any target concept $c \in C_n$, any probabilistic distribution D on X^n and any $m \ge m_A(\varepsilon, \delta)$, $\mathrm{err}_{A,c,\varepsilon,D}(m)$ is at most δ. By Corollary 13, we have

$$\mathrm{err}_{A,c,\varepsilon,D}(m) \le 2 \cdot \Lambda_{C_n}(2m, \frac{\varepsilon m}{2}) \cdot 2^{-\frac{\varepsilon m}{2}}.$$

Then, applying Theorem 19, we have

$$\mathrm{err}_{A,c,\varepsilon,D}(m) \le 2 \cdot \binom{\frac{\varepsilon m}{2} + 2n - 1}{2n - 1} \cdot 2^{-\frac{\varepsilon m}{2}}$$

$$\le \left(\binom{\frac{\varepsilon m}{2} + 2n}{2n} \cdot 2^{-\frac{\varepsilon m}{4}} \right) \cdot \left(2 \cdot 2^{-\frac{\varepsilon m}{4}} \right).$$

The second term is at most δ since $m \ge \frac{4}{\varepsilon} \log \frac{2}{\delta}$. Since $m \ge \frac{39n}{\varepsilon}$, Lemma 20 implies that

$$\binom{\frac{\varepsilon m}{2} + 2n}{2n} \cdot 2^{-\frac{\varepsilon m}{4}} \le \binom{21.5n}{2n} \cdot 2^{-9.75n} \le 2^{(H(2/21.5) \cdot 21.5 - 9.75)n},$$

where $H(2/21.5) \cdot 21.5 - 9.75 = -0.15065 \cdots < 0$. Thus, the first term is at most 1.

□

References

1. M. Anthony, N. Biggs, and J. Shawe-Taylor. The learnability of formal concepts. In *Proceedings of the 3rd Workshop on Computational Learning Theory*, pages 246–257, 1990.

2. A. Blumer, A. Ehrenfeucht, D. Haussler, and M. K. Warmuth. Learnability and the Vapnik-Chervonenkis dimension. *Journal of the Association for Computing Machinery*, 36(4):929–965, Aug. 1989.

3. A. Ehrenfeucht, D. Haussler, M. Kearns, and L. G. Valiant. A general lower bound on the number of examples needed for learning. In *Proc. Conference on Learning*, pages 110–120, 1988.

4. D. Haussler, N. Littlestone, and M. Warmuth. Predicting $\{0, 1\}$-functions on randomly drawn points. In *Proceedings of the 29th Annual IEEE Symposium on Foundations of Computer Science*, pages 100–109. IEEE, 1988.

5. E. Maeda. Private communications.

6. B. K. Natarajan. *Machine Learning: A Theoretical Approach*. Morgan Kaufmann, San Mateo, 1991.

7. N. Pippenger. Information theory and complexity of boolean functions. *Mathematical Systems Theory*, 10:129–167, 1977.

Complexity of Computing Vapnik-Chervonenkis Dimension

Ayumi Shinohara

Research Institute of Fundamental Information Science,
Kyushu University 33, Fukuoka 812, Japan
ayumi@rifis.sci.kyushu-u.ac.jp

Abstract. The Vapnik-Chervonenkis (VC) dimension is known to be the crucial measure of the polynomial-sample learnability in the PAC-learning model. This paper investigates the complexity of computing VC-dimension of a concept class over a finite learning domain. We consider a decision problem called the discrete VC-dimension problem which is, for a given matrix representing a concept class $\mathcal{F}$ and an integer K, to determine whether the VC-dimension of $\mathcal{F}$ is greater than K or not. We prove that (1) the discrete VC-dimension problem is polynomial-time reducible to the satisfiability problem of length J with $O(\log^2 J)$ variables, and (2) for every constant C, the satisfiability problem in conjunctive normal form with m clauses and $C \log^2 m$ variables is polynomial-time reducible to the discrete VC-dimension problem. These results can be interpreted, in some sense, that the problem is "complete" for the class of $n^{O(\log n)}$ time computable sets.

1 Introduction

The PAC learnability due to Valiant [8] is to estimate the feasibility of learning a concept probably approximately correctly, from a reasonable amount of examples (polynomial-sample), within a reasonable amount of time (polynomial-time). It is well-known that the Vapnik-Chervonenkis Dimension (VC-dimension) which is a combinatorial parameter of a concept class plays the key role to determine whether the concept class is polynomial-sample learnable or not [2, 3, 5].

This paper settles a complexity issue on VC-dimension of a concept class over a finite learning domain. We remark that the complexity of computing VC-dimension is of independent interest from the polynomial-time learnability, since it is not directly related to the running time of learning algorithms.

Linial et al. [3] showed that the VC-dimension of a concept class over a finite learning domain can be computed in $n^{O(\log n)}$ time, where n is the size of a given matrix which represents the concept class. Nienhuys-Cheng and Polman [6] gave another $n^{O(\log n)}$-time algorithm, although they have not analyzed its running time. On the other hand, Linial et al. [3] pointed out that the decision version of the problem called the *discrete VC-dimension problem* may have some connection with the problem of finding a minimum dominating set in a tournament, which is shown by Megiddo and Vishkin [4] to be a kind of "complete" problem for the class of $n^{O(\log n)}$ time computable sets.

Along this line, we show that the discrete VC-dimension problem is also "complete" for the class of $n^{O(\log n)}$ time computable sets in the same sense. That is, we give the following two reductions: (1) The discrete VC-dimension problem is reducible in polynomial time to the satisfiability problem of a boolean formula of length J with $O(\log^2 J)$ variables. (2) On the other hand, for every constant C, the satisfiability problem in conjunctive normal form with m clauses and $C \log^2 m$ variables is polynomial-time reducible to the discrete VC-dimension problem. Therefore we can interpret that the discrete VC-dimension problem is one of the natural problems which seem to be neither NP-complete, nor in P.

2 Preliminaries

For a matrix M, let M_{ij} denote the element on row i and column j of M, and the size of M is the number of elements in M. The length of a boolean formula ψ, denoted by $|\psi|$, is the total number of variable occurrences in ψ. For a boolean formula ψ, we denote $[\psi, 1] = \psi$ and $[\psi, 0] = \neg\psi$. For any integers $i \geq 1$ and $t \geq 1$, let $b(i, t)$ denote the t-th binary digit of $(i - 1)$, that is, $i = \sum_{t=1}^{\lceil \log i \rceil} 2^{t-1} \cdot b(i, t) + 1$. For example, $b(7, 1) = 0$, $b(7, 2) = 1$, and $b(7, 3) = 1$.

Let U be a finite set called a *learning domain*. We call a subset f of U a *concept*. A concept f can be regarded as a function $f : U \rightarrow \{0, 1\}$, where $f(x) = 1$ if x is in the concept and $f(x) = 0$ otherwise. A *concept class* is a nonempty set $\mathcal{F} \subseteq 2^U$. We represent a concept class $\mathcal{F}$ over a finite learning domain U, by a $|U| \times |\mathcal{F}|$ matrix M with $M_{ij} = f_j(x_i)$. Each column represents a concept in $\mathcal{F}$. For a $\{0, 1\}$-valued matrix M, let $\mathcal{F}_M$ denote the concept class which M represents.

Definition 1. We say that $\mathcal{F}$ *shatters* a set $S \subseteq U$ if for every subset $T \subseteq S$ there exists a concept $f \in \mathcal{F}$ which *cuts T out of S*, i.e., $T = S \cap f$. The *Vapnik-Chervonenkis dimension* of $\mathcal{F}$, denoted by VC-dim($\mathcal{F}$), is the maximum cardinality of a set which is shattered by $\mathcal{F}$.

Lemma 2. *[5] For any concept class $\mathcal{F}$, VC-dim($\mathcal{F}$) $\leq \log|\mathcal{F}|$.*

By this lemma, Linial et al. [3] immediately claimed that a simple algorithm which enumerates all possible sets to be shattered shall terminate in $n^{O(\log n)}$ time, where n is the size of a given matrix.

Definition 3. [3] The *discrete VC-dimension problem* is, given a $\{0, 1\}$-valued matrix M and integer $K \geq 1$, to determine whether VC-dim$\mathcal{F}_M \geq K$ or not.

Definition 4. [4] The classes $\text{SAT}_{\log^k n}$ and $\text{SAT}_{\log^k n}^{\text{CNF}}$ for $k \geq 1$ are defined as follows:

(1) A set L is in $\text{SAT}_{\log^k n}$ if there exists a Turing machine M, a polynomial $p(n)$, and a constant C, such that for every string I of length n, M converts I within $p(n)$ time into a boolean formula Ψ_I (whose length is necessarily less than $p(n)$) with at most $C \log^k n$ variables, so that $I \in L$ if and only if Ψ_I is satisfiable.
(2) The definition of $\text{SAT}_{\log^k n}^{\text{CNF}}$ is essentially the same as that of $\text{SAT}_{\log^k n}$ except that the formula Ψ_I is in conjunctive normal form.

From the definitions, it is easy to see that for each $k \geq 1$,

$$P \subseteq \text{SAT}_{\log^k n}^{\text{CNF}} \subseteq \text{SAT}_{\log^k n} \subseteq NP.$$

3 Discrete VC-dimension Problem is in $SAT_{\log^2 n}$

In this section, we show that the discrete VC-dimension problem is polynomial-time reducible to the satisfiability problem of a boolean formula of length J with $O(\log^2 J)$ variables.

Theorem 5. *The discrete VC-dimension problem is in $SAT_{\log^2 n}$.*

Proof. Let M be an $m \times r$ matrix and K be an integer. By Lemma 2, we can assume that $K \leq \log r$ without loss of generality. Moreover, we can also assume that $m = 2^l$ for some integer l; if $m < 2^l$ for $l = \lceil \log m \rceil$, then we enlarge M by duplicating the last row of M until the row size reaches 2^l. It is easy to see that the size of the enlarged matrix M' is less than twice as large as that of the original matrix M, and VC-dim($\mathcal{F}_{M'}$) = VC-dim($\mathcal{F}_M$).

We now construct a boolean formula Ψ_M which contains $K \cdot l$ variables v_{kt} ($1 \leq k \leq K$, $1 \leq t \leq l$) as follows:

$$\Psi_M = \bigwedge_{s=1}^{2^K} \bigvee_{j=1}^{r} \beta_{sj},$$
$$\beta_{sj} = \bigwedge_{k=1}^{K} [\alpha_{kj}, b(s, k)] \qquad (1 \leq s \leq 2^K, \ 1 \leq j \leq r),$$
$$\alpha_{kj} = \bigvee_{i \in \{i | M_{ij}=1\}} \psi_{ki} \qquad (1 \leq k \leq K, \ 1 \leq j \leq r),$$
$$\psi_{ki} = \bigwedge_{t=1}^{l} [v_{kt}, b(i, t)] \qquad (1 \leq k \leq K, \ 1 \leq i \leq m).$$

Note that the length of Ψ_M is

$$|\Psi_M| \leq l \cdot m \cdot K \cdot r \cdot 2^K$$
$$\leq \log m \cdot m \cdot \log r \cdot r \cdot 2^{\log r}$$
$$< n^2 \log^2 n,$$

where $n = m \cdot r$ is the size of the given matrix M. Also note that Ψ_M can be constructed in polynomial time with respect to n.

Let $U = \{x_1, x_2, \ldots, x_m\}$ be the learning domain and $\mathcal{F}_M = \{f_1, f_2, \cdots, f_r\} \subseteq 2^U$ be the concept class which M represents. We will show that the formula Ψ_M is satisfiable if and only if $\mathcal{F}_M$ shatters a set $S \subseteq U$ of cardinality K.

For a formula ψ and a truth assignment σ to the variables of ψ, let $\sigma(\psi)$ denote the truth value of ψ evaluated under σ. We denote truth values by 0 and 1. For each assignment σ, we define a set $S_\sigma \subseteq U$ as follows:

$$S_\sigma = \{x_{\langle \sigma, k \rangle} \mid 1 \leq k \leq K\}, \quad \text{where } \langle \sigma, k \rangle = \sum_{t=1}^{l} 2^{t-1} \cdot \sigma(v_{kt}) + 1.$$

It should be noticed that the cardinality of S_σ is not always equal to K, since there may be two distinct k_1 and k_2 with $\langle \sigma, k_1 \rangle = \langle \sigma, k_2 \rangle$ in general.

We now show through a sequence of equivalences that an assignment σ satisfies Ψ_M if and only if $|S_\sigma| = K$ and S_σ is shattered by $\mathcal{F}_M$.

First, for any $k \in \{1, \ldots, K\}$ and any $i \in \{1, \ldots, m\}$,

$\sigma(\psi_{ki}) = 1$

$\iff \sigma([v_{kt}, b(i,t)]) = 1$ for each $t \in \{1, \ldots, l\}$

$\iff \sigma(v_{kt}) = \begin{cases} 1 & \text{if } b(i,t) = 1 \\ 0 & \text{if } b(i,t) = 0 \end{cases}$ for each $t \in \{1, \ldots, l\}$

$\iff b(i,t) = \sigma(v_{kt})$ for each $t \in \{1, \ldots, l\}$

$\iff \sum_{t=1}^{l} 2^{t-1} \cdot b(i,t) = \sum_{t=1}^{l} 2^{t-1} \cdot \sigma(v_{kt})$

$\iff i = \langle \sigma, k \rangle.$

Next, for any $k \in \{1, \ldots, K\}$ and any $j \in \{1, \ldots, r\}$,

$\sigma(\alpha_{kj}) = 1$

$\iff \sigma(\psi_{ki}) = 1$ and $M_{ij} = 1$ for some $i \in \{1, \ldots, m\}$

$\iff i = \langle \sigma, k \rangle$ and $x_i \in f_j$

$\iff x_{\langle \sigma, k \rangle} \in f_j.$

For an integer $s \in \{1, \ldots, 2^K\}$, the s-th subset $S_\sigma^{[s]}$ of S_σ is defined by $S_\sigma^{[s]} = \{x_{\langle \sigma, k \rangle} \mid b(s,k) = 1, 1 \leq k \leq K\}$. For example, $S_\sigma^{[1]} = \emptyset$, $S_\sigma^{[5]} = \{x_{\langle \sigma, 3 \rangle}\}$ and $S_\sigma^{[6]} = \{x_{\langle \sigma, 1 \rangle}, x_{\langle \sigma, 3 \rangle}\}$. Then, for any $s \in \{1, \ldots, 2^K\}$ and any $j \in \{1, \ldots, r\}$,

$\sigma(\beta_{sj}) = 1$

$\iff \sigma([\alpha_{kj}, b(s,k)]) = 1$ for each $k \in \{1, \ldots, K\}$

$\iff \sigma(\alpha_{kj}) = \begin{cases} 1 & \text{if } b(s,k) = 1 \\ 0 & \text{if } b(s,k) = 0 \end{cases}$ for each $k \in \{1, \ldots, K\}$

$\iff \begin{cases} x_{\langle \sigma, k \rangle} \in f_j & \text{if } b(s,k) = 1 \\ x_{\langle \sigma, k \rangle} \notin f_j & \text{if } b(s,k) = 0 \end{cases}$ for each $k \in \{1, \ldots, K\}$

$\iff \{x_{\langle \sigma, k \rangle} \mid b(s,k) = 1, 1 \leq k \leq K\} \subseteq f_j$ and
$\{x_{\langle \sigma, k \rangle} \mid b(s,k) = 0, 1 \leq k \leq K\} \subseteq U - f_j$

$\iff S_\sigma \cap f_j = S_\sigma^{[s]}$ and
$b(s, k_1) \neq b(s, k_2)$ implies $\langle \sigma, k_1 \rangle \neq \langle \sigma, k_2 \rangle$ for any $k_1, k_2 \in \{1, \ldots, K\}.$

Finally, we get the following equivalence:

$\sigma(\Psi_M) = 1$

$\iff \sigma(\bigvee_{j=1}^{r} \beta_{sj}) = 1$ for any $s \in \{1, \ldots, 2^K\}$

$\iff$ for each $s \in \{1, \ldots, 2^K\}$,
there exists $f_j \in \mathcal{F}_M$ with $S_\sigma \cap f_j = S_\sigma^{[s]}$ and
$b(s, k_1) \neq b(s, k_2)$ implies $\langle \sigma, k_1 \rangle \neq \langle \sigma, k_2 \rangle$

$\iff k_1 \neq k_2$ implies $\langle \sigma, k_1 \rangle \neq \langle \sigma, k_2 \rangle$, and
for each $s \in \{1, \ldots, 2^K\}$ there exists $f_j \in \mathcal{F}_M$ with $S_\sigma \cap f_j = S_\sigma^{[s]}$

$\iff |S_\sigma| = K$ and S_σ is shattered by $\mathcal{F}_M$

Thus the formula Ψ_M is satisfiable if and only if VC-dim$\mathcal{F}_M \geq K$. □

4 Discrete VC-dimension Problem is $\text{SAT}^{\text{CNF}}_{\log^2 n}$-hard

This section shows that every set in $\text{SAT}^{\text{CNF}}_{\log^2 n}$ is reducible to the discrete VC-dimension problem in polynomial time, i.e., the problem is $\text{SAT}^{\text{CNF}}_{\log^2 n}$-hard.

Theorem 6. *Every* $L \in SAT^{CNF}_{\log^2 n}$ *is polynomial-time reducible to the discrete VC-dimension problem.*

We use the following lemma in the proof of Theorem 6.

Lemma 7. *Let* $\mathcal{F}$ *be a concept class over a learning domain* U, *and* S *be a subset of* U *with* $|S| = d \geq 2$. *If* S *is shattered by* $\mathcal{F}$, *then for any two distinct* x *and* y *in* S, *the number of concepts which contain exactly one of either* x *or* y *is at least* 2^{d-1}, *i.e.,*

$$\left|\{f \in \mathcal{F} \mid f(x) \neq f(y)\}\right| \geq 2^{d-1}.$$

Proof. Let $\mathcal{F}_{\bar{x}y} = \{f \in \mathcal{F} \mid f(x) = 0, f(y) = 1\}$, and $\mathcal{F}_{x\bar{y}} = \{f \in \mathcal{F} \mid f(x) = 1, f(y) = 0\}$. Then $\{f \in \mathcal{F} \mid f(x) \neq f(y)\} = \mathcal{F}_{\bar{x}y} \cup \mathcal{F}_{x\bar{y}}$, and $\mathcal{F}_{\bar{x}y} \cap \mathcal{F}_{x\bar{y}} = \emptyset$. It is easy to see that if S is shattered by $\mathcal{F}$ then the set $S - \{x, y\}$ is shattered by both $\mathcal{F}_{\bar{x}y}$ and $\mathcal{F}_{x\bar{y}}$. By Lemma 2, $|S - \{x, y\}| \leq \log|\mathcal{F}_{\bar{x}y}|$ and $|S - \{x, y\}| \leq \log|\mathcal{F}_{x\bar{y}}|$. Thus $|\mathcal{F}_{\bar{x}y}| \geq 2^{d-2}$ and $|\mathcal{F}_{x\bar{y}}| \geq 2^{d-2}$, which yield $\left|\{f \in \mathcal{F} \mid f(x) \neq f(y)\}\right| = |\mathcal{F}_{\bar{x}y}| + |\mathcal{F}_{x\bar{y}}| \geq 2^{d-2} + 2^{d-2} = 2^{d-1}$. □

Proof of Theorem 6. Let $L \in \text{SAT}^{\text{CNF}}_{\log^2 n}$. Then there is a constant C_L and a polynomial $p_L(n)$ such that every string I of length n can be reduced in $p_L(n)$ time to a boolean formula in conjunctive normal form with at most $C_L \log^2 n$ variables, whose satisfiability coincides with the membership $I \in L$. Therefore we have only to show that, for any C, there is a polynomial-time reduction from the satisfiability problem in conjunctive normal form with at most $C \log^2 n$ variables to the discrete VC-dimension problem. Let $\Psi = E_1 \wedge \cdots \wedge E_m$ $(m \geq 2)$ be a boolean formula where each E_i is a disjunction and the total number of distinct variables occurring in Ψ is not greater than $C \log^2 m$. Without loss of generality, we can assume that m is a power of 2. We can also assume that the number of variables is exactly $C \log^2 m$, and let us rename them, for convenience, with double indices v_{st} $(1 \leq s \leq \log m, 1 \leq t \leq C \log m)$. We first construct a matrix M_Ψ which has $(m^C + 1) \log m$ rows and $m^2 + m(\log m - 1)$ columns, and then prove that VC-dim$(\mathcal{F}_{M_\Psi}) = 2 \log m$ if and only if Ψ is satisfiable.

The learning domain U corresponding to Ψ is defined as $U = X \cup Y$ with $X \cap Y = \emptyset$, where $Y = \{y_u \mid 1 \leq u \leq \log m\}$ and $X = \{x_{sl} \mid 1 \leq s \leq \log m, 1 \leq l \leq m^C\}$. Let $X_s = \{x_{sl} \in X \mid 1 \leq l \leq m^C\}$ for each $s \in \{1, \ldots, \log m\}$, and let $X^{[k]} = \bigcup_{s \in \{s \mid b(k,s)=1\}} X_s$ for each $k \in \{1, \ldots, m\}$. The i-th subset $Y^{[i]}$ of Y is defined by $Y^{[i]} = \{y_u \in Y \mid b(i, u) = 1\}$ for each $i \in \{1, \ldots, m\}$.

The concept class $\mathcal{F} \subseteq 2^U$ is defined as the union of distinct subclasses $F_1, \ldots, F_m$, and G. Here, the structure of G depends only on the number m:

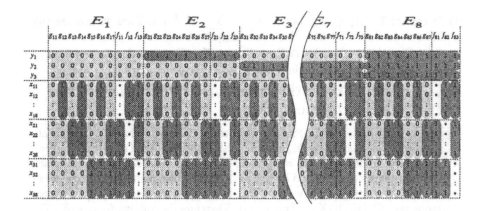

Fig. 1. Structure of the matrix M_Ψ reduced from a boolean formula $\Psi = E_1 \wedge E_2 \wedge \cdots \wedge E_8$ with $C = 1$. In this case, $K = 2\log 8 = 6$. The only elements marked '*' depend on the structure of each clause E_i in Ψ.

$$G = \{g_{ik} \mid 1 \le i \le m, \ 1 \le k \le m-1\}, \quad \text{where } g_{ik} = Y^{[i]} \cup X^{[k]}.$$

On the other hand, each concept in F_i reflects the structure of the clause E_i in Ψ:

$$F_i = \{f_{ij} \mid 1 \le j \le \log m\}, \quad \text{where } f_{ij} = Y^{[i]} \cup (X - X_j) \cup X_j^*(E_i) \quad \text{with}$$

$$X_j^*(E_i) = \left\{ x_{jl} \in X_j \ \middle|\ \begin{array}{l} E_i \text{ contains a positive literal } v_{jt} \text{ with } b(l,t) = 1, \text{ or} \\ E_i \text{ contains a negative literal } \neg v_{jt} \text{ with } b(l,t) = 0 \\ \text{for some } t \in \{1, \ldots, C\log m\} \end{array} \right\}.$$

Figure 1 illustrates the structure of the matrix M_Ψ.

Clearly the cardinality of learning domain, i.e., the row size of the matrix M representing $\mathcal{F}$ is

$$|U| = |X| + |Y| = m^C \cdot \log m + \log m = (m^C + 1)\log m,$$

and the cardinality of the concept class $\mathcal{F}$, i.e., the column size of M is

$$|\mathcal{F}| = |G| + |F_1| + \cdots + |F_m| = m(m-1) + m \cdot \log m.$$

Moreover, it is easy to see that M_Ψ can be constructed in polynomial time with respect to the length of given formula Ψ.

Now we prove that if the formula Ψ is satisfiable then VC-dim$(\mathcal{F}) = 2\log m$. For an assignment σ which satisfies Ψ, we consider the set $S_\sigma = Y \cup X_\sigma$ with

$$X_\sigma = \{x_{s,\langle\sigma,s\rangle} \in X \mid 1 \le s \le \log m\}, \quad \text{where } \langle\sigma,s\rangle = \sum_{t=1}^{C\log m} 2^{t-1} \cdot \sigma(v_{st}) + 1.$$

It is clear that $|S_\sigma| = |Y| + |X_\sigma| = 2\log m$. We will show that S_σ is shattered by $\mathcal{F}$, i.e., for every $T \subseteq S_\sigma$, there exists an $f \in \mathcal{F}$ with $S_\sigma \cap f = T$. Let $i_T = \sum_{y_u \in T \cap Y} 2^{u-1} + 1$. It is easy to see that $i_T \in \{1, \ldots, m\}$ and $T \cap Y = Y^{[i_T]}$. According to $T \cap X_\sigma = X_\sigma$ or not, we have the following two cases.

(1) In case of $T \cap X_\sigma \subsetneq X_\sigma$: Let $k_T = \sum\limits_{x_s,\langle \sigma, s \rangle \in T \cap X_\sigma} 2^{s-1} + 1$. Then we can see that

$k_T \in \{1, \ldots, m-1\}$ and $T \cap X_\sigma = X^{[k_T]} \cap X_\sigma$. Therefore the concept $g_{i_T, k_T} \in G \subseteq \mathcal{F}$ cuts T out of S_σ as follows:

$$g_{i_T, k_T} \cap S_\sigma = (Y^{[i_T]} \cup X^{[k_T]}) \cap (Y \cup X_\sigma) = (Y^{[i_T]} \cap Y) \cup (X^{[k_T]} \cap X_\sigma)$$
$$= (T \cap Y) \cup (T \cap X_\sigma) = T.$$

(2) In case of $T \cap X_\sigma = X_\sigma$: Since σ satisfies Ψ, the disjunction E_{i_T} in Ψ is also satisfied by σ. That means E_{i_T} contains either positive literal v_{st} with $\sigma(v_{st}) = 1$, or negative literal $\neg v_{st}$ with $\sigma(v_{st}) = 0$, for some s and t. Let us take such an s (not necessarily unique), and let $j_T = s$. Then by the definition of $\langle \sigma, j_T \rangle$, we see $b(\langle \sigma, j_T \rangle, t) = \sigma(v_{j_T, t})$ for each t. Thus $x_{j_T, \langle \sigma, j_T \rangle}$ is included in $X^*_{j_T}(E_{i_T})$, and moreover, $X^*_{j_T}(E_{i_T}) \cap X_\sigma = \{x_{j_T, \langle \sigma, j_T \rangle}\}$. Therefore the concept $f_{i_T, j_T} \in F_{i_T} \subseteq \mathcal{F}$ cuts T out of S_σ as follows:

$$f_{i_T, j_T} \cap S_\sigma = (Y^{[i_T]} \cup (X - X_{j_T}) \cup X^*_{j_T}(E_{i_T})) \cap (Y \cup X_\sigma)$$
$$= (Y^{[i_T]} \cap Y) \cup ((X - X_{j_T}) \cap X_\sigma) \cup (X^*_{j_T}(E_{i_T}) \cap X_\sigma)$$
$$= (T \cap Y) \cup (X_\sigma - \{x_{j_T, \langle \sigma, j_T \rangle}\}) \cup \{x_{j_T, \langle \sigma, j_T \rangle}\}$$
$$= (T \cap Y) \cup X_\sigma = (T \cap Y) \cup (T \cap X_\sigma) = T.$$

In each case, T is shown to be cut out of S_σ by some concept in $\mathcal{F}$. Therefore S_σ is shuttered by $\mathcal{F}$.

Now we show the converse. Suppose that VC-dim$(\mathcal{F}) = 2 \log m$. Then there is a set $S \subseteq U$ of cardinality $2 \log m$ which is shattered by $\mathcal{F}$.

Claim 1 *S contains exactly one element from each X_s $(1 \leq s \leq \log m)$, and all elements from Y.*

Proof of Claim 1. Case $m = 2$: The learning domain is $U = \{y_1\} \cup X_1$ and the concept class is $\mathcal{F} = \{f_{11}, g_{11} f_{21}, g_{21}\}$. Since $|\mathcal{F}| = 4$ and $g_{11}(x) = g_{21}(x) = 0$ for any $x \in X_1$, no two elements from X_1 can be included in S which is to be shattered by $\mathcal{F}$. Moreover, since $|Y| = |\{y_1\}| = 1$, the claim holds.

Case $m \geq 3$: Let $s \in \{1, \ldots, \log m\}$ be fixed arbitrarily, and x_1, x_2 be distinct elements in X_s. Suppose that S contains both x_1 and x_2. Then by Lemma 7,

$$\left| \{ h \in \mathcal{F} \mid h(x_1) \neq h(x_2) \} \right| \geq 2^{2 \log m - 1} = \frac{m^2}{2}.$$

On the other hand, let us consider a concept $h \in \mathcal{F}$ with $h(x_1) \neq h(x_2)$. Since $g_{ik}(x_1) = g_{ik}(x_2) = b(k, s)$ for any $g_{ik} \in G$, the concept h is not in G. Moreover, since $f_{ij}(x_1) = f_{ij}(x_2) = 1$ for any $f_{ij} \in F_1 \cup \cdots \cup F_m$ with $j \neq s$, thus h must be one of the concepts from $\{f_{1s}, f_{2s}, \ldots, f_{ms}\}$. Therefore

$$\left| \{ h \in \mathcal{F} \mid h(x_1) \neq h(x_2) \} \right| \leq \left| \{ f_{1s}, f_{2s}, \ldots, f_{ms} \} \right| = m,$$

which yields a contradiction since $\frac{m^2}{2} > m$ for any $m \geq 3$. Thus S can contain at most one element from X_s for each $s \in \{1, \ldots, \log m\}$. Since $|S| = 2 \log m$ and $|Y| = \log m$, the set S must contain exactly one element from each X_s and all elements from Y. $\square$

Therefore for each $s \in \{1, \ldots, \log m\}$, there is a unique $l = l(s) \in \{1, \ldots, m^C\}$ such that $x_{s,l(s)} \in S$, and we can assume that $S = Y \cup X_{(l)}$, where $X_{(l)} = \{x_{s,l(s)} \mid 1 \leq s \leq \log m\}$. Let σ_S be an assignment corresponding to S with

$$\sigma_S(v_{st}) = b(l(s), t) \cdot (1 \leq s \leq \log m, \ 1 \leq t \leq C \log m).$$

Now we show that σ_S satisfies all disjunctions E_i in Ψ. Let $i \in \{1, \ldots, m\}$ be fixed arbitrarily. Since S is shattered by $\mathcal{F}$, for the subset $T_i = Y^{[i]} \cup X_{(l)}$ of S there is a concept $h_i \in \mathcal{F}$ with $S \cap h_i = T_i$. Since $S \cap h_i = (Y \cap h_i) \cup (X_{(l)} \cap h_i)$, the concept h_i must satisfy the following two conditions:

(1) $Y \cap h_i = Y^{[i]}$.
(2) $X_{(l)} \cap h_i = X_{(l)}$.

Note that no concept in G satisfies the condition (2), and no concept in $F_{i'}$ with $i' \neq i$ satisfies the condition (1). Therefore such an $h_i \in \mathcal{F}$ is in F_i, and thus we can assume $h_i = f_{ij}$ for some $j \in \{1, \ldots, \log m\}$. The above condition (2) requires that f_{ij} contains all elements from $X_{(l)}$. Especially, remark that $x_{j,l(j)} \in X_{(l)}$ is included in f_{ij} for the above j. By the definition of f_{ij}, the element $x_{j,l(j)}$ is in $X_j^*(E_i)$. Thus the clause E_i satisfies either (a) or (b):

(a) E_i contains a positive literal v_{jt} with $b(l(j), t) = 1$.
(b) E_i contains a negative literal $\neg v_{jt}$ with $b(l(j), t) = 0$.

By the definition of σ_S, we see $\sigma_S(v_{jt}) = 1$ in case of (a), and $\sigma_S(v_{jt}) = 0$ in case of (b). In each case, $\sigma_S(E_i) = 1$. Therefore σ_S satisfies every disjunction E_i in Ψ. Thus Ψ is satisfiable. $\qquad \square$

5 Conclusion

We showed that the discrete VC-dimension problem is in $\mathrm{SAT}_{\log^2 n}$ and $\mathrm{SAT}_{\log^2 n}^{\mathrm{CNF}}$-hard. Therefore we may interpret that the discrete VC-dimension problem is, in some sense, "complete" for the class of $n^{O(\log n)}$ time computable sets. It remains open that the discrete VC-dimension problem is in $\mathrm{SAT}_{\log^2 n}^{\mathrm{CNF}}$, or $\mathrm{SAT}_{\log^2 n}$-hard.

As a dual to the VC-dimension, Romanik [7] defined the *testing dimension* of a concept class $\mathcal{F}$ as the *minimum* cardinality of a set $S \subseteq U$ which is *not* shattered by $\mathcal{F}$. We can see that testing dimension problem is also in $\mathrm{SAT}_{\log^2 n}$, by a similar reduction in the proof of Theorem 5. It is open whether the problem is $\mathrm{SAT}_{\log^2 n}^{\mathrm{CNF}}$-hard or not. It is also interesting to evaluate the complexity of computing another various dimensions of the class of *multi-valued functions* introduced in [1].

Acknowledgments

The author would like to thank Satoru Miyano and Hiroki Arimura for helpful discussions.

References

1. S. Ben-David, N. Cesa-Bianchi, and P.M. Long. Characterizations of learnability for classes of $\{0, \ldots, n\}$-valued functions. In *Proc. 5th Annual Workshop on Computational Learning Theory*, pages 333–340, 1992.

2. A. Blumer, A. Ehrenfeucht, D. Haussler, and M.K. Warmuth. Learnability and the Vapnik-Chervonenkis dimension. *JACM*, 36(4):929–965, 1989.

3. N. Linial, Y. Mansour, and R.L. Rivest. Results on learnability and the Vapnik-Chervonenkis dimension. *Information and Computation*, 90:33–49, 1991.

4. N. Megiddo and U. Vishkin. On finding a minimum dominating set in a tournament. *Theoretical Computer Science*, 61:307–316, 1988.

5. B.K. Natarajan. *Machine Learning — A Theoretical Approach*. Morgan Kaufmann Publishers, 1991.

6. S.H. Nienhuys-Cheng and M. Polman. Complexity dimensions and learnability. In *Proc. European Conference on Machine Learning, (Lecture Notes in Artificial Intelligence 667)*, pages 348–353, 1993.

7. K. Romanik. Approximate testing and learnability. In *Proc. 5th Annual Workshop on Computational Learning Theory*, pages 327–332, 1992.

8. L.G. Valiant. A theory of the learnable. *CACM*, 27(11):1134–1142, 1984.

ε-approximations of k-label spaces

(Extended Abstract)

Susumu Hasegawa, Hiroshi Imai and Masaki Ishiguro

Department of Information Science, University of Tokyo
Hongo, Bunkyo-ku, Tokyo 113, Japan

Abstract. In learning a geometric concept from examples, examples are mostly classified into two, positive examples and negative examples which are contained and not contained, respectively, in the concept. However, there exist cases where examples are classified into k classes. For example, clustering a concept space by the Voronoi diagram generated by k points is a very common tool used in image processing and many other areas. We call such a space a k-label space. The case of positive and negative examples corresponds to the 2-label space. In this paper, we first extend the ε-approximation for the 2-label space originally considered by Vapnik and Chervonenkis [10] (see also [1, 5]) to that for the k-label space. The generalized ε-approximation is then applied to the randomized algorithm for the assignment problem by Tokuyama and Nakano [9] to obtain tighter bounds. The ε-approximation combined with the capacity of a k-label space generated by Voronoi diagrams gives bounds for learning noisy data for such a k-label space.

1 Introduction

Geometric concepts are quite powerful to represent real-world complicated concepts by virtue of their fertile background. A concept is a subset of the space of appropriate dimensions, and any point in the space is classified into positive and negative according as it is contained in the concept or not. In learning such a geometric concept from examples, we assume in most cases that positive and negative points (examples) are given. In this standard case, examples consists of 2 classes. However, there exist cases where examples are classified into more than 2 classes. For example, clustering the concept space by the Voronoi diagram (e.g., see [2]) generated by k points is a very common geometric tool employed in learning problems encountered in learning geometric concepts in image processing and many other areas. In this general case, examples are classified into k labels, and the space is partitioned into k subsets by the labels. We call this space a k-label space. While the standard case of 2-label spaces consisting of positive and negative examples is well studied, the k-label space ($k \geq 3$) has not been well investigated. In this paper, we will clarify several fundamental properties of k-label spaces from the viewpoint of computational learning theory.

In the original 2-label space, the ε-approximation and ε-net proposed by Vapnik and Chervonenkis [10] and others [1, 5] play a central role in discussing the learning complexity of the space. We will generalize the ε-approximation to the k-label space as follows. A concept space S is a pair $(X, \mathcal{R})$, where X is a set and $\mathcal{R}$ is a set of subsets of X. A probabilistic concept space is a triple $(X, \mathcal{R}, P)$, where $(X, \mathcal{R})$ is a concept space and P is a probability measure on X. An m-sequence $\bar{y} \in X^m$ of X is an ϵ-approximation for a concept space $(X, \mathcal{R}, P)$, if

$$\left| \frac{\text{Member}(\bar{y}, R)}{m} - P(R) \right| < \epsilon \tag{1}$$

for every concept $R \in \mathcal{R}$, where $\text{Member}(\bar{y}, R)$ is the number of elements of R in $\bar{y}$. Haussler and Welzl [5] defined an ϵ-approximation for only a finite set X and a non-probabilistic concept space. In this paper, we use this term for more generalized spaces.

A concept $R \in \mathcal{R}$ is regarded as a function from X to $\{0, 1\}$. We generalize it to a label function L from X to $\{0, \ldots, k-1\}$. A label space is defined as a triple $(X, \mathcal{L}, P)$, where $\mathcal{L}$ is a set of label functions and P is a probability measure on X. In Section 3 we describe the definition of an ϵ-approximation for a label space and give the probability that a random sampling gives an ϵ-approximation.

As applications of this result, we demonstrate that the analysis of the randomized algorithm for the task assignment problem considered by Tokuyama and Nakano [9] can be strengthened. Also, by combining the result for generalized ϵ-approximation with the capacity [4, 3] of k-label space generated by Voronoi diagrams, we can give bounds for learning such k-label spaces from noisy data, as in the case of original ϵ-approximation for the 2-label space.

2 Range spaces, ϵ-nets and the primary shatter functions

In this section we will review the analysis of ϵ-approximation by a random sampling. We state the results in terms of the primary shatter function of a concept space [6]. The primary shatter function $\pi_{\mathcal{R}}(m)$ of a concept space $(X, \mathcal{R})$ is defined by

$$\pi_{\mathcal{R}}(m) = \max_{A \subseteq X |A| \leq m} |\{R \cap A, |R \in \mathcal{R}\}|. \tag{2}$$

We restate the result of [10] in terms of the primary shatter function.

Theorem 1 (Vapnik and Chervonenkis) *Let $S = (X, \mathcal{R}, P)$ be a concept space, c_1, d constants such that $\pi_{\mathcal{R}}(m) \leq (c_1 m)^d$, and $c_2 = 1 + \frac{1}{e}$ and $\bar{Y}$ be a random variable denoting an element of X^m obtained by m independent draws under P. Then if*

$$m \geq \max \left(\frac{4}{\epsilon^2} \ln \frac{4}{\delta}, \frac{16 c_2 d}{\epsilon^2} \ln \frac{32 c_1 c_2 d}{\epsilon^2} \right), \tag{3}$$

$\bar{Y}$ forms an ϵ-approximation for S with probability at least $1 - \delta$. □

A finite subset T of X is called *an ϵ-net for $S = (X, \mathcal{R}, P)$* if,

$$\forall R \in \mathcal{R} \quad P(R) > \epsilon \rightarrow T \cap R \neq \emptyset. \tag{4}$$

Similar to ϵ-approximations, ϵ-nets can be obtained by random sampling. Replacing $\sum_{i=0}^{d} \binom{m}{i}$ by $(c_1 m)^d$ in the proof of Lemma A2.2 in [1]; the following theorem holds.

Theorem 2 (Blumer et al.) *Let $S = (X, \mathcal{R}, P)$ be a concept space, c_1, d constants such that $\pi_{\mathcal{R}}(m) \leq (c_1 m)^d$, $c_2 = 1 + \frac{1}{e}$, $c_3 = \ln 2$ and $\bar{Y}$ be a random variable denoting an element of X^m obtained by m independent draws under P. Then if*

$$m \geq \max \left(\frac{1}{c_3 \epsilon} \ln \frac{1}{\delta}, \frac{4c_2 d}{c_3 \epsilon} \ln \frac{16 c_1 c_2 d}{c_3 \epsilon} \right) \tag{5}$$

the set of distinct elements of $\bar{Y}$ forms an ϵ- net for S with probability at least $1 - \delta$. $\square$

If the probability of an event is positive, the event exists, and taking δ close to 1, we obtain the following.

Corollary 1 *Let S, d, c_1, c_2, c_3 be same as above. Then there exist ϵ-nets for S of size m such that*

$$m = \left\lceil \frac{4 c_2 d}{c_3 \epsilon} \ln \frac{16 c_1 c_2 d}{c_3 \epsilon} \right\rceil . \; \square \tag{6}$$

3 Label spaces

In this section label spaces and ϵ-approximations for label spaces are defined. We will analyze the probability that random samplings construct ϵ-approximations for label spaces.

3.1 The definition of a label space

A k-label space (simply a label space) is a pair $S = (X, \mathcal{L})$, where X is a set and $\mathcal{L}$ is a set of functions from X to $\{0, 1, \ldots, k - 1\}$. Each function $L \in \mathcal{L}$ is called a label function. Further, for each $L \in \mathcal{L}$, a region $L^{[i]} \subseteq X$ is defined by:

$$L^{[i]} = \{ x \in X \mid L(x) = i \} . \tag{7}$$

Similar to the case of concept spaces, probabilistic label spaces are also defined. A probabilistic label space is a triple $S = (X, \mathcal{L}, P)$ where $(X, \mathcal{L})$ is a label space and P is a probabilistic measure on X. For a label space $(X, \mathcal{L})$ and a subset A of X, $\mathcal{L}|_A$ denotes the set of label functions $\{ L|_A : A \rightarrow \{0, \ldots, k - 1\} | L \in \mathcal{L} \}$, where $L|_A$ is the restriction of L to A. We call $(A, \mathcal{L}|_A)$ the subspace of $(X, \mathcal{L})$ induced by A.

The notion of the primary shatter functions is extended to label spaces. The primary shatter function $\pi_{\mathcal{L}}(m)$ of a label space $S = (X, \mathcal{L})$ is defined by

$$\pi_{\mathcal{L}}(m) = \max_{A \subseteq X, |A| \leq m} | \{ L|_A : A \rightarrow \{0, \ldots, k - 1\} | L \in \mathcal{L} \} | \tag{8}$$

The symmetric difference $L_1 \oplus L_2 \subseteq X$ of two labels L_1 and L_2 is defined by $L_1 \oplus L_2 = \{x \in X | L_1(x) \neq L_2(x)\}$. Given two label spaces $S_1 = (X, \mathcal{L}_1)$ and $S_2 = (X, \mathcal{L}_2)$, and a label function L, concept spaces $S_3 = (X, \mathcal{L}_1 \oplus \mathcal{L}_2)$ and $S_4 = (X, L \oplus \mathcal{L}_1)$ are defined by

$$\mathcal{L}_1 \oplus \mathcal{L}_2 = \{L_1 \oplus L_2 | L_1 \in \mathcal{L}_1, L_2 \in \mathcal{L}_2\} \tag{9}$$

$$L \oplus \mathcal{L}_1 = \{L \oplus L_1 | L_1 \in \mathcal{L}_1\}. \tag{10}$$

The primary shatter functions of S_3 and S_4 are bounded by those of S_1 and S_2. That is, the following lemma holds.

Lemma 2 *Let $S_1 = (X, \mathcal{L}_1)$ and $S_2 = (X, \mathcal{L}_2)$ be label spaces. Then $\pi_{\mathcal{L}_1 \oplus \mathcal{L}_2}(m)$ $\leq \pi_{\mathcal{L}_1}(m) \cdot \pi_{\mathcal{L}_2}(m)$ and $\pi_{\mathcal{L}_1 \oplus L}(m) \leq \pi_{\mathcal{L}_1}(m)$.*

Proof:

$$
\begin{aligned}
\pi_{\mathcal{L}_1 \oplus \mathcal{L}_2}(m) &= \max_{A \subseteq X, |A| \leq m} |\{(L_1 \oplus L_2) \cap A | L_1 \in \mathcal{L}_1, L_2 \in \mathcal{L}_2\}| \\
&= \max_{A \subseteq X, |A| \leq m} |\{(L_1|_A) \oplus (L_2|_A) | L_1 \in \mathcal{L}_1, L_2 \in \mathcal{L}_2\}| \\
&\leq \max_{A \subseteq X, |A| \leq m} |\{L_1|_A | L_1 \in \mathcal{L}_1\}| \cdot \max_{A \subseteq X, |A| \leq m} |\{L_2|_A | L_2 \in \mathcal{L}_2\}| \\
&= \pi_{\mathcal{L}_1}(m) \cdot \pi_{\mathcal{L}_2}(m) \tag{11}
\end{aligned}
$$

Therefore the first inequality is proved. The second inequality can be proved by applying the first one with $\mathcal{L}_2 = \{L\}$. $\square$

Similar to the case of the concept space, we can define an ϵ-approximation for the k-label space.

Definition 1 $\bar{Y} \in X^m$ *is an ϵ-approximation for a k-label space $S = (X, \mathcal{L}, P)$ if*

$$\sum_{i=0}^{k-1} |P(L^{[i]})m - \text{Member}(\bar{Y}, L^{[i]})| < \epsilon m , \tag{12}$$

for every label function $L \in \mathcal{L}$. $\square$

Here the distance between $(p_0 m, \ldots, p_{k-1} m)$ and $(m_0, \ldots, m_{k-1})$, where $p_i = P(L^{[i]})$ and $m_i = \text{Member}(\bar{Y}, L^{[i]})$, is measured with L_1 norm. This is natural and suitable when X is a finite set of points and P is a uniform distribution on X. Then, the ϵ-approximation assures that the total discrepancy between the original k-label space and an approximate k-label space found from $\bar{Y}$ is smaller than ϵm. Here, it is crucial to sum up the discrepancy, as in L_1 norm, for each label. In fact, this is a main key in the application discussed in Section 4.

By a random sampling an ϵ-approximation for a label space can be obtained with high probability.

Theorem 3 *Let $S = (X, \mathcal{L}, P)$ be a probabilistic k-label space ($k \geq 3$), c_1, d (for simplicity, in this version, we assume $k \leq d$) constants such that $\pi_{\mathcal{L}}(m) \leq$*

$(c_1 m)^d$, $c_2 = 1 + \frac{1}{e}$ and $\bar{Y} \in X^m$ be a random variable denoting an m sequence of X obtained by m independent draws under P. If

$$m \geq \max \left(\frac{42}{\epsilon^2} \ln \frac{2}{\delta}, \frac{127 c_2 d}{\epsilon^2} \ln \frac{506 c_1 c_2 d}{\epsilon^2} \right) \tag{13}$$

then $\bar{Y}$ is an ϵ-approximation for S with probability at least $1 - \delta$. □

This theorem is proved in the next subsection.

3.2 The proof of the main theorem

First we give two auxiliary lemmas.

Lemma 3 Let $S = (X, \mathcal{R}, P)$ be a concept space, c_1, d constants such that $\pi_\mathcal{R}(m) \leq (c_1 m)^d$ and $\bar{Y}$ be a random variable denoting an m sequence of X obtained by m independent draws under P. Then, for any ϵ $(0 < \epsilon < \frac{1}{4})$, $\delta > 0$ and $q > 0$, if

$$m \geq \max \left(\frac{2}{t\epsilon} \ln \frac{1}{\delta}, \frac{2 c_2 d}{t\epsilon} \ln \frac{4 c_1 c_2 d}{t\epsilon} \right), \tag{14}$$

where $t = q \ln 2 - \frac{4}{3}$ and $c_2 = 1 + \frac{1}{e}$, then with probability at least $1 - \delta$ the following relation holds.

$$\forall R \in \mathcal{R} \quad P(R) < \epsilon \rightarrow Member(\bar{Y}, R) < q\epsilon m . \tag{15}$$

Proof.

Let $A(R), B(R) \subseteq X^m$ and $C(R) \subseteq X^{2m}$ and $\mathcal{R}_\epsilon \subseteq \mathcal{R}$ be as follows:

$$A(R) = \{ \bar{y} \in X^m | Member(\bar{y}, R) > q\epsilon m \} \tag{16}$$

$$B(R) = \{ \bar{y}' \in X^m | Member(\bar{y}', R) = 0 \} \tag{17}$$

$$C(R) = \{ \bar{y} \cdot \bar{y}' \in X^{2m} | \bar{y} \in A(R), \bar{y}' \in B(R) \} \tag{18}$$

$$\mathcal{R}_\epsilon = \{ R \in \mathcal{R} | P(R) < \epsilon \} . \tag{19}$$

Further let $\bar{Y}'$ be a random variable denoting an m sequence of X obtained by independent drawing under P. By using a conditional probability

$$P^{2m} \left(\bar{Y} \cdot \bar{Y}' \in \bigcup_{R \in \mathcal{R}_\epsilon} C(R) \right) = \sum_{\bar{y} \in \bigcup_{R \in \mathcal{R}_\epsilon} A(R)} P^m (\bar{Y} = \bar{y}) \cdot P^m \left(\bar{Y}' \in \bigcup_{R \in \mathcal{S}} B(R) \mid \bar{Y} = \bar{y} \right) \tag{20}$$

holds, where $\mathcal{S}$ is a set of $R \in \mathcal{R}$ which satisfy $\bar{y} \in A(R)$. Since for all $R \in \mathcal{R}_\epsilon$

$$P^m \left(\bar{Y}' \in B(R) \right) = (1 - P(R))^m > (1 - \epsilon)^m > \exp \left(-\frac{\epsilon m}{1 - \epsilon} \right) \tag{21}$$

holds, we can obtain

$$P^m \left(\bar{Y}' \in \bigcup_{R \in \mathcal{S}} B(R) \mid \bar{Y} = \bar{y} \right) > \exp \left(-\frac{\epsilon m}{1 - \epsilon} \right) > \exp \left(-\frac{4 \epsilon m}{3} \right) \tag{22}$$

where the last inequality is obtained since $\epsilon < \frac{1}{4}$. Therefore equation (20) is replaced by

$$P^{2m}\left(\bar{Y}\cdot\bar{Y}'\in\bigcup_{R\in\mathcal{R}_\epsilon}C(R)\right) > \exp\left(-\frac{4\epsilon m}{3}\right)\cdot\sum_{\bar{y}\in\bigcup_{R\in\mathcal{R}_\epsilon}A(R)}P^m(\bar{Y}=\bar{y})$$

$$= \exp\left(-\frac{4\epsilon m}{3}\right)P^m\left(\bar{Y}\in\bigcup_{R\in\mathcal{R}_\epsilon}A(R)\right) . \quad (23)$$

Since the left hand side of equation (23) is the probability that a vector $\bar{Y}\cdot\bar{Y}'$ of length $2m$ has elements of R more than $q\epsilon m$ in the first half and no such elements in the last half, similar to Lemma A2.2 of [1] we can obtain:

$$P^{2m}\left(\bar{Y}\cdot\bar{Y}'\in\bigcup_{R\in\mathcal{R}_\epsilon}C(R)\right) < (2c_1 m)^d\cdot 2^{-q\epsilon m} . \quad (24)$$

Now the equations (23) and (24) yield:

$$P^m\left(\bar{Y}\in\bigcup_{R\in\mathcal{R}_\epsilon}A(R)\right) < \exp\left(\frac{4\epsilon m}{3}\right)P^{2m}\left(\bar{Y}\cdot\bar{Y}'\in\bigcup_{R\in\mathcal{R}_\epsilon}C(R)\right)$$

$$< (2c_1 m)^d\exp\left(-(q\ln 2 - \frac{4}{3})\epsilon m\right) . \quad \square \quad (25)$$

An ϵ-net assures that any concept with sufficiently large probability has at least one sample point. To the contrary Lemma 3 assures that any concept with sufficiently small probability does not have a large number of points.

Lemma 4 *For any label space $S = (X,\mathcal{L},P)$ and fixed $L\in\mathcal{L}$ the following holds:*

$$P^m\left(\bar{Y}\in X^m : \sum_{i=0}^{k-1}|P(L^{[i]})m - Member(\bar{Y}, L^{[i]})| > \epsilon m\right) < 2^k\exp\left(-\frac{\epsilon^2 m}{8}\right) .$$
$$(26)$$

Proof:

Denote $P(L^{[i]})$ by p_i $(i = 0,\ldots,k-1)$. The distribution considered here is a multinomial distribution $M_k(m; p_0, p_1,\ldots,p_{k-1})$ such that the probability that event i occurs x_i times in m trials is given by

$$p(x_0, x_1,\ldots,x_{k-1}) = \frac{m!}{x_0! x_1!\cdots x_{k-1}!}p_0^{x_0}p_1^{x_1}\cdots p_{k-1}^{x_{k-1}} \quad (27)$$

where $x_i \geq 0$ $(i = 0, \ldots, k-1)$ and $x_0 + x_1 + \cdots + x_{k-1} = m$. Then, for $r > 0$,

$$\Pr(\sum_{i=0}^{k-1} |p_i m - x_i| \geq r) = \sum_{\substack{r' \geq r \\ \sum_{i=0}^{k-1} |p_i m - x_i| = r' \\ x_i \geq 0, x_0 + \cdots + x_{k-1} = m}} p(x_0, \ldots, x_{k-1})$$

$$= \sum_{\emptyset \neq I \subset \{1,\ldots,k\}} \left[\sum_{r' \geq r} \sum_{\substack{0 \leq x_i < p_i m \ (i \in I), \ p_i m \leq x_i \leq m \ (i \notin I) \\ \sum_{i \in I}(p_i m - x_i) = r'/2, \ \sum_{i \notin I}(x_i - p_i m) = r'/2}} p(x_0, \ldots, x_{k-1}) \right]$$

$$\leq \sum_{\emptyset \neq I \subset \{1,\ldots,k\}} \left[\sum_{r' \geq r} \sum_{\sum_{i \in I}(p_i m - x_i) = r'/2, \ \sum_{i \notin I}(x_i - p_i m) = r'/2} p(x_0, \ldots, x_{k-1}) \right]$$

$$\equiv \sum_{\emptyset \neq I \subset \{1,\ldots,k\}} P(m; p_0, \ldots, p_{k-1}; I, r)$$

$$\tag{28}$$

For fixed I, $P(m; p_0, \ldots, p_{k-1}; I, r)$ may be bounded from above by using bounds for the binomial distribution as follows. The binomial distribution $B(m; p)$ is $M_2(m; p, 1-p)$, and denote the probability that event 0, corresponding to the probability p, occurs at most $mp - r$ by $P(m, p, r)$. For $I \subseteq \{0, 1, \ldots, k-1\}$, define $p(I)$ to be $\sum_{i \in I} p_i$. Then, for fixed I, $P(m; p_0, \ldots, p_{k-1}; I, r)$ is bounded by $P(m, p(I), r/2)$. By the Chernoff bound [7], $P(m, p, r/2)$ is further bounded by

$$P(m, p, r/2) \leq \exp\left(-\frac{r^2}{8mp}\right). \tag{29}$$

Hence, (28) is bounded by

$$\sum_{\emptyset \neq I \subset \{1,\ldots,k\}} \exp\left(-\frac{r^2}{8mp(I)}\right) \leq 2^k \exp\left(-\frac{r^2}{8m}\right) \tag{30}$$

Setting $r = \epsilon m$, the lemma follows. $\square$

Now we turn to the proof of Theorem 3. We define a concept space S' as $S' = (X, \mathcal{L} \oplus \mathcal{L}, P)$. By Lemma 2, $\pi_{S'}(m) \leq (c_1 m)^{2d}$. Therefore by Corollary 1 if

$$m \geq m_1 = \left\lceil \frac{8c_2 d}{c_3 \epsilon_1} \ln \frac{32 c_1 c_2 d}{c_3 \epsilon_1} \right\rceil, \tag{31}$$

there exit ϵ_1-nets of size m_1. Let T be one of such ϵ_1-nets. We define an equivalence relation '$\simeq_T$' for $\mathcal{L}$ by $L_1 \simeq_T L_2 \Leftrightarrow L_1|_T = L_2|_T$. Further the representative element for each equivalence class is chosen arbitrary. Let $U(L)$ denote the representative element of the class which L is included in. Since T is an ϵ_1-net for S',

$$P(U(L) \oplus L) < \epsilon_1 \qquad \text{for} \quad \forall L \in \mathcal{L}. \tag{32}$$

By triangular inequality, for any $L \in \mathcal{L}$,

$$\sum_{i=0}^{k-1} \left| P(L^{[i]})m - \text{Member}(\bar{Y}, L^{[i]}) \right|$$

$$\leq \sum_{i=0}^{k-1} \left| P(L^{[i]})m - P(U(L)^{[i]})m \right| + \sum_{i=0}^{k-1} \left| P(U(L)^{[i]})m - \text{Member}(\bar{Y}, U(L)^{[i]}) \right|$$

$$+ \sum_{i=0}^{k-1} \left| \text{Member}(\bar{Y}, U(L)^{[i]}) - \text{Member}(\bar{Y}, L^{[i]}) \right|$$

$$\leq 2P(L \oplus U(L))m + \sum_{i=0}^{k-1} \left| P(U(L)^{[i]})m - \text{Member}(\bar{Y}, U(L)^{[i]}) \right|$$

$$+ 2\text{Member}(\bar{Y}, L \oplus U(L)) . \tag{33}$$

The first term of the right-hand side is always less than $2\epsilon_1 m$. By Lemma 4 with the fact that the number of distinct equivalent classes is less than $(c_1 m_1)^d$, the probability that the second term is greater than $\epsilon_2 m$ for some representative element $U(L)$ is less than

$$(c_1 m_1)^d 2^k \exp\left(-\frac{\epsilon_2^2 m}{8}\right) \leq (2c_1 m)^d \exp\left(-\frac{\epsilon_2^2 m}{8}\right) . \tag{34}$$

where it should be noted that we have assumed $k \leq d$. Therefore, if

$$m \geq \max\left(\frac{16}{\epsilon_2^2} \ln \frac{1}{\delta_1}, \frac{16c_2 d}{\epsilon_2^2} \ln \frac{32c_1 c_2 d}{\epsilon_2^2}\right) , \tag{35}$$

with probability at least $1 - \delta_1$ the second term is less than $\epsilon_2 m$ for all $U(L)$. Further, applying Lemma 3 to the concept space S', if

$$m \geq \max\left(\frac{2}{t\epsilon_3} \ln \frac{1}{\delta_2}, \frac{2c_2 d}{t\epsilon_3} \ln \frac{4c_1 c_2 d}{t\epsilon_3}\right) , \tag{36}$$

then with probability at least $1 - \delta_2$ the third term is less than $2q\epsilon_3 m$ for all $L \in \mathcal{L}$. From equations (31), (35) and (36) with $\epsilon_1 = \frac{\epsilon_2^2}{2c_3}$, $\epsilon_3 = \frac{\epsilon_2^2}{8t}$ and $\delta_1 = \delta_2 = \frac{\delta}{2}$, if

$$m \geq \max\left(\frac{16}{\epsilon_2^2} \ln \frac{2}{\delta}, \frac{16c_2 d}{\epsilon_2^2} \ln \frac{64c_1 c_2 d}{\epsilon_2^2}\right) \tag{37}$$

with probability at least $1 - \delta$ the right hand side of equation (33) is at most $(2\epsilon_1 + \epsilon_2 + 2q\epsilon_3)m$. For sufficiently large q,

$$(2\epsilon_1 + \epsilon_2 + 2q\epsilon_3)m = \left(1 + \left(\frac{1}{c_3} + \frac{q}{4\left(qc_3 - \frac{4}{3}\right)}\right)\epsilon_2\right)\epsilon_2 m \leq 2.81\epsilon_2 m. \tag{38}$$

Now setting $\epsilon = 2.81\epsilon_2$ we obtain the theorem. $\square$

4 An application of ϵ-approximations

In this section we consider an algorithm for a geometric problem called a splitter finding problem proposed by Tokuyama and Nakano [9]. This problem corresponds to the traditional task assignment problem. We here analyze it using the ϵ-approximation for a label space, and give tight bounds.

4.1 The definition of the problem and some notations

Let $A(b) = \{a_0 + b, \ldots, a_{k-1} + b\}$ be a set of k affinely independent points in R^{k-1} with a parameter $b \in R^{k-1}$. Label function $\lambda(x; b) : R^{k-1} \rightarrow \{0, \ldots, k-1\}$ is defined by

$$\lambda(x; b) = \min\{ i \mid \forall j : d_{L_2}(x, a_i + b) \leq d_{L_2}(x, a_j + b) \}. \tag{39}$$

Let Q be an n points set in R^{k-1}. For simplicity, we assume that no k points of Q are on the boundary of the partition obtained by any label function λ. $S_\lambda = (Q, \mathcal{L}_\lambda|_Q)$ is a label space, where $\mathcal{L}_\lambda = \{\lambda(x; b) | b \in R^{k-1}\}$. We call this label space as the splitter space.

Lemma 5 $\pi_{S_\lambda}(m) \leq \binom{m+k}{k}$

Proof Let $A = \{(a_0, \ldots, a_{k-1}) | a_i \in N \cup \{0\}, \ a_0 + \cdots + a_{k-1} = m\}$. For any $(a_0, \ldots, a_{k-1}) \in A$ and m points set Q, there exists one and only one splitter λ such that $\forall i \ |\lambda^{[i]} \cap Q| = a_i$ [9]. Hence, there exits an one-to-one mapping from $\mathcal{L}|_Q$ to A and

$$\left| \mathcal{L}|_Q \right| = |A| = \binom{m+k}{k}. \ \square \tag{40}$$

Corollary 6 *For* $k \geq 3$ *and* $m \geq 2$, $\pi_{S_\lambda}(m) \leq (em)^k$.

Proof:

$$\pi_{S_\lambda}(m) \leq \binom{m+k}{k} \leq \frac{(m+k)^k}{k!} \leq \left(\frac{e(m+k)}{k}\right)^k \leq (em)^k \ \square \tag{41}$$

Based on the splitter space we give the definition of the problem called λ-splitter finding problem:

Problem SF:
Given a set Q of n points in R^{k-1} and k integers q_i $(0 \leq i \leq, k-1)$ such that $\sum_{i=0}^{k-1} q_i = n$, find a splitter function $\lambda_t \in \mathcal{L}_\lambda$ (we call it the target splitter) such that for each i, $\lambda^{[i]}$ contains exactly q_i points of Q. $\square$

We express an instance of problem SF as $SF(Q, q_0, \ldots, q_{k-1})$. Tokuyama and Nakano [9] showed that this problem is equivalent to the traditional task assingment problem and found a deterministic algorithm (we call it algorithm D) which starts any splitter and find the target splitter by inserting n points incrementally. Further based on the algorithm D they gave a randomized algorithm which we call an algorithm R:

Algorithm R:
Phase A:

1. Draw m points independently according to a uniform distribution over Q. We regard the resulting vector as a set of m points in R^{k-1} and denote it as Q'.
2. Apply an algorithm D to a problem $SF(Q', \frac{q_0 m}{n}, \ldots, \frac{q_{k-1} m}{n})$ with any splitter as a start. (To simplify the problem we assume that each $\frac{q_i m}{n}$ is an integer.) The resulting splitter is denoted by λ_m.

Phase B:

1. Apply D to the objective problem $SF(Q, q_0, \ldots, q_{k-1})$ with λ_m as a starting splitter. ☐

Before analyzing this algorithm, we give some notations and lemmas here.

Definition 2 *Let λ_1 and λ_2 be splitters and Q be a set of points in R^{k-1}, then $excess_Q(\lambda_1, \lambda_2)$ is defined as:*

$$excess_Q(\lambda_1, \lambda_2) = \sum_{i=0}^{k-1} \left| \left| Q \cap \lambda_1^{[i]} \right| - \left| Q \cap \lambda_2^{[i]} \right| \right| \tag{42}$$

☐

Using an excess, the time complexity of algorithm D is expressed as follows:

Lemma 7 (Tokuyama, Nakano [9]) *Algorithm D finds a target splitter λ_t for any problem $SF(Q, q_0, \ldots q_{k-1})$ with λ_s as a starting splitter, within $ck^2 h \ln n$ time, where c is some constant, $n = |Q|$ and $h = \frac{excess_Q(\lambda_s, \lambda_t)}{2}$*

☐

4.2 An analysis of algorithm R

We analyze algorithm R by using an ϵ-approximation for the label space $\mathcal{L}_\lambda$. Since Phase A solves the subproblem of size m, it takes at most $ck^2 m \ln m$ time. By Theorem 3 and Corollary 6, for

$$m \geq \max \left(\frac{42}{\epsilon^2} \ln \frac{2}{\delta}, \frac{127 c_2 k}{\epsilon^2} \ln \frac{506 c_2 e k}{\epsilon^2} \right) \tag{43}$$

Q' forms an ϵ-approximation for the label space $(Q, \mathcal{L}_\lambda, D)$ with probability at least $1 - \delta$, where D is the uniform distribution over Q. If Q' is an ϵ-approximation for $(Q, \mathcal{L}_\lambda, D)$,

$$\epsilon > \sum_{i=0}^{k-1} \left\| \frac{\left| \lambda_m^{[i]} \cap Q \right|}{n} - \frac{\left| \lambda_m^{[i]} \cap Q' \right|}{m} \right\|$$

$$= \sum_{i=0}^{k-1} \left\| \frac{\left| \lambda_m^{[i]} \cap Q \right|}{n} - \frac{q_i}{n} \right\| \tag{44}$$

holds, since $\lambda_m^{[i]} \cap Q' = \frac{q_i m}{n}$. Therefore,

$$\text{excess}_Q(\lambda_m, \lambda_t) = \sum_{i=0}^{k-1} \left\| \left| \lambda_t^{[i]} \cap Q \right| - \left| \lambda_m^{[i]} \cap Q \right| \right\|$$

$$= \sum_{i=0}^{k-1} \left| q_i - \left| \lambda_m^{[i]} \cap Q \right| \right| < \epsilon n \tag{45}$$

holds. Phase B completes within $O(k^2 (\epsilon n) \ln n)$ time. Now setting $\epsilon = c(k/n)^{1/3}$ and $\delta = c'(k^{1/3} n^{2/3})^{-k}$ for some constants c, c', and setting m to the value of the righthand side of (43) for these ϵ and δ, we obtain the following lemma.

Lemma 8 *Algorithm R solves the problem within $O(k^{\frac{7}{3}} n^{\frac{2}{3}} \ln^2 n)$ time with probability $1 - c'(k^{1/3} n^{2/3})^{-k}$.*

This completes a rigorous tail distribution analysis of the randomized algorithm. This coincides well with the analysis on the expected running time given in [9], and thus strengthens it to guarantee the running time with high probability. This becomes possible through the use of the extended ϵ-approximation.

5 Concluding Remarks

Hasegawa [3] discusses the k-label space induced by the Voronoi diagram generated by k points (called k-Voronoi space), and shows that for m points in the d-dimensional space the primary shutter function of this k-label space is $O(m^{dk})$. This defines the capacity or dimension of this k-Voronoi space.

In the case of 2-label space, the ϵ-net gives bounds for PAC learning, where a concept which is consistent with the given positive and negative examples is constructed as an approximation concept. When noisy data are given as examples, a consistent concept to the given examples does not exist any more and the ϵ-net is not useful. In such cases ϵ-approximation can be used to guarantee the approximate learnability.

In a similar way to this, the generalized ϵ-approximation can be used to the problem of learning the k-Voronoi space from noisy data. More detailed discussions concerning this will be given somewhere else.

Acknowledgment

The work by the second author was supported in part by the Grant-in-Aid of the Ministry of Education, Science and Culture of Japan.

References

1. A. Blumer, A. Ehrenfeucht, D. Haussler, and M. Warmuth. Learnability and Vapnik Chervonenkis dimension. *Journal of the ACM*, 36(4):929-965, October 1989.
2. H. Edelsbrunner. *Algorithms in Combinatorial Geometry*. Springer-Verlag, 1987.
3. S. Hasegawa. *A Study on ϵ-net and ϵ-approximation*. Mater's Thesis, Department of Information Science, University of Tokyo, 1993.
4. D. Haussler and P. Long. A generalization of Sauer's Lemma. Technical Report UCSC-CRL-90-15, University of California at Santa Cruz, 1991.
5. D. Haussler and E. Welzl. Epsilon-Nets and Simplex Range Queries. *Discrete and Computational Geometry*, 2:127-151, 1987.
6. J. Matoušek, E. Welzl and L. Wernisch. Discrepancy and ϵ-approximations for bounded VC-dimension *Proc. 32nd Annual Symposium on Foundations of Computer Science*: 424-430, 1991.
7. P. Raghavan. Lecture Notes on Randomized Algorithms. *IBM Research Report RC 15340*, IBM Research Division, 1990.
8. N. Sauer. On the density of families of sets. *J. Combinatorial Theory (A)*, 13:145-147, 1972.
9. T. Tokuyama and J. Nakano. Geometric Algorithms on an Assignment Problem. *Proc. 7th ACM Symposium on Computational Geometry*: 262-271, 1991.
10. V. N. Vapnik and A. Ya. Chervonenkis. On the uniform convergence of relative frequencies of events to their probabilities. *Theory of Probability and Its Applications*, 16:264-280, 1971.

Exact Learning of Linear Combinations of Monotone Terms from Function Value Queries

Atsuyoshi Nakamura   Naoki Abe

C & C Research Laboratories, NEC Corporation
4-1-1 Miyazaki Miyamae-ku, Kawasaki 216, Japan
E-mail:atsu@ibl.cl.nec.co.jp and abe@ibl.cl.nec.co.jp

Abstract

We investigate the exact learning of real-valued functions on $\{0,1\}^n$ represented by a weighted sum of a number of monotone terms using queries for the values of the target function at assignments of the learner's choice. When all coefficients are nonnegative, we show that these functions are learnable with at most $(n - \lfloor \log_2 k \rfloor + 1)k$ queries, where n is the number of variables and k is the number of terms in the target. We also prove a lower bound of $\Omega(\frac{nk}{\log_2 k})$ on the number of queries necessary for learning this class, so no algorithm can reduce the number of queries dramatically. In the general case, namely, when the coefficients vary over the reals, we show the number of queries required for exact learning of k-term subclass is upper bounded by $q(n, \lfloor \log_2 k \rfloor + 1)$ and is lower bounded by $q(n, \lfloor \log_2 k \rfloor)$, where $q(n,l) = \sum_{i=0}^{l} \binom{n}{i}$. The latter bounds are shown by generalizing Roth and Benedek's technique for analyzing the learning problem for k-sparse multivariate polynomials over $GF(2)$ [8].

1 Introduction

We consider the learning problem for functions that are representable as a weighted sum of a number of *monotone terms*, which are products of Boolean variables. The learning model we study in this paper is exact learning (identification) using *function value queries*, namely queries that ask for the value of the target function at an assignment. The function value query is the *membership query* when the target function is a Boolean function.

Our work was motivated by the question of whether DNF is learnable if the number of satisfied terms is given in addition to the value of the DNF formula. Note that this corresponds to the case in which the target function is the simple sum of the terms. As we can get more information this way, one can imagine that the amount of computation and the number of queries necessary for learning DNF from this type of queries may be significantly less. The motivation for considering a weighted sum of monotone terms, which is more general than a sum of monotone terms, is derived from the idea that it is natural for each term to have a different weight according to its importance. Note that linear combinations of monotone terms are very expressive because all real-valued functions on $\{0,1\}^n$ are representable in this form.

The results we obtained on exact learnability of linear combinations of monotone terms are as follows. If all coefficients are restricted to be nonnegative, this class can be learned exactly using a number of function value queries bounded above by $(n - \lfloor \log_2 k \rfloor + 1)k$, where n and k are the numbers of variables and terms in the target, respectively. We also obtain a lower bound of order $\Omega(\frac{nk}{\log_2 k})$ so no algorithm can reduce the number of queries dramatically. In the general case, namely, when the coefficients vary over the reals, the same upper and lower bounds on the number of required queries that are known to hold for the problem of learning multivariate polynomials over GF(2) can be shown to hold for this class : it is upper bounded by $q(n, \lfloor \log_2 k \rfloor + 1)$ and is lower bounded by $q(n, \lfloor \log_2 k \rfloor)$, where $q(n, l) = \sum_{i=0}^{l} \binom{n}{i}$. These bounds are obtained by generalizing Roth and Benedek's technique they used to analyze the learning problem for multivariate polynomials over GF(2).

As we noted earlier, the problem of exactly learning a function defined as an operator applied on a set of (monotone) terms can be viewed as the problem of identifying the set of terms using that operator as information source. This is similar in spirit to the PAC learning model with additional information proposed by Kakihara and Imai [6]. Together with previously known results due to various authors on the complexity of learning monotone DNF and multivariate polynomials over GF(2), the results in this paper imply that the summation ('+') is significantly more valuable than the logical OR ('∧') or the exclusive OR ('⊕'). For example, for identifying a set of k monotone terms, the progression of required numbers of queries for the three operators goes roughly as: $O(n^k)$ for '∧,' $O(n^{\log k})$ for '⊕' and $O(nk)$ for '+.' Note that the number of queries polynomial in both n and k is only for '+'. For identifying a set of read-k terms, the number of required queries for '∧,' '⊕' and '+' seems to decrease roughly as 'exponential in n', $O(n^{\log k})$, and $O(n^2 k)$. These comparisons are described in details in Section 5.

2 Preliminaries

Let A_n denote the set of Boolean variables $x_1, x_2, ..., x_n$. A *monotone term* is the product of variables in a subset of A_n, and it can be considered to be a function from $\{0, 1\}^n$ to $\{0, 1\}$. Since the product of variables belonging to the empty set is 1, 1 is a monotone term. We let MT denote the set of all monotone terms. For each term $t \in MT$, $v(t)$ denotes the set of variables in t. We define the function class $\mathcal{F}$, the set of *linear combinations of monotone terms*, as follows : $\mathcal{F} = \{\sum_{t \in MT} a_t t : a_t \in \Re\}$. For each $F \in \mathcal{F}$, $T(F)$ denotes the set of terms with nonzero coefficients in F. A function $F \in \mathcal{F}$ is said to be k-*term* if $|T(F)| \leq k$. A function $F \in \mathcal{F}$ is *read-k* if each variable is contained in at most k terms of $T(F)$. We say that a function $F \in \mathcal{F}$ is *inductive-read-k* if, for each $m \in \{1, 2, ..., n\}$, there are at most k terms of $T(F)$ which contain x_m and no variables in $\{x_{m+1}, x_{m+2}, ..., x_n\}$. We also let k-*term-$\mathcal{F}$*, *read-k-$\mathcal{F}$* and *inductive-read-k-$\mathcal{F}$* denote the k-term, read-k and inductive-read-k subclasses of $\mathcal{F}$, respectively. $Q(n, k)$ denotes the class of subsets consisting of at most k elements of A_n. We use the notation 1_S to denote the assignment in which all variables belonging to S are assigned 1 and all others are 0 . We also use 0_S which is similarly defined. We abbreviate 1_{A_n} and 0_{A_n} by 1 and 0, respectively.

3 The Case with Nonnegative Coefficients

In this section, we present our positive learnability result for linear combinations of monotone terms with nonnegative coefficients via function value queries.

In this case, there is a simple learning algorithm that dose fairly well. The algorithm is based on a method for learning monotone DNF [1, 4]. In particular, it uses as a subprocedure the same method for finding one monotone term with a nonzero coefficient that is used in [1, 4]. Suppose $F(1_S) > 0$. This means that there is a term t with a nonzero coefficient such that $v(t) \subseteq S$. For every variable $x \in S$, ask for the value of $F(1_{S-\{x\}})$ and if $F(1_{S-\{x\}}) > 0$ then remove x from S. After this procedure, $F(1_S) > 0$ and $F(1_{S-\{x\}}) = 0$ for all $x \in S$. This means that F contains the term $\prod_{x \in S} x$ with coefficient $F(1_S)$. If $F(1_{A_n}) = 0$ then $F \equiv 0$, otherwise we can find a term t with a nonzero coefficient a by the process described above using exactly n function value queries. Then we can find another term with a nonzero coefficient by applying the same procedure to the function $F - at$. By repeating this procedure as many times as there are terms with a nonzero coefficient, we can identify F. The number of function value queries used in this algorithm is $nk + 1$, where k is the number of terms with a nonzero coefficient. But this algorithm does not fully use the information available to it, because $F(1_{S-\{x\}}) > 0$ contains the two cases $F(1_{S-\{x\}}) = F(1_S)$ and $F(1_{S-\{x\}}) < F(1_S)$, and the behavior of the algorithm is the same in these cases. $F(1_{S-\{x\}}) = F(1_S)$ means that every term t satisfying $v(t) \subseteq S$ does not contain x, and $F(1_{S-\{x\}}) < F(1_S)$ means that there are terms t_1, t_2 satisfying $v(t_1), v(t_2) \subseteq S$ such that $x \notin v(t_1)$ and $x \in v(t_2)$. Procedure Identify (Figure 1) uses this information and when the latter case occurs, it divides its work into two parts. The first stage finds every term t_1 satisfying $v(t_1) \subseteq S - \{x\}$ and the second one finds every term t_2 satisfying $x \in v(t_2) \subseteq S$ making use of the information obtained in the first stage, i.e., the algorithm uses the value of $F - H'$ instead of F, where H' is the subformula of F found in the first stage. The number of function value queries used in procedure Identify varies from $n - \lfloor \log_2 k \rfloor + 3k - 2$ to $(n - \lfloor \log_2 k \rfloor + 1)k$ depending on the exact form of F.

Theorem 3.1 *If F is a linear combination of monotone terms with nonnegative coefficients, then procedure 'Identify' learns F exactly using at most $(n - \lfloor \log_2 |T(F)| \rfloor + 1)|T(F)|$ function value queries in time $O(n|T(F)|^2)$.*

(Proof) Let S be a subset of A_n and H be a subformula of F, i.e., the value of every nonzero coefficient of H is the same as that of F. We define the set $T'(S, H)$ as follows:

$$T'(S, H) = \{t : v(t) \subseteq S, t \in T(F - H)\}.$$

We show that if Claim 3.1 stated below holds, then the output H of Identify must equal F. If $F = 0$, procedure Sub-identify is never called because $a = F(1) = 0$, and $H = 0$ is output. If $F \neq 0$, Sub-identify is called with the input parameter quadruple $(A_n, n, a, 0)$ satisfying Relation 1, so we see by Claim 3.1 that $0 + \sum_{t \in T'(A_n, 0)} a_t t = \sum_{t \in T(F)} a_t t = F$ is output.

procedure Identify return(H)

H : a linear combination of monotone terms with nonnegative coefficients

begin

    $a := F(1)$

    if $a = 0$ then return(0) else return(Sub-identify(A_n,n,a,0))

end

end procedure

procedure Sub-identify(S,i,a,H) return(H_{out})

S : a subset of A_n

i : an element of $\{1, 2, ..., n\}$

a : a positive real number

H, H_{out} : linear combinations of monotone terms with nonnegative coefficients

begin

    while $i > 0$

        $b := F(1_{S-\{x_i\}}) - H(1_{S-\{x_i\}})$

        if $b = a$ then $S := S - \{x_i\}$

        else if $0 < b$ then exit while

        $i := i - 1$

    end while

    if $i = 0$ then return($H + a \prod_{x \in S} x$)

    else begin

        $H' := $ Sub-identify($S - \{x_i\}$,$i - 1$,b,H)

        return(Sub-identify(S,$i - 1$,$a - b$,H'))

    end else

end

end procedure

Figure 1: Procedure Identify

Claim 3.1 *Suppose that $|T'(S, H')| \geq 1$ and a quadruple (S, i, a, H) satisfies Relation 1. Then, Sub-identify(S, i, a, H) outputs $H + \sum_{t \in T'(S,H)} a_t t$.*

Relation 1 *A quadruple (S, i, a, H) is said to satisfy Relation 1 if the following two conditions are satisfied.*

    1. $F(1_S) - H(1_S) = a$

    2. $\forall t \in T'(S, H), S \cap \{x_k : k > i\} = v(t) \cap \{x_k : k > i\}$

Note that after the while-loop, in which S and i are updated, the quadruple (S, i, a, H) still satisfies Relation 1, because, for every $j > i$, if variable x_j is removed from S in the while-loop, then x_j is not contained in any term of $T'(S, H)$ (when $b = a$), and if x_j is not removed in the while-loop, then x_j is contained in all terms of $T'(S, H)$ (when $b = 0$).

First, we prove this claim in the case of $|T'(S, H)| = 1$. Let t_0 be the only term contained in $T'(S, H)$. In this case, $i = 0$ holds after the while-loop, because $0 < b < a$ is never satisfied in the while-loop. When $i = 0$, Relation 1 is satisfied if and only if $a_{t_0} = a, v(t_0) = S$. So, $H + a\prod_{x \in S} x = H + \sum_{t \in T'(S,H)} a_t t$ is output.

Now assume that Claim 3.1 holds when $|T'(S,H)| < l \ (l \geq 2)$. We then show that it also holds when $|T'(S,H)| = l$. Assume the input quadruple (S, i, a, H) satisfies Relation 1 and $|T'(S,H)| = l$. Since $T'(S,H)$ contains more than two terms, there exists a j such that both terms having and not having the variable x_j exist. For such j, $0 < b = F(1_{S-\{x_j\}}) - H(1_{S-\{x_j\}}) < a$ holds. In this case, $i \neq 0$ holds after the while-loop, and hence the two recursive calls to Sub-identify are made. The input quadruple $(S - \{x_i\}, i - 1, b, H)$ of the first recursive call satisfies Relation 1 because $b = F(1_{S-\{x_i\}}) - H(1_{S-\{x_i\}})$ and x_i is not in $v(t)$ for all $t \in T'(S - \{x_i\}, H)$. Since $|T'(S - \{x_i\}, H)| < l$ holds also, by the inductive hypothesis, $H + \sum_{t \in T'(S-\{x_i\},H)} a_t t$ is output and set to H'. Therefore,

$$F(1_S) - H'(1_S) = F(1_S) - H(1_S) - \sum_{t \in T'(S-\{x_i\},H)} a_t t = a - b$$

holds. Note that x_i is contained in every term in $T'(S, H')$, because $T'(S, H') = T'(S, H) - T'(S - \{x_i\}, H)$. Thus, Relation 1 is satisfied by the input quadruple $(S, i-1, a-b, H')$ of the second recursive call, and $|T'(S, H')| < l$ holds, so by the inductive hypothesis, $H' + \sum_{t \in T'(S,H')} a_t t = H + \sum_{t \in T'(S,H)} a_t t$ is output. This completes the proof of Claim 3.1.

Next, we calculate the number of function value queries made by this algorithm. Procedure Identify itself makes only one query (for the assignment 1). The number of queries made by Sub-identify equals the number of times b is calculated, which occurs once in each iteration of the while-loop, and thus essentially for each update of i. If we form a history of the updates on i in Sub-identify by branching every time two recursive calls are made, and growing a single branch by one step every time i is updated, then we obtain a tree of depth n and width $|T(F)|$. Hence, the number of queries, which is equal to the number of nodes of this tree, is at most $n|T(F)|$. More precisely, since at the root the width is one and it takes at least $\log_2 |T(F)|$ steps for the width to reach $|T(F)|$, the bound can be improved to $(n - \lfloor \log_2 |T(F)| \rfloor + 1)|T(F)|$, which is tighter especially when $|T(F)|$ is exponentially large in n.

The running time of this algorithm is determined by the calculation of the value b. To calculate b, $H(1_{S-\{x_i\}})$ has to be calculated, that is to say, for each term $t \in T(H)$, it must be determined whether t is satisfied by $1_{S-\{x_i\}}$ or not. This can be done by finding out whether t is satisfied by 1_S and whether $x_i \in v(t)$, and this takes a constant number of steps by keeping the information of whether t is satisfied by 1_S. Therefore, b can be calculated in $O(T(H))$ time, so the algorithm runs in time $O(n|T(F)|^2)$. $\square$

In order to learn inductive-read-k linear combinations of monotone terms with nonnegative coefficients, the naive algorithm described at the beginning of this section needs $n(n - \lfloor \log_2 k \rfloor)k$ function value queries in the worst case. Algorithm Identify improves upon this bound and reduces the coefficient of n^2 from k to $\frac{1}{2}k$. In the case of read-once, we can further reduce the required number of queries: the coefficient of n^2 can be made $\frac{1}{4}$ by modifying the algorithm so as not to query $F(1_{S-\{x_i\}})$ if a term $t \in T(F)$ which contains the variable x_i has been already found.

Theorem 3.2 *If F is an inductive-read-k linear combination of monotone terms with nonnegative coefficients, then algorithm 'Identify' learns F exactly using at most $\frac{1}{2}((n-\lfloor \log_2 k \rfloor)(n-\lfloor \log_2 k \rfloor +1)+2)k$ function value queries. In particular, when F is read-once, 'Identify' with a minor modification learns F with at most $\frac{1}{4}n^2 + n + 1$ queries.*

(Proof) Let F be inductive-read-k. Every second recursive call finds at most k terms because F is inductive-read-k. Therefore, from the proof of Theorem 3.1, the number of function value queries made by algorithm Identify is at most $1+\sum_{i=1}^{n-\lfloor \log_2 k \rfloor -1}(n-\lfloor \log_2 k \rfloor +1-i)k + \sum_{i=1}^{\lfloor \log_2 k \rfloor +1} 2^{i-1} = \frac{1}{2}((n-\lfloor \log_2 k \rfloor)(n-\lfloor \log_2 k \rfloor +1)+2)k$.

Let F be read-once and $|T(F)| = l$. If the algorithm does not query $F(1_{S-\{x_i\}})$ when a term $t \in T(F)$ which contains the variable x_i has been already found, then the number of function value queries decreases by at least $\sum_{i=1}^{l-2} i$. In order to learn F exactly, Identify uses at most $1+\sum_{i=0}^{l-1}(n-i)$ function value queries, so the modified algorithm needs at most $-l^2 + (n+2)l \le \frac{1}{4}n^2 + n + 1$ function value queries. $\square$

Next, we give a lower bound on the number of function value queries needed for exact learning of a linear combination of monotone terms with nonnegative coefficients.

Theorem 3.3 *Any exact learning algorithm for k-term linear combinations of monotone terms with nonnegative coefficients must make more than $\frac{nk-1}{\log_2(k^2+1)} = \Omega(\frac{nk}{\log_2 k})$ function value queries, in the worst case.*

(Proof) Consider the following subset of k-term linear combinations of monotone terms with nonnegative coefficients :

$$\mathcal{G} = \{F \in \mathcal{F} : |T(F)| = k \text{ and } \forall t \in T(F), a_t \in \{1, 2, ..., k\}\}.$$

The value of $F \in \mathcal{G}$ for an arbitrary assignment is one of the $k^2 + 1$ values, $0, 1, ..., k^2$. Thus, for any assignment, one of these values is assumed by at least $\frac{1}{k^2+1}$ of the functions in $\mathcal{G}$ that are consistent with the answers for the queries made so far. If an adversary selects such a value as the response to the current query, at least $\frac{1}{k^2+1}$ of the consistent functions remain. When any algorithm identifies the target function, there must not be two distinct functions that are consistent with the answers to the queries made up to that point. Therefore, the number of queries l necessary in this case must satisfy

$$\left(\frac{1}{k^2+1}\right)^l |\mathcal{G}| < 2. \tag{1}$$

Now, $|\mathcal{G}|$ is bounded from below by the product of the number of combinations of k terms over n variables $\binom{2^n}{k}$, and the number of coefficient assignments to a fixed set of k terms, i.e., k^k. Since $\binom{2^n}{k} \geq \left(\frac{2^n}{k}\right)^k$ holds, we get $|\mathcal{G}| \geq 2^{nk}$. Therefore, by (1), we must have

$$l > \frac{nk - 1}{\log_2(k^2 + 1)}.$$

□

4 The General Case

In the previous section, we considered exact learnability of the subclass of $\mathcal{F}$ with nonnegative coefficients and derived the result that the number of function value queries needed is polynomial in the number of variables and the number of nonzero coefficient terms. However, this is impossible for the whole class $\mathcal{F}$, when we allow real coefficients. In this case, we show exact learnability of two subclasses, k-term-$\mathcal{F}$ and inductive-read-k-$\mathcal{F}$, by generalizing the method Roth and Benedek [8] developed for multivariate polynomials over GF(2), and get the same bounds on the number of queries needed to identify them. We summarize our results as a theorem below.

Theorem 4.1 *Let k-term-$\mathcal{F}$ and inductive-read-k-$\mathcal{F}$ be as defined in Section 2.*

1. *There exists an algorithm that exactly learns k-term-$\mathcal{F}$, making function value queries at assignments 0_S for all $S \in Q(n, \lfloor \log_2 k \rfloor + 1)$, running in time $O(kn|Q(n, \lfloor \log_2 k \rfloor + 1)|)$.*

2. *There exists an algorithm that exactly learns [1] inductive-read-k-$\mathcal{F}$ making function value queries at assignments 0_S for all $S \in Q(n, \lfloor \log_2 k \rfloor + 2)$, running in time $O(kn|Q(n, \lfloor \log_2 k \rfloor + 2)|)$.*

3. *Any exact learning algorithm for k-term-$\mathcal{F}$ must make more than $|Q(n, \lfloor \log_2 k \rfloor)|$ function value queries, in the worst case.*

4.1 The Generalization of the Technique Developed by Roth and Benedek

Let K be a commutative field with identity elements $0, 1$ of addition and multiplication, respectively. In this subsection, we generalize the technique developed on the vector space $GF(2)^l$ over $GF(2)$ to the vector space K^l over K.

We define a total order on the set 2^{A_n} as follows :

$$S_1 \leq S_2 \Leftrightarrow \sum_{x_i \in S_1} 2^i \leq \sum_{x_i \in S_2} 2^i.$$

---

[1] Note that inductive-read-k-$\mathcal{F}$ properly contains read-k-$\mathcal{F}$.

Next, we define a $|Q(n,k)| \times 2^n$ matrix $H_{n,k}$ composed only of identity elements of addition and multiplication of K. We use elements of $Q(n,k)$ and 2^{A_n} to specify the rows and columns, respectively. Both the rows and columns of $H_{n,k}$ are arranged in the order '$\leq$' defined above. The (S_1, S_2)-component of $H_{n,k}$ is defined as follows :

$$H_{n,k}[S_1, S_2] = \begin{cases} 1 & \text{if the assignment } 0_{S_1} \text{ satisfies the term } \prod_{x \in S_2} x \\ 0 & \text{otherwise} \end{cases}$$

For $F = \sum_{t \in MT} a_t t$, (where $a_t \in K$ and addition and multiplication are operations on K,) consider the 2^n dimensional column vector f whose S-component $f[S]$ is a_t such that $v(t) = S$. Then, the $|Q(n,k)|$ dimensional column vector g calculated by $g = H_{n,k}f$ is the vector whose S-component $g[S]$ is $F(0_S)$, where $S \in Q(n,k)$.

For $f \in K^n$, $|f|$ denotes the number of nonzero components in f. We define K_k^n as follows:
$$K_k^n = \{f \in K^n : |f| \leq k\}.$$

'Interpol', which essentially parallels the algorithm of the same name developed by Roth and Benedek [8], shown in Figure 2, is a procedure which solves simultaneous linear equations representable by the matrix $H_{n,k}$ with the restriction that the solution is in $K_{2^k-1}^{2^n}$. This is a very efficient algorithm because it runs in time polynomial in n though the number of components of $H_{n,k}$ is $2^n|Q(n,k)|$, if all vectors f, f_+, f_0, f_1 are represented by lists of the indices and values of their nonzero components. 'Interpol' here is almost the same as the original algorithm, but one point of deviation is that $f_1 := f_+ \oplus f_0, f_0 := f_+ \oplus f_1, g_2 \oplus g_0$ are replaced by $f_1 := f_+ - f_0, f_0 := f_+ - f_1, g_2 - g_0$, which makes no change on $GF(2)^l$ over $GF(2)$. The other point is that our version of the algorithm computes not only indices of nonzero components but also their values.

Lemma 4.1 (Generalization of Lemma 3.1 and Theorem 3.1 in [8])
Suppose the mapping $\phi_{H_{m,l}}$ from $K_{2^l-1}^{2^m}$ to $K^{|Q(m,l)|}$ defined by $\phi_{H_{m,l}}(f) = H_{m,l}f$ is one-to-one for all nonnegative integers m,l. Then, upon input of arbitrary nonnegative integers n, k and an arbitrary vector $g \in K^{|Q(n,k)|}$, the behavior of procedure 'Interpol' satisfies all of the following.

1. *If there exists a vector $f \in K_{2^k-1}^{2^n}$ satisfying $g = H_{n,k}f$, it outputs such f (it is unique) and STATUS= "success."*

2. *Otherwise, it outputs STATUS= "failure."*

3. *It runs in time[2] $O(2^k n |Q(n,k)|)$.*

(Proof) This lemma can be proven similarly to Lemma 3.1 and Theorem 3.1 in [8]. □

---

[2] To obtain this time complexity upper bound, we need to represent all vectors f, f_+, f_0, f_1 as lists of the indices and values of their nonzero components.

```
procedure Interpol(n, k, g) return(f,STATUS)
    n, k : nonnegative integers
    g : an element of K^{|Q(n,k)|}
    f : an element of K^{2^n}
    STATUS : "success" or "failure"
    begin
            STATUS := "success"
            if k = 0 then
                    if g = 0 then f := 0 else STATUS := "failure"
            else if n = 0 then f := g
            else begin
                    (f_+,STATUS) := Interpol(n - 1, k, g_+)
                    if STATUS="success" then
                            (f_0,STATUS) := Interpol(n - 1, k - 1, g_0)
                            if STATUS="success" then f_1 := f_+ - f_0
                            if STATUS="failure" or |f_0| + |f_1| ≥ 2^k then
                                    (f_1,STATUS) := Interpol(n - 1, k - 1, g_2 - g_0)
                                    if STATUS="success" then
                                            f_0 := f_+ - f_1
                                            if |f_0| + |f_1| ≥ 2^k then STATUS := "failure"
                                    end if
                            end if
                            if STATUS="success" then f := ( f_0 )
                                                         ( f_1 )
                    end if
            end begin
            return(f,STATUS)
    end begin
end procedure
```

[NOTE]

g_+ : the subvector of g consisting of components with indices smaller than $\{x_n\}$

g_0 : the subvector of g consisting of components with indices no smaller than $\{x_n\}$

g_2 : the subvector of g_+ with indices belonging to $Q(n - 1, k - 1)$.

Figure 2: Procedure Interpol

We define ind-k-K^{2^n} as follows :

$$\text{ind-}k\text{-}K^{2^n} \;=\; \{f : f \text{ has at most } k \text{ nonzero components}$$
$$\text{indexed by elements of } 2^{A_m} - 2^{A_{m-1}}, \text{ for each } m = 1, 2, ..., n\}.$$

Hellerstein and Warmuth [5] proposed and analyzed a learning algorithm for read-k multivariate polynomials over GF(2). The basic structure of 'Read-k-interpol' is based on their algorithm and is improved by using 'Interpol' and generalized for a general vector space K^l over K, i.e., it also computes values of nonzero components.

```
procedure Read-k-interpol(n, k, g) return(f,STATUS)
     n, k : nonnegative integers
     g : an element of K^{|Q(n,k+1)|}
     f : an element of K^{2^n}
     STATUS : "success" or "failure"
     begin
          STATUS:="success"
          if n = 0 then f := g
          else begin
               (f_1,STATUS) := Interpol(n − 1, k, g_2 − g_0)
               if STATUS="success" then
                    (f_0,STATUS) := Read-k-interpol(n − 1, k, g_0)
                    if STATUS="success" then f = ( f_0 )
                                                  ( f_1 )
               end if
          end begin
          return(f,STATUS)
     end begin
end procedure
```

[NOTE]

g_+ : the subvector of g consisting of components with indices smaller than $\{x_n\}$

g_0 : the subvector of g consisting of components with indices not smaller than $\{x_n\}$

g_2 : the subvector of g_+ with indices belonging to $Q(n − 1, k)$.

Figure 3: Procedure Read-k-interpol

Lemma 4.2 *Suppose the mapping* $\phi_{H_{m,l}}$ *from* $K_{2^l-1}^{2^m}$ *to* $K^{|Q(m,l)|}$ *defined by* $\phi_{H_{m,l}}(f) = H_{m,l}f$ *is one-to-one for all nonnegative integers* m, l. *Then, upon input of arbitrary nonnegative integers* n, k *and an arbitrary vector* $g \in K^{|Q(n,k+1)|}$, *the behavior of procedure 'Read-k-interpol' satisfies all of the following.*

1. *If there exists a vector* $f \in$ *ind-*$(2^k − 1)$*-*K^{2^n} *satisfying* $g = H_{n,k+1}f$, *it outputs such* f *(it is unique) and STATUS="success."*

2. *Otherwise, it outputs STATUS="failure."*

3. *It runs in time[3]* $O(2^k n|Q(n, k + 1)|)$.

The proof of this lemma is basically similar to the proof of Corollary in [5], so we only show the next lemma which proves the uniqueness claimed in Lemma 4.2 1.

Lemma 4.3 *Suppose mapping* $\phi_{H_{m,l}}$ *from* $K_{2^l-1}^{2^m}$ *to* $K^{|Q(m,l)|}$ *defined by* $\phi_{H_{m,l}}(f) = H_{m,l}f$ *is one-to-one for all nonnegative integers* m, l. *Let* n, k *be arbitrary nonnegative integers. Then,* $\phi_{H_{n,k+1}}$ *is one-to-one as a mapping from* *ind-*$(2^k − 1)$*-*K^{2^n} *to* $K^{|Q(n,k+1)|}$.

---

[3]The same assumption as in Lemma 4.1 is necessary.

(Proof) Let f, f' be arbitrary distinct elements of ind-$(2^k - 1)$-K^{2^n}. Define m_0 as the maximum $m \in \{1, 2, ..., n\}$ satisfying the condition that there exists $S \in 2^{A_m} - 2^{A_{m-1}}$ such that S-components of f and f' are distinct. Let f_{m_0}, f'_{m_0} be subvectors of f, f' consisting of the components indexed by elements of $2^{A_{m_0}}$. To prove $H_{n,k+1}f \neq H_{n,k+1}f'$, we only have to show $H_{m_0,k+1}f_{m_0} \neq H_{m_0,k+1}f'_{m_0}$ because $H_{n,k+1}, f, f'$ are decomposable to $\begin{pmatrix} H_{m_0,k+1} & X \\ Y & Z \end{pmatrix}, \begin{pmatrix} f_{m_0} \\ h \end{pmatrix}, \begin{pmatrix} f'_{m_0} \\ h \end{pmatrix}$, respectively [8, 7]. We can also decompose $H_{m_0,k+1}$ to $\begin{pmatrix} H_{m_0-1,k+1} & H_{m_0-1,k+1} \\ H_{m_0-1,k} & O \end{pmatrix}$ (see [8, 7]). Let $\begin{pmatrix} f_0 \\ f_1 \end{pmatrix}, \begin{pmatrix} f'_0 \\ f'_1 \end{pmatrix}$ be the decomposed representations of f_{m_0}, f'_{m_0}, where f_0, f'_0 and f_1, f'_1 consist of the components indexed by elements of $2^{A_{m_0-1}}$ and $2^{A_{m_0}} - 2^{A_{m_0-1}}$, respectively. If $H_{m_0-1,k}f_0 \neq H_{m_0-1,k}f'_0$, then $H_{m_0,k+1}f_{m_0} \neq H_{m_0,k+1}f'_{m_0}$ holds trivially. Assume $H_{m_0-1,k}f_0 = H_{m_0-1,k}f'_0$. Notice $|f_1| \leq 2^k - 1, |f'_1| \leq 2^k - 1$ and $f_1 \neq f'_1$. So, by the assumption of this lemma, $H_{m_0-1,k}f_1 \neq H_{m_0-1,k}f'_1$ holds. Thus, $H_{m_0-1,k}(f_0 + f_1) \neq H_{m_0-1,k}(f'_0 + f'_1)$ holds. This means $H_{m_0,k+1}f_{m_0} \neq H_{m_0,k+1}f'_{m_0}$ because $H_{m_0-1,k}(f_0 + f_1)$ is a subvector of $H_{m_0,k+1}f_{m_0}$. $\square$

4.2 The Case with Real Coefficients

From Lemma 4.1 and 4.2, we only have to show that K satisfies Condition 1 to prove exact learnability of the k-term and inductive-read-k subclass of functions representable by $F = \sum_{t \in MT} a_t t$, where $a_t \in K$ and addition and multiplication are defined on K.

Condition 1 The mapping $\phi_{H_{m,l}}$ from $K_{2^l-1}^{2^m}$ to $K^{|Q(m,l)|}$ defined by $\phi_{H_{m,l}}(f) = H_{m,l}f$ is one-to-one for all nonnegative integers m, l.

In order to show that K satisfies Condition 1, it suffices to show that every $2(2^l - 1)$ set of column vectors of $H_{m,l}$ is linearly independent, which we can show from the next proposition Roth and Benedek used.

Proposition 4.1 *Let H be an $m \times n$ matrix consisting of elements of K and k be a nonnegative integer. Then, the following two conditions are equivalent.*

1. *Every set of $2k$ column vectors is linearly independent.*

2. *The mapping ϕ_H from K_k^n into K^m defined by $\phi_H(f) = Hf$ is one-to-one.*

Roth and Benedek derived the above linear independence in $GF(2)^{2^m}$ over $GF(2)$ from the fact that the matrix $H_{m,l}$ is the parity check matrix of the binary $(m-l-1)$st order Reed-Muller code of length 2^m, the minimum distance of which is 2^{l+1}. The corresponding linear independence in $\Re^{2^m}$ over $\Re$ is implied by the same in $GF(2)^{2^m}$ over $GF(2)$ from the next lemma.

Lemma 4.4 *If $v_1, v_2, ..., v_k \in \{0,1\}^n$ are linearly independent in vector space $GF(2)^n$ over $GF(2)$, then $v_1, v_2, ..., v_k$ are linearly independent in vector space $\Re^n$ over $\Re$.*

(Proof) We prove the contraposition. Assume the existence of k real numbers $a_1, a_2, ..., a_k$ containing at least one nonzero number and satisfying

$$a_1 v_1 + a_2 v_2 + ... + a_k v_k = 0 \qquad (2)$$

on vector space $\Re^n$. Since the v_i's are integral vectors, (2) with irrational coefficients can be reduced to a number of similar equations with rational coefficients. Hence, we can assume that the a_i's are integers without loss of generality. We can also assume that one of $a_1, a_2, ..., a_k$ is odd, because if all of them are even, then each a_i can be replaced by $\frac{a_i}{2}$. If we define $b_i = a_i \bmod 2$, then on vector space $GF(2)^n$,

$$b_1 v_1 + b_2 v_2 + ... + b_k v_k = 0$$

holds and b_i corresponding to an odd number a_i is not zero. $\qquad\square$

We have thus shown that $\Re$ suffices Condition 1.

Lemma 4.5 *For all nonnegative integers n, k, the mapping $\phi_{H_{n,k}}$ from $\Re^{2^n}_{2^k-1}$ to $\Re^{|Q(n,k)|}$, defined by $\phi_{H_{n,k}}(f) = H_{n,k} f$, is one-to-one.*

(Proof of Theorem 4.1)
1. Lemma 4.5 allows us to use Lemma 4.1. Applying Lemma 4.1 with $\lfloor \log_2 k \rfloor + 1$ in place of k gives us the bounds.
2. Applying Lemma 4.2 with $\lfloor \log_2 k \rfloor + 1$ gives us the bounds.
3. This proof is a straightforward generalization of the corresponding result on multivariate polynomials over GF(2) in [8]. Consider the following set of k-term linear combinations of monotone terms: $\mathcal{G} = \{ \prod_{x \in S}(1-x) \prod_{y \in A_n - S} y : S \in Q(n, \lfloor \log_2 k \rfloor) \}$. An arbitrary element $\prod_{x \in S}(1-x) \prod_{y \in A_n - S} y$ of $\mathcal{G}$ assumes value 1 for the assignment 0_S and value 0 for all other assignments. Thus, value 0 for the assignment 0_S satisfying $|S| > \lfloor \log_2 k \rfloor$ contradicts no member of $\mathcal{G} \cup \{0\}$, and value 0 for the assignment 0_S satisfying $|S| \leq \lfloor \log_2 k \rfloor$ contradicts only one member of $\mathcal{G} \cup \{0\}$. So, the number of queries necessary to remove all members but one of $\mathcal{G} \cup \{0\}$ is at least $|\mathcal{G} \cup \{0\}| - 1 = |\mathcal{G}| = |Q(n, \lfloor \log_2 k \rfloor)|$. $\qquad\square$

5 Remarks

In this paper, we studied exact learnability of functions that can be defined as a weighted sum of a set of monotone terms. If instead of a weighted sum, an operator such as '$\vee$' or '$\oplus$' is applied on the terms, then the corresponding function class is DNF or the class of multivariate polynomials over GF(2). Exact learnability of these classes by queries has been studied by many researchers, and as described in the introduction, the starting point of our research was to extend the family of learnable classes by allowing more powerful queries. From this view point, we compare part of our results with other results on the learning problems for DNF and multivariate polynomials over GF(2).

Figure 4 summarizes bounds on the number of function value queries required to identify a set of monotone terms when the operators applied on them are '$\vee$',

'$\oplus$' and '+.' First, we explain past results concerning the '$\vee$' and '$\oplus$' columns and then compare the three columns for each row.

target set T		$\vee_{t\in T}t$	$\oplus_{t\in T}t$	$\sum_{t\in T}t$
k-term	U	$kn+n^{k-1}$ [4]	$n^{\lfloor\log_2 k\rfloor+1}$ [8] Lem. 2.1	$(n-\lfloor\log_2 k\rfloor+1)k$ Th. 3.1
	L	2^{k-1} [1] Th. 2	$(\frac{n}{\lfloor\log_2 k\rfloor})^{\lfloor\log_2 k\rfloor}$ [8]	$\frac{nk-1}{\log_2(k^2+1)}$ Th. 3.3
read-once	U	$\frac{1}{4}n^2+\frac{3}{2}n+2$ ([2] Th.5)	$\frac{1}{4}n^2+\frac{5}{2}n$	$\frac{1}{4}n^2+n+1$ Th. 3.2
	L	?	?	?
read-k $(k\geq 2)$	U	?	$n^{\lfloor\log_2 k\rfloor+2}$ [5] Lem. 3	$\frac{((n-\lfloor\log_2 k\rfloor)(n-\lfloor\log_2 k\rfloor+1)+2)k}{2}$ Th. 3.2
	L	$2^{\lfloor\frac{n}{2}\rfloor}$ [1] Th. 2	?	?

T : a set of monotone terms
U : upper bound
L : lower bound
(Some bounds are not satisfied when $n\leq 1$ or $k\leq 1$.)

Figure 4: Comparison of value of three function classes

Angluin [1] and Gu and Maruoka [4], independently, gave a learning algorithm for monotone DNF using at most nk membership queries and $k+1$ equivalence queries, where n and k are the numbers of variables and terms in the target formula, respectively. Gu and Maruoka [4] also shown that k-term monotone DNF can be learned with at most $nk+n^{k-1}$ $(k\geq 1)$ membership queries and no equivalence queries. Angluin has also shown a result that implies lower bounds for the number of membership queries needed when no other queries are used: $2^{\lfloor\frac{n}{2}\rfloor}$ for (read-twice) monotone DNF, 2^{k-1} for k-term monotone DNF provided $k-1\leq\frac{n}{2}$. In this paper, we have evaluated upper bounds for the number of membership queries needed for read-once monotone DNF and multivariate polynomials over GF(2) including the coefficients ($\approx\frac{1}{4}n^2$) as indicated in Figure 4, which improves somewhat the $O(n^2)$ bound due to Angluin, Hellerstein and Karpinski for read-once monotone formulas [2].

As for multivariate polynomials[4] over GF(2), Roth and Benedek [8], Hellerstein and Warmuth [5], independently, showed exact learnability of k-term subclass using $q(n, \lfloor\log_2 k\rfloor+1)$ membership queries, where $q(n,l)=\sum_{i=0}^{l}\binom{n}{i}$. Roth and Benedek have also shown that the number of membership queries needed to learn the same class is at least $q(n, \lfloor\log_2 k\rfloor)$. Hellerstein and Warmuth [5] proved exact learnability of read-k subclass with at most $q(n, \lfloor\log_2 k\rfloor+2)$ membership queries.

---

[4]Ben-Or and Tiwari [3] showed that k-term subclass of multivariate polynomials over the reals is exactly learnable using at most $2k$ function value queries.

The upper and lower bounds shown in the last column in Figure 4 are part of our results in this paper. Let us now compare the three columns in each row in Figure 4. For read-once formulas, the upper bounds for the three function classes obtained so far are polynomial in n and have no significant difference. For k-term formulas, the upper bounds for all three function classes are also polynomial in n, but the degrees of n progresses as $k - 1$ for '$\vee$', $\lfloor \log_2 k \rfloor + 1$ for '$\oplus$' and 1 for '$+$'. The upper bounds for read-k $\oplus_{i \in T} t$ and $\sum_{i \in T} t$ are polynomial in n although the degrees of n are improved as $\lfloor \log_2 k \rfloor + 2$ for '$\oplus$' and 2 for '$+$', but no algorithm to learn read-k $\vee_{i \in T} t$ using a number of membership queries polynomial in n exists. Notice that the upper bound for $\sum_{i \in T} t$ is polynomial in both n and k, where k is the number of terms, but no such algorithms to learn $\vee_{i \in T} t$ and $\oplus_{i \in T} t$ exist. From consideration above, we can see that $\sum_{i \in T} t$ has the highest value and the second is $\oplus_{i \in T} t$ regarded as information sources to identify the set of monotone terms.

Acknowledgement

We would like to thank Prof. Manfred K. Warmuth of U. C. Santa Cruz and Prof. Lisa Hellerstein of Northwestern University for sharing their work on learning GF(2) polynomials with us.

References

[1] D. Angluin. Queries and concept learning. *Machine Learning*, 2:319–342, 1988.

[2] D. Angluin, L. Hellerstein, and M. Karpinski. Learning read-once formulas with queries. *Journal of the ACM*, 40(1):185–210, January 1993.

[3] M. Ben-Or and P. Tiwari. A Deterministic Algorithm For Sparse Multivariate Polynomial Interpolation. In *Proc. of the 20th Annual ACM Symposium on Theory of Computing*, pages 301–309, 1988.

[4] Q. P. Gu and A. Maruoka. Learning Boolean Functions. *Technical Report of IEICE COMP87-82*, 1988.

[5] L. Hellerstein and M. Warmuth. Interpolating GF[2] polynomials. Unpublished manuscript.

[6] K. Kakihara and H. Imai. Notes on the PAC learning of geometric concepts with additional Information. In *Proc. of Third Workshop on ALT*, pages 252–259, October 1992.

[7] F. J. MacWilliams and N. J. A. Sloane. *The Theory of Error-Correcting Codes. Noth-Holland, Amsterdam*, 1977.

[8] R. M. Roth and G. M. Benedek. Interpolation and approximation of sparce multivariate polynomials over GF[2]. *SIAM J. Comput.*, 20(2):291–314, April 1991.

Thue Systems and DNA —
A Learning Algorithm for a Subclass

Rani Siromoney, D.G.Thomas, K.G.Subramanian and V.R.Dare

Department of Mathematics, Madras Christian College,
Tambaram, Madras 600 059, India.

Abstract. Thue systems are considered to be appropriate for modeling certain kinds of behavior in DNA sequences [6]. We present a polynomial time algorithm for learning a Church-Rosser Thue system when the quotient monoid is finite, making use of membership queries related to the congruential languages, which specify regular languages.

1.Introduction

Tom Head [6] has formalized the effect of restriction enzymes and a ligase that allow DNA molecules to be cleaved and reassociated to produce further molecules. A regularity result is proved in [5] and identification of splicing systems from examples established in [13]. DNA molecules occur as circular strings as well, and extensions are considered in [7,12]. Tom Head has pointed this out in the conclusion, leaving open the problem of characterizing and studying certain other kinds of behavior of DNA sequences. There are DNA sequences,which he calls excision sequences, that have the ability (in the presence of appropriate enzymes) to insert into other DNA sequences in which they appear. Such behavior could be represented in a natural way in terms of Thue systems and the associated congruential languages [6]. Thue systems are string rewriting systems which present quotient monoids, whose members are congruence classes specifying formal languages. The union of finitely many congruence classes of Thue system, called congruential languages has been studied [3,10] and in particular, for Church-Rosser Thue system.

Learning an unknown concept using queries and examples has been widely studied in recent research and efficient learning methods for different domains are available including those for regular sets and context-free languages [1,11,14]. Since for Church-Rosser Thue systems, the word problem is solvable in linear time, we present a learning algorithm for a subclass when the quotient monoid is finite, using membership queries to the congruence classes that specify regular languages. The time taken for the algorithm is polynomial in the number of congruence classes which depend on the rules of the Thue system.

2. Thue-Systems and DNA

Even though several actions and behavior of DNA sequences have been modeled, there are still quite a few which need to be formalized and theoretical studies made. Among these are those of the introns (an intervening sequence, which generates that part of precursor RNA which is excised during transcription and does not form mRNA and therefore does not specify the primary structure of the gene product), transposons (mobile elements which can move in and out of DNA sequences), damage and repair of DNA and retrovirus and reverse transcription.

In a Thue system, each rule u↔v is symmetric, i.e. u can be replaced by v and v can be replaced by u. We feel that a Thue system may be used to model two kinds of behavior found in DNA sequences. One is that of the excision of introns, which is a well-accepted phenomenon during transcription in RNA splicing in eukaryotes. The other is to represent the damage caused during replication or due to natural causes and the repair, sometimes spontaneous, but otherwise as a remedy for a disease caused by the mutation.

DNA sequences can suffer spontaneous and chemical and radiation induced damages like base substitutions, deletions and insertions. Mutation results when damage changes the coding properties of bases. Replication errors by DNA polymerase may cause single base pair changes or deletions or additions of one or a few base pairs and there are specific enzymatic mechanisms to repair damaged sites. As one example, the base cytosine suffers spontaneous deamination to the closely related base uracil which pairs in DNA like thymine, thus a GC base pair becomes an AT base pair when the DNA is replicated. If this change is not repaired, this may give rise to an intolerable rate of mutation. Some DNA repair enzymes recognize and reverse specific damages in DNA.

Comparison of DNA and protein sequences, shows that they are collinear in prokaryotes, but in eukaryotes coding regions may be interrupted. An interrupted eukaryotic gene consists of exons that are spliced together in RNA, removing the introns. The existence of interrupted genes was revealed by experiments to identify the DNA corresponding to a particular mRNA. Many eukaryotic genes are much longer than their mRNA's. How did interrupted genes evolve? What was the original form of genes that today are interrupted? Did the ancestral protein-coding units consist of uninterrupted sequences of DNA, into which introns were subsequently inserted? Or did these genes initially arise as interrupted structures, which have been maintained in this form? Another form of this question is to ask whether the difference between eukaryotic and prokaryotic genes is to be accounted for by the acquisition of introns in the eukaryotes or by the loss of introns from the prokaryotes. These questions are raised in [9]. There are experimental data to show that introns were present in eukaryotes before plants and animals diverged. This suggests that introns may have been present before divergence between higher and lower eukaryotes but does not tell us whether the evolution of the gene occurred by loss of existing introns or gain of further introns. On the other hand, the organization of some genes shows extensive discrepancies between species. In these cases, there must have been extensive removal or insertion of introns during evolution. The best characteristic case is represented by the actin gene which points towards loss of introns from the primordial gene to the current one. A similar and a more dramatic situation may apply to fibrinogen [9].

On the other hand, polymorphisms seem common in genes for a rRNA and tRNA where alternative forms can often be found, with and without introns. In the case of the tRNAs, where all the molecules conform to the same general structure, it seems unlikely that evolution brought together the two regions of the gene. The different regions are involved in the base pairing that gives significance to the structure. So here it may be that the introns were inserted into continuous genes [9].

Though it is well accepted that introns are excised during eukaryotic transcription, there are several questions to be answered in the evolutionary set up due to presence of interrupted genes showing existence of introns in the current eukaryotic genomes, and absence of introns in prokaryotic genomes. These questions motivate the modelling of behavior/presence of introns by special classes of Thue systems, and it is meaningful to consider learning algorithms for Thue systems i.e., to find the Thue rules from examples which means inferring presence or absence of introns from experimental data or evolutionary evidence. This may help towards unraveling some of these unknown mysteries. In formal languages and automata, the higher the class the more difficult it is to obtain efficient learning algorithms. Quite a bit of effort is spent on learning algorithms for DFA. In the case of Thue systems also, we begin by considering specific subclasses for which efficient learning algorithm can be found.

3. Preliminaries

Let A be an alphabet and A^* the free monoid generated by A. The empty string e is the identity element of A^*. $|x|$ denotes the length of the string x. We consider in the sequel only finite Thue systems [3] on a finite alphabet A.

A finite Thue system T on a finite alphabet A, is a finite subset of $A^* \times A^*$. Each element $(u,v) \in T$ is called a rewriting rule. The symmetric relation $\leftrightarrow_T$ is defined as follows: for $(u,v) \in T$ and $x,y \in A^*$, $xuy \leftrightarrow_T xvy$ and $xvy \leftrightarrow_T xuy$. The reflexive, transitive closure of this relation denoted by $\leftarrow^*\rightarrow_T$ is a congruence relation, called Thue congruence generated by T. Two strings w and z are congruent (modT) if and only if $w \leftarrow^*\rightarrow_T z$. The congruence class of w (mod T) is $[w]_T = \{z \in A^* / z \leftarrow^*\rightarrow_T w\}$. The set of all congruence classes of the congruence $\leftarrow^*\rightarrow_T$ forms a monoid, denoted by M_T, which is the quotient of A^* by $\leftarrow^*\rightarrow_T$ and $[e]_T$ is the identity element. Multiplication in M_T is given by $[x]_T \circ [y]_T = [xy]_T$.

If $x,y \in A^*$, $x \leftrightarrow_T y$ and $|x| > |y|$, then $x \to_T y$. This is the reduction relation. The reflexive, transitive closure of $\to_T$ is denoted by $\to^*_T$. If $u \to^*_T v$ for some $u,v \in A^*$, we say that u is reduced to v using T.

A Thue system T on A is said to be length-reducing if for every $(u,v) \in T$, $|u| > |v|$. A string $x \in A^*$ is said to be irreducible (modT) if there is no string y such that $x \to_T y$. Let IRR(T) denote the set of all strings that are irreducible (modT).

Let $RED(T) = A^* - IRR(T)$. Then for a finite, length-reducing Thue system T on A, $RED(T) = A^*.Domain(T).A^*$ where $Domain(T) = \{u/(u,v) \in T\}$.

A Thue system T on A is called Church-Rosser if for every choice of x and y, $x \leftarrow^* \to_T y$ implies that for some z, $x \to^*_T z$ and $y \to^*_T z$, $x,y,z \in A^*$. For a finite Church-Rosser Thue system T on A, (i) every congruence class has a unique irreducible string (modT) (ii) the word problem for T and hence the membership problem, namely "Is $y \in [x]_T$, for $x,y \in A^*$?" is decidable in linear time (iii) it is decidable whether the quotient monoid M_T is finite or not and the cardinality of M_T, denoted card (M_T) = card (IRR(T)), the cardinality of the set IRR(T) and (iv) every congruence class of a string (modT) is a context sensitive language [3,4,10]. A Thue system T on A is called reduced if for every rewriting rule $(u,v) \in T$, $v \in IRR(T)$ and u cannot be reduced using the system $T - \{(u,v)\}$. Two finite Thue systems T and U on A are called equivalent if for all $x,y \in A^*$, $x \leftarrow^* \to_T y$ implies $x \leftarrow^* \to_U y$ and conversely.

We state certain results concerning Church-Rosser Thue systems, needed for our learning algorithm.

Lemma 1 [8]: For any finite Thue system T_1 that is Church-Rosser, there is a unique finite reduced Thue system T_2 that is Church-Rosser and equivalent to T_1. Furthermore, one can effectively construct T_2 from T_1.

Lemma 2 [8]: Let T and T' be two equivalent Thue systems. If T is Church-Rosser and IRR(T) = IRR(T'), then T' is also Church-Rosser.

4. Learning Procedure

We consider any Church-Rosser Thue system T on A for which the quotient monoid M_T is finite. $A = \{a_1, a_2, ..., a_m\}$ is a known fixed finite alphabet.

Let $M_T = \{L_1, L_2, ..., L_n\}$ where each L_i is a congruence class of a string (modT). Then the congruence relation $\leftarrow^* \to_T$ is of finite index and so each congruence class $L_i (1 \leq i \leq n)$ is regular. It is possible to make use of Angluin's method [1] for learning regular sets from membership and equivalence queries. But algebraic properties of a Church-Rosser Thue system T for which M_T is finite, enable us to present an efficient learning procedure with only membership queries, for congruence classes. The congruence relation of a Thue system T with rules (u,v), $u,v \in A^*$, partitions the set $A^*_\bullet$ of words over A into disjoint congruence classes. Any word w over A is always in A^* but is in only one congruence class of T. So, the membership query for congruence classes is meaningful and reasonable.

The unique, reduced Church-Rosser Thue system R equivalent to T is obtained. The learning procedure consists of two parts, first for IRR(R) and then for the Thue system R.

4.1 Procedure for Learning IRR(R)

For any string $w \in A^*$, as input, the oracle answers membership query by producing an n-tuple that contains n-1 zeros and one 1 since $M_R = \{L_1, L_2, ..., L_n\}$. For a Church-Rosser Thue system the word problem is solvable and when M_R is finite, the number n of congruence classes is finite. So, the value of n is known to the oracle. The learner gets this information when he inputs a membership query for the empty string e.

The input is a string $w \in A^*$ and the output is an n-tuple $t(w) = (t_1, t_2, ..., t_n)$ where $t_i = 1$, if $w \in L_i$ and $= 0$ if $w \notin L_i$ $(1 \le i \le n)$. Let p_i be the projection defined by $p_i(x) = x_i$ for any n-tuple $x = (x_1, x_2, ..., x_n)$, $1 \le i \le n$.

Membership queries are made to the oracle for the input strings, starting with the empty string e which is an irreducible string (modR) and continued with strings in $\{e\}.A$ from which irreducible strings (modR) in A are obtained. For $j = 1, 2, ..., n$, the irreducible string in L_j, is the string w_i whenever $w_i \in L_j$ but $w_k \notin L_j$ for $k = 1, 2, ..., i-1$ where $w_1 = e, w_2, w_3, ...,$ are ordered strings according to length (strings of equal length are lexicographically ordered) from the set $\{e\} \cup A \cup \{(IRR(R) \cap A^r).A \mid r \ge 1\}$.

This process is continued recursively by considering strings in $(IRR(R) \cap A^r).A$ $(r \ge 1)$ with the IRR(R) obtained so far. This gives irreducible strings (modR) in A^2, A^3, A^4 and so on. This process terminates when each L_j obtains an irreducible string (modR). We note that if $x = a_1 a_2 ... a_r$ $(a_i \in A)$ is an irreducible string (modR), then $a_1 \in IRR(R) \cap A$, $a_1 a_2 \in IRR(R) \cap A^2, ..., a_1 a_2 ... a_{r-1} \in IRR(R) \cap A^{r-1}$.

The algorithm for forming irreducible strings (modR), terminates when the process for finding irreducible strings (modR) in A^k where $k = \max\{|x| : x \in IRR(R)\}$ is over, since (a) IRR(R) is finite (b) each L_j $(1 \le j \le n)$ contains exactly one irreducible and (c) irreducible strings (modR) are shortest strings in their respective classes $L_1, L_2, ..., L_n$.

Thus we can find the irreducible strings (modR) as (say) $x_1, x_2, ..., x_n$ corresponding to the classes $L_1, L_2, ..., L_n$ i.e. $IRR(R) = \{x_1, x_2, ..., x_n\}$.

4.2 Procedure for Learning Thue System R

To identify the unique, reduced Church-Rosser Thue System R equivalent to the unknown Thue System T, we perform again the membership query as in previous procedure for the strings in the set $((IRR(R).A)-IRR(R))\cup((A.IRR(R))-IRR(R))$.

We form the Thue system

$S = \{(u,v)/u\in((IRR(R).A)-IRR(R))\cup((A.IRR(R))-IRR(R)),v\in IRR(R),$ u and v both belong to L_j for some $j(1\leq j\leq n)\}$ on A.

From S, the reduced Thue system S' equivalent to S on A is obtained and thus we obtain R which is same as S' on A.

4.3 Algorithm for Learning IRR(R)

```
begin
    IRR(R) = Φ
    Input the empty string w₁ = e
    n = number of entries in t(e)
    L₁ = {e}
    IRR(R) = {e}
    N₁ = 1
    For j = 2 to n, initialize
    Lⱼ = Φ
    Nⱼ = 0
    Input strings wᵢ(i = 2,3,...) ordered according to length   (strings of equal length
    are lexicographically ordered)   such that wᵢ ∈ A∪{(IRR(R)∩Aʳ).A | r ≥ 1}
    While Nⱼ = 0 for some j do
    begin
        For j = 1 to n do
        begin
            If pⱼ(t(wᵢ)) = 1 do
            begin
                Lⱼ = Lⱼ ∪ {wᵢ}
                If Nⱼ = 0 do
                begin
                    Nⱼ = 1
                    IRR(R) = IRR(R)∪{wᵢ}
                end
            end
        end
    end
    Output: IRR(R)
end
```

4.4 Algorithm for Learning R

begin
    Input strings $w_i(i = 1,2,3,...)$ ordered according to length
    (strings of equal length are lexicographically ordered) such that
    $w_i \in ((IRR(R).A) - IRR(R)) \cup ((A.IRR(R)) - IRR(R))$
    Initialize : $S = \Phi$
    For $u \in ((IRR(R).A) - IRR(R)) \cup ((A.IRR(R)) - IRR(R))$ *do*
    begin
        For $v \in IRR(R)$ *do*
            begin
                If $p_j(t(u)) = p_j(t(v)) = 1$ for some j $(1 \le j \le n)$,*then*
                $S = S \cup \{(u,v)\}$
            end
    end
    Initialize: $S' = S$
        begin
            For $(u,v) \in S'$, *do*
                begin
                    If $(u_1,v_1) \in S' - \{(u,v)\}$
                            such that $u = \alpha u_1 \beta, \alpha, \beta \in A^*$, *then*
                    $S' = S' - \{(u,v)\}$
                end
        end
        Output : $R = S'$
end

4.5 Correctness of the Learning Algorithm

We establish the correctness of the learning algorithm by showing that S and R are equivalent and S is Church-Rosser. Also, if S' is the reduced Thue system equivalent to S then S' = R.

Lemma 3: $IRR(S) = IRR(R)$

Proof : Since S is length-reducing,

$$
\begin{aligned}
IRR(S) \quad &= A^* - RED(S) \\
&= A^* - A^*.Domain(S).A^* \\
&= A^* - (A^* - IRR(R)) \\
&= IRR(R)
\end{aligned}
$$

Lemma 4: $x \leftarrow^* \rightarrow_S y$ implies $x \leftarrow^* \rightarrow_R y$, for $x,y \in A^*$

Proof: It is enough to prove that $x \leftrightarrow_S y$ implies $x \leftarrow^* \rightarrow_R y$, for $x, y \in A^*$. Suppose $x \leftrightarrow_S y$ holds. Then $x = \alpha u \beta$ and $y = \alpha v \beta$ where either $(u,v) \in S$ or $(v,u) \in S$ and $\alpha, \beta \in A^*$. By the definition of S, u and v both belong to L_i for some $i(1 \leq i \leq n)$ i.e., $u \leftarrow^* \rightarrow_R v$. This shows $x \leftarrow^* \rightarrow_R y$.

Lemma 5: card $(M_S) = n$

Proof : Let $IRR(S) = \{x_1, x_2, ..., x_n\}$. Now, no two x_i's are congruent (modS). The reason is as follows: Suppose $x_i \leftarrow^* \rightarrow_S x_j$, for some i and j, $1 \leq i \leq n$, $1 \leq j \leq n$, $i \neq j$. Then, by lemma 4, $x_i \leftarrow^* \rightarrow_R x_j$. This is not possible since R is Church-Rosser and x_i and x_j are irreducibles (modR). Thus, every congruence class of S has exactly one x_i. Since card $(IRR(S)) = n$, We have card $(M_S) = n$.

Lemma 6: For $x, y \in A^*$, $x \leftarrow^* \rightarrow_R y$ implies $x \leftarrow^* \rightarrow_S y$

Proof: Suppose, for $x, y \in A^*$, $x \leftarrow^* \rightarrow_R y$ holds. Since R is Church-Rosser, there exists exactly one $x_i \in IRR(R)$ such that $x \rightarrow^*_R x_i$ and $y \rightarrow^*_R x_i$. i.e. $x \leftarrow^* \rightarrow_R x_i$ and $x_i \leftarrow^* \rightarrow_R y$. Since every congruence class (modS) contains exactly one irreducible (modS) and $IRR(S) = IRR(R)$, let $x_j \in [x]_S$. This implies $x_j \leftarrow^* \rightarrow_S x$ and hence $x_j \leftarrow^* \rightarrow_R x$, which means $x_j \leftarrow^* \rightarrow_R x_i$. This is impossible since R is Church-Rosser. Hence $j = i$ i.e., $x_i \in [x]_S$. Similarly, we can show $x_i \in [y]_S$. Thus $x_i \leftarrow^* \rightarrow_S x$ and $x_i \leftarrow^* \rightarrow_S y$ together imply that $x \leftarrow^* \rightarrow_S y$.

Theorem 7: R and S are equivalent

Proof follows from lemmas 4 and 6

Theorem 8: S is Church-Rosser.

Proof follows from lemmas 2 and 3 and theorem 7.

Theorem 9: Let S′ be the reduced Thue system equivalent to S. Then S′ = R.

Proof: By lemma 1 and theorem 8, S′ is unique and Church-Rosser. Since S′ is equivalent to S, which in turn is, equivalent to R (by theorem 7) and R is reduced, again by applying lemma 1 once, we obtain S′ = R.

4.6 Time Analysis

We find that the number of strings to be processed to learn $IRR(R)$ is less than or equal to $1 + mn$ where, $m = card(A)$ which is fixed and $n = $ number of congruence classes (modR), since the set F of strings to be processed for learning $IRR(R)$ consists of strings in
$$\{e\} \cup A \cup (IRR(R) \cap A).A \cup (IRR(R) \cap A^2).A \cup ... \cup (IRR(R) \cap A^{k-1}).A$$
where $k = \max\{|x|/x \in IRR(R)\}$. In fact,

card (F) $= 1+(1.m+s_1.m+s_2.m+...+s_{k-1}.m)$ where
s_j = card $(IRR(R) \cap A^j)$, $1 \leq j \leq k$. So,
card (F) $= 1+(1+s_1+s_2+...+s_{k-1})m = 1+(n-s_k)m \leq 1+nm$.

Hence the time taken by the learning algorithm to learn IRR(R) is O(n).

Similarly, the number of strings in $((IRR(R).A)-IRR(R)) \cup ((A.IRR(R))-IRR(R))$
to be processed is bounded by $2n(m-1)$ and hence the time taken for this is O(n)

   The number of rewriting rules in the Thue system S is almost $O(n^2)$ and hence its
construction requires $O(n^2)$ time. For forming S', each of these $O(n^2)$ rewriting rules
has to be compared with each of the remaining rewriting rules in S. Hence the total
time required by the learning algorithm to form S' = R is $O(n^4)$.

4.7 An Example Run

We illustrate with an example, the learning procedure for an unknown, reduced
Church-Rosser Thue system for which M_R is finite. Table 1 gives the irreducible
elements (modR) of the congruence classes i.e., IRR(R) = {e,a,b} in response to the
membership queries.

STRINGS		CLASSES		
		L_1	L_2	L_3
e		1	0	0
A	a	0	1	0
	b	0	0	1

Table 1

From Table 2, Domain(R) is obtained which gives the set of left side of the rules of
the Thue system R. Hence from these two tables the reduced Church-Rosser Thue
system R is obtained as $\{(a^2,a), (ab,a), (b^2,b), (ba,b)\}$. The fixed alphabet A = {a,b}
is known. From R, the congruence classes (modR) are obtained as $L_1 = [e]_R = \{e\}$,
$L_2 = [a]_R = a\{a,b\}^*$ and $L_3 = [b]_R = b\{a,b\}^*$.

STRINGS		CLASSES		
		L_1	L_2	L_3
((IRR(R).A)–IRR(R))	a^2	0	1	0
$\cup$((A.IRR(R))–IRR(R))	ab	0	1	0
	b^2	0	0	1
	ba	0	0	1

Table 2

5. Finiteness of M_T and a Regularity Criterion

In general, it is known [4] that there is no algorithm to decide for an arbitrary Thue system T, whether M_T is finite or not. Also that M_T is finite if IRR(T) is finite. Hence it is of interest to examine cases where M_T is finite. In this section we consider Thue systems for which M_T is finite and obtain conditions for the finiteness of M_T, since in this paper, we have given a learning algorithm for Church-Rosser Thue systems whose quotient monoids are finite. A characterization of the language class corresponding to finite Church-Rosser monadic Thue systems is also presented.

5.1 Monadic and Special Thue Systems

A Thue system T on A is said to be (i) monadic if $(u,v) \in T$ implies $|u| > |v|$ and $v \in A \cup \{e\}$, (ii) special if $(u,v) \in T$ implies $v = e$.

Proposition 10: Let $A = \{a\}$. Let T be a special Thue system given by $\{(a^{m_1},e),(a^{m_2},e),...,(a^{m_n},e)\}$ where $m_1,m_2,...,m_n$ are positive integers ($n \geq 1$). Then, card $(M_T) = $ g.c.d $(m_1,m_2,...,m_n)$.

5.2 Sufficient Conditions Using Notion of Codes

We recall the notions of a code, factorizing code and uniform code [2] and then show that certain conditions on the Thue systems T imply the finiteness of the monoid M_T.

Let A be a finite alphabet. A subset X of A^* is a code if it satisfies the condition:

Whenever $x_1 x_2 ... x_n = y_1 y_2 ... y_m$, $x_i, y_j \in X$, $1 \leq i \leq n$, $1 \leq j \leq m$, then $n = m$ and $x_i = y_i$ for $i = 1,2,...,n$.

A code $X \subseteq A^*$ is called maximal if it is not properly contained in any other code. It is said to be factorizing if there exist two subsets P and Q of A^* and for every $w \in A^*$ there exist unique $p \in P$, $q \in Q$ and $x \in X^*$ such that $w = qxp$. The pair (P,Q) is called a factorization of X.

It is known [2] that X is a finite maximal factorizing code if and only if P and Q are finite.

A code $X \subseteq A^*$ is called a uniform code, if there exists an integer $p \geq 1$ such that $X = A^p$.

Proposition 11: Let T be a Thue system on alphabet $A = \{a_1, a_2, ..., a_m\}$. Then M_T is finite if $X = IRR(T) - \{e\}$ is a code.

Proof: Let $y \in IRR(T)$. If $|y| > 1$, let $y = b_1 b_2 ... b_n$ where $b_i \in A$ ($i \leq i \leq n$). Clearly, each $b_i \in X$. This gives a contradiction to the fact that X is a code as $y \in X$. Hence $|y| \leq 1$. Therefore $X \subseteq A$.

Proposition 12: If T is a special Thue system on A with rules (w_i, e), $i = 1, 2, ..., n$, such that $Domain(T) = \{w_i \mid i = 1, 2, ..., n\}$ is a finite maximal factorizing code, then M_T is finite.

Proof: If (P,Q) is a factorization of Domain(T), then every word $w \in A^*$ factorizes uniquely into $w = qxp$ with $p \in P$, $q \in Q$ and $x \in (Domain(T))^*$. It is clear that $IRR(T) \subseteq QP$. Since P and Q are finite, IRR(T) is finite and so is M_T.

A subset X of A^* is said to be unavoidable if every word $w \in A^*$ whose length is sufficiently long has a factor in X.

Proposition 13: Let T be a finite Thue system on A. If $X = \{u \mid (u,v)$ or $(v,u) \in T$ and $|u| > |v|\}$ is an unavoidable subset of A^*, then M_T is finite.

Proof: Since X is an unavoidable subset of A^*, there exists a smallest natural number r such that every string $w \in A^*$ with $|w| \geq r$ has a factor in X and is reducible by the rules of T. So, every irreducible word z (modT) is such that $|z| < r$. Thus M_T is finite.

Corollary 14: Let T be a finite Thue System on A. If X, as defined in proposition 13, is a uniform code, then M_T is finite.

The corollary is a consequence of proposition 13.

We now consider Church-Rosser monadic Thue systems for which the quotient monoids need not be finite.

It is well-known that every congruence class of a Church-Rosser monadic Thue system is a deterministic context-free language [3]. Generally, it is decidable when a deterministic context-free language is regular. In the following theorem, we obtain conditions underwhich the congruence classes of a Church-Rosser monadic Thue system to be regular, thus characterizing that class. The proof of the theorem is based mainly on the fact that for Church-Rosser monadic Thue systems $[x]_T \circ [y]_T = [xy]_T = [x]_T.[y]_T$ where . stands for the product of the languages and $xy \in IRR(T)$ with $x,y \in A$. But this need not be true if the Thue system fails to be Church-Rosser or monadic as seen by the following examples.

Example 1. Let $T = \{(abc,ab),(b^2c,cb)\}$ be a Thue system on $A = \{a,b,c\}$. T is Church-Rosser but not monadic. $abc \in [ab]_T = [a]_T \circ [b]_T$, but $abc \notin [a]_T.[b]_T$.

Example 2. Let $S = \{(ab,a),(ba,b)\}$ be a Thue system on $A = \{a,b\}$. S is monadic but not Church-Rosser. $a \in [a^2]_S = [a]_S \circ [a]_S$ but $a \notin [a]_S.[a]_S$.

Theorem 15: Let T be a finite Church-Rosser monadic Thue system on $A = \{a_1, a_2, ..., a_n\}$. Then, every congruence class of T is regular if and only if each one of the members of the set $K = \{[e]_T, [a_1]_T, [a_2]_T, ..., [a_n]_T\}$ is regular.

Proof: It is clear that when every congruence class of T is regular, each one of the members of K is regular. We need to prove only the converse.

Let $w \in IRR(T)$ with $w = b_1 b_2 ... b_m$, $b_i \in A$, $1 \le i \le m$ and $m \ge 2$.

Claim: $[w]_T = [b_1]_T . [b_2]_T . .... .[b_m]_T$ where the dot stands for the product of languages.

It suffices to consider the case $m = 2$, noting that the argument can be extended to the case where $m > 2$. Let $w = b_1 b_2$, $b_i \in A$, $i = 1,2$. Clearly $[b_1]_T.[b_2]_T \subseteq [w]_T$. Suppose $[w]_T \subseteq [b_1]_T.[b_2]_T$ is not true. Let y be a shortest string in $[w]_T$ and $y \notin [b_1]_T . [b_2]_T$. That is, $y \to^*_T w$ and $y \notin [b_1]_T.[b_2]_T$. Clearly $y \ne w$. Hence, $y \to_T z$ for some $z \in A^+$ and $z \in [b_1]_T.[b_2]_T$. Let $z = z_1 z_2...z_j...z_k$ where $z_t \in A$, $1 \le t \le k$ and $j \le k$. Assume that $z_1 z_2...z_j \in [b_1]_T$ and $z_{j+1} z_{j+2}...z_k \in [b_2]_T$. Since T is monadic and $y \to_T z$, there is a word $z' \in A^+$ and a rule (z',z'') in T where $z'' \in A \cup \{e\}$ such that by applying the rule on y, z is derived. There arise two cases.

case(i): Suppose $z'' = z_i$ for some $i (1 \le i \le k)$. Then $y = z_1 z_2...z_{i-1} z' z_{i+1}...z_k$. If $i < j$, then $z_1 z_2...z_{i-1} z' z_{i+1}...z_j \in [b_1]_T$ and $z_{j+1} z_{j+2}...z_k \in [b_2]_T$. This implies $y \in [b_1]_T.[b_2]_T$, which is a contradiction.

The possibilities $i > j$ and $i = j$ are similar.

case(ii): Suppose $z'' = e$. Then y must be any one of the following possibilities:

(a) $y = z'z$ (b) $y = zz'$
(c) $y = z_1 z_2...z_i z' z_{i+1}...z_k$ for some pair $(i,i+1)$ where $1 \le i \le k-1$. This means, z' occurs in between some z_i and z_{i+1}.

(a) Suppose $y = z'z$. Then $z'z_1z_2...z_j \in [b_1]_T$ and $z_{j+1}z_{j+2}...z_k \in [b_2]_T$ and so, $y \in [b_1]_T.[b_2]_T$, a contradiction. (b) and (c) are similar.

This proves $[w]_T = [b_1]_T . [b_2]_T$. Hence the claim.

Suppose all the elements of K are regular. Then, by the claim and the fact that the family of regular languages is closed under product, for every $w \in IRR(T)$, the class $[w]_T$ is regular. Since T is Church-Rosser, it is true that for every $x \in A^*$, $[x]_T = [w]_T$ for exactly one $w \in IRR(T)$ with $x \leftarrow^* \rightarrow_T w$. This proves the theorem.

6. Conclusion

In this paper, we have presented a learning algorithm for finite Church-Rosser Thue systems, when the quotient monoids are finite. The algorithm is polynomial in the number of congruence classes and uses membership queries. We are examining learning of Church-Rosser, monadic Thue systems, for which the quotient monoids are infinite.

Acknowledgment: We wish to thank Miss Lisa Mathew for a critical reading of the manuscript and useful suggestions and Miss Anne Stevens for discussions regarding DNA sequences.

Financial support from Fujitsu Laboratories Ltd. is gratefully acknowledged.

References

1. D.Angluin:Learning regular sets from queries and counter examples, Information and Computation 75, 87-106 (1987)

2. J.Berstel, D.Perrin:"Theory of Codes", Academic press, 1985

3. R.Book: Thue systems as rewriting systems, J.Symbolic Computation 3, 39-68 (1987)

4. R.Book, C.O'Dunlaing: Thue congruences and the Church-Rosser property, Semi Group Forum 22, 325-331 (1981)

5. K.Culik II,T.Harju: Splicing semigroups of dominoes and DNA, Discrete Applied Mathematics, 31, 261-277 (1991).

6. T.Head: Formal Language theory and DNA: An analysis of the generative capacity of specific recombinant behaviours. Bulletin of Mathematical Biology, 49(6):737-759 (1987).

7. T.Head: Splicing schemes and DNA in "Lindenmayer Systems (Impacts on Theoretical Computer Science, Computer Graphics, and Developmental Biology)" (G.Rozenberg and A.Salomaa, Eds.), Springer-Verlag 1992, 371-383.

8. D.Kapur,P.Narendran: The Knuth Bendix completion procedure and Thue systems, Siam J.Computing, 14, 1052-1072 (1985).

9. B.Lewin: "GenesIV", Oxford University Press, Delhi 1990.

10. P.Narendran,C.O'Dunlaing,H.Rolletschek:Complexity of certain decision problems about Congruential languages, J. Comp. Syst. Sci. 30, 343-358 (1985).

11. Y.Sakakibara:Learning context - free grammars from structural data in polynomial time, Theor. Comp. Sci. 76, 223-242 (1990).

12. R.Siromoney,K.G.Subramanian,V.R.Dare: Circular DNA and splicing systems, Lecture Notes in Comp.Sci.654, 1992, 260-273.

13. Y.Takada,R.Siromoney: On identifying DNA splicing systems from examples, Lecture Notes in Arti. Int. 642, 1992, 305-319.

14. T.Yokomori: Learning non-deterministic finite automata from queries and counter examples, "Machine Intelligence" Vol.13. (Furukawa, Michie and Muggleton, Eds.), Oxford University Press (in Press).

The VC-dimensions of Finite Automata with n States

Yoshiyasu Ishigami[†]

Department of Mathematics
Waseda University
3-4-1 Okubo, Shinjuku-ku
Tokyo 169, Japan
62m502@cfi.waseda.ac.jp

Sei'ichi Tani

Center for Information Science
Tokyo Woman's Christian University
2-6-1 Zempukuji, Suginami-ku
Tokyo 167, Japan
tani@twcu.ac.jp

Abstract

In this paper, we investigate the VC-dimensions of finite automata. We show that for a fixed positive integer k (1) the VC-dimension of $\mathrm{DFA}_{k,n}$, which is the class of dfas of an alphabet size k whose minimum states dfa has at most n states, is $(k - 1 + o(1))n \log n$, (2) the VC-dimension of $\mathrm{NFA}_{k,n}$, which is the class of nfas of an alphabet size k whose minimum states nfa has at most n states, is $\Theta(n^2)$ and (3) the VC-dimension of $\mathrm{CDFA}_{k,n}$, which is the class of commutative dfas of an alphabet size k whose minimum states commutative dfas has at most n states, is $(1 + o(1))n$. These results are applied to the problems in computational learning theory.

1   Introduction

In this paper, we investigate Vapnick-Chervonenkis dimensions of deterministic finite automata (dfas for short), nondeterministic finite automata (nfas for short) and commutative deterministic finite automata (cdfas for short). Commutative dfas accepts only commutative regular languages. $\mathrm{DFA}_{k,n}$ denotes the class of dfas of an alphabet size k whose minimum states dfa has at most n states and $\mathrm{NFA}_{k,n}$ denotes the class of nfas of an alphabet size k whose minimum states nfa has at most n states. Let Σ_2 be a two-letter alphabet and $f(n)$ be the maximal number of states of the minimal automaton of a subset of $\Sigma_2^{=n}$ where $\Sigma_2^{=n}$ denotes the set of all words of length n. Champarnaud and Pin showed that $1 = \liminf_{n \to \infty} nf(n)/2^n \leq \limsup_{n \to \infty} nf(n)/2^n = 2$ [10] and this result implies that $\mathrm{VC\text{-}dim}(\mathrm{DFA}_{2,n}) \geq 1/2 \cdot n \log n$. Gaizer independently showed that $\mathrm{VC\text{-}dim}(\mathrm{DFA}_{k,n}) = \Theta(n \log n)$ [13]. We decide the asymptotic behavior of $\mathrm{VC\text{-}dim}(\mathrm{DFA}_{k,n})$. We show that for a fixed integer $k \geq 2$, $\mathrm{VC\text{-}dim}(\mathrm{DFA}_{k,n}) = (k - 1 + o(1))n \log n$. We also show that $(k - 1)n^2 \leq \mathrm{VC\text{-}dim}(\mathrm{NFA}_{k,n}) \leq kn^2$.

$\mathrm{CDFA}_{k,n}$ denotes the class of commutative dfas of an alphabet size k whose minimum states cdfas at most n states. (Regarding commutative dfa, see [16].) Abe showed that $\mathrm{VC\text{-}dim}(\mathrm{CDFA}_{k,n}) = \Omega(k + n)$ and $\mathrm{VC\text{-}dim}(\mathrm{CDFA}_{k,n}) =$

[†]Partially supported by the Grant in Aid for Scientific Research of the Ministry of Education, Science and Culture of Japan

$O(kn \log n)$ [1]. We improve Abe's the lower bound by showing that VC-dim(CDFA$_{k,n}$) = $\Omega(k \log n + n)$. We also consider the VC-dimension of cdfas in case when the cardinality of alphabet k is treated as a constant. We show that for a positive integer k, VC-dim(CDFA$_{k,n}$) = $(1 + o(1))n$. The cardinality of alphabet contributes the VC-dimension of dfas but seems not to contribute the VC-dimension of cdfas.

The relationship between the VC-dimension and the computational complexity of learning has been investigated [9, 8, 14]. Maass and Turán [15] investigated — the query complexity — the number of equivalence and membership queries that are needed to exact-learn an unknown concept from some concept class. Let $\mathcal{B}$ be a class of concepts over some finite universe. Maass and Turán showed that the query complexity of learning $\mathcal{B}$ using equivalence and membership queries is $\Omega(\text{VC-dim}(\mathcal{B}))$. They also showed that there exists a concept class such that its VC-dimension and the query complexity of learning it have the same asymptotic behavior.

In this paper, we also consider the query complexity of learning finite automata. We use our lower bound of VC-dimension of DFA$_{k,n}$ and an extension of results of Maass and Turán to show that the lower bounds of query complexity of learning dfa. Angluin [3] showed that there exists a polynomial time algorithm to learn dfas using equivalence and membership queries. Her algorithm uses $O(n)$ equivalence queries and $O(mn^2)$ membership queries where m is the length of maximum length counterexample. On the other hand, Angluin also showed that neither membership query nor equivalence query is good enough to learn dfas in polynomial time [4, 5]. Balcázar et. al [7] showed that it is impossible to reduce the number of equivalence queries more than $O(\log n)$ factor without using exponential number of membership queries in bounded learning model [17]. However it has not been investigated that when a learning algorithm can use $O(n)$ equivalence queries, how many membership queries are needed for learning dfas. We show that the query complexity of learning dfas using equivalence and membership queries is $\Omega(n \log n)$ where n is the number of states of the minimum states dfa of the target dfa. Roughly speaking, our lower bound implies that when a learning algorithm can use $O(n)$ equivalence queries, $\Omega(n \log n)$ membership queries are necessary. We also consider the query complexity of learning nfas. Yokomori [18] showed that nfas are learnable in polynomial time in the number of states of minimum states "dfa". On the other hand, Angluin and Kharitonov[6] showed that under a certain cryptographic assumption nfas is not learnable in polynomial time. But there has been no lower bound of the query complexity of learning the nfas where no unproven assumption is conditioned. We show that the query complexity of learning nfas using equivalence and membership queries is $\Omega(n^2)$ where n is the number of states of the minimum states nfa of the target nfa.

2 Preliminaries

In this paper we follow standard definitions and notations in formal language theory and computational learning theory. In particular, those for finite automata are used without definitions. The reader will find them in standard textbooks (for example [11]).

Let Σ be an alphabet. A *word* is an element of Σ^*, and a *language* is a subset of Σ^*. λ denotes the empty word. The length of a string x is denoted by $|x|$. $\Sigma^{\leq m}$ and $\Sigma^{=m}$ are used to denote the sets $\{x \in \Sigma^* : |x| \leq m\}$ and $\{x \in \Sigma^* : |x| = m\}$ respectively. For $x, y \in \Sigma^*$, $x \cdot y$ denotes the concatenation of x and y. For every alphabet $a \in \Sigma$ and every word $w \in \Sigma^*$, let $T(w \cdot a) = w$ and $last(w \cdot a) = a$.

For any word $w \in \Sigma^*$, $Proj_{[i]}(w)$ is ith alphabet of w, $Pref_{[i]}(w)$ denotes the prefix of w with the length i, and w^R is the reverse of w. For example, $Proj_{[2]}(10110) = 0$, $Pref_{[3]}(10110) = 101$, and $10110^R = 01101$. For any set A and B, $A \triangle B$ denotes $(A \cup B) - (A \cap B)$. Let Σ_k denote $\{a_1, \ldots, a_k\}$ and throughout this paper, we use Σ_k for our k-letter alphabet. For any set S, $|S|$ denotes the cardinality of S. Let $\mathbb{N}$ denote the set of nonnegative integers and $\mathbb{R}$ denote the set of real numbers. For any $n \in \mathbb{N}$, let $[n] = \{i \in \mathbb{N} : 1 \leq i \leq n\}$.

Let $f, g : \mathbb{N} \to \mathbb{R}$ such that $f(n), g(n) > 0$ for large n. We define the order of magnitude symbols O, Ω, Θ, and o as follows:

- $f = O(g)$ if for some $c > 0$ and for large n, $f(n) \leq c \cdot g(n)$,

- $f = \Omega(g)$ if some $c > 0$ and for large n, $f(n) \geq c \cdot g(n)$,

- $f = \Theta(g)$ if $f = O(g)$ and $f = \Omega(g)$,

- $f = o(g)$ if for every $c > 0$ and for large n, $f(n) < c \cdot (n)$.

Representation classes

We use the notion of "representation class"to specify "concepts", which are objectives for learners.

A *representation class* is a five-tuple $(\Gamma, \Sigma, R, \Phi, \rho)$, where

- Γ is an alphabet for representations,

- Σ is an alphabet for concepts,

- R is a subset of Γ^*, called a *representation language*,

- Φ is a mapping from R to concepts over Σ^*, called a *semantic function*, and

- ρ is a function from R to $\mathbb{N}$, called *size function*.

Since we assume $\Gamma = \{0, 1\}$ throughout this paper, we omit the alphabet for representation and write a representation class as (Σ, R, Φ, ρ). For each representation $r \in R$, the set $\Phi(r)$ is the *concept* represented by r. And $\Phi(R)$ is $\{\Phi(r) : r \in R\}$, called a *concept class*.

For any representation r of dfa or nfa, let $L(r)$ denote the language accepted by the finite automaton specified by r. Let R_{dfa} be the set of representation of dfas and R_{nfa} the set of representation of nfas. Φ_{dfa} is a semantic function for dfas that maps every $r \in R_{\mathrm{dfa}}$ to $L(r)$, and ρ_{dfa} is a function that maps every $r \in R_{\mathrm{dfa}}$ to the number of states of the minimum states dfa accepting $\Phi_{\mathrm{dfa}}(r)$. Φ_{nfa} and ρ_{nfa} for nfas are defined similarly. For example, the representation class for dfas of k-letter alphabet is $\mathrm{DFA}_k = (\Sigma_k, R_{\mathrm{dfa}}, \Phi_{\mathrm{dfa}}, \rho_{\mathrm{dfa}})$, where R_{dfa_k} $= \{r : r \text{ is a description of a dfa of the alphabet } \Sigma_k\}$. For any representation class $C = (\Sigma, R, \Phi, \rho)$, let C_n denote $(\Sigma, R_n, \Phi, \rho)$ where $R_n = \{r \in R : \rho(r) \le n\}$.

Vapnik-Chervonenkis dimension

Let $\mathcal{F} \subseteq 2^X$ be a family of sets over the (finite or infinite) universe X.

Definition 2.1 We say that S is *shattered* by $\mathcal{F}$ if for every $S' \subseteq S$, there exists set a $F \in \mathcal{F}$ such that $S' = S \cap F$. In other words, S is shattered by $\mathcal{F}$ if $\{S \cap F : F \in \mathcal{F}\}$ is the entire power set 2^S.

Definition 2.2 The Vapnik-Chervonenkis dimension (VC-dimension) of a family of sets $\mathcal{F}$ is the maximum size of a set shattered by $\mathcal{F}$ (This value is defined to be infinite if no such maximum exists).

For any representation class $C = (R, \Phi, \rho)$, let VC-dimension of C (VC-dim(C)) be the VC-dimension of $\Phi(R)$.

Learnability

Let $C = (\Sigma, R, \Phi, \rho)$ be a representation class over some universe X. In this paper, we deal with a problem to learn an unknown target concept $C_T \in \Phi(R)$ using information about C_T gathered by asking some of the following three types of queries to a teacher:

1. *Membership query* (Mem): The input is a string $x \in \Sigma^*$ and the output is *yes* if $x \in C_T$ and *no* otherwise;

2. *Equivalence query* (Equ): The input is a representation $h \in R$ and the output is *yes* if $\Phi(h) = C_T$ and *no* otherwise. When the answer is *no*, a *counterexample* is also supplied, that is, a string $s \in \Phi(h) \triangle C_T$. The choice of counterexamples is assumed to be arbitrary;

3. *Arbitrary equivalence query* (Arb): The input is a representation $h \in \Gamma^*$ of a set $H \subseteq X$ and the output is *yes* if $H = C_T$ and *no* otherwise. When the answer is *no*, a *counterexample* is also supplied, that is, a string $s \in H \triangle C_T$. The choice of counterexamples is assumed to be arbitrary. Note that it is not required that a hypothesis h is included in R.

Definition 2.3 Let $C = (\Sigma, R, \Phi, \rho)$ be a representation class and let Q be a set of types of queries. An algorithm A is a *learning algorithm for C using Q* if and only if for every $t \in R$, when A is run with an oracle which can answer Q about $\Phi(t)$, and A eventually halts and outputs some $h \in R$ such that $\Phi(h) = \Phi(t)$.

In this paper, we consider {Equ,Mem} and {Arb,Mem} as sets of query types. Here, we define "query complexity". Intuitively, query complexity is the number of queries asked ask by a learning algorithm A in the worst case. More precisely, for any representation class $C = (\Sigma, R, \Phi, \rho)$, any set of query types Q and any learning algorithm A for C using Q, query complexity $\#query_{(A,Q)}$ is defined as follows: For any $t \in R$,

$$\#query_{(A,Q)}(t) = \max\{i \in \mathbb{N} : \text{there is some choice of counterexamples such}$$
$$\text{that } i \text{ is the number of queries asked until } A$$
$$\text{halts and outputs } h \text{ such that } \Phi(h) = \Phi(t)$$
$$\text{when } \Phi(t) \text{ is the target concept to be learned.}\}$$

3   VC-dimension of DFA$_{k,n}$

Let $\text{DFA}_{k,n} = (\Sigma_k, R_{(k,n)\text{-dfa}}, \Phi_{\text{dfa}}, \rho_{\text{dfa}})$, where $R_{(k,n)\text{-dfa}} = \{r \in R_{\text{dfa}_k} : \rho_{\text{dfa}}(r) \le n\}$.

Theorem 3.1

$$\text{VC-dim}(\text{DFA}_{1,n}) = (1 + o(1))n.$$

For the proof of this theorem, see Section 5.

Theorem 3.2 *Let k be a fixed integer such that $k \ge 2$.*

$$\text{VC-dim}(\text{DFA}_{k,n}) = (k - 1 + o(1))n \log n.$$

Proof. In this proof, we use $[k]$ as a k-letter alphabet instead of Σ_k.
(Upper bound) Since $|\Phi_{\text{dfa}}(R_{(k,n)\text{-dfa}})|$ is at most $2^n \cdot n^{kn}/(n-1)!$,
$\text{VC-dim}(\text{DFA}_{k,n}) \le \log(2^n \cdot n^{kn}/(n-1)!) \le (k - 1 + c/\log n)n \log n$ where c is some constant.
(Lower bound) Let $N_1 = n - \lfloor n/\log n \rfloor$ and $N_2 = \lfloor n/\log n \rfloor$. Subsets of $[k]^*$ W_1 and W_2 are defined as follows:

$$W_1 = \left\{ w \in [k]^* : \sum_{i=1}^{|w|} k^{|w|-i} \cdot Proj_{[i]}(w) < N_1 \right\},$$

$$W_2 = \left\{ w \in [k]^* : \exists a \in [k] \text{ such that } \sum_{i=1}^{|w|} k^{|w|-i+1} \cdot Proj_{[i]}(w) + a \ge N_1 \right\}.$$

Let $W = W_1 - W_2$. Note that

$$|W| = \frac{k-1}{k}\left(1 - \frac{1}{\log n}\right)n + o(n)$$

and for any word $w_1, w_2 \in W$, w_1 and w_2 has no common prefix. Let $N_2' = \lfloor\log(N_2 + 2)\rfloor - 1$. We define a set S_k as $\{w \cdot a \cdot 1^i : w \in W, \quad a \in [k], \quad i = 0, 1, \ldots, N_2' - 1\}$. It is easy to check $|S_k| = |W| \cdot k \cdot \lfloor\log N_2\rfloor = (k - 1 + o(1))n\log n$. Therefore, it sufficient to show that S_k can be shattered by $\Phi(R_{(k,n)\text{-dfa}})$.

For every set $S' \subseteq S_k$, we will construct a dfa $M(S') \in R_{(k,n)\text{-dfa}}$ such that $L(M(S')) = S'$. For every set $S' \subseteq S_k$, every word $w \in W$ and every $a \in [k]$, let $S'_{w \cdot a} = S' \cap \{w \cdot a \cdot 1^i : i = 0, 1, \ldots N_2' - 1\}$. We use the symbol $[S'_{w \cdot a}]$ to denote the word with the length N_2' as follows:

$$Proj_{[i]}([S'_{w \cdot a}]) = \begin{cases} 1 & \text{if } w \cdot a \cdot 1^{i-1} \in S'_w, \\ 0 & \text{otherwise.} \end{cases}$$

For every $S' \subseteq S_k$, we define $M(S') = ([k], Q, \delta, q_{init}, F)$ as follows:

- $Q = A \cup B$ is the set of states, where $A = \{q_{(A,w)} : w \in W_1\}$ and $B = \{q_{(B,w)} : w \in \Sigma^{\leq N_2'} - \{\lambda\}\}$,

- $\delta : Q \times \Sigma \to Q$ is the transition function, where

  1. for A and for $w \in W_2$ and $a \in [k]$ such that $\sum_{k=1}^{|w|} k^{|w|-i+1} \cdot Proj_{[i]}(w) + a < N_1$,

  $$\delta(q_{(A,w)}, a) = q_{(A,w \cdot a)},$$

  2. for A and for every $w \in W$ and $a \in [k]$,

  $$\delta(q_{(A,w)}, a) = q_{(B,[S'_{w \cdot a}]^R)},$$

  3. for B and for every word $w \in \Sigma^{\leq N_2'} - \Sigma^{\leq 1}$,

  $$\delta(q_{(B,w)}, 1) = q_{(B,T(w))}$$

- $q_{init} = q_{(A,\lambda)}$ is the initial state, and

- $F = \{q_{(B,w)} : q_{(B,w)} \in B, \, last(w) = 1\}$ is the set of final states.

The transition rules (1), which maps $A \times [k]$ to A, and the transition rules (3), which maps $B \times [k]$ to B, are common to all dfas among $\{M(S') : S' \subseteq S\}$. The transition rules (2), which maps $A \times [k]$ to B, are different to each other among $\{M(S') : S' \subseteq S\}$.

We extends the domain of δ to $Q \times [k]^*$ such that for every $q \in Q$, $a \in [k]$ and $w \in \Sigma^*$, $\delta(q, a \cdot w) = \delta(\delta(q, a), w)$. It is clear that $M(S')$ accepts no word that is not included in S. Let $w \cdot a \cdot 1^i$ be a word such that $w \in W$ and $w \cdot a \cdot 1^i \in S$. By the definition of δ,

$$\delta(q_{init}, w \cdot a \cdot 1^i) = \delta(q_{(A,w)}, a \cdot 1^i)$$
$$= \delta(q_{(B,[S'_{w \cdot a}]^R)}, 1^i)$$
$$= q_{(B, Pref_{[N_2'-i]}([S'_{w \cdot a}]^R))}.$$

By the definition of the final state sets F, $w \cdot a \cdot 1^i$ is accepted by $M(S')$ if and only if $last(Pref_{[N_2'-i]}([S'_{w \cdot a}]^R)) = 1$. Since $last(Pref_{[N_2'-i]}([S'_{w \cdot a}]^R)) = last(Pref_{[i]}([S'_{w \cdot a}])) = Proj_{[i]}([S'_{w \cdot a}])$, $w \cdot a \cdot 1^i$ is accepted by $M(S')$ if and only if $w \cdot a \cdot 1^i \in S'_{w \cdot a}$, that is to say, $L(M(S')) = S'$. Therefore, S is shattered by $\Phi_{dfa}(R_{(k,n)\text{-dfa}})$. $\square$

4 The VC-dimension of NFA$_{k,n}$

Let $NFA_k = (\Sigma_k, R_{nfa_k}, \Phi_{nfa}, \rho_{nfa})$, where R_{nfa_k} is the set of representations of nfas with the alphabet Σ_k. Let $NFA_{k,n} = (\Sigma_k, R_{(k,n)\text{-nfa}}, \Phi_{nfa}, \rho_{nfa})$, where $R_{(k,n)\text{-nfa}} = \{r \in R_{nfa_k} : \rho_{nfa}(r) \le n\}$.

Theorem 4.1 *Let k be a fixed integer such that $k \ge 2$.*

$$(k-1)n^2 \le VC\text{-dim}(NFA_{k,n}) \le kn^2.$$

Proof. (Upper bound) Since $|\Phi_{nfa}(R_{(k,n)\text{-nfa}})|$ is at most $2^{2n} \cdot 2^{kn^2}/n!$, $VC\text{-dim}(NFA_{k,n}) \le \log(2^{2n} \cdot 2^{kn^2}/n!) \le kn^2$.
(Lower bound) Let $S = \{a_1{}^i a_l a_1{}^j : i = 0, 1, \ldots, n-1, j = 0, 1, \ldots, n-1, a_l \in \Sigma_k - \{a_1\}\}$. Note that $|S| = (k-1)n^2$. For any $S' \subseteq S$, we construct a nfa $M_{S'} = (\Sigma_k, Q, Q_{init}, \delta, F) \in R_{(k,n)\text{-nfa}}$ such that $M_{S'}$ accepts any word in S' and no word $S - S'$. Let $Q = \{q_0, q_1, \ldots, q_{n-1}\}$, $Q_{init} = \{q_0\}$, and $F = \{q_{n-1}\}$. For $i = 0, 1, \ldots, n-2$, $\delta(q_i, a_1) = q_{i+1}$. If $a_1{}^i a_l a_1{}^j \in S'$, $\delta(q_i, a_l) = q_{n-1-j}$. It is clear that $M_{S'}$ accepts any word in S' and no word $S - S'$ and S is shattered by $\Phi_{nfa}(R_{(k,n)\text{-nfa}})$. $\square$

5 The VC-dimension of Commutative DFA

In this section, we consider the VC-dimension of commutative dfas. Commutative dfas accepts only commutative regular languages (see [16]). Let $CDFA_{k,n}$ denote $(\Sigma_k, R_{(k,n)\text{-cdfa}}, \Phi_{dfa}, \rho_{dfa})$ where $R_{(k,n)\text{-cdfa}}$ is the set of representations of commutative dfas of the alphabet Σ_k whose minimum states dfa has at most n states. Since any 1-letter language is commutative,

$$CDFA_{1,k} = DFA_{1,k}.$$

Therefore the proof of Theorem 3.1 is established by showing that $CDFA_{1,k} = (1 + o(1))n$ in this section.

Theorem 5.1 *For a constant $k = 1, 2, \ldots,$*

$$\text{VC-dim}(\text{CDFA}_{k,n}) = (1 + o(1))n.$$

Proof of this theorem is accomplished by a series of the following lemmata.

Lemma 5.2 *Let k be a fixed positive integer,*

$$\text{VC-dim}(\text{CDFA}_{k,n}) \geq n + k - 1$$

Proof. First, we consider the case of $k = 1$. Let $S = \{a_1^i : i = 0, 1, \ldots, n-1\}$. It is clear that $|S| = n + k - 1$ and S can be shattered by $\Phi_{\text{dfa}}(R_{(1,n)\text{-dfa}})$.

In the rest of this proof, we consider the case of $k \geq 2$. Let $S = S_1 \cup S_2$ where $S_1 = \{a_1{}^i : i = 0, 1, \ldots, n-1\}$ and $S_2 = \{a_j : j = 2, 3, \ldots, k\}$. It is clear that $|S| = n + k - 1$. For any set $S' \subseteq S$, we construct a cdfa $M_{S'} = (\Sigma_k, Q, q_{init}, \delta, F) \in R_{(k,n)\text{-cdfa}}$ such that $M_{S'}$ accepts any word in S' and no word in $S - S'$. Let $Q = \{q_0, q_1, \ldots, q_{n-1}\}$ and $q_{init} = q_0$.

Case 1. Suppose $S' \cap S_1 \neq \emptyset$. Let $F = \{q_i : a_1{}^i \in S' \cap S_1\}$. We define $\delta(q_i, a_1) = q_{i+1}$ for $i = 0, 1, \ldots, n-2$. Let $m = \max\{i : q_i \in F\}$. We define $\delta(q_0, a_i) = q_m$ if $a_i \in S' \cap S_2$.

Case 2. Suppose $S' \cap S_1 = \emptyset$. Let $F = \{q_1\}$. We define $\delta(q_0, a_i) = q_1$ if $a_i \in S' \cap S_2$.

In each case, it is clear that $M_{S'}$ accepts any word in S' and no word in $S - S'$. Therefore S is shattered by $\Phi_{\text{dfa}}(R_{(k,n)\text{-cdfa}})$. $\square$

Next we consider the upper bound of $\text{VC-dim}(\text{CDFA}_{k,n})$. Let $f_k(n) = |\{L(r) : r \in R_{(k,n)\text{-cdfa}}\}|$. By estimating the upper bound of $f_k(n)$, we will obtain the upper bound of $\text{VC-dim}(\text{CDFA}_{k,n})$. Let $A : \mathbb{N}^k \to [2]$ be a 2-coloring of k-dimensional infinite array. For any $x \in \mathbb{N}^k$, x_i denotes the i-th element. For any $x, y \in \mathbb{N}^k$, $x + y$ denotes $(x_1 + y_1, \ldots, x_k + y_k)$. For $x \in \mathbb{N}^k$, a 2-coloring of k-dimensional infinite array $[x \backslash A]$ is defined as follows: $[x \backslash A](y) = A(x + y)$ for every $y \in \mathbb{N}^k$.

Lemma 5.3 *Let k be a fixed integer $k \geq 1$.*

$$f_k(n) = |\{a\ 2\text{-coloring of } k\text{-dimensional array } A : |\{[x \backslash A] : x \in \mathbb{N}^k\}| \leq n\}|.$$

Proof Let $L \subseteq \Sigma_k{}^*$ be a language. For any word $u, v \in \Sigma_k{}^*$, we say that $u \equiv_L v$ if $u \cdot w \in L$ if and only if $v \cdot w \in L$ for every word $w \in \Sigma_k{}^*$. Let $[w]_{\equiv_L}$ denote $\{w' \in \Sigma_k{}^* : w' \equiv_L w\}$. It is well known that if L is regular then $|\{[w]_{\equiv_L} : w \in \mathbb{N}^k\}|$ is finite and $|\{[w]_{\equiv_L} : w \in \mathbb{N}^k\}|$ is equal to the number of the minimum states dfa accepting L. For $L \in \Phi_{\text{dfa}}(R_{(k,n)\text{-cdfa}})$, a 2-coloring of k-dimensional infinite array A_L defined as follows:

$$A_L(x) = \begin{cases} 1 & \text{if } a_1{}^{x_1} \cdots a_k{}^{x_k} \in L \\ 0 & \text{otherwise} \end{cases}$$

For $x, y \in \mathbb{N}^k$, $a_1^{x_1} \cdots a_k^{x_k} \equiv_L a_1^{y_1} \cdots a_k^{y_k}$ implies $[x \backslash A_L] = [y \backslash A_L]$, since for any word $w \in \Sigma_k^*$, $a_1^{x_1} \cdots a_k^{x_k} \cdot w \in L$ if and only if $a_1^{y_1} \cdots a_k^{y_k} \cdot w \in L$. If $a_1^{x_1} \cdots a_k^{x_k} \not\equiv_L a_1^{y_1} \cdots a_k^{y_k}$, $[x \backslash A] \neq [y \backslash A]$, since there exists a word $w = a_1^{w_1} \cdots a_k^{w_k}$ such that $A(x_1 + w_1, \ldots, x_k + w_1) \neq A(y_1 + w_1, \ldots, y_k + w_k)$. Therefore $f_k(n) = |\{A : |\{[x \backslash A] : x \in \mathbb{N}^k\}| \leq n\}|$. $\qquad \square$

Let $B : \mathbb{N}^k \to [n]$ be an n-partition of k-dimensional infinite array. For $x \in \mathbb{N}^k$, an n-partition of k-dimensional infinite array $[x \backslash B]$ is defined as follows: $[x \backslash B](y) = B(x + y)$ for every $y \in \mathbb{N}^k$. Let A be a 2-coloring of k-dimensional infinite array such that $|\{[x \backslash A] : x \in \mathbb{N}^k\}| \leq n$. We define an n-partition of k-dimensional infinite array B_A as follows: For any $x, y \in \mathbb{N}^k$, $B_A(x) = B_A(y)$ if and only if $[x \backslash A] = [y \backslash A]$. This partition satisfies the following condition $(*)$.

$(*)$ $\quad$ For any $x, y \in \mathbb{N}^k$, $B_A(x) = B_A(y)$ implies $[x \backslash B_A] = [y \backslash B_A]$.

Let

$$g_k(n) = \{\text{an } n\text{-partition of } k\text{-dimensional infinite array } B : B \text{ satisfies } (*)\}.$$

Then

$|\{\text{a 2-coloring of } k\text{-dimensional array } A : |\{[x \backslash A] : x \in \mathbb{N}^k\}| \leq n\}| \leq 2^n \cdot g_k(n)$.

In consequence of this fact and Lemma 5.3, the following lemma holds.

Lemma 5.4 $f_k(n) \cdot \leq 2^n \cdot g_k(n)$.

For $x, y \in \mathbb{N}^k$, we say that $x \leq y$ if and only if $x_i \leq y_i$ for every $i \in [k]$ and that $x \geq y$ if and only if $x_i \geq y_i$ for every $i \in [k]$. For $i \in [k]$, $y \in \mathbb{N}^{k-1}$ and $x \in \mathbb{N}$, $r_{i,y}(x)$ denotes $(y_1, \ldots, y_{i-1}, x, y_i, \ldots, y_{k-1})$ and $R_{i,j}$ denotes $\{r_{i,y}(x) : x \in \mathbb{N}\}$.

Lemma 5.5 *Suppose B is an n-partition of k-dimensional infinite array satisfying the condition $(*)$. For $i \in [k]$ and $y \in \mathbb{N}^{k-1}$, $B(r_{i,y}(x_1)) = B(r_{i,y}(x_2))$ and $x_1 < x_2$ implies for every $x > x_2$, $B(r_{i,y}(x)) = B(r_{i,y}(x_1 + k))$ where $k = (x - x_2) \mod (x_2 - x_1)$.*

Proof. Since $B(r_{i,y}(x_1)) = B(r_{i,y}(x_2))$, $B(r_{i,y}(x_1 + j)) = B(r_{i,y}(x_2 + j))$ for every $j \in \mathbb{N}$. Let $l = x_2 - x_1$ and $m = (x - x_1 - k)/l$. $B(x_1 + k) = B(x_2 + k) = B(x_1 + l + 1) = \cdots = B(x_1 + ml + k) = B(x)$. $\qquad \square$

Let $\mathcal{Y}_k(n) \in 2^{\mathbb{N}^k}$ be defined as $\{Y \subseteq \mathbb{N}^k : |y| \leq n$ and for any $x, y \in \mathbb{N}^k$, $x \in Y$ and $x \geq y$ implies $y \in Y\}$. We define the k-dimensional partition function $P_k(n)$ as $|\mathcal{Y}_k(n)|$. For $x, y \in \mathbb{N}^k$, we say that $x \preceq y$ if and only if for any i such that $x_i - y_i > 0$, there exists $j < i$ such that $x_j - y_j < 0$ and that $x \prec y$ if and only if $x \preceq y$ and $x \neq y$. For $S \subseteq \mathbb{N}^k$ and $y \in \mathbb{N}^{k-1}$, let $S_{max}(y) = \max\{x : r_{1,y}(x) \in S\}$ For $i \in [k] - 1$, we define $\vec{0_i} \in \mathbb{N}^{k-1}$ as in follows: the $(i-1)$th element is 1 and the other elements are 0.

Algorithm: YDG
Input: A n-partition satisfying the condition ($*$)
Output: $Y \in \mathcal{Y}_k(n)$ satisfying the condition ($**$)
Begin
$X; = \emptyset; \ Y := \emptyset; \ y := (0, \ldots, 0) \ (\in \mathbb{N}^{k-1});$
Repeat
   $X_{k-1} := X; \ y_{k-2} := 0;$
  Repeat
     $X_{k-2} := X; \ y_{k-3} := 0;$

        $\ddots$

        $X_3 := X; \ y_2 := 0;$
        Repeat
          $X_2 := X; \ y_1 := 0;$
          Repeat
            $X_1 := X;$
            $X := X \cup \{\text{the parts of } n\text{-partition of } B \text{ appearing in } R_{1,y}\};$
            $m := X - X_1;$
            $Y := Y \cup \{r_{1,y}(x) : x = 0, \ldots, m-1\};$
            $y_1 := y_1 + 1;$
          Until $|X| - |X_1| = 0;$
          $y_2 := y_2 + 1;$
        Until $|X| - |X_2| = 0 \ ;$

        $\iddots$

  Until $|X| - |X_{k-2}| = 0;$
  $y_{k-1} := y_{k-1} + 1$
Until $|X| - |X_{k-1}| = 0$
End.

Figure 1: The algorithm YDG

Lemma 5.6 *Let B be an n-partition of k-dimensional infinite array satisfying the condition ($*$). There exists $Y \in \mathcal{Y}_k(n)$ such that $\{B(x) : x \in Y\} = \{B(x) : x \in \mathbb{N}^k\} \subseteq [n]$. Moreover, there exists $Y_B \in \mathcal{Y}_k(n)$ satisfying the following condition ($**$):*

For every $y \in \mathbb{N}^{k-1}$

$$(**) \qquad \left\{ B(x) : x \in Y_B \cap \bigcup_{y' \preceq y} R_{1,y'} \right\} = \left\{ B(x) : x \in \bigcup_{y' \preceq y} R_{1,y'} \right\}.$$

Proof For an n-partition of k-dimensional infinite array B satisfying the condition ($*$), the algorithm YDG (shown in Figure 1) generates $Y \in \mathcal{Y}_k(n)$ satisfying the condition ($**$) when YDG is run with B as an input. By the construction of the algorithm YDG and Lemma 5.5, it is clear that the output Y satisfies $(**)$ and $|Y| \leq n$. This proof is accomplished by showing that $Y \in \mathcal{Y}(n)$. Suppose that there exist $y, z \in \mathbb{N}^{k-1}$ such that $y_i + 1 = z_i$, $y_j = z_j$ for every $j \neq i$ and $Y_{max}(y) < Y_{max}(z)$. There exists $u \in Y \cap \bigcup_{y' \preceq y} R_{1,y}$ such that $B(u) = B(r_{1,y}(Y_{max}(y)))$. Since

$$\left\{ B(u + \vec{0_i}) : u \in Y \cap \bigcup_{y' \prec y} R_{1,y'} \right\} \subseteq \left\{ B(x) : x \in Y \cap \bigcup_{y' \preceq y} R_{1,y'} \right\}$$

and

$$\{ B(u + \vec{0_i}) : u \in Y \cap R_{1,y} \} = \{ B(r_{1,z}(x)) : x = 0, 1, \ldots, Y_{max}(y) \}.$$

On the other hand, $r_{1,y}(Y_{max}(y) + 1) + \vec{0_i} = r_{1,z}(Y_{max}(y) + 1)$. These contradict each other. $\qquad\square$

Suppose $Y \in \mathcal{Y}_k(n)$ is given. We construct an n-partition of k-dimensional infinite array B such that B satisfies the condition $(*)$ and Y satisfies the condition $(**)$. Since it is necessary to hold that B satisfies $(*)$ and Y satisfies $(**)$, for $y \in \mathbb{N}^{k-1}$, we assign some $i \in [k]$ to $r_{1,y}(Y_{max}(y) + 1)$, hereupon for every $x \geq Y_{max}(y) + 2$, $B(r_{1,y}(x))$ is decided by Lemma 5.5. For every $y \in \mathbb{N}^{k-1}$, we are not free to choose an integer in $[k]$ for $r_{1,y}(Y_{max}(y) + 1)$. Suppose for $y, z \in \mathbb{N}^{k-1}$ such that $y_i + 1 = z_i$ and $y_j = z_j$ for every $j \neq i$, $Y_{max}(y) = Y_{max}(z)$ and for every $x \in \bigcup_{y' \prec y} R_{1,y'}$, partition numbers are assigned as satisfying the condition $(**)$. Since $Y_{max}(y) = Y_{max}(z)$, for every $x \in Y \cap R_{1,y}$, $B(r_{1,z}(x))$ has been decided. When some $p \in \{B(x) : Y \cap \bigcup_{y' \preceq y} R_{1,y'}\}$ is assigned to $r_{1,y}(Y_{max}(y) + 1)$, for every $x \in Y \cap \bigcup_{y' \preceq y} R_{1,y}$ $B(x + \vec{0_i})$ is decided. Therefore, for every $x \in \bigcup_{y' \preceq y} R_{1,y}$, $B(x + \vec{0_i})$ is decided. Since

$$R_{1,z} = \{ x + \vec{0_1} : x \in R_{i,y} \}$$

the assignment to $R_{1,z}$ has been decided. Let

$$\alpha_k(n) = |\{ S \subseteq \mathbb{N}^k : \text{For any } x, y \in S \text{ such that } x \not\geq y,$$
$$|\{ z \in \mathbb{N}^k : \exists x \in S \text{ such that } x \geq z \}| \leq n \}|.$$

Then

$$g_k(n) \leq n^{\alpha_k(n)} \cdot P_k(n).$$

Y. Ishigami showed that the following property holds.

Proposition 1 ([12]) *Let k be a fixed integer k.*

$$\alpha_k(n) \leq c_k n^{(k-1)/k}$$

for some constant c_k.

In consequence, we gain the following theorem.

Theorem 5.7 *Let k be a fixed positive integer.*

$$g_k(n) \leq n^{c_k n^{(k-1)/k}} \cdot P_k(n)$$

where c_k is some constant.

Proof of Theorem 5.1. It is known the estimation of partition function as in follows [2, 12]: For a positive integer k, $\log P_k(n) = o(n)$. From this fact, Lemma 5.2, Lemma 5.4 and Theorem 5.7, it is straightforward to derive Theorem 5.1. □

In the rest of this section, we consider the case when the alphabet size is treated as a variable.

Theorem 5.8 *For variables k and n, VC-dim(CDFA$_{k,n}$) = $\Omega(n + k \log n)$.*

Proof. In case when $k < n/\log n$, CDFA$_{k,n}$ can shatter $\{a_1{}^i : i = 0, 1, \dots, n-1\} \cup \{a_j : j = 2, 3, \dots, k\}$ (by Lemma 5.2) and VC-dim(CDFA$_{k,n}$) = $\Omega(n + k \log n)$. In the rest of this proof, we consider the case when $k \geq n/\log n$. Let $S = \{a_1{}^i \cdot a : a \in \Sigma_k - \{a_1\}, i = 0, 1, \dots, b-1\}$, where $b = \lfloor \log n - \log \log n \rfloor$. We construct a 2-coloring of k-dimensional infinite array $A_{S'} : \mathbb{N}^k \to \{0, 1\}$ such that $A_{S'}$ is corresponding to a cdfa accepting any word in S' and no word in $S - S'$ and $|\{[x/A_{S'}] : x \in \mathbb{N}^k\}| \leq n$. Let $\bar{0} = (0, \dots, 0) \in \mathbb{N}^{k-1}$. We can assigns $\{0, 1\}$ to $R_{1, \bar{0}}$ such as all string in $\{0, 1\}^{=b}$ appears in $\{r_{1, \bar{0}}(x) : x < b2^b\}$. For $x \geq b2^b$, assign such as $A_{S'}(x) = A_{S'}(x')$ where $x' = x \bmod b2^b$. For every $S' \subseteq S$ and $a \in \Sigma_k - \{a_1\}$, we use the symbol $[S'_a]$ to denote the word with the length b as follows:

$$Proj_{[i]}([S'_a]) = \begin{cases} 1 & \text{if } 1^{i-1} \cdot a \in S', \\ 0 & \text{otherwise.} \end{cases}$$

For $i \in [k] - \{1\}$, find the smallest $k \in \mathbb{N}$ such that

$$[S'_{a_i}] = A_{S'}(r_{1, \bar{0}}(k)) \cdot A_{S'}(r_{1, \bar{0}}(k+1)) \cdots A_{S'}(r_{1, \bar{0}}(k+b-1)).$$

For every $x \in \mathbb{N}$, assign $A_{S'}(r_{1, \bar{0}}(x+k))$ to $r_{1, \bar{0}_i}(x)$. Thus $|\{x \backslash A_{S'} : x \in \mathbb{N}^k\}| \leq b2^b = o(n)$ and $|S| = \Omega(k \log n)$. □

6 On the Lower Bounds of Query Complexity of Learning Finite Automata

In this section, we consider query complexity of learning finite automata. Maass and Turán investigated the query complexity to learn an unknown concept from a representation class over finite domain.

Proposition 6.1 ([15]) *For any representation class* $C = (\Sigma, R, \Phi, \rho)$, *and any learning algorithm A for C using* {Equ, Mem},

$$\max\{\#query_{(A,\{Equ,Mem\})}(t) : t \in R\} = \Omega(\text{VC-dim}(C)).$$

Proposition 6.2 ([15]) *For any representation class* $C = (\Sigma, R, \Phi, \rho)$, *and any learning algorithm A for C using* {Arb, Mem},

$$\max\{\#query_{(A,\{Arb,Mem\})}(t) : t \in R\} = \Omega(\text{VC-dim}(C)).$$

These properties hold in case of learning a representation class C over some infinite universe if VC-dim(C) is finite. As a corollary of these properties, we have the following lemma.

Lemma 6.3 *For any representation class* $C = (\Sigma, R, \Phi, \rho)$, *there exists no learning algorithm A for C using* {Equ, Mem} *such that for any target* $t \in R$

$$\#query_{(A,\{Equ,Mem\})}(t) = o\left(\text{VC-dim}\left(C_{\rho(t)}\right)\right).$$

Proof. Assume that there exists a learning algorithm A such that
$$\#query_{(A,\{Equ,Mem\})}(t) = o(\text{VC-dim}(C_n))$$
for any $t \in R$. Suppose A tries to learn C_n using {Arb,Mem}. Then for any $t \in R$ A learns C_n with $\#query_{(A,\{Arb,Mem\})}(t) = o(\text{VC-dim}(C_n))$. This contradicts Proposition 6.2. □

From Theorem 3.2, Theorem 4.1, and this lemma, it is straightforward to derive the following two corollary.

Corollary 6.4 *There is no learning algorithm A for* DFA$_k$ *using* {Equ, Mem} *such that for any* $t \in R_{\text{dfa}}$

$$\#query_{(A,\{Equ,Mem\})}(t) \geq (k-1)n \log n,$$

where $n = \rho_{\text{dfa}}(t)$.

Corollary 6.5 *There is no learning algorithm A for* NFA$_k$ *using* {Equ, Mem} *such that for any* $t \in R_{\text{nfa}}$

$$\#query_{(A,\{Equ,Mem\})}(t) \geq (k-1)n^2,$$

where $n = \rho_{\text{nfa}}(t)$.

References

[1] Abe, N.: Learning Commutative Deterministic Finite State Automata in Polynomial Time, *New Generation Computing 8*, 1991, pp.319-335.

[2] Andrews, G.: *The Theory of Partitions*, Addison-Wesley, Massachusetts (1976).

[3] Angluin, D.: Learning regular sets from queries and counterexamples, *Information and Computation*, Vol. 75, 1987, pp.87-106.

[4] Angluin, D.: Queries and concept learning, *Machine Learning*, Vol. 2, 1988, pp.319-342.

[5] Angluin, D.: Negative results for equivalence queries, *Machine Learning*, Vol. 5, 1990, pp.121-150.

[6] Angluin, D. and M. Kharitonov: When Won't Membership Queries Help? in *Proc. of 23nd STOC*, ACM, 1991, pp.444-454.

[7] Balcazar, J. L., J. Diaz, R. Gavalda and O. Watanabe: A Note on the Query Complexity of Learning DFA, *Proc. of the 3rd workshop on Algorithmic Learning Theory*, Japanese Society for AI, 1992, pp.53-62.

[8] Benedek, G. M. and A. Itai: Nonuniform learnability, *International Colloquium on Automata, Languages and Programming 1988*, Lecture Notes in Computer Science 317, Springer-Varlag, pp.82-92.

[9] Blumer, A., A. Ehrenfeucht, D. Haussler, and M. K. Warmuth: Classifying learnable geometric concepts with Vapnik-Chervonenkis dimension, in *Proc. 18th ACM Symp. on Theory of Computing*, 1986, pp.273-282.

[10] Champarnaud, J.-M. and J.-E. Pin: A Maximin problem on finite automata, *Discrete Applied Mathematics 23*, 1989, pp.91-96.

[11] Hopcroft, J. E., and Ullman, J. D., *Introduction to Automata Theory, Languages, and Computation*, Addison-Wesley, Massachusetts (1979).

[12] Ishigami, Y.: A Sperner set generating a k-dimensional Young tableaux, Preprint (1993).

[13] Gaizer, T.: The Vapanik-Chervonenkis dimension of finite automata. Unpublished manuscript (1990).

[14] Maass, T., and G. Turán : On the complexity of learning from counterexamples, in *Proc. 30th IEEE Symp. on Foundations of Computer Science* 1989, pp.262-167.

[15] Maass, T., and G. Turán: On the Complexity of Learning from Counterexamples and Membership Queries, in *Proc. 31st IEEE Symp. on Foundations of Computer Science*, 1990, pp.203-210.

[16] Salomaa, A.: *Theory of Automata*, Pergamon Press, Oxford, 1969.

[17] Watanabe, O.: A formal study of learnability via queries, *International Colloquium on Automata, Languages and Programming, 1990*, Lecture Notes in Computer Science 443, Springer-Varlag.

[18] Yokomori, T. : Learning Non-deterministic Finite Automata from Queries and Counterexamples. To appear in *Machine Intelligence*, Vol.13, Oxford Univ. Press.

Unifying Learning Methods by Colored Digraphs

Kenichi Yoshida, Hiroshi Motoda and Nitin Indurkhya*

Advanced Research Laboratory, Hitachi, Ltd.
Hatoyama, Saitama 350-03, Japan.

Abstract. We describe a graph-based induction algorithm that extracts typical patterns from colored digraphs. The method is shown to be capable of solving a variety of learning problems by mapping the different learning problems into colored digraphs. The generality and scope of this method can be attributed to the expressiveness of the colored digraph representation which allows a number of different learning problems to be solved by a single algorithm. We demonstrate the application of our method to two seemingly different learning tasks: inductive learning of classification rules, and learning macro rules for speeding up inference. We also show that the uniform treatment of the above two learning tasks enables our method to solve complex learning problems such as the construction of hierarchical knowledge bases.

1 Introduction

In recent years, much of machine learning research has concentrated on algorithms for two relatively distinct learning tasks: (1) extracting new knowledge from sample data [19, 23], and (2) reorganizing existing knowledge based on various considerations such as efficiency, operationality, etc. [6].

While several algorithms are available for each of the two learning tasks, some applications require the ability to address both problems simultaneously. For instance, the task of hierarchical knowledge base construction involves both kinds of learning. The inference process at lower levels must be reorganized into more efficient processes at higher levels in the hierarchy, and at the same time knowledge bases at higher levels of abstraction must be induced from sample data at lower levels. Similar problems abound in the field of Robotics. Addressing such applications with existing learning methods requires a careful decomposition of the problem such that each part can be solved by existing algorithms. Hence, such problems have been difficult to solve and existing solutions rely on considerable manual interaction. For example, current methods for generating hierarchical knowledge bases depend on manual guidance [8, 17].

This paper examines methods for solving complex learning problems that demand several modes of learning to be performed. We present a unified method

---

* Present address: Telecom Research laboratories, 770 Blackburn Road, Clayton, Vic 3168, Australia

for extracting new knowledge from sample data as well as reorganizing given knowledge for efficient problem-solving. The method relies on the use of colored digraphs for representing learning problems. Different learning problems are mapped into colored digraphs in a manner such that solving the learning problem is achieved by extracting common patterns in the resulting graph. This enables the use of a *single* graph-based induction algorithm for solving a variety of learning problems. Specifically, our method can be used to 1) learn classification rules from sample data, 2) learn macro rules for speeding up inference, and 3) create a hierarchical knowledge base.

The rest of the paper is organized as follows: Section 2 examines related work, Section 3 outlines the basic idea of graph-based induction, Section 4 describes how different learning tasks can be mapped into colored digraphs, Section 5 examines the algorithm for extracting typical patterns, Section 6 presents experimental results and Section 7 summarizes the key points and discusses future issues.

2   Related Work

Learning strategies that involve the use of both inductive and deductive methods have been proposed before [15, 22, 3, 10, 24]. Most of these methods, however, used separate components for inductive learning and deductive learning, and focus instead on how to combine these two separate learning functions. When problems cannot be decomposed easily, such methods are difficult to use.

Colored digraphs have been used for representation of deductive learning problems [7]. However, the use of this representation to solve other learning tasks, especially problems that require a combination of learning methods, has not been attempted before.

Several algorithms are available for extracting substructures from graph or graph-like data structures [26, 1, 16, 9, 11, 12]. However, most of them [26, 1, 16, 9] cannot extract substructures from a single connected structure, and are not applicable for certain kinds of learning such as knowledge-base abstraction. The pattern extraction procedure in SUBDUE [11, 12] is quite general but can be computationally prohibitive for large problems.

3   Graph-based Induction

We employ the colored digraph for representing learning problems. In a colored digraph representation, each node has topological (edge) information and one or more colors attached to it. The colors are used to associate attributes to each node. Colored digraphs can be viewed as a generalization of the attribute-value representation used in conventional classification learning systems [19, 23].

The central intuition behind our graph-based induction method is as follows: a pattern that appears frequently enough in a colored digraph is worth paying attention to and may represent an important concept in the environment. In other words, the repeated patterns in the input graph represent typical characteristics of the given environment. The extraction of patterns is based solely on

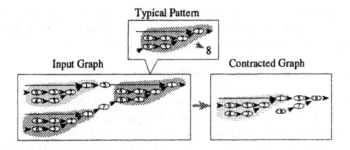

Fig. 1. Graph Contraction By Finding Typical Patterns

finding repetitions of substructures within the input graph. The algorithm analyzes the input colored digraphs, and extracts sets of patterns (each set is called a *view*), in such a way that the patterns in the view can be used to contract the input graph. A key operation in this procedure is graph matching.

Standard graph matching methods check the equivalence of two graphs by considering all possible edge combinations. In our algorithm however, we adopt graph identity as our matching criterion. An important implication of this is that graph matching can be done in polynomial time. This restriction does not seem to limit the class of learning tasks that the algorithm can handle. Figure 1 illustrates pattern matching based on graph identity. Note that though there are three isomorphic subgraphs in the input graph, the algorithm finds only the subgraph in the upper box as a typical pattern, gives this a new identifier color (8 in this case), and contracts the graph such that only the two subgraphs *identical* to the typical pattern are contracted.

Besides the graph contraction procedure, a method to evaluate the contracted graph is also required. This is done by *selection criteria*, based mainly on notions related to graph size, that enable selection of the best contracted graph. These issues are discussed in greater detail in Section 5.

Interpretation of the typical patterns depends on the learning task encoded in the colored digraph. For example, in one interpretation, the typical patterns can be viewed as new knowledge induced from data encoded in the input graph. On the other hand, if the input graph represents inference patterns, then typical patterns may be interpreted as new inference macro-rules. Traditionally, these two learning tasks have been considered separately. However, by encoding them in colored digraphs, it is possible to solve both problems by a single algorithm. We now describe how different learning problems are encoded into the colored digraph representation.

4 Transforming Learning Problems into Colored Digraph Representation

In this section we describe how to map different types of learning problems into colored digraphs. Besides the conventional learning problems such as inductive classification learning and macro-rule learning, we also illustrate how more complex learning problems, that involve a combination of different learning strategies, might be encoded in colored digraphs.

4.1 Conventional Learning Problems

In encoding conventional learning problems from inductive classification learning, speed-up learning, etc, we take advantage of the close connection between colored digraphs and the conventional representations used for these problems. For conventional inductive learning, the nodes represent the attributes of relevant entities, and node color is used to encode attribute values. For conventional deductive learning, the nodes in the graph represent the data referred to or produced by the inference engine. Two colors are used: one to encode the rule used to obtain the data and the other to encode the value or kind of data.

Learning Classification Rules from Sample Data The process of mapping inductive learning problems into colored digraphs is best illustrated by an example. We use the example of learning rules for classifying DNA promoter sequences. In mapping the set of cases into the graph structure, we construct one subgraph for each sequence in the set of cases. The subgraph consists of a root node and a set of leaf nodes. The number of leaf nodes equals the number of attributes (in this case, the number of sequence elements). The color of the root node of the subgraph specifies whether the corresponding sequence represents a promoter sequence or not. The color of the i-th leaf specifies the nucleotide (one of A, T, C, or G) of the i-th position.

We now show that the learning problem is also transformed such that its resolution is equivalent to finding typical patterns in the colored digraph. These typical patterns will consist of a root node and one or more leaf nodes. Note that each extracted pattern is equivalent to a typical DNA sequence (promoter or non-promoter). In order to select "good" typical patterns we need to use some selection criterion that reflects this interpretation of the typical patterns extracted. One such criterion is:

$$\text{minimize} \left[\begin{array}{l} \text{(Number of Nodes in the Contracted Graph)} + \text{(Number} \\ \text{of the different values of Color in the Contracted Graph)}^2 \end{array} \right] \quad (1)$$

The first term of the criterion is intended to encourage a search for typical patterns of the DNA data. The smaller the number of nodes in the contracted graph, the better the typical patterns used. These typical patterns extracted can be interpreted as classification rules. The second term in the selection criterion is a penalty to avoid an excessive number of rules, and ensures that compact solutions are preferred. This ensures that the process of extracting typical patterns is equivalent to the goal of extracting classification rules.

Note that the restriction on graph matching (see section 3) is necessary to map the learning problem correctly. Checking for equivalence between all possible edge combinations would be incorrect, since this would involve comparing different attributes.

The mapping process described above for the DNA sequence example is applicable for learning of all attribute-value type classification systems.

Learning Macro Rules for Speeding up Inference Learning macro rules to make problem-solving more efficient is also a well-studied problem in machine learning. This involves reorganizing prespecified knowledge into a more efficient form.

The colored digraph for this purpose is a type of proof tree. Each node corresponds to a term in the proof process. Each node has two colors. Color1 of a node corresponds to the axiom (or rule) used to prove the term and Color2 refers to the term itself. Here, Color2 is ignored in the matching process, and only the equivalence for Color1 is examined. The ordering of edges from a node encodes the ordering of terms in the proof. For example, in the case of a prolog clause, the ordering of edges from a node indicates the order in the body part of the clause.

The learning problem (finding macro rules to make equation-solving more efficient) is consistent with the goal of finding typical patterns. These patterns can be interpreted as macro rules. This interpretation process for the macro rules is essentially equivalent to the generalization process of EBL [21, 4]. In order to select "good" typical patterns, the following selection criterion can be used:

$$\text{minimize} \left[\begin{array}{l} \text{(Number of Nodes in the Contracted Graph)} + \text{(Number} \\ \text{of different Color1 values in the Contracted Graph)}^2 \end{array} \right] \quad (2)$$

It is the same to the criterion used for solving inductive learning problems. Similar to EBL, the goal here is to make inference efficient by chunking out the intermediate inference. The first term is intended to estimate the effect of chunking. The second term is a penalty for the excessive macro rules.

The mapping process described above for equation-solving is applicable for other macro-rule learning problems of the type discussed in [13].

4.2 Hierarchical Knowledge Base Construction

In this section we illustrate how more complex learning problems can be mapped into colored digraphs. We examine the problem of hierarchical knowledge base construction and show how it can be solved by means of graph-based induction. The solution of this problem calls for a mixture of inductive and deductive learning strategies. Before we discuss how to map this problem into colored digraphs, we briefly describe hierarchical knowledge bases.

Hierarchical Knowledge Bases A hierarchical knowledge base is a method of organizing knowledge about complex systems. Usually the different levels in the hierarchy are abstraction levels. Each abstraction level contains knowledge of the system appropriate at that level. Usually each level has its own *interpretation rules* for manipulating the objects at that level and making inferences about them. The levels are connected together by *reformation rules* that specify how to map knowledge at one level to the level immediately above (or below) it. Given such a hierarchical knowledge base, it is normally used in one of two modes:

Reformation Mode: Given a description at one level, descriptions at other levels are generated.

Inference Mode: In using the knowledge base to make inferences, an appropriate level is selected based on some criterion. Then the descriptions and interpretation rules at that level are used to make inferences.

Construction of such hierarchical knowledge bases is motivated by the fact that the higher levels enable more efficient inferences. The results of these inferences can then be mapped down to the lower levels if needed. This process is considerably more efficient than performing the complete inference at the lower level itself. Creation of hierarchical knowledge bases usually involves creating new abstract level concepts (along with the accompanying interpretation rules for that level) from the lowest level description and rules. Previous approaches to this problem [8, 17] have relied on user-supplied reformation rules.

Construction of Hierarchical Knowledge Bases by Colored Digraphs
In order to create hierarchical knowledge bases, qualitative simulation of the lowest level of the knowledge base is used.

The interpretation rules at the lowest level are used to obtain a qualitative simulation trace. It is this trace that is represented as a colored digraph. Each node corresponds to some physical datum, such as voltage and current at a certain node in the circuit. Two colors are used for each node: Color1 of the node corresponds to the equation used to calculate the value and Color2 is the value itself. Each edge of a node corresponds to a variable in the equation associated with the node. Color2 is ignored in the matching process, but is used later on for extracting interpretation rules from the typical patterns.

We now show how finding typical patterns in the colored digraph is equivalent to constructing the hierarchical knowledge base. The typical patterns are interpreted in two ways: (1) they are viewed as reformation rules which translate the lower-level descriptions into higher-level vocabularies, and (2) they are also used to specify interpretation rules for the higher-level vocabularies. Thus, the process of finding typical patterns in the colored digraph results in automatic generation of the higher levels of the knowledge base. In order to obtain "good" higher levels, the following selection criterion was found useful:

$$\text{minimize} \left[\begin{array}{l} (\text{Number of Nodes in the Contracted Graph}) \\ + (\text{Number of Edges in the Contracted Graph}) \\ + \sum(\text{Number of Nodes in the Pattern}_j)^2 \end{array} \right] \quad (3)$$

The first term represents the amount of data to be handled, the second term the matching cost during inference, and the third term the cost to generate hierarchical descriptions from the input descriptions.

The mapping process illustrated above is applicable for other problems that involve creation of hierarchical knowledge bases based on QSIM [14] type qualitative simulators.

5 An Algorithm for Finding Typical Patterns

After the learning problem is encoded as a colored digraph, the key step is to extract typical patterns. In this section we describe in detail an algorithm, called

CLiP, for performing this task. CLiP is a beam-search algorithm, amenable to parallel implementation, that searches for typical sub-patterns in the colored digraph. The objective is to find typical patterns that help contract the graph. In order to assist in the choice of desirable typical patterns, a selection criterion for comparing alternative contracted graphs is provided to the CLiP algorithm. The search procedure focuses on obtaining a reasonably good solution, not necessarily an optimal one. The method involves three basic operations: *Pattern Modification*, *Pattern Combination* and *View Selection*. It starts with a set of null patterns, one for each view (a view holds promising typical patterns), and iteratively extends these patterns by the first two operations (the old patterns are also retained). In each iteration, the input graph is contracted using patterns in each view and only the good views are retained. The heart of the CLiP procedure involves iteratively performing the following three steps:

Pattern Modification: Each pattern in each view is extended. Figure 2 shows how patterns are modified in this step. First, a view is selected and the graph is contracted using patterns in the selected view. The reduced graph is then analyzed and every possible pattern made up of two linked nodes is considered. These patterns are referred to as temporary patterns. Each such temporary pattern is then expanded based on patterns in the current view, and used to create new views. In the example, three new views, each with two patterns, are generated from the current view. Note that patterns in the parent view are also stored in the new views. As mentioned earlier, the matching procedure is restricted to checking for graph identity. In contrast to matching based on graph isomorphism (an NP-complete problem), matching based on graph identity has a time complexity that is O(Number of Nodes).

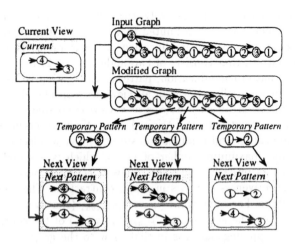

Fig. 2. Pattern Modification

Pattern Combination: Existing views are combined in pairs to obtain new views. All possible combinations are considered.

View Selection: Estimates are obtained for each view as to how much reduction in the size of the graph can be expected after contracting the input graph using patterns in that view, and only those views that rank high are chosen within the allowable number of views.

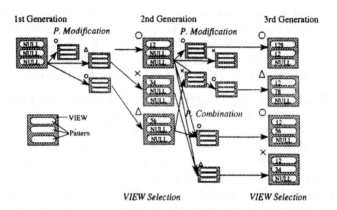

The mark on the top left of each view indicates the result of graph contraction:○ for good, × for bad and △ for intermediate. The size of a mark indicates whether the result is based on actual rewriting of the graph (big) or on estimation (small).

Fig. 3. Evolution of Patterns in a View

Figure 3 illustrates how patterns evolve through generations. In this example, the maximum number of views is limited to 4. In the first generation, starting from null patterns, the pattern modification step generates three views, each containing one pattern that consists of two nodes (*e.g.* 1-2) In succeeding generations, besides pattern modification, pattern combination is also performed. In pattern modification, all patterns in all views are considered in turn as candidates for modification. In each case, a new view is created consisting of the modified pattern appended to the original view. View selection is done after the pattern modification and combination steps. The view selection step selects the best N views (maximum number of views, which is a search parameter; 4 in Figure 3) views in each generation. View selection is based on estimates of the relative effectiveness of views in graph contraction. These estimates are computed as follows: For each of the views $V_{Previous}$ (that were selected in the previous generation), the actual size $C_{Exact}(V_{Previous})$ of the contracted graph that results from applying patterns in that view to the graph is calculated at the beginning of the current generation. For each new view $V_{Current}$ that is generated by pattern modification step, the estimated size $C_{Estimate}(V_{Current})$ of the contracted graph that would result from applying that view is calculated as a perturbation to the actual size of the contracted graph due to the parent view $V_{Previous}$ (see Equation 4). The graph size for views generated by the pattern combination step is estimated as the average of the two views involved in the combination.

The estimated size and the actual size (calculated at the beginning of the next generation) may not agree (see the difference between big and small marks on the top left of each view in Figure 3). However, the estimate is reasonable enough to ensure that good views are usually selected. The estimation process is necessary because computation of the actual size is computationally expensive.

$$C_{Estimate}(V_{Current}) = (1.0 - 0.1 \times F) \times C_{Exact}(V_{Previous}) \qquad (4)$$

$$F = \frac{\text{No. of occurrences of Temp. Pattern } \textcircled{A} \rightarrow \textcircled{B}}{\text{No. of occurrences of Node } \textcircled{B}}.$$

Our procedure for extracting typical patterns is closely related to SUBDUE [11, 12]. The key difference is that we use a more efficient graph-matching procedure.

6 Experimental Results

6.1 DNA Sequence Classification

A set of DNA promoter sequence data [25] was prepared for analyses by graph-based induction. The dataset consisted of 106 sequences, half of which were promoter sequences and the rest were non-promoter sequences. Each sequence had 57 DNA nucleotides (one of A, T, C, or G). This data was encoded in a colored digraph as described earlier, and the CLiP algorithm was used to extract typical patterns. Since each generation creates views that give rise to classification rules, another issue is the selection of views from one particular generation as the final answer. A very good measure of success is the error rate for unseen data. Hence, those views that generate classification rules with minimum leave-one-out error [5], were selected in each generation. The following classification rules were obtained:

$$
\begin{aligned}
F_{06} = A \wedge F_{15} = T \wedge F_{16} = T &\rightarrow positive \\
F_{16} = T \wedge F_{17} = G &\rightarrow positive \\
F_{15} = T \wedge F_{16} = T &\rightarrow positive \\
True &\rightarrow negative
\end{aligned}
$$

Table 1 summarizes the predictive performance of these rules and compares it to previous results. Error-rates were estimated by the leave-one-out method. In Table 1, ID3 [23] and SWAP1 [27] represent rule learning systems, and BP a standard back propagation neural network method with one hidden layer. CLiP outperforms the standard ID3 tree-induction program.

	Previously Reported Methods			CLiP
Method	ID3	SWAP1	BP	
Error/106	19	14	8	14

Table 1. Inductive Learning: Comparison with other classification methods

6.2 First Order Equation Solving

Eighty five inference traces of first-order equation solving were obtained by running a prolog program in which each prolog clause represented an axiom for equation solving. These traces were mapped to colored digraphs as described earlier. The CLiP algorithm extracted a set of macro operators for more efficient equation solving. The macro operators were encoded as prolog clauses, each representing a new theorem for equation solving. Figure 4 illustrates an inference trace before and after learning.

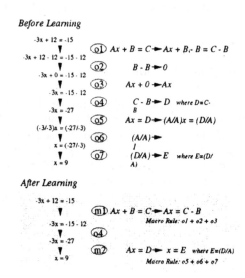

Fig. 4. Use of Learned Macro Operations in Equation Solving

The effectiveness of the learning process can be assessed by computing the *speed-up ratio* which compares the length of inferences before and after learning. The speed-up ratio was computed by adding the generated clauses at the top of the original prolog program, and solving the equations with this new set of rules.

As in the previous task, we need a measure to select a set of views as the final answer. The best measure of success here is the utility of acquired macros. So we select the views which show the best speed-up ability over the training problems. The results (Table 2) indicate, in contrast to a non-selective EBL system, learning by graph-based induction improved problem-solving efficiency.

Method	Non Learning	EBL	CLiP
CPU sec.	100.0	279.9	87.5
Speed-up Ratio	1.00	2.80	0.88

Table 2. Equation Solving: Comparison with EBL

6.3 Circuit Equation Reformation

Qualitative simulation results of an NMOS circuit that calculates a carry in a CPU [18] were used for constructing a hierarchical knowledge base of the circuit. The simulation traces were mapped to colored digraphs and the graph-based induction algorithm successfully extracted the patterns and truth tables corresponding to NOR and NOT functions. It first found an analog version of these concepts and introduction of additional approximation resulted in the logical concepts [2].

We set the maximum number of views at 15 and limited the algorithm to 50 iterations. The minimum graph size was attained in the 25th generation.

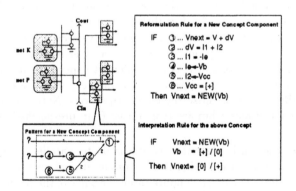

Fig. 5. A Pattern that corresponds to NOT operation

---

[2] See [28] for more details of this approximation procedure.

Unlike the previous two tasks, there is no uniform criterion to measure the usefulness of acquired concepts, and we select views that gives the smallest contracted graph. Figure 5 shows an example of a typical pattern extracted (lower left) and the corresponding reformation/interpretation rules (right). The variables on the right hand side of each equation in the conditional portion of the extracted concept (e.g. V and dV in equation ①) are arranged so that they correspond to the numbers indicated on the edges pointing to each node (e.g. the edges 1 and 2 going into the node ①). The reformation rule says that if there is a set of 6 relations (equations) as shown, it is reasonable to infer that V_{next} can be calculated by some relation NEW from the current value of V_b. The interpretation rule shown, describes a set of relationships between the input V_b and the output V_{next}. These relationships indicate that the concept generated is the NOT operation. In another result from the same experiment, the concept corresponding to the NOR operation was also discovered.

Further experiments were performed using different selection criteria. In this experiment, the third term of the selection criterion (the cost to generate hierarchical descriptions), is multiplied by a factor of 10. The minimum size was attained at the 15th iteration, and the result corresponds to intermediate physical structures, i.e. pullup transistor and pulldown transistor. Using this level as input, the algorithm obtained exactly the same set of concepts as in the previous experiment.

These results indicate that the resultant structure varies depending on the characteristics of the inference system. While an inference system with strong reformation capability can recognize a complex object by one level of abstraction, another inference system with weaker reformation capability needs two levels of abstraction for the same object. In contrast to previous studies on hierarchical knowledge representation, our method automatically considers a number of factors (such as cost of inference) during the construction of hierarchical device representations. It is also able automatically find new abstract level concepts while at the same time re-compiles the lower level information into higher level knowledge structures that involve the abstract concepts.

Achieving equivalent results solely by using conventional deductive learning techniques is difficult. The appropriate operationality criteria and goal concepts to cover the whole range of concepts are difficult to specify. Similarly, conventional inductive learning techniques are also not very useful for this task as they lack the ability to reorganize the circuit equations. Note that all concepts acquired by the CLiP algorithm are reorganized from prespecified circuit equations.

In principle, the CLiP algorithm is amenable to parallel implementation. However, even on conventional hardware, computational requirements of the algorithm are modest. The most compute-intensive parts of the procedure involve graph matching and graph contraction. By using an efficient graph matching procedure, and by using estimators to avoid having to do graph contraction too often, both these stages are well optimized.

For the hierarchical knowledge base creation task, the input graph had 2176 nodes and 2144 edges. Using this graph, the concepts of Analog NOT/NOR were obtained in about 6 minutes of CPU time on a SPARCstation-1. The largest graph we have considered to date involved over 50K nodes and over 50K links, representing an inductive learning problem involving data from the protein identification resource (PIR) [2]. Solving this problem required about 2 days of CPU time on a SPARCstation-1.

7 Discussion

In this paper we have examined the use of colored digraphs in integrating different modes of learning. We have shown how different learning problems can be mapped into colored digraphs. The mapping is achieved such that resolution of the corresponding learning task can be achieved by finding typical patterns in the resultant colored digraph. The use of a common representation allows us to conceive of a *single* algorithm to perform a variety of inductive and deductive learning tasks such as learning classification rules, generating new macro operators, and creating hierarchical knowledge bases by identifying new concepts. The algorithm for extracting typical patterns relies on an efficient matching procedure. The interpretation of the typical patterns depends on the learning task being solved. For classification learning, they are interpreted as classification rules; in speed-up learning, they are viewed as macro operators; and in hierarchical knowledge base construction, typical patterns help obtain both reformation rules as well as interpretation rules by identifying new concepts.

Empirical results indicate the viability of our graph-based induction method for solving a variety of learning problems. While it is reasonable to expect that methods for solving a specific learning problem might do better in certain specialized representations, our ultimate goal is to devise a unified learning mechanism and towards this end the colored digraph representation is more promising. However, to accomplish our ultimate goal, further research is necessary. For example, we have not addressed the issue of defining selection criteria is a principled fashion. We still need to select this criterion carefully for each task. Methods to convert other learning problem into colored digraph representation still have room for investigation. For example, although the application to the PRODIGY [20] type control rule learning appears to be straightforward, experimental results have not yet been analyzed.

Acknowledgements:

We would like to thank Sholom Weiss, Haym Hirsh, Pat Langley and Hari Narayanan for helpful comments on the research. We would also like to thank Ayumi Shinohara for providing the PIR data.

References

1. J. R. Anderson and P. J. Kline. A Learning System and Its Psychological Implications. In *IJCAI-79*, pages 16–21, 1979.
2. S. Arikawa, S. Miyano and A. Shinohara. Knowledge Acquisition from Amino Acid Sequences by Learning Algorithms. In *JKAW92*, pages 109–128, 1992.

3. A. P. Danyluk. The Use of Explanation for Similarity-Based Learning. In *IJCAI-87*, pages 274–276, 1987.

4. G. DeJong and R. Mooney. Explanation-Based Learning : An Alternative View. *Machine Learning*, pages 145–176, 1986.

5. B. Efron. The jackknife, the bootstrap and other resampling plans. In *SIAM*, 1982.

6. T. Ellman. Explanation-Based Learning: A Survey of Programs and Perspectives. *ACM Computing Surveys*, 21(2):165–220, 1989.

7. O. Etzioni. STATIC : A Problem-Space Compiler for PRODIGY. In *AAAI-91*, pages 533–540, 1991.

8. B. Falkenhainer and K. D. Forbus. Compositional modeling : finding the right model for the job. *Artificial Intelligence*, 51:95–143, 1991.

9. N. S. Flann and T. G. Dietterich. A stduy of explanation-based methods for inductive learning. *Machine Learning*, pages 187–226, 1989.

10. H. Hirsh. Combining Empirical and Analytical Learning with Version Spaces. In *ML-89*, pages 29–33, 1989.

11. L. B. Holder. Empirical Substructure Discovery. In *ML-89*, pages 133–136, 1989.

12. L. B. Holder, D. J. Cook, and H. Bunke. Fuzzy Substructure Discovery. In *ML-92*, pages 218–223, 1992.

13. R. E. Korf. Macro-operators : A weak method for learning. *Artificial Intelligence*, pages 35–77, 1985.

14. B. Kuipers. Qualitative Simulation. *Artificial Intelligence*, 29:289–338, 1986.

15. M. Lebowitz. Integrated Learning: Controlling Explanation. *Cognitive Science*, 10:219–240, 1986.

16. R. Levinson. A Self-Organizing Retrieval System for Graphs. In *AAAI-84*, pages 203–206, 1984.

17. Z. Y. Liu and A. M. Farley. Shifting Ontological Perspectives in Reasoning about Physical Systems. In *AAAI-90*, pages 395–400, 1990.

18. C. Mead and L. Conway. *Introduction to VLSI Systems*. Addison-Wesley Publishing Company, 1980.

19. R. S. Michalski. A theory and methodology of inductive learning. *Artificial Intelligence*, 20:111–161, 1983.

20. S. Minton. Quantitative Results Concerning the Utility of Explanation-Based Learning. *Artificial Intelligence*, 42:363–391, 1990.

21. T. M. Mitchell, R. M. Keller and S. T. Kedar-Cabelli. Explanation-Based Generalization : A Unifying View. *Machine Learning*, pages 47–80, 1986.

22. M. Pazzani, M. Dyer, and M. Flowers. The Role of Prior Causal Theories in Generalization. In *AAAI-86*, pages 545–550, 1986.

23. J. R. Quinlan. Induction of Decision Trees. *Machine Learning*, 1:81–106, 1986.

24. P. S. Rosenbloom and J. Aasman. Knowledge Level and Inductive Uses of Chunking (EBL). In *AAAI-90*, pages 821–827, 1990.

25. G. G. Towell, J. W. Shavlik and M. O. Noordewier. Refinement of Approximate Domain Theories by Knowledge-Based Neural Networks. In *AAAI-90*, pages 861–866, 1990.

26. S. A. Vere. Induction of Relational Productions in the Presence of Background Information. In *IJCAI-77*, pages 349–355, 1977.

27. S. M. Weiss and N. Indurkhya. Reduced Complexity Rule Induction. In *IJCAI-91*, pages 678–684, 1991.

28. K. Yoshida and H. Motoda. Automatic Knowledge Reformation. In *JKAW92*, pages 263–293, 1992.

A Perceptual Criterion for Visually Controlling Learning

Masaki Suwa and Hiroshi Motoda

Advanced Research Laboratory, Hitachi Ltd.,
2520, Hatoyama, Saitama, 350-03, Japan

Abstract. Acquiring search control knowledge of high utility is essential to reasoners in speeding up their problem-solving performance. In the domain of geometry problem-solving, the role of "perceptual chunks", an assembly of diagram elements many problems share in common, in effectively guiding problem-solving search has been extensively studied, but the issue of learning these chunks from experiences has not been addressed so far. Although the explanation-based learning technique is a typical learner for search control knowledge, the *goal-orientedness* of its chunking criterion leads to produce such search control knowledge that can only be used for directly accomplishing a target-concept, which is totally different from what perceptual-chunks are for. This paper addresses the issues of acquiring domain-specific perceptual-chunks and demonstrating the utility of acquired chunks. The proposed technique is that the learner acquires, for each control decision node in the problem-solving traces, a chunk which is an assembly of diagram elements *that can be visually recognizable and grouped together* with the control decision node. *Recognition rules* implement this chunking criterion in the learning system PCLEARN. We show the feasibility of the proposed technique by investigating the applicability and cost-effective utility of the learned perceptual chunks in the geometry domain.

1 Introduction

Acquiring search control knowledge (SCK) of high utility is essential to reasoners in speeding up their problem-solving performance. We have been investigating in the domain of geometry problem-solving what kind of SCK should be learned and how it should be learned from the problem-solving episode of a problem.

Let's look at the two problems in Fig.1 to make the research issues of this paper clear. The first problem is to prove that $\angle BAC = \angle DEC$, where the conditions of Fig.1 are given. We can solve it only after coming up with appropriate auxiliary-lines, CF and DF with a new point F such that A, C, F are on a line and $AC = CF$, because those additional lines enable us to use the midpoint condition $BC = CD$ as one of the antecedent conditions for proving congruence of triangles and consequently it prompts us to use the given condition $AB = DE$ to prove $\triangle DEF$ is an isosceles. The second problem, the more difficult one, is to prove $\triangle AQR$ is an isosceles. We can solve it similarly if we come up with auxiliary-lines like APE, CE and DE with a new point E such that A, P, E

are on a line and $AP = PE$, because those lines enable us to use the two given conditions, $BP = PC$ and $AB = CD$ to prove that $\triangle ABP$ and $\triangle ECP$ are congruent and $\triangle CDE$ is an isosceles.

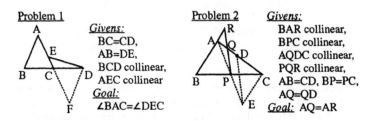

Fig. 1. Two auxiliary-line problems in geometry

The important thing is that although both problems have quite different goal-structures from each other, they need to be solved by utilizing some of the given conditions in exactly the same way with similar auxiliary-lines. In other words, if such SCK as shown in the upper left of Fig.5 is provided to a reasoner, it can solve both problems in an analogous way as human experts do; the SCK tells that if there is a midpoint condition $XY = YZ$ with X, Y, Z being on a line, then recall (or find) another collinear line VYW passing through Y such that $VY = YW$ in order to prove that $\triangle XYW$ is congruent to $\triangle ZYV$. Preferably selecting to apply this SCK in reaction to the above midpoint condition, although the condition may be possibly used as one of the antecedent conditions of other geometry theorems, can lead the reasoner to the solutions in both problems.

This kind of SCK corresponds to what is called "perceptual-chunks" in the research field of cognitive science where the role of perceptual-chunks in effectively guiding problem-solving has been studied [3][4][11][6]. A perceptual-chunk is a chunk of diagram elements which many problems share in common as a portion of the whole diagram. Recalling an appropriate perceptual-chunk at a control decision node during problem-solving processes and applying the macro-operator attached to the perceptual-chunk contributes much to controlling problem-solving search effectively. The most important characteristic of search-control by perceptual-chunks is that it is *not goal-oriented*; applying a perceptual-chunk is meant for locally recognizing it in the diagram of the problem, *without caring* whether or not its application will directly accomplish the goal of the problem. In spite of the beneficial role of perceptual-chunks, however, the past studies have not addressed the issue of learning the set of effective perceptual-chunks from experiences.

The explanation-based learning (EBL) module of PRODIGY system [8] is a typical system that learns SCK from experiences. It explains why a selection at a control decision node has led the reasoner to a target concept and chunks only the features relevant to the explanation as the antecedent conditions of the learned SCK. The *goal-orientedness* of its chunking criterion produces the kind of SCK that is used for directly accomplishing a target concept, which is totally different from perceptual-chunks. In this sense, we cannot expect EBL systems to learn perceptual-chunks [12].

This paper addresses the issue of learning perceptual-chunks from an experience. Its basic concept is to chunk the **diagram elements that can be visually recognizable and grouped together** with the diagram elements corresponding to each control decision node of the problem-solving traces. *"Recognition rules"* implement this chunking criterion in the learning system PCLEARN. Each of them is domain-specific knowledge which describes the necessary conditions for each domain object to be visually recognizable. Notice that the chunked area does not always include the goal of the problem because the goal is not always recognizable (i.e. too far away) at the local control decision node. This distinguishes PCLEARN chunking facility from the EBL learner which chunks all the paths to the target concept of the problem, in most cases the goal of the problem.

In the second section, we show the basic notion of recognition rules and its use for determining the area to be chunked out. In the third section, we discuss the feasibility of the recognition rules as a perceptual criterion, by presenting some experimental data on the operationality and cost-effective utility of the learned perceptual-chunks when the solver uses them as search control knowledge. In the fourth section, we discuss the thrust of this perceptual-chunking method in terms of comparisons with other methods as well as its generality.

2   The PCLEARN chunking module

2.1   Recognition rules

PCLEARN, after solving a problem[1], learns a perceptual chunk with macro-operator information for each control decision node[2] of the given problem-solving traces. For the purpose of chunking diagram elements that can be visually recognizable and grouped together for each control decision node, we need to provide a criterion for determining what is "visually recognizable". The use of recognition rules determines it.

Recognition rules themselves are a semantic representation of how human experts visualize domain objects. More precisely speaking, each rule itself describes the necessary conditions for a domain object to be "visually recognizable", consisting of the recognizabilities of other related domain objects and some additional conditions.

Figure 2 is the set of recognition rules in the domain of geometry. Points, segments, triangles and angles are the domain objects in this domain. The first

---

[1] In case of auxiliary-line problems as shown in Fig.1, PCLEARN needs to be taught about how to draw lines in order to solve it. Note that once it learns a chunk, the chunk can be used to find auxiliary lines.

[2] A control decision node is the one that is a member of the successful proof tree and also to which there were more than one tested domain rules some of which have been successfully applied and others were not.

rule states that a point X is always recognizable when a segment XY is found recognizable because X is a constituting member of XY. In general, when an object is already found recognizable and we want to prove the recognizability of another object which is a structurally constituting member of the former object, we do not need any additional conditions. The first three rules in Fig.2 belong to this category. On the other hand, when we prove the recognizability of an object from the other objects which structurally composes that target object, we need some (sometimes no) additional conditions. For example, when we prove the recognizability of segment XY from the recognizabilities of the two points X and Y, an additional condition is needed, i.e. the segment XY actually has to exist in the problem space(corresponding to $exist(s(X,Y))$ in Fig.2), or two segments $s(X,Z)$ and $s(Z,Y)$ have to be on the same line for another point Z (corresponding to $collinear(X,Z,Y)$ in Fig.2). The last two recognition rules are the examples where no additional condition is needed by chance, although they belong to this category.

```
recognizable(X):- recognizable(s(X,Y)).
recognizable(s(X,Y)):- recognizable(a(X,Y,Z)).
recognizable(s(X,Y)):- recognizable(tr(X,Y,Z)).
recognizable(s(X,Y)):- recognizable(X), recognizable(Y), exist(s(X,Y)).
recognizable(s(X,Y)):- recognizable(X), recognizable(Y), collinear(X,Z,Y).
recognizable(a(X,Y,Z)):- recognizable(s(X,Y)), recognizable(s(Y,Z)).
recognizable(tr(X,Y,Z)):-
        recognizable(s(X,Y)), recognizable(s(Y,Z)), recognizable(s(Z,X)).

where s(X,Y) -- segment XY,  tr(X,Y,Z) -- triangle XYZ,  a(X,Y,Z) -- angle XYZ
       The literals underlined are additional conditions.
```

Fig. 2. The set of recognition rules in geometry

2.2 Algorithm for perceptual-chunking

For the purpose of chunking perceptual-chunks for each control decision node, PCLEARN first identifies all the recognizable domain objects when the objects included in the domain rule[3] that has been successfully applied to the control decision node are supposed to be recognizable, and enumerates all the recognizable features of the above objects, to finally obtain a perceptual chunk which is the assembly of the recognizable objects with the recognizable features.

Step 1: Picking up recognizable objects. The first step is to enumerate all the recognizable objects relevant to a control decision node. The procedures are

1. to assert that all the objects which appear as the arguments of the literals in the SAR are recognizable, and
2. to enumerate all the objects which can be proved as recognizable, using **recognition rules.**

---

[3] This rule is denoted as SAR (Successfully Applied Rule) in this paper.

Figure 3 is a successful proof tree of the Problem 1 in Fig.1. The underlined nodes are the control decision nodes. Here, the learning process for the control decision node, $AC = CF$, is illustrated. The SAR for this control decision node is Cong-by-2Side-1Ang. First, the objects appearing in this SAR, $s(b,c)$, $s(a,c)$, $s(c,d)$, $s(f,c)$, $a(b,c,a)$, $a(d,c,f)$, $tr(a,b,c)$ and $tr(f,c,d)$, are asserted to be recognizable. Then, by use of the recognition rules, the following objects, a, b, c, d, f, $s(a,b)$, $s(d,f)$, $s(b,d)$, $s(a,f)$, $a(b,a,c)$, $a(b,a,f)$, $a(a,b,c)$, $a(a,b,d)$, $a(d,f,c)$, $a(d,f,a)$, $a(f,d,c)$, $a(f,d,b)$, $a(b,c,f)$ and $a(a,c,d)$ are justified to be recognizable.

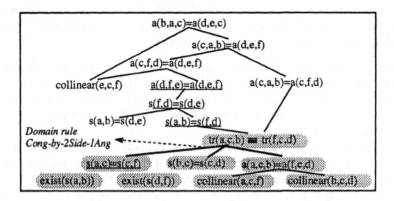

Fig. 3. The successful proof tree of Problem 1 of Fig.1

Step 2: Enumerating recognizable features. The second step is to derive and pick up from the problem-solving traces all the recognizable features of the above recognizable objects. The procedures are

1. the literals appearing in the SAR are picked up as recognizable,
2. the literals of the **additional conditions** which appeared in the recognition rules used successfully for proving the recognizability of objects in Step 1 are picked up as recognizable, and
3. all the features that can be derived from the above recognizable literals using domain rules are regarded as recognizable.

What we have obtained so far is the derivation tree (the third procedure of Step 2), and importantly the derivation tree itself represents a piece of macro-operator information that can be applied to the same control decision node in future problems; the lowest nodes of the tree are the IF-part of the macro-operator and the other nodes are the THEN-part. If we notice that the macro-operator has been derived only from the recognizable features that has been determined by use of recognition rules, the significant role of recognition rules in chunking the macro-operator may be clear.

Let's look at the example case of learning from the control decision node $AC = CF$ in Fig.3. The recognizable literals to be picked up before the derivation process are shown in Fig.3 as the nodes colored grey, out of which the literals that have been incorporated as a result of using recognition rules (the 2nd of Step 2) are $collinear(a, c, f)$, $collinear(b, c, d)$, $exist(s(a, b))$, $exist(s(d, f))$. The first two, $collinear(a, c, f)$ and $collinear(b, c, d)$, have been picked up because they appeared in the recognition rules used for proving the recognizability of the object $s(a, f)$ and $s(b, d)$ respectively. Out of these four, the last two will not be used in the derivation process and therefore removed off from the macro-operator.

Notice here that due to the existence of some additional conditions in the set of recognition rules, the learned macro-operator becomes more specific[4] than the SAR itself. In case of the above example, incorporating the two collinearities has been significant in obtaining a perceptual-chunk of the two congruent triangles located in a completely point-symmetry (the upper left of Fig.5), which is more specific than the two merely congruent triangles.

Step 3: Generalizing. The final step is to generalize each node of the acquired derivation tree by dissolving the bindings of the variables of the used domain rules. The generalized tree itself represents a macro-operator that has been learned for the control decision node. In case of the above example, the one in the upper left in Fig.5 is acquired.

We call this sort of macro-operator a perceptually-chunked macro-operator because the recognition rules work as a perceptual criterion for determining the area to be chunked out, just as human experts might do visually. Also in this sense, we can say that chunking by use of recognition rules has a connotation of visually controlling learning processes.

3 Experimental Results

In this section, we made some experiments for showing the feasibility of PCLEARN perceptual-chunking in comparison with the EBL system that learns SCK.

The typical EBL system that learns SCK is PRODIGY [8]. It implements selective learning by providing four kinds of meta-level target-concept, "succeeds", "fails", "sole-alternative", and "goal-interference". However, in the domain of geometry problem-solving, learning from "fails", "sole-alternative", and "goal-interference" will not lead to useful knowledge[5]. Therefore, we compare the PCLEARN system with the EBL system that learns from "succeeds" in which the goal-node of the problem is specified as the target-concept and the learner acquires so-called *preference rule* by explaining why the selection of a domain rule at each control decision node contributed to achieving the goal.

---

[4] This specificity directly influences much to the operationality of the learned perceptual-chunks. In this sense, recognition rules play a crucial role in determining the chunked area.

[5] In this domain, there may be no positive reason why a choice leads to a failure, and there may be no problem-solving phenomenon corresponding to sole-alternative and goal-interference.

3.1 Generality of the learned knowledge

Whether or not the learned macro-operator represents a general and meaningful perceptual chunk depends upon how frequently it is learned from various problems. We made an experiment in which each of the 20 geometry problems (shown in Appendix A) were separately solved and learned, in both learners i.e. PCLEARN and the EBL system.

Table 1 shows how many times the same macro-operator has been learned from the 20 problems for both learners. PCLEARN learns the same macros much more frequently (i.e. more than twice), while it is quite rare that the EBL learner acquires the same macro more than twice. In other words, the macro-operators learned by EBL tend to be more specific to the goal-structure of the original problems. This is mainly because *goal-orientedness* of the EBL chunking criterion does not suit for the nature of the geometry domain where there is little consistency in goal-structure across problems but rather much consistency in visual/perceptual chunks across problems. In this respect, PCLEARN is superior to EBL systems as a method of learning domain-specific perceptual chunks in this domain.

Table 1. The frequencies of macro-operators being learned

Frequency	The number of macro-operators	
	PCLEARN	EBL
1	43	86
2	9	7
3	5	0
4	3	1
more than 4	4	0
total	64	94

3.2 Cost-effective utility of the learned knowledge

In order to evaluate the feasibility of PCLEARN as a learner for speeding up problem-solving performances, we evaluated the cost-effective utility of each of the learned macro-operators throughout its use over many problems according to the following formula,

$$Utility = TotalSavings - TotalMatchCosts, \tag{1}$$

where *TotalSavings* is the cumulative cpu-time benefit that results from applying the macro-operator frequently, so-called *re-ordering effect* [7], and *TotalMatchCosts* is the cumulative time cost spent in testing to apply the macro-operator in vain over frequent testings during many problems. Every time a problem is solved, the above costs and benefits are measured for each macro-operator, as a result of which the utility of each macro-operator is updated.

The cpu-time benefit of a macro-operator is calculated, when it was applicable to a problem, by subtracting the costs relevant to the use of the macro-operator from the corresponding costs in solving the same problem without any macro-operators[6]. The former cost tends to become large by the existence of macro-operators because it potentially causes the solver to try more matchings in searching for applicable knowledge, while the latter may include the cost spent in vain for producing some irrelevant branches of nodes at the decision node before applying the successful domain rule, which the solver would have avoided by use of the macro-operator. The tradeoff between these two effects influences the benefit of each macro-operator.

We do not separate a training session from a testing one. Instead, going through the 20 problems, each one is solved using only a limited number[7] of macro-operators that have been already learned and have scored high utility values *up to* that point. The problem order is assigned so that the problems become progressively larger.

In Fig.4, the cumulative problem-solving cost is plotted against the number of problems solved up to that point, for both cases of using PCLEARN macro-operators ("with-PCLEARN-macro" system) and the EBL macro-operators ("with-EBL-macro" system), together with the case of the system without any macro-operators ("without-macros" system) for reference. The cost of the "with-PCLEARN-macro" system becomes worse for the first several problems than the "without-macro" system due to the increased matchings by use of the macro-operators, but after going through eleven problems, the slope of the cumulative cost gradually begins to become smaller than that of the "without-macro" system, i.e. learning effects by use of macro-operators begin to appear. On the other hand, the cost of the "with-EBL-macro" system is larger by the cost of using macro-operators than that of the "without-macros" system, which indicates that only a few macro-operators were useful out of the learned ones.

The results of Tables 2, 3 and 4 can be an explanation of why the EBL macro-operators can exhibit little learning effect in geometry domain. Table 2 shows, for both of the PCLEARN and EBL macro-operators, the total costs of testing to apply macro-operators at control choice nodes. The average cost per one testing of the EBL macro-operators is about double the cost per one testing of the PCLEARN macro-operators. Table 3 shows the applicability of both macro-operators. The PCLEARN macro-operators scored higher percentage of successful applications, i.e. higher applicability, in this domain than the EBL macro-operators. The lower applicability of the EBL macro-operators may be an explanation of their lower learning effect. Table 4 shows why the EBL macro-operators cost much more in testing and have lower applicability. The av-

---

[6] The results of solving all the problems without macro-operators are provided in advance.

[7] In this experiment, the top 25 macro-operators are selected as available every time each problem is solved.

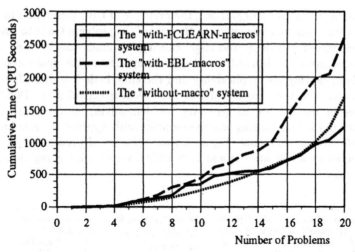

Fig. 4. The performance results of PCLEARN and the EBL learner

erage sizes[8] and the standard deviations of the applied macro-operators weighed with the frequency of applications are compared with those of all the learned macro-operators. For the PCLEARN macro-operators, the distributions of the sizes of the learned macro-operators and the applied ones are not very different from each other, which tells that PCLEARN produces macro-operators with appropriate size in terms of applicability. On the contrary, however, for the EBL macro-operators, the distribution of the sizes of the learned macro-operators is extremely shifted towards larger side than that of the applied ones, which indicates that the EBL method tends to produce too large macro-operators in terms of applicability.

Table 2. The cost of testing both PCLEARN and EBL macro-operators

	Total (msec)	Frequency of testing	Cost per one testing(msec)
PCLEARN	288,733	1,146	252
EBL	976,963	1,910	511

Table 3. The applicability of both PCLEARN and EBL macro-operators

	Frequency		Percentage (%)
	testing	applying	
PCLEARN	1,146	39	3.4
EBL	1,910	12	0.6

Table 4. The sizes of the learned and applied macros

	All the macros		The applied macros	
	Average	Standard Deviation	Average	Standard Deviation
PCLEARN	3.3	1.4	2.7	1.3
EBL	5.0	1.8	3.2	1.2

---

[8] For simplicity, we define that the size of a macro-operator is equal to the number of its preconditions, reflecting the ease of finding appropriate instantiations of the preconditions.

To sum up the above observations, the principle of EBL techniques to chunk the path from a control choice node to the goal of the problem tends to make the size of the learned knowledge too large, causing too much cost in using it, and also tends to make the learned knowledge too specific to the goal-structure of the problem, causing low applicability and therefore little learning effect in future problems. On the contrary, use of the "recognition rules" in PCLEARN is effective in the domain of geometry problem-solving in the sense that it contributes much to acquiring macro-operators with appropriate size and better cost-effective performance. Figure 5 shows the set of perceptually-chunked macro-operators which have scored high utility values after going through all the problems.

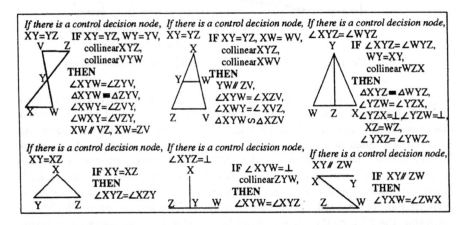

Fig. 5. The perceptual-chunks learned by PCLEARN and their macro-operator information

4 Discussions

4.1 Difference from operationality criterion

Since the PCLEARN system chunks a partial structure of the problem-solving traces, some readers may associate its chunking criterion with the concept of operationality criterion [9] that has been discussed with EBL. Here we have to discuss that both have different connotations from each other; both ideas come from different purposes for chunking, different semantics, and the different ways in using chunking criteria.

First, the purpose of PCLEARN chunking is to use the learned knowledge for parsing the whole problem space into some meaningful subparts, which helps a solver avoid traversing irrelevant paths of inference as shown in Koedinger's CD model [4]. On the other hand, the purpose of learning by use of operationality criterion is to acquire operational and usable knowledge, e.g. in recognition task.

This difference in purpose brings about the difference of semantics; the semantic of PCLEARN chunking is to chunk out what is "visually grouped" in the light of human being's way of visualizing domain objects, while the semantic of learning by use of operationality criterion is to chunk out such a portion of the obtained explanation tree that is "described only by operational predicates (i.e. descriptive predicates) and not by functional predicates", in recognition task. Here one may regard what is represented by descriptive predicates as equal to what is chunked by use of recognition rules, and also think that PCLEARN is an extension of operationality criterion. But it is not correct; distinguishing functional and descriptive concepts (e.g. "liftable" v.s. "with-hand" in the example of CUP [13]) is one thing and grouping the represented concepts into some chunks visually is another. The difference between both learners may not be clear yet enough in the above example, because the concept of a single cup is not so complicated as to make humans feel like parsing it into sub-concepts. But suppose an object with more complicated configurations which a human reasoner tries to parse into smaller parts when he thinks of the object. In this example, the explanation tree of the whole object would be represented by many predicates, some of which are functional and others are descriptive. Here, the most important thing is to parse the whole object into several subparts from a viewpoint of "visual grouping", not from a view point of differentiating between functional and descriptive predicates. In this example, it is clear that the criteria of both chunking methods will bring about different learning results.

In order to implement the above purposes and semantics of learning, both chunking methods employ different criteria in different ways. EBL systems are provided in advance with the list of operational (descriptive) predicates that will work in target tasks, and chunks such a portion of the obtained explanation tree that is represented by those operational predicates. However, we cannot determine in advance what predicates in the explanation tree are visually grouped at each of the local control decision nodes, i.e. in other words, it must be dependent on each control decision node. Thus, the PCLEARN system dynamically determines it at each control decision node, using knowledge about how humans visualize each domain object in relation to other objects.

4.2 Comparison with other chunking methods

Other kind of chunking methods to be mentioned are SOAR [5] and ACT theory [1], in which the learners chunk the problem-solving processes relevant to satisfying the subgoals the solvers have established. We call these methods, including EBL, *goal-oriented* chunking because they share the view that the learner should chunk the goal-oriented problem-solving processes toward accomplishing subgoals and/or target-concepts. Compared to these works, the PCLEARN's chunking criterion is new in the sense that it is not *goal-oriented* but rather a bottom-up perception-oriented chunking mechanism. In this respect, we call it *perceptual* chunking.

4.3   Generality of the perceptual chunking

We have shown the feasibility of the PCLEARN chunking method in geometry problem-solving domain. In this subsection, we discuss the class of problems to which this learning method can be applicable.

The general notion of PCLEARN chunking method is that the learner acquires a useful set of perceptual chunks which help the solver parse problem diagrams into several subparts, contributing much to focusing only on some key inference steps to generate a solution plan. This kind of abstract planning is important especially in domains that have been studied in the research area of diagrammatic reasoning, e.g. not only geometry problem solving [11][6][4] but also the task of reasoning about interactions with spatial configurations[10], the task of reasoning in blocks worlds with spatial distribution [2] and so on.

The key process underlying the acquisition of perceptual chunks is "visual grouping", and the criterion for controlling "visual grouping" is the set of recognition rules. Therefore, the applicability of the PCLEARN chunking mechanism in a domain depends upon whether or not we can provide a useful set of recognition rules in the target domain. Although the specification itself of the rules is general in the sense that we have only to enumerate the domain objects and to describe what conditions are needed for each domain object to be recognizable or accessed, we have to wait for future researches for actual implementation in other domains.

5   Conclusion

We proposed a new mechanism of learning from experiences domain-specific perceptual-chunks that can be used as search control knowledge. Its basic concept is that the learner acquires, for each control decision node in the problem-solving traces, a chunk which is an assembly of diagram elements *that can be visually recognizable and grouped together* with the control decision node. Its chunking criterion is quite different from that of the other learners that chunk the goal-oriented problem-solving processes of the solver, e.g. explanation-based learners. *Recognition rules*, domain-specific knowledge describing the necessary conditions for domain objects to be "visually recognizable", implement this chunking criterion in PCLEARN. The set of rules can be regarded as *a semantic representation of human's visual actions* in seeing objects and its effective use as a guide of learning processes is demonstrated here. The use of the perceptual-chunks obtained *semantically* guides problem-solving processes toward relevant paths of inference just as human beings do visually, although PCLEARN syntactically does sentential searches in logic for applicable perceptual-chunks.

In the domain of geometry problem-solving, a typical domain where diagrammatic reasoning is effective, we made some experiments of measuring the effectiveness of the learned perceptual chunks and showed that the chunks learned by PCLEARN can work as search control knowledge with higher applicability and better cost-effective utility, compared to conventional learners like EBL systems. The results suggest that the chunking criterion proposed are acceptable and useful in this domain.

We expect this method to be a good approach to providing domain-specific perceptual-chunks for reasoners, as the first step for implementing hybrid reasoning systems which use diagrammatic and non-diagrammatic knowledge effectively for problem-solving and learning. The second step for this, one of our future works, will be to improve the problem-solving performance of PCLEARN by reducing the cost of searching for applicable perceptual-chunks in some way analogous to human's way of controlling search by use of visual scan on diagrams.

References

1. J. R. Anderson. *The Architecture of Cognition*. Harvard University Press, London, England, 1983.
2. Barwise J. and J. Etchemendy. Hyperproof: Logical reasoning with diagrams. In *Working Notes of AAAI Stanford Spring Symposium on Diagrammatic Reasoning*, pages 80–84, March 1992.
3. M. Y. Kim. Visual reasoning in geometry theorem proving. In *Proceedings of IJCAI-89*, pages 1617–1622, 1989.
4. K. R. Koedinger and J. R. Anderson. Abstract planning and perceptual chunks: Elements of expertise in geometry. *Cognitive Science*, 14:511–550, 1990.
5. J. E. Laird, A. Newell, and P. S. Rosenbloom. SOAR: An architecture for general intelligence. *Artificial Intelligence*, 33(1):1–64, 1987.
6. T. McDougal and K. Hammond. A recognition model of geometry theorem-proving. In *Proceedings of the Fourteenth Annual Conference of the Cognitive Science Society*, pages 106–111, 1992.
7. S. Minton. Quantitative results concerning the utility of explanation-based learning. *Artificial Intelligence*, 42:363–391, 1990.
8. S. Minton, J. G. Carbonell, C. A. Knoblock, D. R. Kuokka, O. Etzioni, and Y. Gil. Explanation-based learning: A problem solving perspective. *Artificial Intelligence*, 40:63–118, 1989.
9. D. J. Mostow. Machine transformation of advice into a heuristic search procedure. In *Machine Learning : An Aritificial Intelligence Approach*. CA: Tioga Press, 1983.
10. N. H. Narayanan. Reasoning visually about spatial interactions. In *Proceedings of IJCAI-91*, pages 360–365, 1991.
11. M. Suwa and H. Motoda. Acquisition of Associative Knowledge by the Frustration-Based Learning Method in an Auxiliary-line Problem. *Knowledge Acquisition*, 1:113–137, 1989.
12. M. Suwa and H. Motoda. Learning perceptually-chunked macro-operators. In *Proceedings of the international workshop on Machine Intelligence*, 1992. (to appear in Machine Intelligence volume 1, Oxford University Press).
13. P. H. Winston, T. O. Binford, B. Katz, and M. Lowry. Learning physical descriptions from functional definitions, examples and precedents. In *Proceedings of AAAI-83*, pages 433–439, 1983.

A Geometry problems for experiments

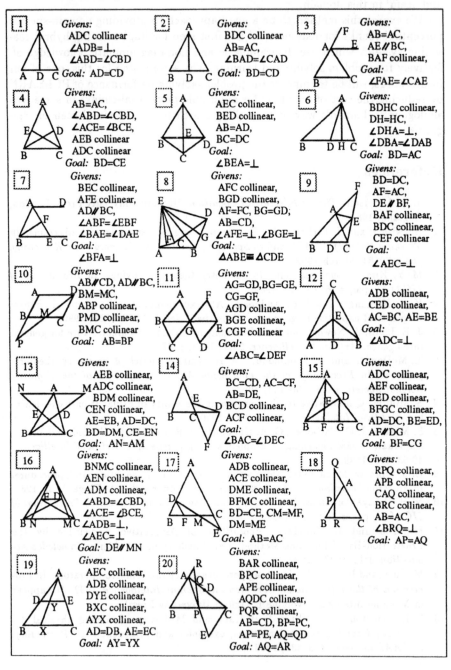

1 B
Givens:
ADC collinear
∠ADB=⊥,
∠ABD=∠CBD
A D C Goal: AD=CD

2 A
Givens:
BDC collinear
AB=AC,
∠BAD=∠CAD
B D C Goal: BD=CD

3 Givens:
/F
A E AB=AC,
AE∥BC,
BAF collinear,
Goal:
B C ∠FAE=∠CAE

4 A
Givens:
AB=AC,
∠ABD=∠CBD,
E D ∠ACE=∠BCE,
AEB collinear
B C ADC collinear
Goal: BD=CE

5 A
Givens:
AEC collinear,
BED collinear,
E AB=AD,
B D BC=DC
C Goal:
∠BEA=⊥

6 A
Givens:
BDHC collinear,
DH=HC,
∠DHA=⊥,
B D H C ∠DBA=∠DAB
Goal: BD=AC

7 Givens:
BEC collinear,
AFE collinear,
A D AD∥BC,
F ∠ABF=∠EBF
∠BAE=∠DAE
B E C Goal:
∠BFA=⊥

8 Givens:
AFC collinear,
E BGD collinear,
D AF=FC, BG=GD,
AB=CD,
F C G ∠AFE=⊥,∠BGE=⊥
A B Goal:
△ABE≣△CDE

9 Givens:
F BD=DC,
AF=AC,
A DE∥BF,
E BAF collinear,
BDC collinear,
B D C CEF collinear
Goal:
∠AEC=⊥

10 Givens:
A D AB∥CD, AD∥BC,
BM=MC,
B M C ABP collinear,
PMD collinear,
P BMC collinear
Goal: AB=BP

11 Givens:
A F AG=GD,BG=GE,
CG=GF,
AGD collinear,
B BGE collinear,
G E CGF collinear
C D Goal:
∠ABC=∠DEF

12 C
Givens:
ADB collinear,
E CED collinear,
AC=BC, AE=BE
A D B Goal:
∠ADC=⊥

13 Givens:
N A M AEB collinear,
ADC collinear,
BDM collinear,
E D CEN collinear,
AE=EB, AD=DC,
B C BD=DM, CE=EN
Goal: AN=AM

14 Givens:
A BC=CD, AC=CF,
E AB=DE,
D BCD collinear,
B C ACF collinear,
Goal:
F ∠BAC=∠DEC

15 A Givens:
ADC collinear,
AEF collinear,
F D BED collinear,
BFGC collinear,
B F G C AD=DC, BE=ED,
AF∥DG
Goal: BF=CG

16 Givens:
A BNMC collinear,
AEN collinear,
E D ADM collinear,
∠ABD=∠CBD,
∠ACE=∠BCE,
B N M C ∠ADB=⊥,
∠AEC=⊥
Goal: DE∥MN

17 A Givens:
ADB collinear,
ACE collinear,
D DME collinear,
C BFMC collinear,
B F M BD=CE, CM=MF,
E DM=ME
Goal: AB=AC

18 Q Givens:
RPQ collinear,
A APB collinear,
CAQ collinear,
P BRC collinear,
AB=AC,
B R C ∠BRQ=⊥
Goal: AP=AQ

19 A Givens:
AEC collinear,
D E ADB collinear,
Y DYE collinear,
BXC collinear,
AYX collinear,
B X C AD=DB, AE=EC
Goal: AY=YX

20 R Givens:
A Q BAR collinear,
D BPC collinear,
APE collinear,
B P C AQDC collinear,
PQR collinear,
E AB=CD, BP=PC,
AP=PE, AQ=QD
Goal: AQ=AR

Learning Strategies Using Decision Lists

Satoshi Kobayashi

Department of Computer Science and Information Mathematics
The University of Electro-Communications

1-5-1, Chofugaoka, Chofu, Tokyo 182, Japan
e-mail:satoshi@cs.uec.ac.jp

Abstract

This paper presents a new technique for improving the efficiency of problem solvers by using trace data. The learning schema of this kind is usually called *speedup learning*. Although this learning paradigm has been formalized by [Nat89] based on the Probably Approximately Correct Learning model[Val84], his algorithm does not use operator sequence information of traces at all. We give a new algorithm which makes use of operator sequence information, and its effectiveness is shown by experiments in several domains. We also show theoretical backgrounds of the algorithm.

1 Introduction

This paper presents a new technique for improving the efficiency of problem solvers by using traces of correct solutions. The learning schema of this kind is usually called *speedup learning*. Although this learning paradigm is formalized by [Nat89] based on the Probably Approximately Correct Learning model[Val84], his algorithm does not use operator sequence information of traces at all. The main purpose of our research is to give a new algorithm which makes use of operator sequence information, and to show its effectiveness.

In this paper, we deal with a problem which is defined by a quadruple $< S, I, G, O >$, where S, I, G, and O are a state space, an initial state space, a goal state space, and a set of operators. A solution of an initial state s is an operator sequence which transforms s into a state in G. Many researches on speedup learning deal with the problem of learning a subspace of S for each operator o where o should be applied. This learning problem is recognized as a concept learning problem of classifying each state to the corresponding class of the operator. [Nat89] has formalized speedup learning problem along this outline. However, his work does not take operator sequence information of trace data into account.

Our proposal is to make use of operator sequence information of traces for learning an appropriate pre-condition of each operator. Intuitively speaking, the proposed algorithm is based on the idea that similar problem instances might be solved in a similar way. The algorithm searches similar continuous subsequences in the given traces and collects them together to obtain appropriate pre-conditions of operators.

In this paper, we use the tree pattern language to represent pre-conditions of operators. In chapter 2, we introduce the tree pattern language, a decision list over tree patterns, and a decision list strategy which is used as a control mechanism for applying operators. This is an application of a decision list over tree patterns[KHO93] to speedup learning. Then we propose an algorithm for learning decision list strategies from trace data using operator sequence information. Experiments in two domains, circuit design and 8-puzzle, are presented in chapter 4 and the effectiveness of the algorithm is shown. Then we show theoretical backgrounds of the algorithm in chapter 5.

2 Preliminaries

In this paper, we denote the complementary set of a set S by $\overline{S}$, the set of all natural numbers by N, and a composite function $\overbrace{f \circ f \circ \cdots \circ f}^{n}$ of a function f by f^n.

2.1 Decision List over Tree Patterns

Let $\Sigma = \{a,b,c,...\}$ be a finite alphabet, representing *functors*, associated with a mapping *arity* from Σ to nonnegative integers, and $V = \{X,Y,Z,...\}$ be a countable set of symbols disjoint from Σ whose elements are called *variables*.

Definition 1 *A tree pattern on* $\Sigma \cup V$ is defined recursively as follows[ASO91].

(1) t is a tree pattern if t is a 0-ary functor (i.e. $arity(t) = 0$).

(2) a variable is a tree pattern.

(3) an ordered tree $f(t_1, ..., t_n)$ is a tree pattern if f is a n-ary functor, and $t_i, ..., t_n$ are tree patterns.

The size of a tree pattern t is defined as the length of the string representing it and is written $| t |$. A tree pattern which has no variables is called *a tree*. $TP(\Sigma)$ denote the set of all tree patterns on $\Sigma \cup V$.

A *substitution* is a homomorphism θ from $TP(\Sigma)$ to $TP(\Sigma)$ such that $\theta(f) = f$ for any 0-ary functor $f \in \Sigma$. We write $p \preceq q$ if there exists a substitution θ such that $p = \theta(q)$.

Definition 2 A tree pattern p defines the following language.

$$L(p) = \{t \in TP(\Sigma) \mid t \preceq p\}$$

Let S be a set of tree patterns. The tree pattern tp^* that satisfies the next condition is called *a least general generalization of S* and is written $lgg(S)$.

$$tp^* \in \{q \mid \forall p \in S \ (p \preceq q)\} \land \forall tp \in \{q \mid \forall p \in S \ (p \preceq q)\} \ (tp^* \preceq tp)$$

The least general generalization $lgg(S)$ can be calculated in polynomial time with respect to the size of S[Plo70].

Let C be an alphabet representing classes of objects. *A decision list over tree patterns* is a list of tuples,

$$(tp_1, c_1), (tp_2, c_2), ..., (tp_r, c_r), (tp_{r+1}, c_{r+1})$$

where tp_i and $cp_i(1 \leq i \leq r+1)$ are elements of $TP(\Sigma)$ and C respectively. tp_{r+1} is a tree pattern composed of only one variable X itself. A decision list over tree patterns dl defines a function f_{dl} from $TP(\Sigma)$ to C as follows. When we are given a tree pattern t, test for $t \preceq tp_i$ from $i = 1$ to $i = r+1$. Let i^* be the minimum integer such that $t \preceq tp_{i^*}$, then output c_{i^*}.

For the decision list shown below,

$$(a(X, X), c_1), (a(X, Y), c_2), (b(d(X), Y), c_3), (X, c_4),$$

the values of $f_{dl}(a(f,g))$ and $f_{dl}(b(d(c),e(Z)))$ are c_2 and c_3 respectively.

The length $|\ dl\ |$ of decision list dl is defined as the number of tuples in dl. This is a natural extension of decision lists[Riv87] over tree patterns.

2.2 Decision List Strategy

A problem is a quadruple $(TP(\Sigma), I, G, O)$, where $TP(\Sigma)$, I, G, and O are a state space, an initial state space such that $I \subseteq TP(\Sigma)$, a goal state space such that $G \subseteq TP(\Sigma)$, and a set of operators, respectively. An operator is a function from $TP(\Sigma)$ to $TP(\Sigma)$.

A problem instance of a problem P is a state $s \in I$. *A solution for a problem instance s* is an operator sequence $o_n \cdots o_2 \cdot o_1$ such that $o_n \cdots o_2 \cdot o_1(s) \in G$.

We call a function f from $TP(\Sigma)$ to O, *a strategy for a problem P*.

Definition 3 Let $P = (TP(\Sigma), I, G, O)$ be a problem, and f be a strategy for P, then a function $\hat{f}$ from $TP(\Sigma)$ to $TP(\Sigma)$ is defined as follows.

$$\forall s \in TP(\Sigma) \ (\hat{f}(s) = (f(s))(s))$$

$\hat{f}(s)$ gives the next state of s when we apply operators according to the strategy f.

Definition 4 Let $P = (TP(\Sigma), I, G, O)$ be a problem, and f be a strategy for P, then *a finite domain $FD(f)$ of f* is defined as follows.

$$FD(f) = \{s \in (I - G) \mid \exists k \ \hat{f}^k(s) \in G\}$$

Then we define $FD^*(f)$ recursively as follows.

(1) each $s \in FD(f)$ is an element of $FD^*(f)$.

(2) if $s \in FD^*(f)$ and $\hat{f}(s) \notin G$, then $\hat{f}(s) \in FD^*(f)$.

A trace of a state $s \in FD^*(f)$ according to a strategy f is given as a sequence,

$$s \overset{f(s)}{\rightarrow} \hat{f}(s) \overset{f(\hat{f}(s))}{\rightarrow} \hat{f}^2(s) \overset{f(\hat{f}^2(s))}{\rightarrow} \cdots \overset{f(\hat{f}^k(s))}{\rightarrow} \hat{f}^{k+1}(s) \in G$$

where, $\forall i \le k \; (\hat{f}^i(s) \notin G)$. Here, $f(\hat{f}^k(s))f(\hat{f}^{k-1}(s)) \cdots f(\hat{f}(s))f(s)$ is called *a solution of s according to f* and is written $sol(s, f)$. In case of $s \notin FD^*(f)$, $sol(s, f)$ denotes an infinite operator sequence, $\cdots f(\hat{f}^2(s))f(\hat{f}(s))f(s)$. The length of a trace t is defined as the number of operators applied in it, and is written $|t|$.

We define $SOL(S, f) = \{sol(s, f) \mid s \in S\}$.

A sequence of traces, $t_1, t_2, ...$, is called *a positive presentation of traces for a strategy f* if $\forall s \in FD(f) \; \exists i \; (t_i$ is a trace of $s)$ holds.

Definition 5 *A decision list strategy for a problem $P = (TP(\Sigma), I, G, O)$ is a decision list over tree patterns $dl = (tp_1, o_1), (tp_2, o_2), ..., (tp_k, o_k), (X, o_{k+1})$ which satisfies the next conditions.*

(1) $k \in N \cup \{0\}$

(2) $\forall i \le k \; (tp_i \in TP(\Sigma) \wedge o_i \in O) \; \wedge \; o_{k+1} \in O$

For any problem P, the set of all decision list strategies for a problem P is denoted by $DLS(P)$. $DLSF(P)$ is defined as a set of functions, $\{f_{dl} \mid dl \in DLS(P)\}$.

3 A Learning Algorithm

In this section, we propose a learning algorithm $\mathbf{M_1}$ which learns a decision list strategy from trace data. In the algorithm, trace data are collected as a set of tuples, (s, o), where s is a state in the given trace set and o is the operator applied to s.

Let T be a set of traces whose ith element is represented as,

$$s_{n(i,1)} \overset{o_{m(i,1)}}{\rightarrow} s_{n(i,2)} \overset{o_{m(i,2)}}{\rightarrow} \cdots \overset{o_{m(i,k_i)}}{\rightarrow} g \in G \;\; (i = 1, 2, ..., K, i.e. \mid T \mid = K)$$

Then, we define

$$D(T) = \{(s_{n(i,j)}, o_{m(i,j)}) \mid 1 \le i \le K, 1 \le j \le k_i\}.$$

Let us consider a set D of tuples (s, o) such that $s \in TP(\Sigma)$ and $o \in O$. A decision list strategy dl is said to *be consistent with D* if the next condition holds.

$$\forall (s, o) \in D \; (f_{dl}(s) = o)$$

We define $pat(D) = \{s \mid (s, o) \in D\}$ and $op(D)$ is defined as follows.

$$op(D) = \begin{cases} an\ element\ \hat{o} \in \{o \mid (s, o) \in D\} & if\ |\{o \mid (s, o) \in D\}| = 1 \\ undefined & otherwise \end{cases}$$

The learning algorithm $\mathbf{M_1}$ receives a positive presentation of traces, $t_1, t_2, ...,$ from a teacher, and outputs a sequence of decision list strategies, $dl_1, dl_2, ...$ Let us consider the ith stage of $\mathbf{M_1}$ and let T_i be a set of traces which $\mathbf{M_1}$ has at the ith stage. If dl_{i-1} is consistent with $D(T_i)$, then $\mathbf{M_1}$ outputs dl_{i-1}, otherwise, $\mathbf{M_1}$ executes the algorithm called $\mathbf{A_1}$, and outputs its answer.

The algorithm $\mathbf{A_1}$ is shown bellow.

Input: A set of traces T whose ith element is represented as,

$$s_{n(i,1)} \overset{o_{m(i,1)}}{\rightarrow} s_{n(i,2)} \overset{o_{m(i,2)}}{\rightarrow} \cdots \overset{o_{m(i,k_i)}}{\rightarrow} g \in G\ (i = 1, 2, ..., K, i.e.\ |T| = K)$$

Output: a decision list over tree patterns consistent with $D(T)$

```
begin
    L := null list;
    l_max := max{|t| | t ∈ T};
    D := D(T);
    while(D ≠ φ)do
        flag := NO;
        for l = 1 to l_max do
            C := classify(D, l);
            if |C| = 1 then do      (Here, C = {C_1})
                insert (X, op(C_1)) at the end of L;
                D := D - C_1;
                flag := YES;
                break;
            end
            for r = 1 to |C| do     (Here, C = {C_1, C_2, ..., C_|C|})
                if ∀p (op(C_r) ≠ op(C_p) ⟹ ∀s ∈ pat(C_p) (s ∉ L(lgg(pat(C_r))))) ...(*)
                then do
                    insert (lgg(pat(C_r)), op(C_r)) at the end of L;
                    D := D - {(s, o) ∈ D | s ∈ L(lgg(pat(C_r)))};
                    flag := YES;
                    break;
                end
            end
            if flag = YES then break;
        end
        if flag = NO then fail;
    end
    output L;
end
```

classify(D, l)

begin

$C := \phi;$

$p := 0;$

while$(D \neq \phi)$**do**

$p := p + 1;$

$C_p := \phi;$

insert C_p into C;

delete one element $(s_{n(i,j)}, o_{m(i,j)})$ from D and insert it into C_p;

$l_1 := min\{k_i - j, l - 1\};$

forall $(s_{n(i',j')}, o_{m(i',j')}) \in D$ **do**

$l_2 := min\{k_{i'} - j', l - 1\};$

if $l_1 \neq l_2$ **then** continue;

if $\forall k \in \{0, 1, ..., l_1\}$ $(o_{m(i,j+k)} = o_{m(i',j'+k)})$ **then do**

insert $(s_{n(i',j')}, o_{n(i',j')})$ into C_p;

delete $(s_{n(i',j')}, o_{n(i',j')})$ from D;

end

end

end

return C;

end

In the algorithm above, the subroutine $classify(D, l)$ classifies a given set D of tuples (s, o) into mutually disjoint subsets based on l-length suffix of $sol(s, f^*)$, where f^* is a target strategy.

4 Experimental Results

4.1 An Experiment in Circuit Design Domain

The problem we deal with in this subsection is converting multi-level AND-OR logic into multi-level NOR-NOR logic. For example, a solution for a problem instance $and(or(a,b), or(c,d))$ is $nor(nor(a,b), nor(c,d))$. Operators are represented in Prolog Language and some of them are shown in Figure 1.

```
equiv_func_main(X,X) :-           equiv_func(and(X,Y),nor(neg(X),neg(Y))).
    specification(X),
    !.                            equiv_func(and(X,Y),and(Z,W)) :-
                                      equiv_func(X,Z),
equiv_func_main(X,Y) :-               equiv_func(Y,W).
    equiv_func(X,Z),
    equiv_func_main(Z,Y).         equiv_func(neg(X),neg(Y)) :-
                                      equiv_func(X,Y).
equiv_func(X,X).

equiv_func(neg(neg(X)),X).
```

Figure 1: Circuit Design Domain Knowledge

In this experiment, each Boolean formula is expressed as a term, and a SLD resolution sequence is used as a trace. We mean that a selected subgoal and a rule applied to it are regarded as a state and an applied operator, respectively.

We gave 10 traces of training examples to the learning algorithm, and tested the obtained decision list strategy on 10 unknown problems. The obtained decision list strategy could solve all unknown problems correctly. A part of training examples and unknown problems are shown in Table 1.

Training Problem Instance	Solution
and[x,y]	nor[neg(x),neg(y)]
and[or[a,b],or[c,d]]	nor[nor[a,b],nor[c,d]]
and[x,y,neg(z),w,s,t]	nor[neg(x),neg(y),z,neg(w),neg(s),neg(t)]
Test Problem Instance	Solution
and[x,neg(y)]	nor[neg(x),y]
and[or[neg(a),neg(b),c],or[x,neg(y)]]	nor[nor[neg(a),neg(b),c],nor[x,neg(y)]]
and[or[neg(x),a],or[y,b],or[neg(e),neg(f)]]	nor[nor[neg(x),a],nor[y,b],nor[neg(e),neg(f)]]

Table 1: Training and Test Examples

4.2 An Experiment in 8-puzzle Domain

In 8-puzzle domain, the effective strategy called the Korf's macro table[Kor85] is well known. In this experiment, each state is represented as a tree structure of $st(X0, X1, ..., X8)$, where each X_i denotes the position of the tile i (X_0 denotes the position of blank). We obtained the target decision list strategy by transforming the Korf's macro table into a decision list form automatically. The length of the target decision list is 230.

We selected 300 problems at random, solved them according to the target decision list, and gave the traces to the learning algorithm. The obtained decision list strategy, denoted by dl^*, has the length of 298. The coverage of 1000 unknown problems chosen at random were tested for dl^*. The result is that dl^* has 100% coverage for the test problems. The trace set given to the algorithm includes 4959 different states, which are only 2.7 percent of all problem instances. The coverage indicates the effectiveness of the proposed algorithm.

5 Theoretical Results

In this section, we show theoretical backgrounds for the proposed algorithm.

5.1 Target Class of Strategies

Definition 6 Let x and y be strings on an alphabet O. Then a relation $\trianglelefteq$ over strings on O is defined as

$$x \trianglelefteq y \quad \text{iff.} \quad \exists s \in O^* \; (s \cdot x = y).$$

If $x \not\trianglelefteq y$ and $y \not\trianglelefteq x$ hold, we write $x \otimes y$. Let S be a set of strings. Then a *common suffix* of S is defined as

$$cs(S) \;=\; z^* \text{ such that } z^* \in \{y \mid \forall x \in S\ (y \trianglelefteq x)\}\ \wedge$$
$$\forall z \in \{y \mid \forall x \in S\ (y \trianglelefteq x)\}\ (z \trianglelefteq z^*).$$

Let s be a string on O and l be a positive integer, then *l-length suffix $suf(s,l)$* of s is defined as follows.

$$suf(s,l) = \begin{cases} \text{a string } s' \text{ such that } \exists x \in O^*(s = x \cdot s' \wedge \mid s' \mid = l) & \text{if } \mid s \mid \geq l \\ s & \text{otherwise} \end{cases}$$

Then we have the following lemmas.

Lemma 1 Let s_1 and s_2 be strings on O. If $s_1 \trianglelefteq s_2$ and $l \in \mathbf{N}$, then $suf(s_1, l) \trianglelefteq suf(s_2, l)$ holds.
Proof
Because it is straightforward to prove this lemma, we omit the proof. □

Lemma 2 Let S be a set of strings on O. If $s \in S$ and $\mid cs(S) \mid \leq l$, then $cs(S) \trianglelefteq suf(s, l)$ holds.
Proof
$cs(S) \trianglelefteq s$ holds for any $s \in S$, by the definition of $cs(S)$. Then we have

$$\begin{aligned} cs(S) \;&=\; suf(cs(S), l) \quad (\text{ because } \mid cs(S) \mid \leq l\ ) \\ &\trianglelefteq\; suf(s, l) \quad\quad\ \ (\text{ by Lemma 1 }) \end{aligned}$$

□

Definition 7 Let $P = (TP(\Sigma), I, G, O)$ be a problem., and $dl = (tp_1, o_1), (tp_2, o_2),$ $..., (tp_n, o_n)$ be a decision list strategy for P. Then we define $U(dl, i)$ and $CS(dl, i)$ for each $i = 1, ..., n$ as follows.

$$\begin{aligned} U(dl, i) \;&=\; L(tp_i) \cap \overline{\cup_{j < i} L(tp_j)} \\ CS(dl, i) \;&=\; cs(SOL(U(dl, i) \cap FD^*(f_{dl}), f_{dl})) \end{aligned}$$

Then we define the target class of strategies $DLSF^*(P)$.

Definition 8 Let $P = (TP(\Sigma), I, G, O)$ be a problem. Then subclasses of $DLS(P)$ and $DLSF(P)$ are defined as follows.

$$\begin{aligned} DLS^*(P) \;&=\; \{dl \in DLS(P) \mid \forall\, 1 \leq i < j \leq\mid dl \mid\ (CS(dl, i) \otimes CS(dl, j))\} \\ DLSF^*(P) \;&=\; \{f_{dl} \mid dl \in DLS^*(P)\} \end{aligned}$$

The target strategies used in section 4 can be proved to belong to this class.

5.2   Convergence in the Limit

Definition 9 Let $P = (TP(\Sigma), I, G, O)$ be a problem. A relation $\simeq$ between strategies is defined as follows.

for any pair of strategies f_1, f_2, $f_1 \simeq f_2$ iff $FD^*(f_1) \subseteq FD^*(f_2) \wedge$
$$\forall s \in FD^*(f_1) \, (f_1(s) = f_2(s))$$

Definition 10 Let $P = (TP(\Sigma), I, G, O)$ be a problem, and f be a strategy such that $f \in DLSF(P)$. Let us consider a learning algorithm M which takes a positive presentation of traces for f as an input and outputs a sequence of decision list strategies, $dl_1, dl_2, \ldots$. We say M *infers* f *in the limit*, if there exists $n \in N$ such that $\forall k \geq n \, (dl_k = dl_{k+1} \wedge f \simeq f_{dl_k})$. We say M *infers* $F \subseteq DLSF(P)$ *in the limit*, if M infers every $f \in F$ in the limit.

Let dl be a decision list strategy, then *a suffix length $sufLen(dl)$ of a decision list dl* is defined as follows.

$$sufLen(dl) \; = \; max\{| \, CS(dl, i) \, | \; | \; 1 \leq i \leq | \, dl \, |\}$$

Let $f^* \in DLSF^*(P)$ be a strategy, and $DL(f^*)$ be a set

$$\{dl \in DLS^*(P) \; | \; f^* \simeq f_{dl}\}.$$

We define the value

$$l_{min}(f^*) = min\{sufLen(dl) \; | \; dl \in DL(f^*)\}.$$

First, we prove the following lemmas.

Lemma 3 Let T be a set of traces according to a strategy f^*, l be a positive integer, and $C = \{C_1, C_2, \ldots, C_r\}$ be the output of $classify(D(T), l)$. Then the next sentence holds.

$$\forall 1 \leq i \leq r \forall p, q \in pat(C_i) \, (suf(sol(p, f^*), l) = suf(sol(q, f^*), l))$$

Proof
This lemma is easy to prove, so we omit the proof. $\square$

Lemma 4 Let T be a set of traces according to a strategy f^*, l be a positive integer, and $C = \{C_1, C_2, \ldots, C_r\}$ be the output of $classify(D(T), l)$. Then the next sentence holds.

$$\forall 1 \leq i \leq r \forall p \in pat(C_i) \, (suf(sol(p, f^*), l) \trianglelefteq cs(SOL(pat(C_i), f^*)))$$

Proof
By the definition of $suf(s, l)$, we have the next sentence for each i.

$$\forall p \in pat(C_i)(suf(sol(p, f^*), l) \trianglelefteq sol(p, f^*))$$

Then we can easily derive the next sentence for each i from Lemma 3 and the definition of $cs(S)$.

$$\forall p \in pat(C_i)\ (suf(sol(p, f^*), l) \trianglelefteq cs(SOL(pat(C_i), f^*)))$$

□

Lemma 5 Let P be a problem and T be a finite set of traces according to a strategy $f^* \in DLSF^*(P)$, then the learning algorithm $\mathbf{A_1}$ with the input T always outputs a decision list strategy dl consistent with $D(T)$ and the value of l in the first for loop does not exceed $l_{min}(f^*)$.

Proof

We can regard the $while$ loop section of the algorithm as a subroutine with input D. It is sufficient for us to show that, for any input D, the $while$ loop section outputs a decision list consistent with D and the value of l in the first for loop does not exceed $l_{min}(f^*)$. The claim is proved by the induction of the number of elements in D, written rn.

In case of $rn = 1$, $|C|$ must be 1. Therefore, the claim holds obviously.

Suppose the claim holds in case of $rn \leq k - 1$ and consider the case $rn = k$. There exists a decision list strategy $dl^* = (tp_1^*, o_1^*), (tp_2^*, o_2^*), ..., (tp_m^*, o_m^*)$ such that

$$dl^* \in DL(f^*) \wedge l_{min}(f^*) = sufLen(dl^*)$$

because $f^* \in DLSF^*(P)$.

Let us consider the output of $classify(D, l)$, for each $l = 1, ..., l_{min}(f^*)$. If the second if condition (denoted by (*) in the program) is satisfied for some $i = 1, ..., l_{min}(f^*) - 1$, the claim holds obviously. Therefore, we consider the output $C = \{C_1, ..., C_r\}$ of $classify(D, l_{min}(f^*))$.

First, we prove that, for any i, there exists j such that $pat(C_i) \subseteq U(dl^*, j)$. It is obvious that $\forall p \in pat(C_i) \exists j(p \in U(dl^*, j))$ holds for each i. Therefore, it is sufficient to prove that for any i and for any tree patterns $p, q \in pat(C_i)$,

$$p \in U(dl^*, j_1) \wedge q \in U(dl^*, j_2) \Rightarrow j_1 = j_2 \tag{1}$$

holds. Let us assume that, for some i, there exist tree patterns $p, q \in pat(C_i)$, such that $p \in U(dl^*, j_1)$, $q \in U(dl^*, j_2)$ and $j_1 \neq j_2$. Here, we have

$$CS(dl^*, j_1) \trianglelefteq suf(sol(p, f^*), l_{min}(f^*)) \tag{2}$$

because $sol(p, f^*) \in SOL(U(dl^*, i) \cap FD(f^*), f^*)$, $|CS(dl^*, i)| \leq l_{min}(f^*)$ and Lemma 2. We can also prove that

$$CS(dl^*, j_2) \trianglelefteq suf(sol(q, f^*), l_{min}(f^*)) \tag{3}$$

in a similar way. We have

$$suf(sol(p, f^*), l_{min}(f^*)) \trianglelefteq cs(SOL(pat(C_i), f^*)) \tag{4}$$
$$suf(sol(q, f^*), l_{min}(f^*)) \trianglelefteq cs(SOL(pat(C_i), f^*)) \tag{5}$$

by Lemma 4. (2), (3), (4), (5) leads to the following relations.

$$CS(dl^*, j_1) \trianglelefteq cs(SOL(pat(C_i), f^*))$$
$$CS(dl^*, j_2) \trianglelefteq cs(SOL(pat(C_i), f^*))$$

Then we can conclude that either $CS(dl^*, j_1) \trianglelefteq CS(dl^*, j_2)$ or $CS(dl^*, j_1) \trianglelefteq CS(dl^*, j_2)$ holds. This contradicts the fact that $CS(dl^*, j_1) \otimes CS(dl^*, j_2)$. Therefore, (1) holds.

Let $j^* = min\{j \mid \exists i \, (pat(C_i) \subseteq U(dl^*, j))\}$, and i^* be an integer such that $pat(C_{i^*}) \subseteq U(dl^*, j^*)$. Here, we can easily derive $lgg(pat(C_{i^*})) \preceq tp_{j^*}^*$ from the definition of i^* and j^*. Threfore, we have

$$\exists o \in O \, \forall s \in pat(D) \, (s \in L(lgg(pat(C_{i^*}))) \Rightarrow f^*(s) = o)$$

This means that the value of l in the first *for* loop does not exceed $l_{min}(f^*)$ at the first loop of *while* loop, and that some elements are removed from D and a tuple (tp, o) is inserted into L. Obviously, (tp, o) is consistent with the removed elements. Let us denote the set of removed elements by D'.

In the rest of *while* loop, we can conclude that the algorithm outputs a decision list strategy which is consistent with $D - D'$, and that the value of l does not exceed $l_{min}(f^*)$, by the induction hypothesis. Therefore, the claim holds in case of $rn = k$.

This completes the proof. □

Theorem 1 For any given problem $P = (TP(\Sigma), I, G, O)$, the algorithm M_1 infers $DLSF^*(P)$ in the limit.

Proof

Let f^* be a target strategy. The length of decision list strategies which $\mathbf{M_1}$ outputs is bounded by $\sum_{i=1}^{l_{min}(f^*)} \mid O \mid^i$ because of Lemma 5.

There exists some $k \in \mathbf{N}$ such that the trace set at the kth stage $T_k = \{t_1, t_2, ..., t_k\}$ includes all possible operator sequences seq such that $\mid seq \mid \leq l_{min}(f^*)$ as continuous subsequences.

Let $dl = (tp_1, o_1), ..., (tp_m, o_m)$ be any decision list strategy which $\mathbf{M_1}$ outputs after the kth stage. Then, by the definition of k, each $U(dl, i)$ must include some state s which appears in T_k.

Let n be the maximum size of states in $pat(D(T_k))$. The number of possible tree patterns which include a tree whose size is less than or equal to n is bounded by $(\mid \Sigma \mid +n + 1)^n$.

Therefore, the number of all possible decision lists which $\mathbf{M_1}$ outputs after the kth stage is bounded by $((\mid \Sigma \mid +n + 1)^n + 1)^{\sum_{i=1}^{l_{min}(f^*)} \mid O \mid^i}$. So, at some stage, the guess converges to the correct decision list dl^* such that $f^* \simeq f_{dl^*}$ because $\mathbf{M_1}$ always outputs decision list strategies consistent with given traces. This completes the proof. □

Note that the class of strategies which $\mathbf{M_1}$ outputs is not equivalent to the target class of strategies $DLSF^*(P)$. A decision list dl which $\mathbf{M_1}$ outputs does not always satisfy the condition $\forall 1 \leq i < j \leq \mid dl \mid (CS(dl, i) \otimes CS(dl, j))$.

It is easy to show that the time required for updating conjectures is polynomial with respect to the size of given traces.

5.3 Sample Complexity of the Algorithm

In the experiments in section 4, the proposed algorithm requires only small number of training examples. The reason why the algorithm works so efficiently is analyzed formally in this subsection. The analysis is based on the PAC model introduced by [Val84].

Let $P = (TP(\Sigma), I, G, O)$ be a problem and n be a positive integer. We define $DOM^n(P) = \{t \in I - G \mid |t| \leq n\}$. By $Prob^n$, we denote a distribution function over $DOM^n(P)$. Here, n is called *a length parameter*.

Let f and g be a strategy for a problem P. A subset $f \Delta g$ of $DOM^n(P)$ is defined as follows.

$$f\Delta g = \{s \in DOM^n(P) \cap FD(f) \mid sol(s,f) \neq sol(s,g)\}$$

We define $dif(f,g) = \sum_{s \in f \Delta g} Prob^n(s)$. Then we say g is *an ϵ approximation of f* if $dif(f,g) \leq \epsilon$ holds. We say g is *an ϵ bad hypothesis for f* if g is not an ϵ approximation of f.

We assume that the teacher picks up a problem instance s at random with the distribution $Prob^n$ and gives $sol(s, f^*)$ to the learner, if $s \in FD(f^*)$, otherwise, gives an empty trace.

Let $k - DLS^*(P)$ be a class of decision list strategies defined as follows.

$$k - DLS^*(P) = \{dl \in DLS^*(P) \mid sufLen(dl) \leq k\}$$

Then, $k - DLSF^*(P)$ is defined as a set of strategies,

$$k - DLSF^*(P) = \{f_{dl} \mid dl \in k - DLS^*(P)\}.$$

Theorem 2 Let $P = (TP(\Sigma), I, G, O)$ be a problem, n be a length parameter, and $f^* \in k - DLSF^*(P)$ be a target strategy. The algorithm $\mathbf{A_1}$ with an input of more than $N_1 = \frac{1}{\epsilon} \log \frac{N_2}{\delta}$ traces according to f^* which are chosen at random with probability $Prob^n$, outputs an ϵ approximation of the target strategy with probability larger than $1 - \delta$. Here, $N_2 = ((|\Sigma| + n + 1)^n + 1)^{\sum_{i=1}^{k} |O|^i}$.

Proof

The number of tree patterns consistent with some tree whose size is less than or equal to n is bounded by $(|\Sigma| + n + 1)^n$. The length of decision lists which M_1 outputs is bounded by $\sum_{i=1}^{k} |O|^i$ by Lemma 5. Therefore, the number of possible decision lists which M_1 outputs is less than $N_2 = ((|\Sigma| + n + 1)^n + 1)^{\sum_{i=1}^{k} |O|^i}$.

For any decision list dl such that f_{dl} is an ϵ bad hypothesis for the target strategy f^*, the probability that f_{dl} is consistent with more than N_1 traces chosen at random with the distribution $Prob^n$ is at most $(1 - \epsilon)^{N_1} \leq \frac{\delta}{N_2}$. Therefore, the probability that the algorithm $\mathbf{A_1}$ with an input of more than N_1 traces outputs an ϵ bad hypothesis is less than δ because the number of hypotheses which A_1 outputs is bounded by N_2. This completes the proof. $\square$

We can conclude that, for a fixed k, the sample complexity of $k - DLSF^*(P)$ is polynomial with respect to $\frac{1}{\epsilon}$, $\frac{1}{\delta}$, and n.

6   Related Research

[KHO93] has introduced a concept of *decision list over tree patterns* and its learnability has been studied in it. They have shown that the class *k-node-DLTP*, which is a class of decision lists over tree patterns of length at most k, is not PAC learnable in polynomial time if $NP \neq RP$ and that the class *DLRTP*, which is a class of decision lists over tree patterns whose tree patterns are all *regular*, is learnable in the limit with polynomial updating time. In this paper, we show that another subclass $DLSF^*(P)$ is learnable in the limit with polynomial updating time.

Although [Nat89] has formalized speedup learning as a concept learning of a pre-condition for each operator, his work does not take operator sequence information of traces into account at all. This paper shows experimentally and theoretically that the use of such operator sequence information is very effective for speedup learning.

[Tad91] has presented a formal approach to *macro operator learning*. He defined two biases called *a sparse solution space bias* and *a macro table bias*, and proposed learning algorithms effective for domains satisfying these biases. While his work is limited to macro operator learning, our work pays attention to *precondition learning*. The coupling of these methods or frameworks seems to be a more promising approach for speedup learning.

Other many researches on Explanation Based Learning[MKKC86] deal with the speedup of general prolog programs ([Coh90][Lai92], etc.). The proposed learning algorithm in this paper cannot be applied to learning complex strategies as described in those works, because such target strategies often do not exist in $DLSF^*(P)$. Although the tree pattern language does not have enough power to represent complex preconditions of operators, we showed a new direction of speedup learning algorithms in this simple framework.

7   Conclusion and Future Works

Although we show the effectiveness of the use of operator sequence information of traces, the proposed algorithm have some problems. First, the proposed algorithm is not an incremental one. Therefore, the time required for updating conjectures increases, as the number of given traces grows. Second, the proposed framework cannot deal with proof *trees* as traces. For speedup learning of general prolog programs, our framework must be extended so as to deal with proof trees as traces. The third problem is that the proposed method cannot learn macro operators. The combination of macro operator and precondition learning is a more promising approach for speedup learning. We are now working on these problems.

Acknowledgments

The author would like to thank Dr. Takashi Yokomori, University of Electro-Communications, for his helpful suggestions and comments. He is also grateful to anonymous referees, Dr. Tatsuya Akutsu, Mechanical Engineering Laboratory, and Noriyuki Tanida, University of Electro Communications, for their helpful comments, which greatly improved the correctness of the paper.

References

[ASO91]   H. Arimura, T. Shinohara, and S. Otsuki. Polynomial Time Inference of Unions of Tree Pattern Languages. In *Proc. of 2nd Workshop on Algorithmic Learning Theory*, pages 105–114, 1991.

[Coh90]   W. W. Cohen. Learning Approximate Control Rules of High Utility. In *Proc. of International Conference on Machine Learning'90*, pages 268–276, 1990.

[KHO93]   S. Kobayashi, K. Hori, and S. Ohsuga. Learning Decision Lists over Tree Patterns and Its Application. In *Proc. of 13th International Joint Conference on Artificial Intelligence*, 1993. to appear.

[Kor85]   R. E. Korf. Macro Operators: A Weak Method for Learning. *Artificial Intelligence*, 26:35–77, 1985.

[Lai92]   P. Laird. Dynamic Optimization. In *Proc. of International Workshop on Machine Learning'92*, pages 263–272, 1992.

[MKKC86]   T. M. Mitchell, R. M. Keller, and S. T. Kedar-Cabelli. Explanation-Based Generalization : A Unifying View. *Machine Learning*, 1:47–80, 1986.

[Nat89]   B. K. Natarajan. On Learning from Exercises. In *Proc. of Computational Learning Theory'89*, pages 72–87, 1989.

[Plo70]   G. D. Plotkin. A Note on Inductive Generalization. *Machine Intelligence*, 5:153–163, 1970.

[Riv87]   R. L. Rivest. Learning Decision Lists. *Machine Learning*, 2:229–246, 1987.

[Tad91]   P. Tadepalli. A Formalization of Explanation-Based Macro-Operator Learning. In *Proc. of IJCAI'91*, pages 616–622, 1991.

[Val84]   L. G. Valiant. A Theory of Learnable. *Comm. Assoc. Computing Machinery*, 27:1134–1142, 1984.

A Decomposition Based Induction Model for Discovering Concept Clusters from Databases

Ning Zhong   Setsuo Ohsuga
Research Center for Advanced Science and Technology
The University of Tokyo
4-6-1 Komaba, Meguro-ku, Tokyo 153, Japan

Abstract. This paper presents a *decomposition based induction* model for discovering concept clusters from databases. This model is a fundamental one for developing DBI which is one of sub-systems of the GLS discovery system implemented by us. A key feature of this model is the formation of concept clusters or sub-databases through analysis and deletion of noisy data in decomposing databases. Its development is based on the concept of Simon and Ando's near-complete decomposability that has been most explicitly used in economic theory. In this model, the process of discovering concept clusters from databases is a process based on incipient hypothesis generation and refinement, and many kinds of learning methods are cooperatively used in multiple learning phases for performing multi-aspect intelligent data analysis as well as multi-level conceptual abstraction and learning, so that a more robust, general discovery system can be developed.

1   Introduction

Knowledge discovery in databases (KDD) is becoming an important topic in AI and is attracting the attention of leading researchers in databases[10]. This topic is different from traditional researches of machine learning, though it uses their results [8, 16, 6, 11]. In previous researches of machine learning, learning from examples, formating concept from instances have been investigated by many researchers. For example, Feigenbum's EPAM, Fisher's COBWEB, Lebowitz's NUIMEM, Cheeseman's AutoClass, Gennari, Langley and Fisher's CLASSIT as well as Quinlan's ID3, Cendrowska's PRISM etc. are some well-known systems or algorithms [6, 11, 3, 1, 4]. These systems or algorithms have led to some successes and provide a good background for our research. However, they are not designed to perform multi-aspect data analysis as well as multi-level conceptual abstraction and learning [21]. In particular, when they are used for discovering knowledge from complex large-scale databases, some of their limits become apparent. One of main reasons for this is that databases have several features different from other learning objects. For example, databases are not always complete but contain uncertain and incomplete data; databases for discovering knowledge are not always static but dynamic; databases are generally very large and complex and so on [19, 16].

In order to discover concept clusters from real databases, we have developed a *decomposition based induction* model. In this model, these features of real databases stated above are considered. This model is a fundamental one for developing DBI which is one of sub-systems of the GLS discovery system implemented by us [16, 14, 18, 20]. Its development is based on the concept of Simon and Ando's near-complete decomposability that has been most explicitly used in economic theory as a technique to study and evaluate the dynamics of systems of great size and complexity [12]. This technique is based on the idea that in many large systems all variables can somehow be clustered into a small number of groups [2]. This idea is of course quite general, and has, at least indirectly, been productive in disciplines other than economics.

A key feature of our approach is the formation of concept clusters or sub-databases through analysis and deletion of noisy data in decomposing a database. That is, the decomposition of databases is also a kind of aggregation and abstraction, and some minor factors (or noises) are neglected in decomposing. As the result of decomposing a database, concept clusters or sub-databases are formed, and they are used as background information for further learning in the next learning phase of GLS [16, 17]. Three kinds of database spaces, the instance space, the probability space and the learning space, are defined for these purposes. Using this technique, two kinds of noisy data, those whose probability values are much smaller than other related values and those which are irrelevant elements, can be reasonably treated.

Since the uncertainty and incompleteness of data in databases, the concept clusters discovered from databases are only hypotheses, which must be refined (evaluated/modified). Also, because databases are not always static but dynamic, new hypotheses are generated when data change in databases. Therefore, the refinement for the discovered hypotheses must been done. In order to achieve our purpose, we adopt the process of discovering concept clusters from databases based on incipient hypothesis generation and refinement (evaluation & modification) as shown in Figure 1 [7]. In this process, many kinds of learning methods are used in multiple learning phases for performing multi-aspect intelligent data analysis as well as multi-level conceptual abstraction and learning, so that a more robust, general discovery system can be developed.

In the following, we will describe the details of our approach. Section 2 gives the definitions of three kinds of database spaces and the method for creating the probability space as a preparation of decomposing databases. Section 3 gives the method of decomposing databases. Section 4 describes how to refine the discovered concept clusters by means of the learning space for processing

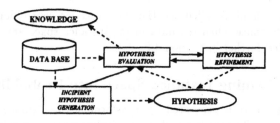

Figure 1: The process of knowledge discovery from databases

the perturbation of databases. Finally, Section 5 discusses a further extension which makes our approach more useful, and gives concluding remarks including a summary of the features of our approach as well as some future research subjects.

2 Database Spaces

As a preparation of decomposing databases, this section first defines three kinds of database spaces. Then the method of creating the probability space of databases is described.

2.1 Three Kinds of Database Spaces

In our approach, three kinds of database spaces, the instance space, the probability space and the learning space, are defined. These database spaces involve such data about analyzing object characteristics, the relations among original data, state changes and so on.

First, the *instance space* is the space for storing instances describing object problems as ordinary databases. Second, the *probability space* is the space for storing distributions describing the distributed status of data in the instance space. Using the probability space, we can gain an insight into the relations among physical entities in the instance space and what are major (and minor) factors for describing object problems. By means of the probability space, we can omit minor factors (or noises) so that we finally achieve the purpose of decomposition and aggregation of databases. The decomposed probability space provides background information for forming concept clusters and further generating *classification knowledge with hierarchical models* by HML (Hierarchical Model Learning), which is another subsystem of GLS [17], in a knowledge-base. The method of creation of the probability space will be discussed in Section 2.2.

Finally, in order to learn and trace dynamically changing conditions in the probability space along with the change of data in the instance space (i.e., to add, to delete or to update), the history of the change of probability space must be recorded. Additionally, some decisions must be made so that the optimal information can be selected for further knowledge discovery. In our model, we also define a mechanism called the *learning space* for controlling and learning new probability space when the data in the instance space are changed. This will be discussed further in Section 4.2.

2.2 Transforming Instance Space into Probability Space

The probability space is generated by transforming the instance space into the probability space. The basis for this transformation is to create a *Probability Distribution Matrix (PDM)*. By applying analysis, evaluation and learning on the PDM, the purpose of decomposition and aggregation of databases can be achieved.

There are many kinds of methods for creating the PDM, depending on their purposes. For our application of knowledge discovery, the dependency relations between any two attributes are considered, their probability distributions are calculated and recorded in a PDM. We present here one of two methods developed by us. Let $\mathbf{a} = \{a_1, a_2, ..., a_n\}$ and $\mathbf{b} = \{b_1, b_2, ..., b_m\}$ be the sets of different values of any two attributes in the instance space. Using conditional probability, we have

$$p(x_i|x_j) = \frac{p(x_i \cap x_j)}{p(x_j)} \qquad x_i, x_j \in \mathbf{a}, \mathbf{b}. \tag{1}$$

From this we define p_{ij}, the probability distributions, to be $p(x_i|x_j)/N$, where N is the number of attributes. These p_{ij} constitute the entities of the PDM. The PDMs generated by above method have the following characteristics:

- The elements p_{ij} of PDM ≤ 1.

- $\sum_{j=1}^{n} p_{ij} = 1$

- PDM is a square matrix.

- $Max_i \ \lambda_i \leq 1$ for the maximal eigenvalue of PDM.

The reason why we use this method of creating the PDM is that *classification knowledge*, one of two kinds of knowledge that we intend to discover from databases [19], is the dependency relations between attributes, $a_i \rightarrow b_j$ and $b_j \rightarrow a_i$, where $x \rightarrow y$ means that y depends on x, and a_i, b_j are attributes. By means of the PDM created in Eq.(1), we can know the distribution of this dependency. Then, major correlated items can be aggregated by near-complete decomposition as shown in Section 3. The results of decomposing can be processed in a distributed way for forming concept clusters and further generating *classification knowledge with hierarchical models* by hierarchical model learning in the next learning phase [17].

Discovering and analyzing data dependencies is of primary importance among the goals of knowledge discovery in databases [21]. The real-lift cause-effect relationships assume the form of data dependencies in such databases. Many useful properties of data are hidden in data dependencies. Ziarko used a rough set-based technique for identifying and analyzing dependencies in data. Piatetsky-Shapiro developed a tuple-oriented algorithm and knowledge is discovered by data dependency analysis [10]. In our approach, the distributions of data dependencies are focused and concept clusters are discovered by decomposing a database.

3 Decomposing Databases

In this section, we first overview Simon and Ando's theory of near-complete decomposability and then our method of decomposing databases is introduced. Two methods for forming the diagonal PDM and an algorithm of decomposing databases are described.

3.1 Near-Complete Decomposability

We first define *completely decomposable matrix* and *nearly completely decomposable matrix* developed by Simon and Ando [12, 2]. Let us consider the formula:

$$P = P^* + \varepsilon C, \tag{2}$$

where P and P^* are PDM of order n when the PDM is decomposable. P^* is called *completely decomposable matrix*. P is called *nearly completely decomposable matrix*. P^* is equal to

$$P^* = \begin{bmatrix} P_1^* & & & & \\ & \ddots & & & \\ & & P_I^* & & \\ & & & \ddots & \\ & & & & P_N^* \end{bmatrix}, \tag{3}$$

where the P_I^* are square sub-matrices, and the remaining elements, not displayed, are all zero. Let $n(I)$ be the order of P_I^*, then $n = \sum_{I=1}^{N} n(I)$. ε is a very small real number and is called the maximum degree of coupling between sub-systems P_{II} of P. ε is equal to

$$\varepsilon = \max_{i_I} \left(\sum_{J=1, J \neq I}^{N} \sum_{j=1}^{n(J)} p_{i_I j_J} \right). \tag{4}$$

C is a square matrix of the same order as P^* and is called the error matrix. C is equal to

$$C = \sum_{J=1, J \neq I}^{N} \sum_{j=1}^{n(J)} c_{i_I j_J} = \frac{1}{\varepsilon} \sum_{J=1, J \neq I}^{N} \sum_{j=1}^{n(J)} p_{i_I j_J}. \tag{5}$$

A criterion for near-complete decomposability is

$$\varepsilon < \frac{[1 - \max_I |\lambda^*(2_I)|]}{2}, \tag{6}$$

$$| \lambda^*(2_I) | \leq \min\{ \max_{1 \leq \mu, \rho \leq n(I)} \frac{1}{2} \sum_{i=1}^{n(I)} \nu_{i_I}^*(1_I) \mid \frac{p_{i_I \mu_I}^*}{\nu_{\mu_I}^*(1_I)} - \frac{p_{i_I \rho_I}^*}{\nu_{\rho_I}^*(1_I)} \mid,$$

$$\max_{1 \leq \mu, \rho \leq n(I)} \frac{1}{2} \sum_{i=1}^{n(I)} | p_{\mu_I i_I}^* - p_{\rho_I i_I}^* | \}. \tag{7}$$

Eq.(7) was defined by Bauer et al. [2] and is called the upper bound of the second largest eigenvalue of nonnegative indecomposable matrices. $\lambda^*(2_I)$ is the second largest eigenvalue of P_I^*, that is, we use $\lambda^*(i_I)$, $i = 1,...,n(I)$, to denote the eigenvalues of P_I^*, we suppose that they are ordered so that for our method of creating the PDM

$$\lambda^*(1_I) = 1 > |\lambda^*(2_I)| \geq |\lambda^*(3_I)| \geq \ldots \geq |\lambda^*(n(I)_I)|. \tag{8}$$

3.2 Selectivity of Proper Attributes in Decomposition

The experience of applying near-complete decomposability in economic theory shows that it is both powerful and interesting. However, our purpose is quite different from Simon and Ando's one. Our purpose is to form concept clusters or sub-databases through decomposing a database on a PDM. As stated in Section 1, one of the features of databases is that databases for discovering knowledge are not always complete but contain uncertain and incomplete data. In particular, the collected data in business databases typically reflect the uncontrolled real world, where many different causes overlap, and many patterns are likely to exist simultaneously. The patterns are likely to have some uncertainty [9]. Therefore, the formed concept clusters or sub-databases are only a representation of incipient structure of some concepts and must be refined by further learning in the next learning phase (see Figure 1). On the other hand, the formation of concept clusters is also a kinds of aggregation and abstraction for databases. That is, primary factors for describing some concepts are aggregated by selecting proper attributes in decomposition, and noises are neglected. There are two kinds of noises: minor elements and irrelevant elements.

The *minor elements* are those whose probability values are much smaller than other related values (i.e., εC in Eq.(2) & (11)). For example, in the *breast cancer database* [5], if bare-nuclei = 1 then the probabilities of malignant and benign are 0.02 and 0.98 respectively. Thus bare-nuclei = 1 cannot be used as one of the conditions of malignant cancer. This kind of minor factors (or call noises) can be neglected by Simon and Ando's method as described in Section 3.1.

The *irrelevant elements* are defined as follows: let P^* be represented in the form of Eq.(9),

$$P^* = \begin{bmatrix} P_1^* & & & & \\ & P_2^* & & & \\ & & \ddots & & \\ & & & P_{N-1}^* & \\ G^* & & & & P_N^* \end{bmatrix}, \tag{9}$$

then G^* is the *irrelevant elements*. Comparing Eq.(3) with Eq.(9), we see that the difference is only G^*. G^* is a sub-matrix whose number of rows is equal to the number of rows of P_N^* and not all elements are zero. For example, in the *breast cancer database* [5], if mitoses $= 3$ then the probabilities of malignant and benign are both nearly 0.5, mitoses $= 3$ cannot be used to differentiate malignant from benign. That is, mitoses $= 3$ is useless for knowledge discovery and cannot be classified into a cluster (i.e., it is a part of G^*). G^* can be considered as noisy data (i.e., the irrelevant elements) and should be omitted although its probability value may be fairly large. Thus we can get Eq.(10) from Eq.(9),

$$P_{sub}^* = \begin{bmatrix} P_1^{*'} & & & \\ & P_2^{*'} & & \\ & & \ddots & \\ & & & P_{N-1}^{*'} \end{bmatrix}, \tag{10}$$

and we can define P' which is a sub-matrix of P by means of Eq.(11) and call it a *nearly decomposable matrix*,

$$P' \simeq P_{sub}^* + \varepsilon C. \tag{11}$$

When F_{sub}^* is got from P^* (see Figure 2), the probability values in P_{sub}^* must be normalized by the following Eq.(12),

$$p_{ij}^{*'} = \frac{p_{ij}^*}{1 - \sum_{m=k+1}^n e_{im}}, \tag{12}$$

so that $\sum_{j=1}^k p_{ij}^{*'} = 1$. Where $i, j = 1 \sim k$. The deleted elements as irrelevant ones are e_{im} and $m = k+1 \sim n$.

Our tests of some real databases such as the *breast cancer database* [5] show that the PDM of real databases which can be represented in the form of Eq.(9) is fairly widespread. In other words, near-complete decomposability of databases which can be represented in the form of Eq.(3) is a special case of no irrelevant elements existing in databases. Eq.(11) is an important extension for satisfying the selectivity of proper attributes in decomposing a database. By means of this extension, our method of decomposing databases can process two kinds of noisy data: those whose probability values are much smaller than other related values (i.e., εC in Eq.(2) & (11)) and those which are the irrelevant elements (i.e., G^* in Eq.(9)), and the purpose of the decomposition and aggregation can be achieved. The result of decomposing a database can be used as background information for further formation of concept clusters as described in Section 3.4.

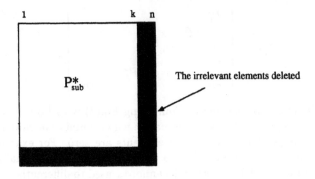

Figure 2: Forming P^*_{sub}

3.3 Two Methods for Forming the Diagonal PDM

Forming the diagonal PDM is valuable as pre-processing before decomposing the PDM. The following two methods are used for forming the diagonal PDM, according to the cases in which a criterion of forming the diagonal PDM is given by user or not.

3.3.1 Forming the Diagonal PDM by a Special Attribute

Forming the diagonal PDM by a special attribute is a supervised method. Before forming the diagonal PDM, an attribute is chosen by the user. The different values of the chosen attribute are used as a criterion of forming the diagonal PDM. In other words, the values of the chosen attribute serve for clustering the values of other attributes. This method is effective for the databases in which an attribute can be chosen as a criterion of forming the diagonal PDM.

More formally, let $\mathbf{c} = \{c_1, c_2, ..., c_n\}$ be the value set of the attribute $\mathbf{c}$ and be chosen as a criterion of forming the diagonal PDM. Let a PDM be created from the attribute $\mathbf{a}$ and $\mathbf{c}$. Select an element a_{ij} in PDM. Then the dependency probabilities between c_k and a_{ij}, $p(c_k|a_{ij})$, are included in PDM. That is, $p(c_k|a_{ij})$ is the probability of c_k for the given a_{ij} and $k = 1,..., n$.

Let a_{ij} be assigned to the cluster corresponding to c_i with the maximal $p(c_i|a_{ij})$. If there is such a case as the same max-value $p(c_m|a_{ij}) = p(c_n|a_{ij})$ for c_m and c_n, then compare their $p(a_{ij}|c_m)$ and $p(a_{ij}|c_n)$ which are the dependency probabilities of a_{ij} for given c_m and c_n respectively. Let $p(a_{ij}|c_m)$ be larger than $p(a_{ij}|c_n)$, then a_{ij} is assigned to the cluster corresponding to c_m. If there are also at least two clusters that contain c_m and c_n with the maximal $p(a_{ij}|c_m) = p(a_{ij}|c_n)$, then a_{ij} is assigned to the cluster corresponding to c_m with the minimal m.

The diagonal PDM shown in Figure 3 can be obtained by the above method and can be decomposed by the algorithm stated in Section 3.4. These diagonal

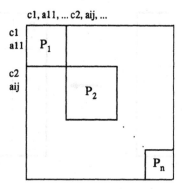

Figure 3: Forming the diagonal PDM by a special attribute

sub-matrices thus obtained have the following two features: there is at least one c_i in a diagonal sub-matrix and the number of the diagonal sub-matrices is less than or equal to the number of different values of the attribute **c**.

3.3.2 Forming the Diagonal PDM by the Optimum Decomposition

Forming the diagonal PDM by the optimum decomposition is an unsupervised method. That is, forming the diagonal PDM is automatically done by the seven steps as follows:

step1 : Select one row from a PDM using order of ascending row number.

step2 : Aggregate the elements larger than the *threshold value* in the selected row into the left side of a PDM. Here, the *threshold value* is the one for distinguishing the meaningful elements from noises. The *threshold value* can be adjusted in decomposing as described in Section 3.4.

step3 : Select a point, by which the PDM is divided into two sub-matrices P_1' and P_2', in the diagonal line (see Figure 4(a)), and calculate ε in Eq.(4).

step4 : Calculate ε' in Eq.(4) for the case when a row in P_1' was moved into P_2' or a row in P_2' was moved into P_1'.

step5 : If the value of $\varepsilon' < \varepsilon$ exists, then reform the diagonal matrix so that it has the smaller ε'.

step6 : Execute repeatedly *step4* and *step5* until no smaller value ε' exists. In this moment, the sub-diagonal matrix P_1 is formed as shown in Figure 4(b).

step7 : P_2 is regarded as new PDM and execute repeatedly *step1-step6* for P_2.

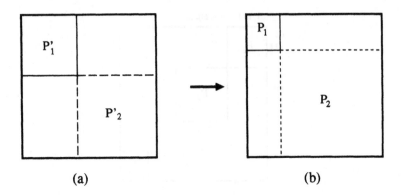

Figure 4: Forming the diagonal PDM by the optimum decomposition

3.4 An Algorithm for Decomposing Databases

An algorithm for decomposing databases is divided into the following five operations:

1. Adjust the *threshold value*. As described in Section 3.3, in the process of forming the diagonal matrix, the elements larger than the *threshold value* are moved into the diagonal sub-matrices and the elements smaller than the *threshold value* are moved outside the diagonal sub-matrices. When the *threshold value* is too small, however, it is possible that no nearly/near-completely decomposable matrix can be obtained. Therefore, the *threshold value* should be repeatedly adjusted so that a nearly/near-completely decomposable matrix (i.e., Eq.(2) or Eq.(11)) can be obtained as far as possible.

2. Form the diagonal matrix. The diagonal matrix is formed by the methods described in Section 3.3.

3. Decompose the diagonal PDM by the following seven steps:

   step1 : Let all points in the diagonal line be the possible cut-points.

   step2 : Select a cut-point with the minimal ε calculated in Eq.(4) as real cut-point from the possible cut-points. By using the selected cut-point, the PDM is divided into two diagonal sub-matrices, and let upper one be called P_i, lower one be called P_k, $i \neq k$.

   step3 : Calculate respectively the second largest eigenvalues of all current diagonal sub-matrices.

   step4 : If the minimal ε satisfies Eq.(6), then record the real cut-point selected in *step2*.

   step5 : Execute repeatedly *step2-step4* for P_i and P_k until the minimal ε satisfying Eq.(6) can be found.

step6 : If the near-complete decomposition is successful, then goto *step7*, else delete the irrelevant elements by using the method described in $\cdot4$ below of this algorithm. Then execute *step2-step6* for the new PDM in which the irrelevant elements have been deleted.

step7 : Create the matrix P^* (or P^*_{sub}) and the error matrix C.

4. Delete the irrelevant elements by the following four steps:

step1 : Let δ be the sum of elements of off-diagonal sub-matrices in a row, that is, $\delta_{i_I} = \sum_{J=1, J \neq I}^{N} \sum_{j=1}^{n(J)} p_{i_I j_J}$. Record the numbers of the row with $\delta = \varepsilon$. Record the numbers of the column which contain the elements larger than the *threshold value* in off-diagonal sub-matrix of the recorded rows.

step2 : Select the column with the maximal sum of the probability from the recorded columns.

step3 : Execute repeatedly *step1-step3* until no row with $\delta = \varepsilon$ exists except the rows related with the selected columns (i.e., the row-column pairs).

step4 : If the number of the selected columns is more than the number of the recorded rows, then delete all rows with $\delta = \varepsilon$ and the related columns (i.e., the row-column pairs), else delete all selected columns and the related rows.

5. Form concept clusters. The result of learning in the probability space is applied back to the instance space and the job of decomposing the instance space is thus completed. As the result of decomposing the instance space, the concept clusters are formed.

4   Refinement

4.1   Perturbation of Databases

In Section 1, we have described that databases for discovering knowledge are not always static but dynamic. This means that databases change often, for example, some data in databases are often added, updated or deleted. We deal with this problem as the perturbation problem of databases, that is, how the probability space changes when the instance space changes.

In general, the perturbation problem is to find how the eigenvalues and eigenvectors of a square matrix P change, when the matrix changes by $P \rightarrow P + E$ [13]. In our application, two perturbation problems must be considered. One is to decompose the original PDM matrix P, $P \rightarrow P^* + \varepsilon C$ ($P' \rightarrow P^*_{sub} + \varepsilon C$); the other one is to modify the original matrix P to $P^{\#}$ when the data in the instance space are changed, $P^{\#} \rightarrow P^{\#*} + \varepsilon^{\#} C^{\#}$ ($P^{\#'} \rightarrow P^{\#*}_{sub} + \varepsilon^{\#} C^{\#}$). In these two kinds of conditions, Eq.(6) & (7) must be used for evaluating decomposable suitability.

We also use $\nu^{*T}(i_I)$ and $\upsilon^*(i_I)$, respectively, to denote the left and right eigenvectors of $P^*(P^*_{sub}, P^{\#*} or P^{\#*}_{sub})$ corresponding to $\lambda^*(i_I)$. Similarly for P ($P', P^{\#}$ or $P^{\#'}$), the left and right eigenvectors associated with $\lambda(i_I)$ will be denoted $\nu^T(i_I)$ and $\upsilon(i_I)$ respectively. For the matrices P ($P', P^{\#}$ or $P^{\#'}$) and

$P^*(P^*_{sub}, P^{\#*} or P^{\#*}_{sub})$, we shall make repeated use of the scalars (i = 1,...,n(I), I=1,...,N)

$$s(i_I) = \nu^T(i_I)v(i_I), \qquad s^*(i_I) = \nu^{*T}(i_I)v^*(i_I). \tag{13}$$

These scalars are called the *condition numbers* of P (P', $P^\#$ or $P^{\#'}$) and $P^*(P^*_{sub}, P^{\#*} or P^{\#*}_{sub})$, respectively. The condition number is an important concept in perturbation theory [13]. We distinguish two kinds of condition numbers: *ill-conditioned* and *well-conditioned*. A condition number is called *ill-conditioned* if the eigenvalues of a matrix are very sensitive to small changes in the data, and *well-conditioned* if the eigenvalues are comparatively insensitive. For our application, the condition number $s^*(i_I), i \neq 1$, expresses the sensitivity of the eigenvalue $\lambda^*(i_I)$ to perturbations of the aggregate P^*_I ($P^{*'}_I$). Wilkinson gives some examples of matrices that are ill-conditioned with respect to certain eigenvalues. These matrices have condition numbers of the order of 10^{-7} to 10^{-19} [13].

4.2 Learning Space

As shown in Section 2.1, the *learning space*, one of three kinds of database spaces, is a space that records probability distributions and errors for learning, describing and controlling the PDM. It is necessary to define the learning space for processing the perturbation problem of databases, so that the result of decomposing database is not affected by smaller change of data in the instance space. That is, if the change of data in the instance space is not larger than the max-error, then record only $P^{\#*}(P^{\#*}_{sub})$ and $\varepsilon^\# C^\#$ in the learning space, and do not change the decomposed PDM. This method is reasonable because there were errors in the result of decomposing database from the beginning, and the results of learning in the learning space may be either an increase or decrease of errors of the PDM. Therefore, the learning process in the learning space is only the accumulation of errors along with the change of data in the instance space. Only when the accumulation of errors exceeds its upper bound, the result of learning in the learning space affects the decomposition of the probability/instance space, that is, the result of learning in the learning space is transmitted into the probability space and the decomposition of the probability/instance space is done again.

5   Discussion & Concluding Remarks

Up to now, we briefly presented a *decomposition based induction* model for discovering concept clusters from databases. Furthermore, we have also developed several auxiliary techniques for making our approach more useful [15]. First, attribute oriented clustering can be performed as a step of pre-processing before creating the probability space. Its objective is the quantization of continuous values and conceptual abstraction (generalization). In the real world, there are many real-valued attributes as well as symbolic-valued attributes. When creating the probability space in our approach, continuous values must be quantized. On the other hands, in order to discover the better knowledge, conceptual abstraction and generalization by using background knowledge are also necessary before creating the probability space. Second, learning is used in the creation of the probability space. Its objective is to improve the performance of our approach. Generally speaking, databases for discovering knowledge are very large and complex. Therefore, the performance of our approach must be considered in its implementation. The probability space is the main operating object in our approach. Thus, it is important to improve the performance of PDM. In general, the maximal number of individual data that can be included in the instance space is $\prod_{i=1}^{m} n_i$ and the number of the probability space is $(\sum_{i=1}^{m} n_i)^2$. Here, m is the number of attributes, n is the number of different data values in each attribute. In creating the probability space, the performance of the probability space can be improved by using background knowledge and learning. Since have limited space, a more general discussion on these issues is not given here, and the details about them refer to [15].

Main features of the *decomposition based induction* model can be summarized as follows: (1) It can either work in the environment of supervised learning or unsupervised learning according to whether a criterion of forming the diagonal PDM is assigned by user or not. That is, when supervised learning is used, concept clusters are formed by a special attribute based diagonalization and decomposition. On the other hand, when unsupervised learning is used, sub-databases are formed by the optimum decomposition. (2) Decomposition is not only done by clustering, but also by the omission of two kinds of noisy data: those whose probability values are much smaller than other related values and those which are the irrelevant elements. (3) Multiple learning phases such as attribute oriented clustering, creating the probability space, near-complete decomposition, refinement, are provided. (4) In order to acquire better knowledge and improve the performance of the discovery mechanism, background knowledge can be used in the procedures of knowledge discovery.

This model is a fundamental one for developing DBI which is one of sub-systems of the GLS discovery system implemented by us [19, 16]. Our experience with a prototype of our approach implemented in a knowledge-based system KAUS and a application to a breast cancer database shows that the features of databases as shown in Section 1 can be reasonably processed by means of above features of our approach. We also use another sub-system called HML (Hierarchical Model Learning) for generating *classification knowledge with hierarchical models* from the discovered concept clusters by DBI, and managing/refining them in a knowledge-base [17].

Several problems are also left for further investigations. Among them are the completion of the learning space, and the concept description to the result of decomposing in unsupervised learning, to implement our approach in the distributed environment and to apply it to more real problems and application fields for further testing and demonstration.

References

[1] J. Cendrowska. PRISM: An Algorithm for Inducing Modular Rules. *International Journal of Man-Machine Studies*, Vol.27, (1987) 349-370.

[2] P.J. Courtois. *Decomposability - Queueing and Computer System Applications*. ACADEMIC PRESS, (1977).

[3] J.H. Gennari, P. Langley, and D. Fisher. Models of Incremental Concept Formation. *Artificial Intelligence*, Vol.40(No.1-3), (1989) 11-61.

[4] M. Lebowitz. Categorizing Numeric Information for Generalization. *Cognitive Science*, Vol.9(No. 3), (1985) 285-308.

[5] O.L. Mangasarian and W.H. Wolberg. Cancer diagnosis via linear programming. *SIAM news*, Vol.23(No.5), (1990) 1-18.

[6] R.S. Michalski, J.G. Carbonell, and T.M. Mitchell. *Machine Learning - An Artificial Intelligence Approach*, volume Vol.1 Vol.2 Vol.3. Tioga Publishing Company, (1983 1986 1990).

[7] S. Ohsuga. AI Paradiam and Knowledge. *Proc. Symposium of New Generation Knowledge Processing*, (1992) 1-21.

[8] G. Piatetsky-Shapiro. Knowledge Discovery in Real Databases: A Report on the IJCAI-89 Workshop. *AI magazine*, Vol.11(No.5), (1991) 68-70.

[9] G. Piatetsky-Shapiro. Discovery,Analysis, and Presentation of Strong Rules. *Piatetsky-Shapiro and Frawley (eds.) Knowledge Discovery in Databases*, (AAAI Press and The MIT Press, 1991) 229-248.

[10] G. Piatetsky-Shapiro and W.J. Frawley (eds.). *Knowledge Discovery in Databases*. AAAI Press and The MIT Press, (1991).

[11] J.W. Shavlik and T.G. Dietterich (eds.). *Readings in Machine Learning*. MORGAN KAUFMANN PUBLISHERS, INC., (1990).

[12] H.A. Simon and A. Ando. Aggregation of Variables in Dynamic Systems. *Econometrica*, Vol.29, (1961) 111-138.

[13] J.H. Wilkinson. *The Algebraic Eigenvalue Problem*. Oxford Univ. Press, (1965).

[14] N. Zhong and S. Ohsuga. A Consideration to Automatic Decomposition of Databases. *Proc. the 5th Annual Conference of JSAI*, (1991) 185-188.

[15] N. Zhong and S. Ohsuga. Discovering Concept Clusters by Decomposing Databases. *manuscript*.

[16] N. Zhong and S. Ohsuga. GLS - A Methodology for Discovering Knowledge from Databases. *Proc. the 13th International CODATA Conference entitled "New Data Challenges in Our Information Age"*, (1992).

[17] N. Zhong and S. Ohsuga. Hierarchical Model Learning in Knowledge Discovery in Databases. *Proc. International Conference on Intelligent Information Processing & System (ICIIPS'92)*, (1992) 420-423.

[18] N. Zhong and S. Ohsuga. Attribute Calculation in Knowledge Discovery in Databases. *(to appear in) Journal of Japanese Society for Artificial Intelligence*, Vol.9 No.1, (1994).

[19] N. Zhong and S. Ohsuga. Knowledge Discovery and Management in Integrated Use of KB and DB. *Ohsuga et al. (eds.) Information Modelling and Knowledge Bases: Foundations, Theory and Application*, (IOS Press, Amsterdam, 1992) 480-495.

[20] N. Zhong and S. Ohsuga. An Integrated Calculation Model for Discovering Functional Relations from Databases. *Proc. 4th International Conference on Database and Expert Systems Applications (DEXA'93) published in the Lecture Notes in Computer Science series*, (Springer-Verlag, Heidelberg, 1993).

[21] W. Ziarko. The Discovery, Analysis, and Representation of Data Dependencies in Databases. *Piatetsky-Shapiro and Frawley (eds.) Knowledge Discovery in Databases*, (AAAI Press and The MIT Press, 1991) 195-209.

Algebraic Structure of some Learning Systems

Jean-Gabriel GANASCIA

LAFORIA - Institut Blaise Pascal, Université Pierre et Marie CURIE
Tour 46-0, 4 Place Jussieu, 75252 Paris, CEDEX, FRANCE
ganascia@laforia.ibp.fr

Abstract: The goal of this research is to define some general properties of representation languages, e.g. lattice structures, distributive lattice structures, cylindric algebras, etc. to which generalization algorithms could be related. This paper introduces a formal framework providing a clear description of version space. It is of great theoretical interest since it makes the generalization and comparison of many machines learning algorithms possible. Moreover, it could lead to reconsider some aspects of the classical description of version space. In this paper, the scope of investigation will be restricted to lattices — i.e. to cases where there exists one and only one generalization for any set of examples — and in particular to Brouwerian lattices. It is shown that a particularly interesting case covered by this restriction is the product of hierarchical posets which is equivalent to the conjunction of tree structured or linearly ordered attributes.

1. Introduction

In the past, very little attention has been paid to the mathematical structure of objects involved in machine learning processes. For example, the notion of linearly ordered or tree structured attributes which is currently used in Machine Learning is not related to well defined mathematical properties. In fact, most of the time, machine learning mechanisms rely on the introduction of an ordering relation which is linked to the notion of generality or to the notion of subsumption. The problem is that this ordering relation restricts the range of the mathematical framework which can structure the representation language.

The traditional artificial intelligence approach defines knowledge representation languages before defining the properties of those languages. In the case of machine learning there is no precise definition of the language properties which are required by the learning algorithms. This leads to some confusion since the limitations of representation languages and the limitations of algorithms which manipulate expressions are not clearly distinguished. For instance, in the case of ID3-like induction systems, it appears that the attribute-value representation is a particular case of some more general representation language which could easily extend these systems. However, the classical description of the algorithm does not facilitate extensions to more general languages since it is limited to representation languages based on attribute-value structural descriptions.

The goal here is to define some general properties of the representation languages, e.g. lattice structures or distributive lattice structures, on which generalization algorithms could be based. It is of great theoretical interest, since it makes the generalization and comparison of many machine learning algorithms possible. However, it should not be confused with learnability, be this Gold [1], Valiant [2] or other learning paradigms. The present goal is not to define general limitations of learning mechanisms but to relate machine learning algorithms to the mathematical properties of the manipulated objects. On the other hand, there have been some attempts to define a general learning framework using the notion of version space, but recent studies show that, in practice, this framework is not usable. (See 3 or 4).

This paper only deals with lattices. It means that each set of descriptions has one and only one least general generalization. This restriction covers many applications in machine learning; for instance, it covers all the ID3-like systems [5, 6], but it does not cover the case where matching is multiple, i.e. where a first order logic is required (Cf. [7]). In these cases, the notion of cylindric algebra has to be introduced. It could be seen as a generalization of the present work, and the principles on which it relies are similar to those presented here.

2. Introduction to Version Space

Introduced by T. Mitchell [8], version space has been seen as a general framework in which every machine learning algorithm could be described as a search algorithm. In this framework, similarity-based learning — SBL — could be summarized as follows (Cf. [8]).

"Given:
- a language in which to describe instances
- a language in which to describe generalizations
- a matching predicate that matches generalizations to instances
- a set of positive and negative training instances of a target generalization to be learned
Determine: generalizations within the provided language that are consistent with the presented training instances (i.e. plausible descriptions of the target generalization)."

It is assumed that the generalization space is ordered with the relation "is more general than", noted $\leq_g$, which is defined by: $G1 \leq_g G2$ if and only if $\{i \in I \mid M(G1,i)\} \supseteq \{i \in I \mid M(G2,i)\}$ where $M(G,i)$ means that the generalization G matches the instance i, M being the matching predicate.

The set of all consistent hypotheses is defined by two sets, the set of *maximally specific generalizations*, noted S-set, and the set of *maximally general generalizations*, noted G-set. Adding positive and negative instances of the target concept leads to increase the S-set and to decrease the G-set. The algorithm stops when the S-set equals the G-set or when some inconsistency arises.

Practical and theoretical studies [9, 3] have shown that this framework was not actually usable. The first reason is that the number of examples required to ensure the convergence of the algorithm can be exponential with the problem size. The second is that the size of the G-set can also become exponential, even in some trivial cases like the one given in [3].

For the sake of clarity, let us recall the examples given in [3]. On the one hand, Haussler says that if the instance space X is defined by the Boolean attributes A_1, A_2, ..., A_n, and if the target concept h is supposed to be $A_1 = true$, we need more than 2^{n-2} positive examples and more than 2^{n-2} negative examples, even if the hypothesis space is restricted to pure conjunctive concepts. The reason is that there are 2^{n-2}

positive examples such that $A_1 = true$ and $A_2 = true$, so if we want to distinguish $A_1 = true$ from $A_1 = true$ & $A_2 = true$, we need more than 2^{n-2} positive examples. The same argument can be applied to negative examples. Therefore, the number of examples needed is exponential with the number of attributes.

On the other hand, let us suppose that X is always defined by the Boolean attributes A_1, A_2, ...A_n and that there is one positive example Q (true, true, ..., true) and n/2 negative examples:

(false, false, true, true, true, ..., true, true, true)
(true, true, false, false, true, ..., true, true, true)
...
...
(true, true, true, true, true, ..., true, false, false)

If we assume that the target concept h is a pure conjunctive hypothesis consistent with the positive example Q, then

(1) h is of the form $A_{i1} = true$ & $A_{i2} = true$ & ... & $A_{ij} = true$, for some $\{A_{i1}, A_{i2}, ..., A_{ik}\}$ such that $\{A_1, A_2, ..., A_n\} \supseteq \{A_{i1}, A_{i2}, ..., A_{ik}\}$ and

(2) h must contain the following atoms:

either the atom $A_1 = true$ or the atom $A_2 = true$ to exclude the first counter-example
either the atom $A_3 = true$ or the atom $A_4 = true$ to exclude the second counter-example
...
...
either the atom $A_{n-1} = true$ or the atom $A_n = true$ to exclude the last counter example.

Therefore, it is easy to show that the maximally general concept which meets both (1) and (2) is a disjunctive normal form containing at least $2^{n/2}$ conjunctions. It follows that the size of the G-set is exponential with the number of counter-examples.

Many solutions have been proposed to solve the difficulties encountered. One, for instance, suggested modifying the learning bias (Cf. [10, 11]). Another proposed to consider only a list of negative instances to represent the G-set [4]. It is also possible to provide new ad-hoc representations of the version space [12] or to decompose the generalization language on a product of attributes [13], etc. It appears that all these solutions are restricted to particular cases, for instance to the cases where the S-set is conjunctive or/and where the generalization space is a *lattice* — i.e. each set of instances has one and only one generalization.

I suggest that, the learning problem being stated as above, it is possible to get a better formalization than the one proposed by the classical version space. It is just necessary to add the hypothesis that the instance language is ordered by the generality relation. In this way, the S-set and the G-set are related to the instance language and not to the generalization language which is just used to compute an efficient generalization. In this framework, we do not have to make the S-set and the G-set converge since it is only necessary to have a maximally general generalization consistent with the instances, i.e. a G-set.

To clarify these ideas, let us limit ourselves to the case where the description language is a *Brouwerian lattice* (see appendix) and let us formalize the notions of S-set and G-set in this context.

3. Formalization of the Learning Problem

It is possible to formulate the learning problem as it was introduced by T. Mitchell (see above) using elementary lattice theory notions (see appendix). To do so let us suppose that given a set $E = \{e_1, e_2,..., e_n\}$ of positive instances and a set $CE = \{ce_1, ce_2,..., ce_m\}$ of negative instances, a concept C has to be learned.

We shall assume that the positive instances, e_i, and the negative instances, ce_j, are described as points of a representation space $\mathfrak{R}$ which is ordered by the generality relation $\leq_g$. We shall also assume that $\mathfrak{R}$ is a Brouwerian lattice which means (see appendix) that for each pair $\{a, b\}$ there exists a *least upper bound* of a and b, noted $(a \vee b)$, a *greatest lower bound* of a and b, noted $(a \wedge b)$ and a pseudo-complement of a related to b, noted $(a : b)$.

3.1. An Example

To make the presentation easier to understant, here is an example which will illustrate all the abstract notions presented in the paper.

This example is drawn from [12] and [13] and involves playing cards. The description language uses three tree-structured attributes, *rank, oddity, color*:

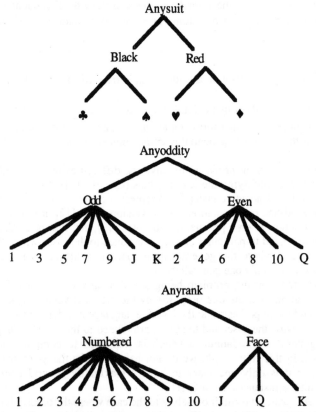

Let us suppose there are two positive examples, (7 ♠) and (King ♠). The least general generalization is the greatest lower bound of (7 ♠) and (King ♠), i.e. it is the greatest element E such that $E \leq_g$ (7 ♠) and $E \leq_g$ (King ♠).

In other words, E = (7 ♠) ∧ (King ♠) = (7 ♠) OR (King ♠). In the case of classical version space, we should have E = (oddity = Odd) & (color = ♠). Using only distributive lattices and tree structured properties it is possible to obtain E = (rank = Anyrank) & (oddity = Odd)à & (color = ♠) & R where R is a logical disjunction:

E = (((rank=7) ∨ (oddity=7) ∨ (color=♠)) ∧ ((rank=K) ∨ (oddity=K) ∨ (color=♠)))

E = ((color = ♠) ∨ (((rank = 7) ∨ (oddity = 7)) ∧ ((rank = K) ∨ (oddity = K))))

E = ((color = ♠) ∨ (rank = Anyrank) ∨ (oddity = Odd) ∨ (((rank = 7) ∨ (oddity = 7)) ∧ ((rank = K) ∨ (oddity = K))) because due to tree-structured attribute properties we have

(rank = 7) = (rank = 7) ∨ (rank = anyrank),
(rank = K) = (rank = K) ∨ (rank = anyrank),
(oddity = 7) = (oddity = 7) ∨ (oddity = Odd) and
(oddity = K) = (oddity = K) ∨ (oddity = Odd).

It makes E = ((color = ♠) & (rank = Anyrank) & (oddity = Odd) & R where R is a logical disjunction of the form :

R = ((rank = 7) ∨ (oddity = 7)) ∧ ((rank = K) ∨ (oddity = K)).

Remark that the least general generalization of a and b is noted (a ∧ b) which corresponds to the logical disjunction (a OR b) and not to the logical conjunction (a & b). This surprising effect is due to the fact that the ordering relation is that of generalization and not implication or subsumbtion.

As we shall see in the following, the least general generalization keeps a lot of information which makes it very similar to a factorization process. Here, the learning mechanism which has to forget some information is related to the search procedure, not to the generalization mechanism. In fact, generalization is just used to store information and to define the search space properly.

3.2. Maximally Specific Generalization

As stated above, the *maximally specific generalization*, i.e. the S-set, is the least general generalization of the positive instances which is consistent with the negative instances. In other words, it is the greatest lower bound of E which means

that: $S = \bigwedge_{u=1}^{n} e_u$

Using the properties of the glb — *greatest lower bound* — it is obvious that $\forall i \in [1, n]$ $S \leq_g e_i$, so S generalizes all the positive instances. Moreover, if there exists another generalization it is lower — i.e. more general — than S which means that S is the least general generalization of the examples.

However, the negative instances are not involved in this definition of S. In fact, due to the properties of lattices, a negative instance is covered by S if and only if it is covered by at least one positive instance. This shall now be proved.

Definition: an element $a \neq \emptyset$ is said to be ∧-irreducible (resp. ∨-irreducible) iff $b \wedge c = a$ (resp. $b \vee c = a$) implies b = a or c = a.

Lemma: if P is ∨-irreducible [resp. ∧-irreducible] in a distributive lattice L then:

$$P \leq \bigvee_{i=1}^{k} x_i \text{ implies } \exists i \in [1, k] \text{ such that } P \leq x_i.$$

$$\left[\text{resp. } P \geq \bigwedge_{i=1}^{k} x_i \text{ implies } \exists i \in [1, k] \text{ such that } P \geq x_i. \right]$$

Proof: $P \leq \bigvee_{i=1}^{k} x_i$ implies $P = P \wedge \bigvee_{i=1}^{k} x_i = \bigvee_{i=1}^{k} (x_i \wedge P)$ (because the lattice is distributive)

P being ∨-irreducible, then $\exists i \in [1, k]$ such that $P = P \wedge x_i$.

[the proof is similar for ∧-irreducible elements]

Remark: as mentioned in the appendix, every Brouwerian lattice is distributive.

Theorem: if $\exists j \in [1, m]$ such that $ce_j \geq S$ then $\exists i \in [1, n]$ such that $ce_j \geq e_i$. In other words, if some negative instance of C is covered by S, then it is also covered by some positive instance of C.

Proof: the above lemma just needs to be applied. Two cases have to be studied: either ce_j is ∧-irreducible or not. In the first case, only the above lemma needs to be applied with $P = ce_j$ and $x_i = e_i$. In the second case, ce_j is the conjunction of

∧-irreducible elements, so $ce_j = \bigwedge_{i=1}^{p} a_i^j$ the a_i^j being ∧-irreducible elements.

Therefore it is possible to apply the above theorem to the a_i^j, since it is obvious that

$a_i^j \geq S$ which means that $a_i^j \geq \bigwedge_{u=1}^{n} e_u$.

Being a least general generalization, S is mainly restricted to the positive instances.

Example: the previous example shows how it works to compute the least general generalization, i.e. to compute S. Note that, using the classical version space generalization — i.e. E = Odd & ♠ & Anyrank — should lead to some contradiction with (9 ♠) as a counter-example, while the present generalization — i.e. E = ((color = ♠) ∨ (rank = Anyrank) ∨ (oddity = Odd) ∨ ((rank = 7) ∨ (oddity = 7)) ∧ ((rank = K) ∨ (oddity = K))) — should not.

Here, the generalization is essentially viewed as a factorization. However, as we shall see in the following, the description language is a distributive lattice so it may involve a product of distributive lattices, each one being associated with some hierarchy. In this case, the generalization may correspond to what people call "climbing a hierarchy of predicates".

3.3. Maximally General Generalization

As described above, the *maximally general generalization* G — the so-called G-set — can be a huge formula, even in trivial cases. Here the G-set is defined using lattice operators so as to provide some useful representation of the version space.

In the appendix, the difference operator has been defined, noted /, which is the dual of the pseudo-complement. More precisely, the difference b/a is the least x — i.e. the more general — such that $(a \lor x) \geq b$. The considered lattices being Brouwerian, it is obvious that there exists a difference for each pair of elements. Using the difference operator, it is thus possible to define the G-set by the following formula:

$$G = \bigwedge_{i=1}^{n} e_i \bigg/ \bigwedge_{j=1}^{m} ce_j$$

To prove that G corresponds effectively to the G-set, we have to prove the following three points:

1- G covers all the positive instances, i.e. $\forall i \in [1,n]\ G \leq e_i$.

2- G does not cover any negative instance, i.e. $\forall j \in [1,m]\ \neg\,(G \leq ce_j)$.

3- G is the most general expression satisfying 1 and 2.

1- G covers all the positive instances

Proof: by definition, G is the least element such that $G \lor \bigwedge_{j=1}^{m} ce_j \geq \bigwedge_{i=1}^{n} e_i$.

It follows that $G \leq \bigwedge_{i=1}^{n} e_i$ because if is were not the case, $\bigwedge_{i=1}^{n} e_i$ would be a solution. Therefore $\forall i \in [1, n]\ G \leq e_i$.

2- G does not cover any negative instance

It is possible to prove that if G covers a negative instance, then this instance is also covered by S.

Proof: let us assume that there exists a negative instance, ce_k which is covered by G, i.e. such that $G \leq ce_k$.

Using distributivity we obtain $G \lor \bigwedge_{j=1}^{m} ce_j = \bigwedge_{j=1}^{m} (G \lor ce_j)$

But as $G \le ce_k$, if follows that

$$\bigwedge_{j=1}^{m}(G \vee ce_j) = \bigwedge_{j=1, j \ne k}^{m}(G \vee ce_j) \wedge ce_k \ge \bigwedge_{i=1}^{n} e_i$$

which means that $S \le ce_k$.

3- G is the most general expression which satisfies 1 and 2
Proof: it follows naturally from the definition of the difference.

The preceding formula make possible the computation of G, but it is very inefficient. As Hirsh [4] shows, it is sufficient to provide some representation of the version space in order to:
1- test if an instance does or does not belong to the version space and
2- update the version space.
Hirsh proposes to memorize a list of negative instances. Here, the proposed solution is similar: it is to memorize the least general generalization of the positive instances, i.e. S, and the least general generalization of the negative instances. In other words we just have to memorize:

$$\bigwedge_{j=1}^{m} ce_j \text{ and } \bigwedge_{i=1}^{n} e_i$$

It is interesting to show that general considerations on the mathematical structure of the objects manipulated by learning algorithms lead to some practical consequences. Moreover, it appears that this solution is better than the solution proposed by Hirsh because we only have to memorize the generalization of the negative instances and not the list of all the instances. Even if generalization keeps a lot of information, it is more efficient to consider only the generalization of negative instances.

Example: let us consider again the previous example and let us suppose there are two positive instances (7 ♠) and (K ♠) and two negative instances (7 ♣) and (Q ♠). The least general generalization of the example is:
E = ((color = ♠) ∨ (rank = Anyrank) ∨ (oddity = Odd) ∨
            (((rank = 7) ∨ (oddity = 7)) ∧ ((rank = K) ∨ (oddity = K)))
while the least general generalization of the counter-example is:
CE = ((color = Black) ∨ (rank = Anyrank) ∨ (oddity = Anyoddity) ∨
(((rank = 7) ∨ (oddity = 7) ∨ (color = ♣)) ∧ ((rank = Q) ∨ (oddity = Q) ∨ (color = ♠))).
Now, let us suppose we want to generate some conjunctive hypothesis h belonging to the version space. It is the least general conjunction G such that $G \vee CE \ge E$, i.e. G = E/CE.
In our example, neither G = (color = ♠) nor G = (oddity = Odd) are solutions, so there is only one conjunctive solution which is (color = ♠) ∨ (oddity = Odd), i.e. (color = ♠) & (oddity = Odd). As we can see, in the case of conjunctions of attributes it is very easy to compute the solution using this formalism. It is also easy to check that no conjunctive solution is allowed and to compute disjunctive solutions.

4. Structure of Lattices

So far we have seen how to circumscribe the version space in $\Re$. We shall now study the exploration of $\Re$ and its structure. Let us remember that $\Re$ is a distributive lattice. Using classical results from lattice theory it is possible to define a particular class of distributive lattices for which the complexity of the exploration is linear with the number of *points*, i.e. with the number of symbols belonging to $\Re$. These lattices, as we shall see, will play a central role in the structure of $\Re$. This section introduces this class of lattices and goes on to show how it is possible to structure $\Re$.

Definition: A subset A of a poset X is said to be ∧-**closed** if and only if $\forall a \in A$, $\forall x \ x \leq a \Rightarrow x \in A$.

Theorem: the free distributive lattice generated with n symbols, $\varnothing$ and I, is dually isomorphic to the ring of all ∧-closed subsets of $\{1, 2, ..., n\}$.

Proof: See [14].

Definition: A poset X is said to be **hierarchical** if and only if $\forall (a, b) \in X^2$ $a \geq b$ or $a \leq b$ or $\{x/ \ x \geq a \text{ and } x \geq b\} = \{\}$. It means that $a \vee b = \textbf{if } a \geq b \textbf{ then } a$ **else if** $a \leq b$ **then** b **else** I.

Example: all three attributes *Color*, *Oddity* and *Rank* are built on hierarchical posets since the conjunction of any two values is either one of the two values of the empty set. For instance, (color = ♠) ∨ (color = ♣) = ⊥ while (color = ♠) ∨ (color = black) = (color = black).

Theorem: Any element belonging to a distributive lattice L built on a hierarchical poset X is a greatest lower bound of elements of the poset X.

Proof: See [15, 16].

Therefore, in the case of distributive lattices built on hierarchical posets, only the greatest lower bounds of elements of the hierarchical posets have to be considered. In other words, using hierarchical posets the search space, i.e. the portion of the distributive lattice $\Re$ which has to be explored, is restricted to the ring of all elements of the poset X on which $\Re$ is built. In practice, hierarchical posets correspond to attributes whose values are disjointed, linearly ordered or, more generally, partially ordered through a hierarchy.

Hierarchical posets correspond to hierarchies of propositions. It is not a sufficient representation since it was not possible to be restricted to just one attribute at a time. To increase the representational power, therefore, the product of distributive lattices built on hierarchical posets can be introduced. Since it has been proved that the product of distributive lattices is a distributive lattice, the resulting relation lattice $\Re$ is just the product of n hierarchical posets.

This structure could be interpreted in terms of machine learning as a set of attributes, each of which being a hierarchical attribute, i.e an attribute whose values belong to a hierarchy. This is exactly the structure of the representation that is used by T. Mitchell in [8] and in many other papers in the literature (see for instance [3, 4]). It can also be seen as an extension of the attribute-value vectors as they are used in the classical TDIS algorithms like ID3 [6], CN2 [17], etc. The use of tree structured attributes is a particular case of this structure, so the previous example falls within this framework. One of the main advantages of this structure is that it allows a direct introduction of knowledge through hierarchies, whether this knowledge be composed of exclusive values, ordered attributes or general hierarchies.

5. Conclusion

The aim of this paper was to give a clear presentation of the proposed formal framework which is a nice formulation of version space. Other research (See [15, 16, 19]) based on the same formalism allows the classical Top Down Induction Techniques to be generalized and extended with respect to learning strategies and representation. In a nutshell, it allows an approximation of the G-set to be computed with k-CNF formulas. Therefore it makes it possible to establish strong links between the version space and the Top Down Induction Techniques.

However it is not only of theoretical interest and the next step is to use it in a practical manner along three lines.

The first is to use it to test different learning strategies. It will be implemented so as to enable an easy introduction of many different learning strategies, thus allowing classical "ID-like" or "CN-like" strategies to be evaluated and compared with new ones.

The second concerns the ability to modify the learning bias and to express it as part of knowledge. In the case of the classical version space, the generalization language is given at the very beginning when the S-set and the G-set are computed, while in the new framework it only needs to be given when the generalization step occurs. In this way it is possible to transform dynamically that part of the learning bias which is related to the generalization language.

The third is related to the extension of the proposed framework to cylindric algebra in order to take into account cases where the reduction to attribute-value language is not possible (Cf. [14, 21, 22]).

As far as the theoretical interest is concerned, it involves examining the exact role of Brouwerian logic on which the new model is based. The introduction of Brouwerian logic was in fact a surprising spin-off considering that the only objective was to get a non atomic lattice as a description space. This question is directly related to the fact that distributive complemented lattices are necessarily atomic which means that the introduction of classical negation leads to a Boolean representation. Since the aim was to justify attribute-value representation, a simplification operation has to be introduced without using complementation; the choice was pseudo-complementation. It now seems interesting to investigate the links between induction and intuitionism in more depth. Concerning this point, my intuition is that it could shed light on the philosophical debate related to induction since Hempel's classical objections could be refuted [18].

References

1. Gold: Language identification in the limit, in Information and control, Vol. 10, pp. 447-474 (1967).
2. L.G. Valiant: A theory of the learnable, in Comm. ACM 27, (11), 1984, pp. 1134-1142.
3. D. Haussler: Quantifying Inductive Bias: AI Learning Algorithms and Valiant's Learning Framework, in Artificial Intelligence 36, pp. 177-221, (1988).
4. H. Hirsh: Polynomial-Time Learning with Version Space, in the proceedings of the 9th International Conference on Machine Learning, pp. 117-122, (1992).
5. R. Quinlan: Learning efficient classification procedures, in Machine Learning: an artificial intelligence approach, Michalski, Carbonell & Mitchell (eds.), Morgan Kaufmann, p. 463-482, (1983).

6. R. Quinlan: The effect of noise on concept learning, in Machine Learning: an artificial intelligence approach, Vol. II, Michalski, Carbonell & Mitchell (eds.), Morgan Kaufmann, p. 149-166, (1986).

7. S. Muggleton, C. Feng: Efficient Induction of Logic Programs, Proceedings of the First Conference on Algorithmic Learning Theory, Tokyo, Japan: Ohmasha. (1990).

8. T. Mitchell: Generalization as search, in Artificial Intelligence 18, pp. 203-226, (1982).

9. A. Bundy, B. Silver, D. Plummer: An analytical comparison of some rule-learning programs, in Artificial Intelligence 27, (1985).

10. T.Mitchell: The need for Biases in Learning Generalizations, internal report Rutgers University, 1980, reprint in Readings in Machine Learning, Morgan Kaufman.

11. P. Utgoff: Shift of bias for inductive concept learning, in Machine Learning: an artificial intelligence approach, Vol. II, Michalski, Carbonell & Mitchell (eds.), Morgan Kaufmann, (1986).

12. J. Nicolas: Une représentation efficace pour les espaces des versions, in Proceedings of the Journées Francophones d'Apprentissage, (1993).

13. C. Carpineto: Trading off Consistency and Efficiency in Version-Space Induction, in the proceedings of the 9th International Conference on Machine Learning, pp. 43-48, (1992).

14. G. Birkhoff: Lattice theory, Third Edition, American Mathematical Society, Providence, RI (1967).

15. J.-G. Ganascia: An algebraic formalization of CHARADE, LAFORIA, Internal report, (1992).

16. J.-G. Ganascia: TDIS: an Algebraic Formalization, in the Proceedings of International Jointed Conference on Artificial Intelligence, (1993).

17. P. Clark, T. Niblett: The CN2 Induction Algorithm, in Machine Learning, 3, 1987, pp. 261-283.

18. C. Hempel: Aspects of Scientific Explanation and other Essays in the Philosophy of Science, The Free Press, A Division of Macmillan Publishing Co. Inc., New-York, (1965).

19. R. Wille: Knowledge Acquisition by methods of formal concept analysis, Technische hochschule Darmstadt, Fachbereich Mathematik, preprint N° 1238, (1989).

20. G. Grätzer: General Lattice Theory, Academic Press, New-York, (1978).

21. H. Rasiowa: An algebraic Approach to Non-Classical Logics, Studies in Logic and the Foundations of Mathematics, North-Holland, Amsterdam-London, (1974).

22. L. Henkin, D. Monk, A. Tarski: Cylindric Algebras, Studies in Logic and the Foundations of Mathematics, Vol. 64, North-Holland Publihing Company, Amsterdam, (1971).

Appendix

This is an introduction to some basic notions of lattice theory. More details can be found in [14, 20, 21].

Definition: A **poset** is a partially ordered set, i.e. a set with a partial ordering relation ≤.

Definition: A **lattice** is a poset E such that there exists a least upper bound and a greatest lower bound for each pair (a, b) of elements of E. The least upper bound is noted (a ∨ b) and the greatest lower bound (a ∧ b).

Definition: A lattice is distributive if and only if it satisfies (1) or (2):

(1) $\forall$(x, y, z)   x ∧ (y ∨ z) = (x ∧ y) ∨ (x ∧ z)

(2) $\forall$(x, y, z)   x ∨ (y ∧ z) = (x ∨ y) ∧ (x ∨ z)

Definition: The **pseudocomplement** of a **relative** to b is the greatest element x such that $a \wedge x \leq b$. It is noted "$a{:}b$".

Since we need to simplify, to subtract, i.e. to build the difference between two descriptions, the dual of the relative pseudocomplement will defined and named "difference".

Definition: The difference "b/a" is the dual of the pseudocomplement of a relative to b. In other words, it is the least x such that $a \vee x \geq b$.

Definition: a **Brouwerian lattice** is a lattice where there exists a difference "b/a" for each pair of elements (a, b).

Theorem: Every Brouwerian lattice is distributive.

Proof: See [14].

Theorem: Every finite distributive lattice is a Brouwerian lattice See [14].

Proof: See [14].

Induction of Probabilistic Rules Based on Rough Set Theory

Shusaku Tsumoto and Hiroshi Tanaka

Department of Informational Medicine

Medical Research Institute,Tokyo Medical and Dental University

1-5-45 Yushima, Bunkyo-ku Tokyo 113 Japan

TEL: +81-03-3813-6111 (6159) FAX: +81-03-5684-3618

E-mail:tsumoto@tmd.ac.jp,tanaka@tmd.ac.jp

Abstract

Automated knowledge acquisition is an important research issue in machine learning. There have been proposed several methods of inductive learning, such as ID3 family and AQ family. These methods are applied to discover meaningful knowledge from large database, and their usefulness is in some aspects ensured. However, in most of the cases, their methods are of deterministic nature, and the reliability of acquired knowledge is not evaluated statistically, which makes these methods ineffective when applied to the domain of essentially probabilistic nature, such as medical one. Extending concepts of rough set theory to probabilistic domain, we introduce a new approach to knowledge acquistion, which induces probabilistic rules based on rough set theory(PRIMEROSE) and develop an program that extracts rules for an expert system from clinical database, based on this method. The results show that the derived rules almost correspond to those of the medical experts.

1 Introduction

One of the most important problems in developing expert systems is knowledge acquisition from experts. While there have been developed a lot of knowledge acquistion tools to simplify this process, it is still difficult to automate this process. In order to resolve this problem, many methods of inductive learning, such as induction of decision trees[1, 14], AQ method[9, 10] and rough set theory[13], are introduced and applied to discover knowledge from large database, and their usefulness is in some part ensured. However, their methods are of the deterministic nature, and,in most of the cases,the acquired knowledge is not evaluated statistically.Hence they are not applicable in their present form to probabilistic domain, such as medical domain.

Extending the concepts of rough set theory to probabilistic domain, we introduce a new approach to knowledge acquistion, which we call Probabilistic

Rule Induction MEthod based on ROugh SEt theory(PRIMEROSE) and develop an program that is based on this method to extract rules for an expert system from clinical database. It is applied to three medical domain:headache and facial pain, where one of the authors previously developed an expert system,called RHINOS(Rule-based Headache and facial pain INformation Orgranizing System), meningitis, and cerebrovascular diseases. The results show that the derived rules almost correspond to those of the medical experts.

The paper is organized as follows:in section 2, we discuss about probabilistic rules in RHINOS2. In section 3,the elementary concepts of rough set theory are introduced,and several characteristics are discussed. Section 4 presents our new method,PRIMEROSE for induction of RHINOS-type rules. Section 5 gives experimental results. Finally,in section 6 we compare our work with induction of decision trees and AQ method.

2   Probabilistic Rules in RHINOS2

RHINOS is an expert system which diagnoses the causes of headache or facial pain from manifestations. RHINOS2 is the second version of RHINOS, in which we refine the previous diagnosing model. For the limitation of the space, in the following, we only discuss about the acquisition of inclusive rules,which are used for differential diagnosis. For further information,refer to [3, 6, 7].

Inclusive rule consists of several rules,which we call positive rules. The premises of positive rules are composed of a set of manifestations specific to a disease to be included for the candidates of disease diagnoses. If a patient satisfy one set of the manifestation of a inclusive rule, we suspect the corresponding disease with some probability. These rules are derived by asking the following questions in relation to each disease to the medical experts:*1.a set of manifestations by which we strongly suspect a corresponding disease. 2.the probability that a paitent has the disease with this set of manifestations:SI(Satisfactory Index) 3.the ratio of the number the patients who satisfy the set of manifestations to that of all the patients having this disease:CI(Covering Index) 4.If sum of the derived CI(tCI) is equal to 1.0 then end. If not, goto 5. 5.For the patients suffering from this disease who do not satisfy all the collected set of manifestations, goto 1.* An inclusive rule is described by the set of manifestations, and its satisfactory index. Note that SI and CI are given experimentally by medical experts.

Formally, we can represent each positive rule as tuple: $\langle d, R_i, SI_i(, CI_i)\rangle$, where d denotes its conclusion, and R_i denotes its premise. The inclusive rule is described as: $\langle\{\langle d, R_1, SI_1(, CI_1)\rangle, \cdots, \langle d, R_k, SI_k(, CI_k)\rangle\}, tCI\rangle$. where total CI(tCI) is defined as the sum of CI of each rule with the same conclusion:$\sum_i CI_i$.

Let us show an example of the inclusive rule of common migraine(tCI=0.9), which is composed of the following three rules:

If history:paroxysmal, jolt headache:yes, nature: throbbing or persistent, prodrome:no, intermittent symptom:no, persistent time: more than 6 hours, and location: not eye, then we suspect common migraine(SI=0.9).

If history:paroxysmal, jolt headache:yes, nature: throbbing or persistent, prodrome:no, intermittent symptom:no, and location: not eye, then we suspect common migraine with probability(SI=0.8).

If history:sudden, jolt headache:yes, nature: throbbing or persistent, and prodrome:no, then we suspect common migraine with probability(SI=0.5).

3 Rough Set Theory

3.1 Elementary Concepts of Rough Set Theory

Rough set theory is one of the most important approach in inducive learning. It is developed and rigorously formulated by Pawlak[13]. This theory can be used to acquire certain sets of attributes which would contribute to class classification and can also evaluate how precisely these attributes are able to classify data. For the limitation of space, we mention only what we need in relation to our reasoning strategy.

Table 1 is a small example of database which collects the patients who complained of headache. First,let us consider how an attribute "loc" classify the headache patients' set of the table. The set whose value of the attribute "loc" is equal to "who" is $\{2,8,11,13,14,16,17,20\}$(In the following,the numbers represent each record number). This set means that we cannot classify $\{2,8,11,13,14,16,17,20\}$ further solely by using the constraint R ="loc=who". This set is defined as indiscernible set over relation R and described as follows: $IND(R)=U/R= \{2,8,11,13,14,16,17,20\}$ (U denotes the total set of database). In this set, $\{2,11,14,16,17\}$ suffer from muscle contraction headache("m.c.h."), $\{8\}$ suffers from intracranial mass lesion("i.m.l.") and $\{13,20\}$ suffer from common migraine("common"). Hence we need other additional attributes to classify this set of patients as to their diseases. Using this concept, we can evaluate the classification power of each attribute. For example, "prod=1" is specific to the case of classic migraine ("classic"). We can also extend this indiscernible relation to multivariate cases, such as $IND("loc=who"$ and $"M2=1")= \{2,6,11\}$. Moreover,we can take not only an attribute-value set as a relation, but also an attribute itself as a relation. For example, $IND("loc")= \{IND("loc=who"),IND("loc=lat"), IND("loc=occ")\}=\{\{2,8,11,13,14,16,17,20\},\{3,4,9\}, \{1,5,6,7,10,12,15,18,19\}\}$. By using these basic concepts, several topological sets and measures for these sets can be defined as:

Definition 1 (Elementary concepts of Rough Set Theory) *Let R be an equivalence relation and X be the subset of U.*

$$
\begin{aligned}
&\text{(R-)positive region} &&Posi_R(X) = \bigcup\{Y \in U/R : Y \subseteq X\} \\
&\text{(R-)possible region} &&Poss_R(X) = \bigcup\{Y \in U/R : Y \cap X \neq \phi\} \\
&\text{(R-)boundary region} &&Bound_R(X) = Poss_R(X) - Posi_R(X) \\
&\text{(R-)accuracy measure} &&\alpha_R(X) = \frac{card\ Posi_R(X)}{card\ Poss_R(X)} \\
&\text{(R-)roughness measure} &&\rho_R(X) = 1 - \alpha_R(X)
\end{aligned}
$$

Table 1: an Example of Database

No.	loc	nat	his	prod	jolt	nau	M1	M2	class
1	occ	per	per	0	0	0	1	1	m.c.h.
2	who	per	per	0	0	0	1	1	m.c.h.
3	lat	thr	par	0	1	1	0	0	common
4	lat	thr	par	1	1	1	0	0	classic
5	occ	rad	acu	0	0	0	1	1	m.c.h.
6	occ	per	sub	0	1	1	0	0	i.m.l.
7	occ	per	acu	0	1	1	0	0	sah
8	who	per	chr	0	1	0	0	0	i.m.l.
9	lat	thr	par	0	1	1	0	0	common
10	occ	per	per	0	0	0	1	1	m.c.h.
11	who	per	per	0	0	0	1	1	m.c.h.
12	occ	per	per	0	0	0	0	0	m.c.h.
13	who	thr	par	0	1	0	0	0	common
14	who	per	per	0	0	0	1	0	m.c.h.
15	occ	per	per	0	0	0	0	0	psycho
16	who	per	per	0	0	0	1	1	m.c.h.
17	who	rad	acu	0	0	0	0	0	m.c.h.
18	occ	per	per	0	0	0	0	0	m.c.h.
19	occ	per	per	0	0	0	0	0	m.c.h.
20	who	thr	par	0	1	1	0	0	common

The above abbreviations stand for the following meanings:
loc:location,nat:nature,his:history,prod:prodrome,jolt:jolt headache
nau:nausea,M1:tenderness of M1,M2:tenderness of M2,who:whole,occ:
occular,lat:lateral,per:persistent,thr:throbbing,rad:radiating,
par:paroxysmal,sub:subacute,chr:chronic progressive,1=Yes,0=No
m.c.h.:muscle contraction headache,common:common migraine
classic:classic migraine,psycho:psychogenic pain.

(In [13],Pawlak does not use the above term "possible region". He refers to our possible region as *upper* approximation of X, and he rarely use it. In this paper, we focus on this region in order to deal with probabilistic domain. In contrast with the word "positive", we use "possible" which reflects the intuitional meaning of an *upper* approximation.)

For example, the set whose class is "m.c.h." is composed of $\{1,2,5,10,11,12,14, 16,17,18,19\}$. Let us take this set as X. We take the following relations R_i so that $U/R_i \cap X \neq \phi$:

$R_1 = [(loc,who),(nat,per),(his,per),(prod,0),(jolt,0),(nau,0), (M1,1),(M2,1)]$
$R_2 = [(loc,who),(nat,per),(his,per),(prod,0),(jolt,0),(nau,0), (M1,1),(M2,0)]$
$R_3 = [(loc,who),(nat,rad),(his,acu),(prod,0),(jolt,0),(nau,0), (M1,0),(M2,0)]$
$R_4 = [(loc,occ),(nat,per),(his,per),(prod,0),(jolt,0),(nau,0), (M1,0),(M2,0)]$
$R_5 = [(loc,occ),(nat,per),(his,per),(prod,0),(jolt,0),(nau,0), (M1,1),(M2,1)]$
$R_6 = [(loc,occ),(nat,rad),(his,acu),(prod,0),(jolt,0),(nau,0), (M1,1),(M2,1)]$

where $[(A,a),(B,b),....]$ denotes a conjunction: "A=a" and "B=b" and ... Let R denote $R_1 \cup R_2 \cup R_3 \cup R_4 \cup R_5 \cup R_6$. That is,$U/R=\{\text{IND}(R_1),\text{IND}(R_2),\text{IND}(R_3), \text{IND}(R_4),\text{IND}(R_5),\text{IND}(R_6)\}= \{\{2,11,16\},\{14\},\{17\},\{12,15,18,19\},$

$\{1,10\}, \{5\}\}$. The positive region of "m.c.h." over the relation R is U/R - $\{12,15,18,19\}=\{1,2,5,10,11,14,16,17\}$. And then its possible region is $\{1,2, 5,10,11,12,14,15,16,17,18,19\}$, which includes one case for "psycho":$\{15\}$. Moreover, we can derive $\{12,15,17,19\}$ as the boundary region, and the accuracy measure is 8/11.

3.2 Reduction of Knowledge

Reduction of Knowledge is a method to examine the independencies of the attributes iteratively and extract the minimum indispensable part of equivalence relations, based on the above-mentioned elementary concepts.

For the limitation of the space, we only mention about the definition of *consistent rules* and their knowledge reduction. For further details, see [13].

Definition 2 (Definition of a consistent rule) *Let R be an equivalence relation and X be the subset of U. $R \Rightarrow X$ is called a consistent rule when $Posi_R(X)$ is given by:*

$$Posi_R(X) = \bigcup\{Y \subseteq X, R \Rightarrow Y\}$$

Definition 3 (Class-consistency based Reduction of Knowledge) *If an attribute a is satisfied with the following equation:*

$$Posi_{R-\{a\}}(X) = Posi_R(X),$$

then we say that a is dispensable in R, and a can be deleted from the relation R. We refer to an equivalent relation without any dispensable attribute as a minimal reduct, which is denoted by $min(R)$. □

For the above example,since $Posi_{R_1}(X) = \{2,11,16\}$, $R_1 \Rightarrow X$ is a consistent rule of X. Here, when we delete an attribute "prod", $R_1 - \{prod\}$ is equal to$[(loc,who),(nat,per),(his,per),(jolt,0),(nau,0), (M1,1),(M2,1)]$ and $IND(R_1 - \{prod\}) = IND(R_1)$. Therefore,$Posi_{R_1-\{prod\}}(X) = Posi_{R_1}(X)$. We can delete this attribute. Applying this method iteratively, we derive a minimum equivalent relation: $min(R_1)=[(loc,who),(M2,1)]$.

Using these elementary concepts and knowledge reduction technique, we can apply rough set theory to inductive discovery of classification rules. In [13], this application to deterministic domain is discussed in detail.

3.3 Characteristics of Rough Set Theory

The rough set theory clarify topological characteristics of the classes over combinatorial patterns of the attributes. These characteristics of rough set theory are precisely discussed in [13, 16, 20].

However, there are also the following problems when we apply the method to inductive learning for real-world domain: 1) When the numbers of samples or attributes are huge, computational cost is very high. 2) This method does not provide statistical evaluation of the derived rules. That is, we cannot evaluate bias, or the degrees of overfitting of the induced rules. Moreover, this method does not give how to evaluate *a priori* probability of the classes, while experts know this kind of knowledge,such as the frequency of each disease. It is also critical information for diagnosis. 3)In the original work, rules are induced only when IND is a subset of positive region. However, it is too strict in real world, especially in probabilistic domain. As a matter of fact, it is possible that these positive regions might be too narrow. For the above example, in the case of "psycho", the positive region is an empty set {}, so we should use a possible region for diagnosis. Moreover, when some new data are added and the revision is required, it could happen that they contradicts the rules derived before. For the above example, let us consider a situation when the following case(the 21th case) is given. The attributes are: "loc=who,nat=per,his=per,prod=0,jolt=0,nau=0,M1=1 and M2=1", and its class is "diseases of cervical spine". Since this case satisfies the abovementioned relation R_1, IND(R_1) is changed from {2,11,16} into {2,11,16,21}. Then IND(R_1) is no longer a subset of the positive region of "m.c.h." and the positive region reduces to {1,5,10,14,17}.

Therefore when we apply rough set theory to automated knowledge acquisition, the following extentions are required: 1) For rule induction, it is insufficient to consider only the rules covering positive region. We have to focus on the possible region. A positive region is thought to be the specific subset of a possible region. 2) we need introduce some measures to evaluate statistical characteristics. In the next section, we intoduce a new program for automated knowledge acquistion which adds the above two features to the basic rough set theory.

4 PRIMEROSE

4.1 Definition of Probabilistic Rules

We extend the definition of consistent rules to probabilistic domain. For this purpose, we use the definition of inclusive rules discussed in Section 2.

Definition 4 (Definition of Probabilistic Rules) *Let R_i be an equivalence relation and X denotes one class, which is the subset of U. A probabilistic rule of X is defined as a tuple, $< X, R_i, SI(R_i, X), CI(R_i, X) >$ where $R_i, SI,$ and CI are defined as follows.*
R_i is a conditional part of a class X and defined as:

$$R_i \quad s.t. \quad IND(R_i) \bigcap IND(X) \neq \phi$$

SI and CI are defined as:

$$SI(R_i, X) = \frac{card\ \{(IND(R_i) \bigcup IND^c(R_i)) \bigcap (IND(X) \bigcup IND^c(X))\}}{card\ \{IND(R_i) \bigcup IND^c(R_i)\}}$$

$$CI(R_i, X) = \frac{card\ \{(IND(R_i) \bigcup IND^c(R_i)) \bigcap (IND(X) \bigcup IND^c(X))\}}{card\ \{IND(X) \bigcup IND^c(X)\}}$$

where $IND^c(X)$ or $IND^c(R_i)$ consists of unobserved future cases of a class X or those which satisfies R_i, respectively. □

A total rule of X is given by $R = \bigvee_i R_i$, and then total CI(tCI) is defined as: $\text{tCI}(R,X) = \text{CI}(\bigvee_i R_i, X)$. In the following subsection, we discuss about the extension of reduction and the estimation of SI and CI. Since SI and CI include unobserved cases, it is necessary to estimate these measures only from the training samples. We introduce new method for this estimation.

4.2 Estimation of CI and CI

Since $IND(R_i) \bigcap IND^c(X) = \phi, IND^c(R_i) \bigcap IND(X) = \phi$, SI and CI can be simplified as follows:

$$SI(R_i,X) = \frac{card\ IND(R_i \cap X) + card\ IND^c(R_i \cap X)}{card\ IND(R_i) + card\ IND^c(R_i)}$$

$$CI(R_i,X) = \frac{card\ IND((\cup R_i) \cap X) + card\ IND^c((\cup R_i) \cap X)}{card\ IND(X) + card\ IND^c(X)}$$

Since the above formulae include unobserved cases, we are forced to estimate these measures from the training samples. Note that these estimations have the same problem as that of error rates of discriminant function in multivariate analysis[8]. As discussed in that area, it is very difficult to derive unbiased error rates in an analytic form even if we assumes normal model with equal covariance matrices as a probabilistic model. Hence,recently,several resampling methods,such as cross validation method, the bootstrap method are introduced to overcome this problem[2, 8]. Moreover, for categorical analysis, it is pointed out in [1] that the bootstrap method performs much worse than cross validation method since resampling from the empirical disributional function is biased toward underestimation of true error rate. So we use cross validation method to estimate the unbiased total CI(utCI).

Cross-validation method for error estimation is performed as following: first, the whole training samples $\mathcal{L}$ are split into V blocks: $\{\mathcal{L}_1, \mathcal{L}_2, \cdots, \mathcal{L}_V\}$. Second, repeat for V times the procedure in which we induce rules from the training samples $\mathcal{L} - \mathcal{L}_i (i = 1, \cdots, V)$ and examine the error rate err_i of the rules using $\mathcal{L}_i$ as test samples. Finally, we derive the whole error rate err by averaging err_i over i, that is, $err = \sum_{i=1}^{V} err_i / V$ (this method is called V-fold cross-validation). Therefore we can use this method for estimation of CI by replacing the calculation of err by that of CI, and by regarding test samples as unobserved cases.

One of the problem is how to choose the value of V. However, in order to avoid the overestimate of CI, we can safely choose 2-fold cross-validation. For precise information about this problem,please refer to [1].

4.3 Cluster-based Reduction of Knowledge

In this subsection, we extend the concept of reduction to probabilitic domain. Instead of consistency based reduction, we delete an attribute when the deletion does not make SI change. First, we define a primitive cluster.

Definition 5 (a Primitive Cluster) *Let the set of whole attributes denote* $E = \{a_i\}$, *whose cardinality is equal to* p. *And let* $|R_i|$ *denote the number of the attributes included in the relation* R_i. *A primitive cluster is defined as:*

$$Prim_{R_i}(X) = \bigcap_i^p IND(a_i = v_i) = IND(R_i)$$

such that $|R_i| = p$ *and* $IND(R_i) \cap IND(X) \neq \phi$ *where* v_i *denotes the value of an attribute* "a_i" □

We also define the partial order of relation R_i. Let $A(R_i)$ denote the set whose elements are the attributes included in R_i. If $A(R_i) \subseteq A(R_j)$, then we represent this relation as $R_i \preceq R_j$.

Using these notations, we define cluster-based reduction method as follows:

Definition 6 (Cluster-based Reduction of Knowledge) *If an attribute* a *is satisfied with the following equation:*

$$Poss_{R_i - \{a\}}(X) = Poss_{R_i}(X) = Prim_R(X) \quad (R_i \preceq R, |R| = p)$$

then we say that a *is dispensable in* R_i, *and* a *can be deleted from* R_i. □

This knowledge reduction technique deletes dependent variables which does not change the possible region. That is, the possible region of each relation is invariant over this process, which means that we fix the probabilistic nature of the indiscernible set. We discuss the advantage of this method in Section 6. Note that this definition includes deterministic cases where SI is always equal to one. That is, an indiscernible set of a rule whose SI is equal to one corresponds to the positive region of a class.

The original Pawlak's method ignores an inconsistent part such as R_4 in the example of Section 3.1, and executes reduction of the consistent parts. So this approach is similar to AQ method as discussed in Section 6. Since our approach does not ignore these inconsistent samples, even the derived minimal reducts of positive regions can be different from those induced by the ordinary method. In the next subsection, we introduce algorithms for rule induction based on the methods discussed in Section 3 and in Section 4.2.

4.4 Algorithm for PRIMEROSE

Algorithms for rule induction can be derived by embedding rough set theory concept into the algorithms discussed in Section 2. An algorithm for induction of inclusive rules is described as follows:

1)Using all attributes,calculate all equivalent relation $\{R_i\}$ which covers all of the training samples, that is, calculate $\{R_i | \bigcup IND(R_i) = U\}$.
2)For each class d_j, collect all the equivalent relation R_i such that $IND(R_i) \cap IND(d_j) \neq \phi$. For each combination, calculate its possible region.
3)Calculate $SI(R_i, d_j)$.

Table 2: Experimental Results

Domain	Method	Rules	Conditions	Accuracy	Experts' Accuracy
headache	CART	11	40	72.5%	95%
(116/116)	AQ15	19	38	62.4%	
	PRIMEROSE	25	70	89.6%	
meningitis	CART	9	25	74%	99%
(100/100)	AQ15	14	26	75%	
	PRIMEROSE	20	55	85%	
CVD	CART	13	47	72.6%	90%
(137/124)	AQ15	12	26	71.0%	
	PRIMEROSE	15	35	82.5%	

(A/B) denotes (Training samples/Test samples).

4)Apply probabilistic reduction of knowledge to each relation R_i until SI is changed(Minimize the components of each relation). If several candidates of minimalization are derived, connect each with disjunction.

5)Collect all the rules,perform the cross validation method to estimate utCI for each d_j.

5 Results

We apply PRIMEROSE to headache(RHINOS's domain), meningitis, and cerebrovascular diseases as shown in Table 2. For estimation of SI and CI,we use 2-fold cross validation. In each row of CART,we assign the numbers of leafs to the column of those of rules, and the numbers of nodes to the column of those of conditions. The results of CART are those after pruning.

6 Discussion

As shown in Table 2, in probabilistic domain, PRIMEROSE can almost perform much better than CART or AQ15 method. In this section, we discuss about the differences between PRIMEROSE and those existing methods.

6.1 Comparison with AQ15

AQ is an inductive learning system based on incremental STAR algorithm[9]. This algorithm selects one seed from positive examples and starts from one "selector"(attribute). It adds selectors incrementally until the "complexes" (conjunction of attributes) explain only positive examples. Since many complexes can satisfy these positive examples, according to a flexible extra-logical crite-rion,AQ finds the most preferred one.

It would be surprising that the complexes supported only by positive exam-ples corresponds to the positive region. That is, the rules induced by AQ is equivalent to consistent rules. As a matter of fact, AQ's star algorithm can be reformulated by the concepts of rough set theory except for constructive gener-alization rules[9] as follows:

1)Select a seed e from a class d(positive examples).
2)Generate a bounded star $G(e|U-IND(d),m)$ of the seed e against negative examples($U-IND(d)$) with no more than m elements, where U denotes all of the training samples(using INDUCE method).
3)In the obtained star, find a description D with the highest preference accord-ing to the following criterion: $LEF_i =< (-card(IND(R_i) \cap (U - IND(d))), \tau_1),$ $(card(IND(R_i) \cap IND(d), \tau_2) >$ where τ_1 and τ_2 are tolerances.
4)If IND(D)=IND(c),then go to step 6.
5)Otherwise IND(d):=IND(d)$-$IND(D),U:=U$-$IND(D) and repeat the whole process from step 1.
6)The disjunction of all generated description D is a complete and continous concept description. That is IND($\bigvee D$)=IND(d).

LEF_i denotes lexicographic evaluation functional and is defined as the follow-ing pair:$< (-negcov, \tau_1),(poscov, \tau_2) >$ where negcov and poscov are numbers of negative and positive examples,respectively,covered by an expression in the star,and where τ_1 and tau_2 are tolerance threshold for criterion $poscov, negcov$ ($\tau \in [0..100\%]$). For the definition of INDUCE method, please refer to [9].

However, as shown in [13], the ordinary rule induction by rough set theory is different from AQ in strategy;Pawlak's method starts from description by total attributes, and then performs reduction to get minimal reducts,that is, rules are derived in a top-down manner. On the contrary, AQ induces in a bottom-up manner. While these approaches are different in strategies, they are often equivalent because of logical consistency, and this difference suggests that when we need the large number of attributes to describe rules, induction based on rough set theory is faster. These computational aspects of AQ method and Pawlak's method can be formulated in terms of matroid theory. In appendix, we show that AQ's INDUCE method which generates a bounded star is a kind of a greedy algorithm for generating a basis of a specific matroid.

One of the important problem of the AQ method is that it does not work well in probabilistic domain [10]. This problem is also explained by matroid theory: inconsistent data do not satisfy the condition of independence, so we cannot derive a basis of matroid in probabilistic domain using the proposed definition, which is the same problem as the Pawlak's method,as discussed in Section 4. Hence it is necessary to change the definition of independence to solve those problems.

As discussed earlier,in PRIMEROSE, we adopt cluster membership as the condition of independence, instead of using class membership. Restricting the probabilistic nature, we can use almost the same algorithm as class-consistency based reduction. Then we estimate the probabilistic nature of the derived rules using some resampling plans, such as cross-validation method in this paper.

This is one kind of solution to the above problems, and the similar approach can also solve the disadvantage of AQ.

6.2 Comparison with CART and ID3

Induction of decision trees, such as CART[1] and ID3[14] is another inductive learning method based on the ordering of variables using information entropy measure or other similar measures. This method splits training samples into smaller ones in a top-down manner until it cannot split the samples, and then pruns the overfitting leaves.

There are many discussions about the problems of this approach[15, 11, 12]. Two of the important problems are about high computational costs of pruning and structural instability. As shown in [4], constructing optimal binary decision trees is NP-complete. In this context, this means that it is difficult to determine which leaves should be pruned. CART uses the combination of cross validation method and minimal cost complexity. The difficulty is to calculate the complexity because we should choose the pruned leaves.

PRIMEROSE method also have the similar problems since reduction technique corresponds to pruning. While reduction technique examines the dependencies of attributes, pruning techniques are mainly based on the trade-off between an accuracy and a structural complexity.

Note that reduction technique only uses topological characteristics of the training samples. And dependencies and independencies of the attributes are important factors, since dependent attributes will not change accuracy of the induced rules. Moreover,as shown in [5], if the attributes are independent and quantitized to k levels, there is no peaking phenomenon of accuracy in the Bayesian context, as discussed in the test-sample accuracy of decision trees.

Hence extracting independent variables is very important in probabilistic domain. These facts suggests that when the attributes are the mixture of dependent and independent ones, PRIMEROSE performs much better.On the other hand, when almost all of the attributes are independent, PRIMEROSE is much worse since we cannot use information about dependencies.

7 Conclusion

We introduce a new approach to knowledge acquistion, PRIMEROSE, and develop an program based on this method to extract rules for an expert system from clinical database. It is applied to three medical domains. The results show that the derived rules almost correspond to those of the medical experts. We are studying formulation of our approach in terms of matroid theory. In future work, we will introduce some common framework of machine learning method based on matroid theory and rough set theory.

References

[1] Breiman,L.,Freidman,J.,Olshen,R.,and Stone,C. *Classification And Regression Trees*. Belmont,CA:Wadsworth International Group, 1984.

[2] Efron B. *The Jackknife,the Bootstrap and Other Resampling Plans*. Pensylvania:CBMS-NSF,1982.

[3] Kimura,M, et al. RHINOS: A Consultation System for Diagnoses of Headache and Facial Pain: RHINOS,*Proc. of IJCAI-85*,393-396,Morgan Kaufmann,1985.

[4] Hyafil,L and Rivest,R.L. Constructing Optimal Binary Decision Trees is NP-complete. *Information Processing Letters*,1976.

[5] Chandrasekaran,B.and Jain,A.K. Quantization complexity and independent measurements *IEEE Trans.Comput.*,**23**,102-106,1974.

[6] Matsumura,Y, et al. Consultation system for diagnoses of headache and facial pain: RHINOS,*Medical Informatics*,11,145-157,1986.

[7] Matsumura,Y, et al. Consultation system for diagnoses of headache and facial pain: RHINOS2,*Proc. of Logic Programming Conference '85*, Spring Verlag,1985.

[8] McLachlan, G.J. *Discriminant Analysis and Statistical Pattern Recognition*, John Wiley and Sons,1992.

[9] Michalski,R.S. A Theory and Methodology of Machine Learning. Michalski,R.S.,Carbonell,J.G. and Mitchell,T.M., *Machine Learning - An Artificial Intelligence Approach*, Morgan Kaufmann,1983.

[10] Michalski,R.S.,et al. The Multi-Purpose Incremental Learning System AQ15 and its Testing Application to Three Medical Domains, *Proc. of AAAI-86*, 1041-1045,Morgan Kaufmann,1986.

[11] Mingers,J. An Empirical Comparison of Selection Measures for Decision Tree Induction. *Machine Learning*,**3**,319-342, 1989.

[12] Mingers,J. An Empirical Comparison of Pruning Methods for Decision Tree Induction. *Machine Learning*,**4**,227-243, 1989.

[13] Pawlak,Z *Rough Sets*,Kluwer Academic Publishers, 1991.

[14] Quinlan, J.R. Induction of decision trees,*Machine Learning*,1,81-106,1986.

[15] Shapiro,G.P.and Frawley, W.J.(eds) *Knowledge Discovery in Datatbases*,AAAI press, 1991.

[16] Slowinski,K et al. Rough sets approach to analysis of data from peritoneal lavage in acute pancreatitis, *Medical Informatics*,**13**,143-159,1988.

[17] Tsumoto,S.and Tanaka,H. Comparing inductive learning methods in the framework of the combination of rough set theory and matroid theory. *to be appeared*,1993.

[18] Welsh,D.J.A. *Matroid Theory*,Academic Press,1976.

[19] Whitney,H. On the abstract properties of linear dependence, *Am.J.Math.*,57,509-533,1935.

[20] Ziarko,W The Discovery,Analysis, and Representation of Data Dependencies in Databases, in:*Knowledge Discovery in Database*,Morgan Kaufmann, 1991.

A   Matroid theory and AQ method

In this appendix, we briefly discuss about relation between AQ method and Pawlak's method in the framework of matroid theory. Full theoretical description will be appeared in [17].

Matroid theory abstracts the important characteristics of matrix theory and graph theory, firstly developed by Whitney[19]. A matroid is defined as an independent space which satisfies the following axiom:

Definition 7 (Definition of a Matroid) *The pair $M(E, \mathcal{J})$ is called a matroid(or an independence space),if*

1) E is a finite set,
2) $\emptyset \in \mathcal{J} \subset 2^E$,
3) $X \in \mathcal{J}, Y \subset X \Rightarrow Y \in \mathcal{J}$,
4) $X, Y \in \mathcal{J}, card(X) = card(Y) + 1 \Rightarrow (\exists a \in X - Y)(Y \cup \{a\}) \in \mathcal{J}$.

If $X \in \mathcal{J}$, it is called **independent**, *otherwise X is called* **dependent**.   □

One of the most important characteristic of matroid theory is that this theory refers to the notion of independence using the set-theoretical sheme. As discussed above in the paper, we also consider the independence of the attributes in terms of rough sets,which uses the set-theoretical framework. Therefore our definition of independence can be also partially discussed using matroid theory.

Here we show that our "rough sets" version of AQ algorithm is equivalent to a greedy algorithm for calculating a specific kind of a matroid, which is dual of the matroid defined by the concept "dispensable" in rough set theory. For simplicity, we assume that one class d_j corresponds to only one relation R_j^* in which all variables are specified to one value,that is, formally,$IND(d_j) = IND(R_j^*)$.

Under this assumption we define a set of the whole attributes as $E = \{a_i\}$ and an independent set $\mathcal{J}$ as $\mathcal{J} = \{a_i | \exists R_k, IND(d_j) \subseteq IND(R_k \cap a_i) \subset IND(R_k)\}$, which satisfies the above axiom. We call this type of matroid,$M(E, \mathcal{J}),AQ$ *matroid*. Since it is important to calculate a basis of a matroid in practice, several methods are proposed. In these methods, we focus on the greedy algorithm. This algorithm can be formulated as follows:

1.$B \leftarrow \phi$
2.Calculate "priority queue" Q using weight function of E.
3.If B is a basis of $M(E, \mathcal{J})$ then stop. Else go to 4.
4.$e \leftarrow first(Q)$, which has a minimum weight in Q.
5. If $B \cup \{e\} \in \mathcal{J}$ then $B \leftarrow B \cup \{e\}$. goto 2.

This greedy algorithm has the following characteristics:

Theorem 1 (Computational Complexity of the Greedy Algorithm)
The complexity of the greedy algorithm is

$$\mathcal{O}(f(\rho(M)) + m \log m)$$

where $\rho(M)$ is equal to a rank of matroid M, m is equal to the number of the elements in the matroid, $|E|$, f represents a function of complexity of a independent test, called independent test oracle. □

Theorem 2 (Optimal Solution by the Greedy Algorithm)
The optimal solution is derived by these algorithm if and only if a subset of the attributes satisfies the axiom of the matriod. □

(For the limitation of the space, the proofs of these theorems are not given in this paper. Please refer to [18].)

Using the above two theorems, we can formally show the relation between AQ method and Pawlak's method, and explain why AQ method does not work well when training samples are incomplete.

As shown in [9], it is possible to show that INDUCE method is equivalent to the greedy algorithm for acquiring a basis of AQ matroid. Therefore,since the independent test depends on the calculus of $IND(a_i)$ and on the comparison of the cardinality between $IND(R_1 \cap R_2)$ and $IND(R_1)$,its complexity is less than $\mathcal{O}(\rho(M) * (n^2 + 2n))$ where n denotes a sample size.

On the other hand, since $\rho(M)$ is the number of independent variables, $m - \rho(M)$ is equal to the number of dependent variables. From the concepts of the matroid theory, if we define a dependent set as $\mathcal{I} = \{a_i | \exists R_k, IND(R_k \cap a_i) = IND(R_k) = IND(d_j)\}$, then $M(E, \mathcal{I})$ satisfies the condition of the dual matroid of $M(E, \mathcal{J})$,and we call $M(E, \mathcal{I})$ Pawlak's matroid. Therefore the algorithm of Pawlak's method is formally equivalent to the algorithm for the dual matroid of AQ matroid, and the computational complexity of Pawlak's method is less than $\mathcal{O}((m - \rho(M)) * (n^2 + 2n) + m \log m)$. From these consideration, if $\rho(M)$ is small, AQ algorithm performs better than Pawlak's one under our assumption. Note that we can extend this result to a case where one class corresponds to many relations, or $IND(d_j) = \bigcup IND(R_j)$, under the completeness assumption.

However, this assumption is very strict as discussed in Section 3. In practice, it is often violated by given training samples. So these greedy algorithm cannot generate an optimal basis, which is equal to the phenomenon that when a seed is choson from a incomplete subset, calculation does not stop until all the attributes are exhausted, and no optimal star cannot be obtained.

Index of Authors

Springer-Verlag
and the Environment

We at Springer-Verlag firmly believe that an international science publisher has a special obligation to the environment, and our corporate policies consistently reflect this conviction.

We also expect our business partners – paper mills, printers, packaging manufacturers, etc. – to commit themselves to using environmentally friendly materials and production processes.

The paper in this book is made from low- or no-chlorine pulp and is acid free, in conformance with international standards for paper permanency.

Lecture Notes in Artificial Intelligence (LNAI)

Vol. 568: H.-J. Bürckert, A Resolution Principle for a Logic with Restricted Quantifiers. X, 116 pages. 1991.

Vol. 587: R. Dale, E. Hovy, D. Rösner, O. Stock (Eds.), Aspects of Automated Natural Language Generation. Proceedings, 1992. VIII, 311 pages. 1992.

Vol. 590: B. Fronhöfer, G. Wrightson (Eds.), Parallelization in Inference Systems. Proceedings, 1990. VIII, 372 pages. 1992.

Vol. 592: A. Voronkov (Ed.), Logic Programming. Proceedings, 1991. IX, 514 pages. 1992.

Vol. 596: L.-H. Eriksson, L. Hallnäs, P. Schroeder-Heister (Eds.), Extensions of Logic Programming. Proceedings, 1991. VII, 369 pages. 1992.

Vol. 597: H. W. Guesgen, J. Hertzberg, A Perspective of Constraint-Based Reasoning. VIII, 123 pages. 1992.

Vol. 599: Th. Wetter, K.-D. Althoff, J. Boose, B. R. Gaines, M. Linster, F. Schmalhofer (Eds.), Current Developments in Knowledge Acquisition - EKAW '92. Proceedings. XIII, 444 pages. 1992.

Vol. 604: F. Belli, F. J. Radermacher (Eds.), Industrial and Engineering Applications of Artificial Intelligence and Expert Systems. Proceedings, 1992. XV, 702 pages. 1992.

Vol. 607: D. Kapur (Ed.), Automated Deduction – CADE-11. Proceedings, 1992. XV, 793 pages. 1992.

Vol. 610: F. von Martial, Coordinating Plans of Autonomous Agents. XII, 246 pages. 1992.

Vol. 611: M. P. Papazoglou, J. Zeleznikow (Eds.), The Next Generation of Information Systems: From Data to Knowledge. VIII, 310 pages. 1992.

Vol. 617: V. Mařík, O. Štěpánková, R. Trappl (Eds.), Advanced Topics in Artificial Intelligence. Proceedings, 1992. IX, 484 pages. 1992.

Vol. 619: D. Pearce, H. Wansing (Eds.), Nonclassical Logics and Information Processing. Proceedings, 1990. VII, 171 pages. 1992.

Vol. 622: F. Schmalhofer, G. Strube, Th. Wetter (Eds.), Contemporary Knowledge Engineering and Cognition. Proceedings, 1991. XII, 258 pages. 1992.

Vol. 624: A. Voronkov (Ed.), Logic Programming and Automated Reasoning. Proceedings, 1992. XIV, 509 pages. 1992.

Vol. 627: J. Pustejovsky, S. Bergler (Eds.), Lexical Semantics and Knowledge Representation. Proceedings, 1991. XII, 381 pages. 1992.

Vol. 633: D. Pearce, G. Wagner (Eds.), Logics in AI. Proceedings. VIII, 410 pages. 1992.

Vol. 636: G. Comyn, N. E. Fuchs, M. J. Ratcliffe (Eds.), Logic Programming in Action. Proceedings, 1992. X, 324 pages. 1992.

Vol. 638: A. F. Rocha, Neural Nets. A Theory for Brains and Machines. XV, 393 pages. 1992.

Vol. 642: K. P. Jantke (Ed.), Analogical and Inductive Inference. Proceedings, 1992. VIII, 319 pages. 1992.

Vol. 659: G. Brewka, K. P. Jantke, P. H. Schmitt (Eds.), Nonmonotonic and Inductive Logic. Proceedings, 1991. VIII, 332 pages. 1993.

Vol. 660: E. Lamma, P. Mello (Eds.), Extensions of Logic Programming. Proceedings, 1992. VIII, 417 pages. 1993.

Vol. 667: P. B. Brazdil (Ed.), Machine Learning: ECML – 93. Proceedings, 1993. XII, 471 pages. 1993.

Vol. 671: H. J. Ohlbach (Ed.), GWAI-92: Advances in Artificial Intelligence. Proceedings, 1992. XI, 397 pages. 1993.

Vol. 679: C. Fermüller, A. Leitsch, T. Tammet, N. Zamov, Resolution Methods for the Decision Problem. VIII, 205 pages. 1993.

Vol. 681: H. Wansing, The Logic of Information Structures. IX, 163 pages. 1993.

Vol. 689: J. Komorowski, Z. W. Raś (Eds.), Methodologies for Intelligent Systems. Proceedings, 1993. XI, 653 pages. 1993.

Vol. 695: E. P. Klement, W. Slany (Eds.), Fuzzy Logic in Artificial Intelligence. Proceedings, 1993. VIII, 192 pages. 1993.

Vol. 698: A. Voronkov (Ed.), Logic Programming and Automated Reasoning. Proceedings, 1993. XIII, 386 pages. 1993.

Vol. 699: G.W. Mineau, B. Moulin, J.F. Sowa (Eds.), Conceptual Graphs for Knowledge Representation. Proceedings, 1993. IX, 451 pages. 1993.

Vol. 723: N. Aussenac, G. Boy, B. Gaines, M. Linster, J.-G. Ganascia, Y. Kodratoff (Eds.), Knowledge Acquisition for Knowledge-Based Systems. Proceedings, 1993. XIII, 446 pages. 1993.

Vol. 727: M. Filgueiras, L. Damas (Eds.), Progress in Artificial Intelligence. Proceedings, 1993. X, 362 pages. 1993.

Vol. 728: P. Torasso (Ed.), Advances in Artificial Intelligence. Proceedings, 1993. XI, 336 pages. 1993.

Vol. 743: S. Doshita, K. Furukawa, K. P. Jantke, T. Nishida (Eds.), Algorithmic Learning Theory. Proceedings, 1992. X, 260 pages. 1993.

Vol. 744: K. P. Jantke, T. Yokomori, S. Kobayashi, E. Tomita (Eds.), Algorithmic Learning Theory. Proceedings, 1993. XI, 423 pages. 1993.

Vol. 745: V. Roberto (Ed.), Intelligent Perceptual Systems. VIII, 378 pages. 1993.

Vol. 746: A. S. Tanguiane, Artificial Perception and Music Recognition. XV, 210 pages. 1993.

Lecture Notes in Computer Science